Pistachio Production Manual

Louise Ferguson | David R. Haviland

Technical Editors

University *of* California
Agriculture and Natural Resources

Publication 3545

To order or obtain ANR publications and other products, visit the ANR Communication Services online catalog at http://anrcatalog.ucdavis.edu or phone 1-800-994-8849. You can also place orders by mail or FAX, or request a printed catalog of our products from

University of California
Agriculture and Natural Resources
Communication Services
2801 Second Street
Davis, CA 95618

Telephone 1-800-994-8849
510-665-2195
E-mail: anrcatalog@ucanr.edu

Publication 3545

ISBN-13: 978-1-60107-877-3

Library of Congress Cataloging-in-Publication Data

Names: Ferguson, Louise, editor. | Haviland, David R., editor.

Title: Pistachio production manual / Louise Ferguson, David R. Haviland, technical editors.

Other titles: Publication (University of California (System). Division of Agriculture and Natural Resources) ; 3545.

Description: Oakland, California : University of California Agriculture and Natural Resources, [2016] | Series: Publication ; 3545 | Includes bibliographical references and index.

Identifiers: LCCN 2016035713 | ISBN 9781601078773 (pbk.)

Subjects: LCSH: Pistachio--California. | Pistachio--Diseases and pests--Control--California. | Pistachio industry--California.

Classification: LCC SB401.P5 P57 2016 | DDC 634/.57409794--dc23
LC record available at https://lccn.loc.gov/2016035713

Editing: Linda Ribera. Design and illustrations: Celeste Rusconi and Robin Walton, except as noted in the captions. Proofreading and indexing: Hazel White. Photo research: Evett Kilmartin. Photo credits are given in the captions. Front cover photo: Evett Kilmartin. Back cover background photo: courtesy of American Pistachio Growers. Back cover inset photos: Louise Ferguson. Chapter opening images: courtesy of American Pistachio Growers.

UC PEER REVIEWED This publication has been anonymously peer reviewed for technical accuracy by University of California scientists and other qualified professionals. This review process was managed by ANR Associate Editors Larry Bettiga, David Haviland, Mary Louise Flint, and Joe Nuñez.

Printed in Canada on recycled paper.

1m-rep-2/18-LR/CR, RW

PRECAUTIONS FOR USING PESTICIDES

Pesticides are poisonous and must be used with caution. READ THE LABEL CAREFULLY BEFORE OPENING A PESTICIDE CONTAINER. Follow all label precautions and directions, including requirements for protective equipment. Use a pesticide only on crops specified on the label. Apply pesticides at the rates specified on the label or at lower rates if suggested in this publication. In California, all agricultural uses of pesticides must be reported. Contact your county agricultural commissioner for details. Laws, regulations, and information concerning pesticides change frequently, so be sure the publication you are using is up to date.

LEGAL RESPONSIBILITY. The user is legally responsible for any damage due to misuse of pesticides. Responsibility extends to effects caused by drift, runoff, or residues.

TRANSPORTATION. Do not ship or carry pesticides together with foods or feeds in a way that allows contamination of the edible items. Never transport pesticides in a closed passenger vehicle or in a closed cab.

STORAGE. Keep pesticides in original containers until used. Store them in a locked cabinet, building, or fenced area where they are not accessible to children, unauthorized persons, pets, or livestock. DO NOT store pesticides with foods, feeds, fertilizers, or other materials that may become contaminated by the pesticides.

CONTAINER DISPOSAL. Dispose of empty containers carefully. Never reuse them. Make sure empty containers are not accessible to children or animals. Never dispose of containers where they may contaminate water supplies or natural waterways. Consult your county agricultural commissioner for correct procedures for handling and disposal of large quantities of empty containers.

PROTECTION OF NONPEST ANIMALS AND PLANTS. Many pesticides are toxic to useful or desirable animals, including honey bees, natural enemies, fish, domestic animals, and birds. Crops and other plants may also be damaged by misapplied pesticides. Take precautions to protect nonpest species from direct exposure to pesticides and from contamination due to drift, runoff, or residues. Certain rodenticides may pose a special hazard to animals that eat poisoned rodents.

POSTING TREATED FIELDS. For some materials, reentry intervals are established to protect field workers. Keep workers out of the field for the required time after application and, when required by regulations, post the treated areas with signs indicating the safe reentry date.

HARVEST INTERVALS. Some materials or rates cannot be used in certain crops within a specific time before harvest. Follow pesticide label instructions and allow the required time between application and harvest.

PERMIT REQUIREMENTS. Many pesticides require a permit from the county agricultural commissioner before possession or use. When such materials are recommended in this publication, they are marked with an asterisk (*).

PROCESSED CROPS. Some processors will not accept a crop treated with certain chemicals. If your crop is going to a processor, be sure to check with the processor before applying a pesticide.

CROP INJURY. Certain chemicals may cause injury to crops (phytotoxicity) under certain conditions. Always consult the label for limitations. Before applying any pesticide, take into account the stage of plant development, the soil type and condition, the temperature, moisture, and wind direction. Injury may also result from the use of incompatible materials.

PERSONAL SAFETY. Follow label directions carefully. Avoid splashing, spilling, leaks, spray drift, and contamination of clothing. NEVER eat, smoke, drink, or chew while using pesticides. Provide for emergency medical care IN ADVANCE as required by regulation.

WARNING ON THE USE OF CHEMICALS

Pesticides are poisonous. Always read and carefully follow all precautions and safety recommendations given on the container label. Store all chemicals in their original labeled containers in a locked cabinet or shed, away from foods or feeds, and out of the reach of children, unauthorized persons, pets, and livestock.

Recommendations are based on the best information currently available, and treatments based on them should not leave residues exceeding the tolerance established for any particular chemical. Confine chemicals to the area being treated. THE GROWER IS LEGALLY RESPONSIBLE for residues on the grower's crops as well as for problems caused by drift from the grower's property to other properties or crops.

Consult your county agricultural commissioner for correct methods of disposing of leftover spray materials and empty containers. **Never burn pesticide containers.**

PHYTOTOXICITY: Certain chemicals may cause plant injury if used at the wrong stage of plant development or when temperatures are too high. Injury may also result from excessive amounts or the wrong formulation or from mixing incompatible materials. Inert ingredients, such as wetters, spreaders, emulsifiers, diluents, and solvents, can cause plant injury. Since formulations are often changed by manufacturers, it is possible that plant injury may occur, even though no injury was noted in previous seasons.

CONTAINER DISPOSAL. Dispose of empty containers carefully. Never reuse them. Make sure empty containers are not accessible to children or animals. Never dispose of containers where they may contaminate water supplies or natural waterways. Consult your county agricultural commissioner for correct procedures for handling and disposal of large quantities of empty containers.

PROTECTION OF NONPEST ANIMALS AND PLANTS. Many pesticides are toxic to useful or desirable animals, including honey bees, natural enemies, fish, domestic animals, and birds. Crops and other plants may also be damaged by misapplied pesticides. Take precautions to protect nonpest species from direct exposure to pesticides and from contamination due to drift, runoff, or residues. Certain rodenticides may pose a special hazard to animals that eat poisoned rodents.

Contents

Acknowledgments

The California pistachio industry began in the last quarter of the twentieth century. With the publication of this first edition of the *Pistachio Production Manual*, produced by the University of California Division of Agriculture and Natural Resources, we would like to acknowledge the pistachio industry pioneers who developed the information synthesized here. From the University of California, first and foremost we acknowledge Dr. Julian Crane of the UC Davis Plant Sciences Department. He and Joseph Maranto, UC Cooperative Extension Farm Advisor, Kern County, first defined the physiological and horticultural parameters of the major cultivar *Pistacia vera*. UC Davis Plant Sciences Department colleagues Drs. Kay Uriu and Steven Weinbaum developed the parameters of pistachio nutrient use. Dr. Vito Polito of the UC Davis Plant Sciences Department determined the pollination and fruit set parameters of *Pistacia vera*. Dr. David Goldhamer of the UC Davis Land, Air and Water Resources Department developed the basis of the current irrigation programs. Dr. Dick Rice of the UC Davis Entomology Department and Walter Bentley of the Statewide IPM Program at the UC Kearney Agricultural Research and Extension Center identified and developed management programs for insect pests. Dr. Themis Michailides of the UC Davis Plant Pathology Department characterized the diseases and designed the basis of the current pistachio disease management program. Dr. Adel Kader of the UC Davis Plant Sciences Department and Dr. James F. Thompson of the UC Davis Biological and Agricultural Engineering Department determined how to harvest, handle, and process pistachios. Dr. Lee Ashworth of the Plant Pathology Department at UC Berkeley was the inventor of a major rootstock, UCBI, and with Earl Surber of Blackwell Land Ranch first identified macronutrient deficiencies. Dr. Karen Klonsky of the UC Davis Agricultural and Resource Economics Department produced the first UC Cost Studies for pistachio production. Robert H. Beede, UC Cooperative Farm Advisor, Kings County, not only cooperated in producing most of the research detailed above but also led the effort in developing practical applications of these results.

Dr. Beth Teviotdale, Plant Pathology Specialist at UC Davis, cooperated with Bob Gross and Karen Reinicke (first and second directors) in forming the first California Pistachio Commission.

From the pistachio industry, we acknowledge industry consultants Carlo Fanucchi and Bob Peterson, nurserymen Jerry Allan, Corky Anderson, Ken Puryear, and Brian Blackwell, pruning and harvesting contractor Erick Nielsen, pest control consultant Gary Weinberger, ranch managers Joe MacIlvaine, Phil Pierre, Kevin Olsen, Dennis Elam, Earl Surber, Don Castle, Donny Rose, Chris and Craig Wylie, Kevin Herman, and Bill Seaman, processing facility managers Jeff Gibbons, Andy Anzaldo, and Johnnie Starling, and, finally, ranch owners Roger Jensen, Joshua Setton, Stewart and Linda Resnick, Chuck Nichols, Robert Keenan, Ali Orandi, and Ali Amin. A special acknowledgment goes to Brenda Hansen of Paramount Farming's grading and processing facility for the thousands of samples she processed for researchers. All of these people generously supplied knowledge, orchards, facilities, services, or labor that produced the results synthesized in this manual.

We gratefully recognize the multiple authors who developed and synthesized the research results presented in this manual. And we acknowledge the patience and meticulous editing of Linda Ribera, UC ANR editor, and the expertise and project coordination of Ann Senuta, UC ANR Director of Publishing and Production, who transformed our collective knowledge into a useful publication.

We apologize for unintentional omissions. It has been an honor to work with those named and unnamed in building the California pistachio industry and producing this publication.

~***Louise Ferguson and David R. Haviland***
Technical Editors
Pistachio Production Manual, First Edition
University of California
Division of Agriculture and Natural Resources

Part 1 Economics

1 The Economic Outlook for California Pistachios

Steven C. Blank

The commercial pistachio industry in California is a relatively young business compared with that of other nut crops, but it has a record of steady expansion. This indicates both that strong consumer demand exists and that the industry appeals to farmers. However, before prospective pistachio producers commit to the long-term investment of significant amounts of money per acre, they should be very familiar with the current and future economic outlook for the industry. The developing domestic and global markets for California pistachios are both dynamic, offering much potential for the industry's future.

A key factor is that pistachio growers in California produce about 98 percent of the total crop in the United States, effectively making up the market in this country (table 1.1). However, approximately 55 percent of the total output is marketed by a single grower and processor. This creates some unique economic challenges and opportunities for pistachio growers in California, which they need to be aware of so that successful strategies can be developed for future production and marketing decisions. Production decisions are made at the farm level, but in the aggregate they have dramatic effects on the marketing decisions of the entire industry due to California's dominant role in the concentrated national market, as well as its prominent role in the global market.

This chapter summarizes the main economic challenges and opportunities facing individual growers and the pistachio industry in California. It also points out some of the economic implications of possible strategies that might be taken by the industry. The future of the pistachio industry in California is not easily forecast because it is in the hands of those making production and marketing decisions in response to the unique domestic and global markets for this commodity.

Recent Trends

The scale of the pistachio industry in the United States has increased dramatically from 1,700 acres in 1977 to 232,653 in 2012 (table 1.1). Total acreage increased 84 percent just in the decade from 2002 to 2012. Domestic pistachio production is increasing for many reasons. In general, research has shown

Table 1.1. Pistachio farms and acreage, 2002 and 2012

Location	Total farms			Total acres		
	2002	2012	% Change	2002	2012	% Change
Arizona	109	77	−29	3,509	3,645	4
California	1,055	1,305	24	121,562	228,248	88
Nevada	8	16	100	85	85	0
New Mexico	126	76	−40	1,350	513	−62
Texas	15	15	0	32	114	256
Utah	7	7	0	31	48	55
United States	1,320	1,496	13	126,569	232,653	84

Source: NASS 2014a.

that tree and vine crops are more profitable than field crops, and thus farmers are gradually shifting more acreage into the production of tree crops, including pistachios (Blank 2008). California's climate and soil types are well suited to pistachio production. Also, recent research shows that pistachios can thrive in cheaper, poorer-quality ground and can be irrigated with water less suitable for other crops. Thus, pistachios are an attractive option for farmers, especially those located in parts of the San Joaquin Valley where soils and water costs make some other crops less competitive.

Production of pistachios in California has continued to expand since 1977, with rising yields per acre and an increasing number of acres planted. The rate of increase in pistachio acreage was slow during the 1980s but rose during the 2000s. As shown in table 1.2, bearing acreage has now expanded to 203,000 acres. This total is about seven times the acreage that existed in 1980. With about 47,000 nonbearing acres at present, pistachio production will clearly expand over the next decade, likely putting downward pressure on prices.

An important characteristic of pistachio production is that the trees are alternate bearing, so yields per acre vary from year to year. The third column of table 1.2 shows this significant effect during 3 consecutive years. For example, between 2012 and 2013 there was a 23 percent decrease in average yield. The scale of the yield volatility is such that total production (shown in the fourth column of table 1.2) has fluctuated widely in recent years, despite steadily increasing acreage. This creates a challenge to the industry because of its implications for pistachio prices and for the industry's ability to serve market segments that are much less volatile over time. For example, most domestic market segments have relatively smooth levels of demand across time, so those markets prefer to have a relatively stable flow of a commodity to them. The supply volatility makes it particularly difficult to develop secondary uses for pistachios—as ingredients in baked goods and other foods, for example—as these uses generally require a stable supply with a consistent cost. As a result, packers mitigate yearly supply fluctuations by carrying over large crop surpluses into small crop years.

The trend in global production of pistachios has paralleled the growth of the industry in the United States. World production of pistachios is now about three times as high as it was just 20 years ago. Until recently, Iran was the leading producer, representing about half of global output. Since 2012, the United States has been the production leader, with about 43 percent of global production and over 55 percent of pistachio supplies traded in 2013 to 2014 (FAS 2015). Other top-producing countries include Syria and Turkey, each accounting for about 10 percent of global production. The need for a warm, desert-type climate limits the number of countries where pistachios can be grown successfully, but the potential exists for significant expansion of acreage worldwide in the future. This potential growth in acreage presents the challenge of expanded competition in global markets, although currently California is in a strong competitive position relative to Iran due to California's higher-quality product and more-developed infrastructure.

Consumption

The demand for pistachios comes from both domestic and export markets. Until recently, California's pistachio sales to the domestic market exceeded sales to the export market. This is typical for a relatively new commodity market that is still growing in size. However, export demand now far exceeds domestic sales (FAS 2015). In particular, China has become a huge customer recently, leading the world in pistachio imports.

Per-capita consumption of pistachios in the United States is very low, but it has expanded gradually. In general, per-capita consumption was fairly stable during the late 1980s and early 1990s, averaging about 0.1 pound per year, before rapidly rising to over 0.2 pound early in the 2000s. The higher annual per-capita consumption of pistachios in recent years

Table 1.2. California pistachio production and market data, 2011–2013

Year	Bearing acreage (ac)	Yield per acre (1,000 lbs)	Utilized production (million lbs in-shell equivalent)	Price ($/lb)	Value of utilized production ($ billion)	Value of utilized production ($/ac)
2011	153,000	2.90	444	1.98	0.879	5,745
2012	182,000	3.03	551	2.61	1.438	7,901
2013	203,000	2.32	470	2.20	1.034	5,094

Source: NASS 2014b.

is likely attributable to nutritional research and advertising of the health benefits of nuts, which has increased consumption of all nuts. Also, in recent years, record crops of pistachios have decreased prices, contributing to higher consumption. However, U.S. consumption of pistachios remains low relative to other nuts. In comparison, consumption per capita of almonds was 1.3 pounds in 2008, and for walnuts it was 0.5 pound. Pistachios compete with other nuts and snack foods for consumer dollars. In the domestic market, pistachios are marketed primarily as in-shell nuts (table 1.3) and are sold for consumption as a snack food. A small portion of pistachios is used as ingredients in candies, baked goods, ice cream, confectioneries, and flavorings, whereas a majority of almonds and walnuts are used as ingredients. Thus, for consumers, pistachios are considered an impulse purchase, similar to many snack foods (NASS 2014b). Impulse purchase items usually need stronger marketing efforts to raise and maintain sales levels, and much uncertainty exists about the long-term growth potential for domestic per-capita consumption of an impulse snack product. Exploring ways to expand the use of pistachios in processed food items could provide additional opportunities for domestic market expansion.

In many commodity markets, export prices "make the market" and thus cannot be overlooked. For pistachios, the key issue is that domestic production is increasing faster than domestic demand (measured by consumption), and therefore exports must continue to increase to absorb the surplus.

Exports

Pistachios have been eaten as a snack in many parts of the world for decades. While domestic consumption has been growing slowly in recent years, so has the value of U.S. exports of pistachios. In recent years the export share of the California pistachio crop has been over 60 percent. The export market is essentially where production in excess of domestic consumption must be sold. Export sales are frequently more profitable for the California industry than are domestic sales. This is good news, considering that the size of the export market for California's pistachios is much bigger than the domestic market simply because there are billions of consumers outside the United States.

Until recently, Western Europe was the leading destination for U.S. exports in terms of value. Now, China and Hong Kong are major destinations for U.S. pistachio exports. Market destinations vary over time as prices offered by the U.S. and Iran attract different customers. Price differences between U.S. and Iranian pistachios on world markets will continue to be a significant factor in sales results, despite the quality advantage of U.S. pistachios.

Iran remains the second-largest producer and exporter of pistachios, accounting for slightly over 40 percent of world exports in recent years. The United States is the largest exporter, with a recent share of about 55 percent of the global export market. The amount exported by the United States and its share of global trade vary from year to year due to the alternate-bearing nature of pistachio trees, but both categories are growing significantly. For example, in the 2006–2007 marketing year U.S. exports of pistachios were 80 million pounds, whereas over the 2013–2014 year the total was 385 million pounds. Table 1.4 shows recent pistachio export performance relative to that of other U.S. tree nut products.

Prices

As with all commodities, the price of pistachios depends both on physical characteristics of the product and on supply and demand factors in the market. Physical characteristics affecting the price of pistachios include, for example, the degree of shell splitting and the color of the kernel. Pistachios enclosed in shells that have not opened are worth less than those that have opened, because the shells must be opened mechanically. In market terms, there are many more factors influencing prices, as noted below.

Table 1.3. Pistachios: Utilization and meat production, 2011–2013

Year	Utilized production (million lbs)		Meat production of nuts sold shelled (million lbs)
	Sold in-shell	Sold shelled	
2011	341	103	51.5
2012	464	87	43.9
2013	379	91	46.0

Source: NASS 2014b.

Table 1.4. U.S. exports of selected tree nut products (shelled basis)

Type of tree nut	Marketing season	Season to date, through May (1,000 lbs)		Year-to-date change (%)
		2012	2013	
pistachios	Sep–Aug	117,440	123,548	5.2
almonds	Aug–Jul	1,168,577	1,068,804	-8.5
walnuts	Aug–Jul	239,052	267,888	12.1
pecans	Oct–Sep	53,945	70,816	31.3

Source: ERS 2013.

Changes in pistachio supplies of other major producing countries have the potential to significantly influence prices for U.S. pistachio growers. For example, in 2004, despite a record high U.S. crop, lower supplies from Syria and Turkey led to increased prices. Historically, however, California's large-crop years have corresponded with years of large global supplies, and prices have remained low for producers (ERS 2013). Note the decreasing prices from 2012 to 2013 shown in the fifth column of table 1.2. The average price in 2013 of $2.20 per pound was only about 84 percent of the average price of $2.61 received by growers in 2012. As a result, the total value of utilized production, which is the quantity used in some way, not stored, during the period (shown in the sixth column of table 1.2), was lower in 2013 ($1.034 billion) than it was in 2012 ($1.438 billion). The risks to individual growers can be seen more easily when looking at gross revenues per acre. As shown in the last column of table 1.2, revenue per acre (expressed as "value of utilized production, $/ac") was significantly higher in 2012 ($7,901) than it was in 2013 ($5,094). In other words, lower total sales revenues per acre for California growers are quite likely in some years. These wide swings in revenues are a major challenge to the industry because individual producers must manage their personal financial risks associated with the underlying market price and yield swings.

When foreign producers of pistachios have a good output year, they often lower their price to sell more to all customers, including customers that also buy from the United States, thus adding downward pressure on the prices that U.S. producers receive (NASS 2014b). Increased price competition between sellers is likely to be the norm in global markets for commodities like pistachios.

In early 2013, the export market was taking higher volumes of U.S. pistachios due to the weak U.S. dollar versus other currencies (ERS 2013). As shown in table 1.4, exports of pistachios and other nuts were better in 2013 than in the previous marketing season, with the exception of almonds. Whenever the U.S. dollar grows stronger relative to other currencies, the prices faced by foreign buyers of California's pistachios and other commodities rise accordingly. Thus far, the export market continues to expand for California pistachios, despite price increases, due partly to the high quality of U.S. exports (ERS 2013). Nevertheless, this source of price risk is part of the challenge to domestic producers trying to decide which commodity market offers the best long-term returns on investments. That question is explored next.

Market Choices

In the simplest of terms, the question raised in an economic assessment of the outlook for a commodity comes down to, which market is more profitable for California growers? The question does not refer to the domestic versus export markets for California pistachios, but rather to the total market for California pistachios versus the market for other commodities that an individual grower might produce. In other words, the economic outlook for pistachios, or any other commodity, must be compared with the outlook for alternative commodity investments to determine which crop a grower should produce. That task is never simple. Luckily, a grower does not have to analyze domestic and export market prospects to answer the outlook question. All a grower needs to evaluate is total profits received per acre from all market sales over time for pistachios and total profits received per acre for alternative crops. This is not an exact science, but it can be done with surprising accuracy, as described in the example in the following sections.

Profitability

The profitability of an existing commodity operation or a hypothetical alternative can be projected using the estimated prices and costs method applied by the University of California in its cost studies. A cost study for California pistachios is available at the UC Davis Cost Studies website, coststudies.ucdavis.edu. As shown in that study, commodity price projections are made when evaluating the market situation. The per-unit costs are taken from estimates of management practices used by typical growers, as described in the cost study for each commodity studied by the University of California. To determine the profitability of a particular operation, the prospective grower should prepare a projected net income statement for it. Net (after-tax) income should be estimated for each year of the operation's life. If the operation involves expanding an established pistachio orchard, for example, the productive life of the expansion could be over 35 years. The task of preparing this data often goes uncompleted, yet it is critical for determining a proposed orchard's true economic viability. Using a computer spreadsheet facilitates the work and will show that income levels will vary widely from year to year during the life of an orchard. In particular, losses will be incurred during the first several years because heavy development and operation costs are incurred and the trees do not yet produce a crop. As the trees mature, yields and sales receipts will increase, only to decline again at the end of the orchard's life. Without preparing income statements for each year, a manager will not know when profits will be earned or at what point the orchard's economic life will end (the point at which profits decline below the minimum level needed for economic viability).

For pistachios, the dynamic market causes profitability to vary as the industry in California grows. For example, in 2004 a University of California cost study found that, on average, pistachios grown in the San Joaquin Valley were not profitable, with net returns above total costs of –$558 per acre. In 2008, the cost study was updated with current production and marketing data, which showed a much-improved profit outlook for pistachio growers in the San Joaquin Valley (Beede et al. 2008).

For example, with the results presented in the "ranging analysis" section of the 2008 pistachio cost study and the average price and yield data for relevant years, the average profit per acre was estimated for 2005 to 2008 for the California industry. The estimates of net returns per acre above total costs were as follows: 2005 = $1,280; 2006 = –$25; 2007 = $690; and 2008 = $325.

The causes of the swings in profitability from year to year are twofold. First, the alternate-bearing nature of the trees means that low-yielding years are likely to generate lower revenues per acre. Second, significant risk factors (discussed in the next section) exist in a market that is geographically concentrated to the extent seen in U.S. pistachios. These risk factors need to be understood and included in any analysis of the economic outlook for pistachios.

Risk Issues

Historical risk analysis is important in selecting between alternative investments. It involves evaluating the expected volatility of income levels over the life of an investment. By using past (or expected) annual income data, a grower can estimate how widely net incomes vary from year to year. This volatility can be compared with average income levels over a time period to get an index of risk (called the coefficient of variation, which is calculated automatically by spreadsheets as the standard deviation of incomes divided by the average of incomes). This index can be compared across alternative investments to indicate relative risk levels.

For pistachios, historical risk levels have been similar to those typical of other tree crops but higher than those for the less profitable field crops. This risk is the result of fluctuations in price, yields, production costs, and market environment. Unfortunately, many of these sources of risk are related. In particular, industry yields and total output are directly linked to prices, as summarized below.

The biggest risk factor is due to the pistachio industry being concentrated in one area. The fact that 98 percent of pistachio production in the United States is concentrated in the southern part of California's San Joaquin Valley means that nearly all orchards are in the same microclimate. Thus, nearly all pistachio orchards are subject to the same production situations (e.g., the same alternate-bearing cycle). This causes supply swings in the same direction for nearly all producers and, hence, the entire industry. Next, these supply swings lead directly to price swings. As a result, nearly all pistachio producers in California have similar annual financial returns per acre in mature orchards. This is very different from the experience of producers in geographically dispersed industries, such as tomatoes; those growers can adopt a geographic diversification strategy to manage weather-based yield swings.

The second major source of risk is the fact that the domestic market for pistachios is volatile—and it may become more so as the large, nonbearing acreage comes into bearing. The California industry can overshoot the quantity demanded over time, thus depressing prices and incomes for periods of years. This situation happened to almond growers in the 1990s, even though that commodity market was much larger and more established compared with the current market for pistachios. In essence, overshooting results from individual decisions within a small industry—particularly when it is a small industry that produces a perennial crop with relatively good market prospects. Growers are attracted to the profit potential for such a crop and they plant new acreage. Unfortunately, it does not take much increase in acreage to expand supplies of the commodity beyond the level needed to satisfy demand. This problem is complicated by the fact that a pistachio tree normally takes 6 or more years to begin producing nuts and more than 9 years to reach full production. This means that farmers have to forecast the market demand over a decade in advance before making the decision to plant new acreage. If there is no coordination of planting decisions within the industry, too many farmers may jump in. This type of overshooting adversely affects all growers because prices drop significantly. For pistachios, there is already evidence of overshooting in the 2012 to 2013 data in table 1.2. Total industry revenues per acre were lower in 2013 than they were in both 2011 and 2012. With 47,000 nonbearing acres entering production in the next 3 to 5 years, the question is, can the market absorb a 20 percent expansion in California's output without large price effects?

The investment decision about whether to produce pistachios should be treated just like any other investment: tally the return (profit) and risk, and compare the two against the same information for alternative investments, such as other crops and putting the money in a guaranteed savings account. The fact that pistachio production is risky does not mean it is a bad investment—all agricultural commodity markets are risky. Each person has unique tolerances for risk and unique return-on-investment requirements. If producing pistachios offers a farmer a profit sufficient to compensate for the associated risks, then it is a good investment—and that assessment is a personal decision.

Opportunities and Strategies

There is no way to determine whether consumers will ultimately prefer pistachios to any other nut or snack food, yet it could happen. This illustrates the type of opportunities that farmers may see in a still-developing market. However, strategies for fulfilling any particular opportunity must be developed and executed before outcomes can be observed. Thus, the pistachio industry in California has many choices facing it now that will determine what opportunities the future will hold.

Choices facing individual farmers are very different from those facing the industry as a whole, but both types of choices need attention. Actions by individual farmers can influence only the supply of pistachios, whereas the industry can influence both demand and supply. The choices are briefly summarized below.

Strategies for Individual Farmers

The choices facing an individual farmer are (1) whether to produce pistachios and, if so, (2) what scale of operation to develop, (3) whether to invest in branding, and (4) whether to invest in industry infrastructure. These choices need to be considered in the order they are listed.

First, is an investment in pistachio production a good choice given alternative uses of the money? The answer depends on the rate of return expected on the investment, as well as the compatibility of pistachio production with the interests and skills of the farm. Part of this assessment will involve risk and the farmer's tolerance for it. As a person's level of risk aversion increases, it is increasingly likely that some degree of investment diversification is needed to reduce the farm's exposure to swings in total income over time. This means that people who are more risk averse should produce more than just pistachios, adding other commodities to their crop portfolio to spread their financial risk across commodity markets.

Second, how many acres of pistachios are a good fit in the total farming operation? This is part of the decision about putting together a crop portfolio to manage farm income fluctuation (risk) over time.

Third, are the operation's pistachios going to be sold as a commodity or as a branded product? Creating and promoting one's own brand are expensive endeavors, but they can generate higher prices. The price of a commodity is determined mostly by the interaction of supply and demand at any point in time, and thus a producer is a "price taker" from the market. A firm that sells a branded product that consumers believe has some unique attributes has some control over the price. Therefore, it can charge more per unit for that product compared with the price of the commodity from which the branded product is made. However, establishing that level of belief in sufficient numbers of consumers is very expensive and rarely possible to accomplish for a small firm, even though some firms have succeeded in selling small amounts of pistachios through the internet. Thus, sometimes farmers consider investing in the industry's efforts at brand development rather than try to establish their own brand.

Finally, can the industry take actions that will improve the financial results for the producer? For example, the scale of advertising and promotion costs associated with raising the level of demand for a commodity often makes it more feasible to undertake such an effort at the industry level, where the costs can be spread among all of the producers that will benefit by such actions. Industry actions can influence both the demand for and the supply of a commodity if done correctly, so they can benefit all producers. However, some industry infrastructure or organization is needed to facilitate industry activities.

Strategies for the California Pistachio Industry

The geographic concentration of growers in the San Joaquin Valley means they can come together to discuss methods for managing their market. Grower meetings and grower associations can facilitate communication that leads to strategy development. This is an opportunity not available to most commodity industries because, compared with pistachio producers, growers are usually much more scattered geographically.

One of the first questions addressed through grower communications could be, is there substitution between nut crops on the part of consumers? In other words, can consumption of pistachios be increased by competing directly with other nut crops for the impulse snack market within the United States? The answer is yes, but the costs and time frame associated with such a strategy must be determined at the industry level. Another approach might be for nut producers to jointly pitch nuts as a nutrition-rich alternative to snack foods and avoid competing against each other.

This latter type of strategy involves demand-enhancing activities that can be successful only if undertaken at the industry level because of the high total costs and marketing skills needed. Individual growers cannot afford to engage in a national marketing campaign, so it is a strategy that must be pursued jointly, with the costs shared by all of the growers who will benefit. Such a campaign can be based on many factors, such as the health benefits of pistachio consumption, but it would require marketing expertise to identify consumer preferences that are served by pistachios rather than by competing snack foods. Even more important, a successful campaign needs to expand consumer perceptions of pistachios in order to expand demand for these nuts beyond the snack market. One example is a marketing effort aimed at expanding food processing uses of pistachios. Such a marketing effort would most likely have a slow, cumulative effect over time, so it would require a long-term perspective as an investment of pistachio industry funds. Such an effort is under way in marketing shelled and broken pistachios and is becoming an increasingly important part of the marketing activities of the leading processor. This bodes well for market prospects.

In addition to influencing demand factors, an industry-wide approach can influence the supply of pistachios. The geographically concentrated nature of the California industry would facilitate efforts aimed at supply control. Many examples exist of this strategy being applied in other commodity markets, including the peanut quota in the Southeast, the adjustable quality and size limits on papayas in Hawaii, and to some extent raisin branding in California. The goal of such strategies is to gain control over total industry supplies of a commodity so that price levels can be controlled somewhat by the industry. The logistics of implementing such a strategy usually determine whether it will succeed, but a concentrated industry has more chance of succeeding than does a geographically diverse industry. The pistachio industry's carry-over program to compensate for the effects of alternate bearing on market supplies is working, thus benefiting all growers.

Pistachio growers showed an awareness of the potential for successful industry-wide strategies when they successfully led an initiative to establish a pistachio federal marketing order. Marketing Order 983 went into effect in 2005 and was designed to increase demand for pistachios and enhance future grower returns through quality control regulations. Funded by assessments on California pistachio handlers, the order established maximum tolerance levels for aflatoxin, mandated testing and certification for aflatoxin, and set minimum nut quality standards (which have since been suspended). Thus, it influenced both pistachio demand and supply.

Summary

The pistachio industry in California is relatively young and, as a result, not yet fully developed, thus creating many challenges and opportunities for industry participants. Although volatile, the average profit per acre for growers is positive and continues to attract additional acreage. This growth in industry size is expected to continue as long as export sales are profitable. The industry's rapidly expanding output is already overshooting the slowly increasing domestic demand for the commodity. Thus, export sales will increase, but those sales have often been less profitable in the past. Therefore, average profits per acre and producer income levels will continue to be volatile for the foreseeable future as marketing efforts in export markets become increasingly important. This calls for risk-reducing efforts on the part of growers, with the most likely strategy being diversification of crop portfolios. Given the limited strategies that individual growers can adopt to improve their profit margins, industry-wide strategies aimed at enhancing pistachio demand and managing market supplies may offer financial benefits to growers. Ultimately, the future of the pistachio industry depends on consumers, and that future would be brighter if pistachios were thought of as more than an impulse snack product. Nonetheless, over the next decade the high quality of California pistachios gives the industry much potential for sales growth, especially in export markets.

References

Beede, R., C. Kallsen, B. Holtz, L. Ferguson, K. Klonsky, and R. DeMoura. 2008. Sample costs to establish and produce pistachios. Low-volume irrigation. San Joaquin Valley. University of California Cooperative Extension.

Blank, S. C. 2008. The economics of American agriculture: Evolution and global development. Armonk, NY: M. E. Sharpe, Inc.

ERS (USDA Economic Research Service). 2013. Fruit and tree nuts outlook. July 26. United States Department of Agriculture, FTS-356. USDA ERS website, http://www.ers.usda.gov/publications/fts-fruit-and-tree-nuts-outlook/fts-356.aspx.

FAS (USDA Foreign Agricultural Service). 2015. Pistachios: World markets and trade. USDA FAS website, http://apps.fas.usda.gov/psdonline/circulars/TreeNuts.pdf.

NASS (USDA National Agricultural Statistics Service). 2014a. 2012 census of agriculture. USDA Census of Agriculture website, http://www.agcensus.usda.gov/Publications/2012/.

———. 2014b. Noncitrus fruits and nuts 2013 summary. USDA NASS website, http://www.usda.gov/nass/PUBS/TODAYRPT/ncit0714.pdf.

2 Production Costs

Karen M. Klonsky, Richard L. DeMoura, Robert H. Beede, Brent A. Holtz, and Louise Ferguson

Costs of producing pistachios can be divided into operating costs, business overhead costs, and ownership costs. Operating costs are typically thought of as the costs of production for an orchard. These include materials (such as fertilizer, water, and pesticides), labor, equipment, and custom services. Without these expenditures a crop will not be produced or harvested. But additional costs are incurred regardless of whether a crop is produced. These include business overhead and ownership costs. Business overhead includes taxes and insurance, while ownership costs refer to the costs of owning the assets involved in the farming operation, including the land, trees, equipment, buildings, and irrigation system. Detailed information for costs associated with establishing and producing pistachios is available from the UC Davis Department of Agricultural and Resource Economics at the UC Davis Cost Studies website, coststudies.ucdavis.edu. The following sections discuss the three categories of costs associated with pistachio production for a representative orchard in the San Joaquin Valley of California.

Operating Costs

Operating costs include the costs of day-to-day cultural practices of an orchard. Material inputs are discussed in the context of their function in pistachio production. Accounting for the costs of labor and equipment use is discussed separately. A calendar of operations for a representative pistachio orchard in the San Joaquin Valley is presented in table 2.1.

Table 2.1. Example of a calendar of operations for pistachio, San Joaquin Valley

Operation	Month
prune: hand-prune and stack	Jan
prune: shred prunings (custom)	Jan
winter sanitation: shake, rake, blow, chop	Jan
winter sanitation: disk (mummies and shredded prunings)	Feb
weed control: winter strip	Feb
fertilize: Zn, B	Feb
irrigate: water, labor	Mar–Oct
insect control: NOW (PCA hangs traps)	Mar
fertilize: N injected through drip system	Apr
disease control: Bot/Botrytis (fungicide)	Jun, Jul
insect control: plant bugs (insecticide)	Apr
weed control: disk middles, 3X	Jun, Jul, Aug
vertebrate control: gopher (gopher bait)	Apr, Aug, Nov
vertebrate control: squirrel (squirrel bait)	May
fertilize: N and K injected through irrigation system	May, Jun
weed control: spot-spray tree row 2X (herbicide)	May, Jul
disease control: Bot/Alternaria (fungicide)	Jun, Jul
fertilize: N and K injected through irrigation system	Jul
insect control: citrus flat mite (sulfur)/plant bug (insecticide)	Jul
insect control: NOW (insecticide)	Aug
leaf sampling (collected by PCA) and analysis (custom)	Aug

Pruning and Winter Sanitation

Trees are hand-pruned during the dormant season, usually from mid-November through February. Tipping cuts on long, 1-year-old fruitwood are made in combination with thinning cuts on larger, older, generally more interior branches. The goal is to distribute fruitwood and allow sufficient light into the canopy center. The prunings are hand-stacked in alternate row middles and shredded commercially by a custom operator, leaving the residue on the orchard floor. The residue is later incorporated into the soil during the winter (February) sanitation disking. The pruning and winter sanitation operations are typically carried out by custom operators but may be done by the grower and farm employees. Mechanical hedging and topping are becoming increasingly common as means of controlling tree growth. However, some hand-pruning, particularly large thinning cuts that allow light into the canopy, is a necessary complement to mechanical topping and hedging. Some growers place a circular belt around the expanding tree canopy to keep branches in a more upright position (See chapter 12, "Pruning Mature, Bearing Trees," for more detailed information.)

Winter sanitation refers to mechanical trunk shaking to remove the overwintering nuts (mummies) not removed during pruning. A hand-poling crew may be sent through to remove the remaining nuts. The berms are blown free of mummy nuts and debris using a tractor-mounted blower supplied by the grower. The row centers are then disked to incorporate the mummies and shredded prunings.

Fertilization

Pistachios are not luxury consumers; they do not consume more nutrients than needed, even if they are available in the soil. As a result, tailoring fertilizer applications to the time of greatest use is becoming more common in pistachio production. Nitrogen (N) is applied through the drip system beginning in late April, during early nut development, and continuing through July. Nitrogen (UN32) is applied alone in April and in liquid blends (10-0-10 and 15-0-5) combined with potassium (K). A total of 200 pounds of nitrogen and 125 pounds of potassium are applied annually; the potassium requirement of pistachios is high (see table 2.1). Zinc (Zn) is foliarly applied (as zinc sulfate 36%) at 50 percent canopy expansion from the end of April through May. Boron (B) may be applied with zinc to optimize flower nutrition at bud swell in February.

Leaf tissue sampling should be done annually to determine needed adjustments in the nutritional program. Leaf samples are taken from nonfruiting spurs in August. One sample per 25 acres is sent to a lab for analysis. A recently developed precision fertilization program based on spring leaf sampling is currently being implemented by a few growers but has not yet become standard practice.

Irrigation

Orchards are irrigated with either a single- or a double-line system. No research is available comparing single- with two-line systems. A properly designed, single-line system will meet the water requirements of mature trees. The selected system should be designed for 0.4 to 0.5 inch (0.03 to 0.04 ac-ft, respectively) of water per acre per day. As an example, four 1-gallon-per-hour emitters per tree deliver 0.45 inch of water per acre per day. Growers in the San Joaquin Valley irrigate with water from wells and from surface sources. Water quality and power requirements can vary considerably, depending on source. (See chapter 6, "Microirrigation Systems and Fertigation," and chapter 13, "Irrigation Management," for more detailed information.)

In our representative orchard (table 2.1), the first irrigation in March replenishes the water in the root zone and prepares the system for use during the season. Irrigation costs include pumping costs, use of an ATV (all-terrain vehicle) to check the emitters, and labor for cleaning the filters, repairing the lines, and monitoring all the emitters. Irrigation in April is primarily for applying fertilizer. The trees are assumed to have a seasonal consumptive water use of 42 acre-inches and the irrigation efficiency is 90 percent. Therefore, a total of 45 acre-inches is applied during the year.

Pest Management

Pesticides and application rates are listed in the UC IPM Pistachio Pest Management Guidelines, ipm.ucanr.edu/PMG/selectnewpest.pistachios.html. Licensed pest control advisers (PCAs) provide the written recommendations required for many pesticides. In addition, PCAs monitor orchards for pest, disease, and nutritional problems. Growers may hire private PCAs or receive the service as part of an agreement with an agricultural chemical and fertilizer company. In our representative orchard, the grower contracts with a private PCA. Adjuvants are recommended for many pesticides for effective control and are an added cost. Pesticide costs may vary by location and grower volume.

Weeds. Pre- and postemergent herbicides are applied as a winter strip spray to the tree row in February following winter sanitation. Weeds in the tree rows during the growing season are controlled with two herbicide spot sprays, in May and July. Each application is assumed to be applied to 33 percent of each berm, or 11 percent of the total acreage. The row middles are disked three times: in April, June, and August. (See chapter 11, "Weed Management," for more detailed information.)

Insects and mites. In mid-March, the PCA hangs traps for navel orangeworm (NOW) in the orchard, one trap per 10 acres. The PCA monitors the traps weekly from mid-March to the end of July. The trap cost is included in the PCA monitoring fee. From July to late August, the field is monitored for NOW damage by observing early-split nuts. During harvest, NOW egg-laying activity is monitored in split nuts and maturing hull tissue. In mid-August, a NOW cover spray is applied. Wettable sulfur is applied in July to control citrus flat mite. In April and July, an insecticide is applied to control plant bugs (such as stink bug, *Lygus*, leaffooted bug, *Phytocoris* spp.). The July citrus flat mite and plant bug sprays can be combined to save on application costs and can assist in NOW control, depending on the time of application. (See chapters 18 to 24, on insect pest management, for more detailed information.)

Diseases. Two foliar fungicide applications are targeted for Botryosphaeria panicle and shoot blight (Bot), Botrytis blossom and shoot blight (Botrytis), and Alternaria late blight (Alternaria). Treatment timing depends on the disease or diseases most prevalent. Rain during bloom may cause Bot or Botrytis infection. Bot can be a serious problem in some areas or years. It is a panicle (nut cluster) and shoot disease that can cause deterioration. Controlling Bot with fungicides and pruning can be very expensive. Pruning out infected clusters and 1-year-old branches during the winter is tedious and expensive, but it is critical to controlling the disease. Fungicide (sometimes applied as two sprays at one-half the top label rate) sprayed at bloom for Botrytis control reduces the primary Bot inoculum. In addition to the removal of infected rachises during pruning, orchards with Bot may require several fungicide applications beginning in April to mid- or late July. (See chapter 26, "Foliar, Fruit, and Branch Diseases," for more detailed information.)

Vertebrate pests. Gophers are baited year-round at 3- to 4-month intervals (e.g., April, August, and November) and squirrels in May. The grower-supplied worker uses an ATV to move around the field for baiting purposes. Baiting can sometimes be incorporated with weed spot spraying to save labor costs. Also, bird damage to the maturing crop can be a major problem in some areas, and control costs are highly variable. (See chapter 27, "Vertebrate Pest Management," for more detailed information.)

Harvest

Pistachio trees typically reach full production by the 12th or 13th year. Two methods are used for commercial harvest: the bulk method and the bin method. The bulk method is now more common, and it is used in this orchard example. The costs are approximately the same for both methods. Pistachios are harvested mechanically using a shaker with tarpaulins supported above the horizontal shaker head. The shaker travels down one side of the tree in unison with a catch frame harvester (receiver) on the opposite side. The shaker hydraulically clamps and vibrates the tree trunk, dislodging the nuts onto the tarpaulins and receiver. The bin harvest system employs 4-by-4-by-3-foot wooden or plastic bins, which are distributed throughout the field and carried four at a time on the receiver. Full bins are dropped back in the row and picked up six at a time by a bin carrier. The bins are delivered to a loading area, where they are dumped into large bottom-dump trailers using a specialized forklift that picks up and clamps the bins and then rotates the nuts into the trailers.

The bulk harvest system, by comparison, uses a large trailer attached to the receiver, which continuously conveys harvested nuts into the trailer until it approaches capacity. A mobile bankout wagon then butts up to the back of the receiver trailer and actuates a lever that transfers the nuts into the bankout wagon by way of a cleated conveyor belt incorporated into the floor of the trailer. This eliminates the need for the harvesting equipment to stop for unloading. The bankout wagon then travels to a loading area. The nuts are dumped onto elevators that deposit the crop into large bottom-dump trailers for hauling to the processor. Upon arrival at the processor, the nuts are weighed, hulled, dried, graded, and packed. (See chapter 17, "Harvesting, Transporting, Processing, and Grading," for more detailed information.)

Yields and Returns

Pistachios are an alternate-bearing crop, having a high yield one year and a low yield the next year. Although an economic yield usually begins in the sixth year, the alternate-bearing cycle begins when the trees are between 10 and 12 years old. An average of the high-low yields cycle is shown in table 2.2. Marketable yields are divided into three categories: split in-shell (unstained and light-stained split in-shell); shelling stock (dark-stained split nuts, undersized nuts, and nuts with adhering hulls, loose kernels and shells, or damaged shells); and closed shell. A typical breakout is 80 percent split in-shell, 5 percent shelling stock, and 15 percent closed shell (see table 2.2).

Table 2.2. Example of annual pistachio yields by category (lb/ac)

Year	Total yield	Split in-shell (80%)	Shelling stock (5%)	Closed shell (15%)
6	600	480	30	90
7	1,300	1,040	65	195
8	2,200	1,760	110	330
9	2,800	2,240	140	420

The total yield and the breakout by category collectively determine a grower's gross return. Blanks and other unmarketable nuts (such as those with insect and vertebrate damage) are not included in the total yield. Shelling stock yields, as they appear in table 2.2, include the weight of the kernel (i.e., the nutmeat) and the shell. However, growers should be aware that processors pay only for the kernel in the shelling stock and closed-shell categories. Prices received by growers for their marketable products vary by category. Usually, there are not enough loose kernels (< 0.5%) to affect a grower's return. Loose kernels are normally grouped with the shelling stock. The split in-shell payment is for the shell plus kernel weight. The payment for closed shell and shelling stock is based on the kernel price. For shelling stock and closed shell, the kernel is assumed to be 50 percent of the total nut weight (shell plus kernel). Therefore, the actual price received by growers for these categories is 50 percent of the kernel price.

Assessments

Under a state marketing order, mandatory assessment fees are established, collected, and administered annually by the California Pistachio Research Board. Currently, growers are charged an assessment on total production to pay for industry research programs.

Labor and Management

Labor rates for machine operators and for general labor include payroll overhead of approximately 36 percent. The overhead includes the employer's share of federal and California state payroll taxes, workers' compensation insurance for orchard and nut crops (California Food and Agricultural Code § 0045), and a percentage for other possible benefits. Workers' compensation costs will vary among growers. Labor time for operations involving machinery is typically 20 percent higher than the operation time for equipment use to account for the extra labor involved in equipment setup, moving, maintenance, work breaks, and field repair.

Some orchards employ a salaried manager. About three-quarters of pistachio growers hire professional management services. The manager makes all production decisions, including cultural practices, action to be taken on pest management recommendations, labor, and approval of invoices payable by the grower. This cost will vary depending on the size of the orchard and benefits provided, such as housing or a pickup truck.

Equipment Operating Costs

Equipment operating costs consist of fuel, regular maintenance, and repairs. Assigning fuel costs to an operation requires knowledge of the fuel use per hour for tractors and the number of hours each tractor is used in the pistachio orchard. Most farms own a mix of used and new equipment. In general, new equipment incurs high up-front costs and low repair costs, while used equipment is paid off but repair costs increase with use. Growers need to keep track of the repair costs for individual pieces of equipment in order to know the true cost of production. Also, for a farm with multiple enterprises, the grower needs to record the number of hours of use for each piece of equipment for the pistachio enterprise so that the equipment costs can be allocated accurately. Some managers want to know the cost of individual farming operations and the costs of one orchard block in comparison with another, so they record the equipment used and hours spent in each block.

Interest on Operating Capital

Some farming operations are self-financing, which means that they do not take out annual operating loans for material, labor, and custom services, and therefore there are no cash interest payments. However, when a farm business self-finances, it ties up capital in the orchard until payment is received for the crop. Although there is not a cash cost, there is an opportunity cost of tying up funds that could have been invested in another way over the time period. Therefore, this cost of capital should be accounted for when assessing profitability. In practice, many farms use a combination of borrowed funds and self-financing.

Business Overhead

Business overhead consists of various cash expenses during the year that are assigned to the whole farm and not to a particular enterprise. For farms involving more than one enterprise, these costs must be allocated to each of the enterprises. This can be done in a variety of ways, but most commonly it is done on a per-acre basis or by the percentage of total farm revenue generated by the enterprise.

Property Taxes

Counties charge a base property tax rate of 1 percent on the assessed value of the property. In some counties, special assessment districts exist and charge additional taxes on property, including equipment, buildings, and improvements, which include orchard trees.

Insurance

Insurance for farm investments varies depending on the assets included and the amount of coverage. Property insurance provides coverage for property loss and is charged on the value of the farm assets, including buildings.

Office Expenses

These expenses include office supplies, telephones, bookkeeping, accounting, legal fees, shop and office utilities, and miscellaneous administrative charges.

Sanitation Services

Sanitation services provide double portable toilets, washbasins, soap, and towels for the orchard. The cost includes delivery and 12 months of weekly service. Facilities must be in compliance with the regulations of the California Department of Industrial Relations, Division of Occupational Safety and Health.

Investment Repairs

Repairs on buildings, roads, and the irrigation system must be allocated to the farming enterprises to correctly calculate profit.

Ownership Costs

Equipment Ownership

For equipment that is owned outright, growers need to know the salvage value for tax and insurance purposes. This information is also needed to value the capital tied up in equipment so that the returns on total investment can be calculated. The principal and interest costs on equipment loans must be included and allocated to the farm enterprises.

Orchard Establishment

Costs to establish the orchard are used to determine depreciation and interest on investment for the production years. Establishment cost is the sum of the costs for land preparation, planting, trees, and cash overhead and production expenses for growing the trees through the first year that pistachios are harvested minus any returns from production in the early years. Regardless of whether the trees are owned or there is still an outstanding loan for establishment costs, the total establishment expenditure is used in calculating returns on investment.

Drip Lines

In our representative orchard, a single drip line is used during the early establishment years and a second drip line is installed once the trees are established. Single drip lines are laid out prior to planting. The labor cost for laying out the line is included in the irrigation system cost. The second line is installed during the fifth year, and the materials and labor are included in the ownership costs.

Irrigation System

Water may be delivered from an irrigation system, pumped groundwater, or a combination of both. The system cost includes a booster pump, filtration, fertilizer injector, and mainlines. Costs will vary depending on the size and location of the orchard and also the water source.

Land

Open cropland values vary widely. Again, the farm may have a cash outlay for land in the form of a loan, or it may own the land outright—or a combination of the two may apply. In either case, the capital tied up in the land must be accounted for.

Buildings, Tools, and Fuel Tanks

Buildings include any used in the farming operation for storage, office operations, or maintenance. Tools include shop tools, hand tools, and miscellaneous field tools such as those used for pruning. Most fuel tanks use gravity feed and are on metal stands. The tanks are set up in a cement containment pad that meets federal, state, and county regulations.

Profitability of the Orchard—the Bottom Line

The cost of pistachio production varies for a variety of reasons. Most notably, the cost attributable to the land and orchard establishment varies depending on location, age of the orchard, and orchard condition. Other factors that may differ widely among farms include the costs of buildings and the irrigation system. When the opportunity cost of the orchard is included, capital recovery costs (ownership costs) represent about one-third of total costs. Harvest costs are only 6 percent of total costs, and cash business expenses (taxes, insurance, and repairs) are another 10 percent of total costs. Cultural costs are about half of total costs. The biggest factor in terms of cultural costs is irrigation, which can typically be over one-third of all cultural costs (37% in our orchard). The next-largest contribution to cultural costs is pest control for insects, diseases, and vertebrates (at 25%), while weed control is only 4 percent of cultural costs. Fertilization accounts for 18 percent of cultural costs, and pruning another 12 percent.

Once all of the inputs and associated costs of production have been identified and quantified, the profitability of the orchard can be calculated. The total revenue minus the operating costs equals the net return above operating costs. If this number is negative, then the orchard did not generate enough income to cover the cultural costs, and the business would have been better off without producing a crop. The positive net returns go, in part, to paying the cash business overhead—primarily taxes, insurance, and sanitation. Once these are also deducted from the gross returns, the remainder is the return on capital investment and management. Then the cash ownership costs of loans, as well as the cash payments to a hired manager, should be deducted. This balance is the return on the owner's capital investment and management input into the operation. If this number is negative, the pistachio orchard production is eroding the equity embodied in the orchard and equipment.

References

University of California, Davis, Department of Agricultural and Resource Economics. Cost Studies website, http://coststudies.ucdavis.edu.

University of California Statewide Integrated Pest Management Program. UC IPM Pistachio Pest Management Guidelines. UC IPM website, http://www.ipm.ucanr.edu/PMG/selectnewpest.pistachios.html.

Part 2 The Tree

3 The Pistachio Tree: Physiology and Botany

Louise Ferguson and Craig E. Kallsen

Origin and History in California

The multiple species of the genus *Pistacia* comprise both the male and female commercial cultivars, as well as the rootstocks they are propagated on. *Pistacia* species are native to latitudes between 40° and 70° in western Asia and Asia Minor. *Pistacia vera*, the species used to produce commercial nuts, and multiple other *Pistacia* species are still growing wild in numerous desert locations in Lebanon, Israel, Syria, Iran, Iraq, India, Kyrgyzstan, Afghanistan, southern Europe, and the desert countries of Asia and Africa. The first pistachio introductions to the United States were by the U.S. Department of Agriculture (USDA) plant exploration service in 1890. The first introductions to California were planted at the Plant Introduction Station in Chico in the northern Sacramento Valley in 1904. This collection was the initial source of the current California pistachio industry's major female and male cultivars, Kerman and Peters, though the latter was a selection identified in a Fresno, California, orchard. The newer patented California cultivars—the male Randy and females Lost Hills and Golden Hills—also have parents from this collection. This collection also provided the two *Pistacia* species, *P. atlantica* and *P. integerrima*, and the hybrids from them that are now the major California pistachio seedling and clonal rootstocks.

The first significant California plantings occurred in the 1960s, with the first recorded commercial harvests in the early 1970s. The California industry's initial rapid development was aided by a climate that provided sufficient winter chill; sustained arid summer heat without rain and with low relative humidity; deep, loamy soils with high boron concentrations; and affordable, available water. As of 2016, pistachio constituted almost 310,000 acres in California's central San Joaquin and northern Sacramento Valleys. The most productive orchards and major concentration of orchards are located on the lower west side of California's San Joaquin Valley.

Historical and current acreage, production, and gross value statistics can be viewed at the Administrative Committee for Pistachios website, acpistachios.org/statistics.shtml.

As can be seen in the average annual historical yields given on the above website, pistachio orchards are alternate bearing, having high and low crops in alternating years. Furthermore, the alternate-bearing cycles in individual orchards appear to become synchronized over the entire southern San Joaquin Valley, resulting in industry-wide high and low nut production in alternate years. This phenomenon is common to many tree crops but much more marked in pistachio. It does not appear to harm tree health or longevity. The physiology of alternate bearing and possible methods of mitigation are discussed in more detail in chapter 28, "Alternate Bearing, Nut Blanking, and Shell Splitting."

Botany

The pistachio of commerce, *Pistacia vera*, is the only edible species among the 11 species in the genus *Pistacia*; all are characterized by their ability to exude turpentine or mastic. Several are referred to as pistachios, but the name is generally reserved for the edible nut of commerce, as only *Pistacia vera* produces a large enough seed to be a commercial nut. *Pistacia vera* is a member of the family Anacardiaceae, as are cashew, mango, poison ivy, oak, pepper tree, and sumac. The hull of pistachio produces urushiol, which causes skin rashes and other allergic reactions. Botanically, pistachio fruit are semidry drupes, the same classification as almonds, peaches, apricots, cherries, and plums. All drupes consist of three parts: a combined exocarp and fleshy mesocarp (which form the fleshy hull), and a hard endocarp (called the shell, which encloses the seed, i.e., the edible kernel) (fig. 3.1). The nut of *Pistacia vera* is different from that of the other smaller-seeded *Pistacia* species in that, as it matures, the shell splits along a suture on one side of the shell. As the shell splits, the enclosing hull remains intact or partially intact (fig. 3.2). In California, at harvest, the nut is shaken from the tree, caught on a catching frame, and quickly transported to the processing plant, where the hull is physically removed before it degrades and stains the shell or dries and adheres to the shell.

Pistachio trees are capable of growing moderately tall (25 to 35 ft), and, if not trained to a trunk as in figure 3.3, they will grow as a bush with multiple trunks. Once the tree is mature both vegetative and reproductive growth is apically dominant, growing or setting fruit primarily at shoot and flower apices. The leaves are large (called imparipinnate compound leaves), grayish green, and have three to five roundish leaflets that are 2 to 3 inches long.

California pistachios are grown on rootstocks that are also within the genus *Pistacia* but are a different species. The current commercial rootstocks were selected for, in order of importance, disease, cold and salinity tolerance, and yield enhancement. Current rootstocks of choice include both seedling and clonal *Pistacia integerrima* and the two reciprocal hybrids of *P. integerrima* and *P. atlantica*. These rootstocks are capable of exploiting shallow soils with lateral spreading roots and deep soils with taproots. There have been multiple observations of wild pistachios surviving, and producing, without irrigation in arid climates. However, to produce commercially viable yields, irrigation is required. Recent research has demonstrated a differential ability among the currently used pistachio rootstock species to exclude sodium from the transpiration stream by sequestration in the vacuoles of rootstock cells. A rootstock with the ability to protect the scion by decreasing sodium transport to it would be valuable for both current orchard production (as pistachios are increasingly planted in saline soils and irrigated with saline groundwater) and future plant improvement programs.

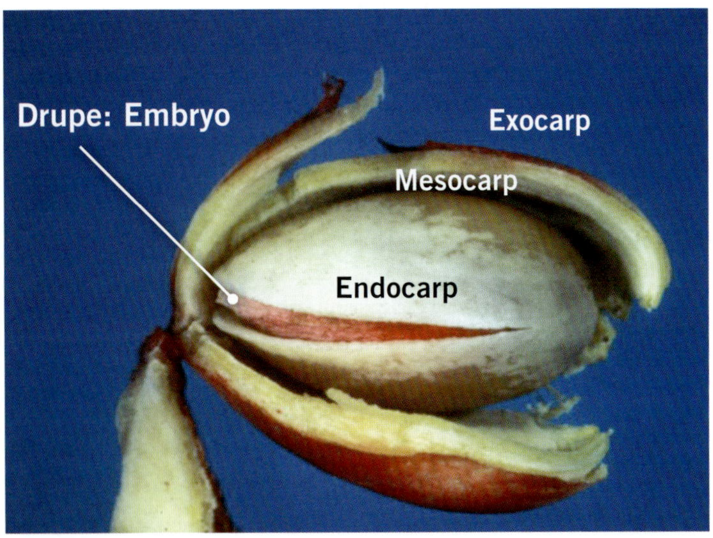

Figure 3.1. A fully mature and naturally ripened pistachio nut (embryo or seed), with the shell (endocarp) naturally split and the hull (internal mesocarp and external exocarp) puffed and turning red. *Photo:* L. Ferguson.

Figure 3.2. A cluster of nuts near harvest. Generally, as harvest approaches, the nuts split and the covering hulls remain largely intact, becoming puffed away from the shell and developing a red blush. *Photo:* B. Holtz.

Figure 3.3. Mature pistachio tree trained to a single trunk, with strong, apically dominant canopy growth habit. *Photo:* L. Ferguson.

Climatic Requirements

Pistachio trees are temperate deciduous trees with physiological adaptations to desert and saline environmental conditions. Areas suitable for pistachio production have long, hot, dry summers with no rainfall, humidities below 50 percent during the growing season, and moderate winters with sustained temperatures in the 40s (°F) and no hard freezes. Since the beginning of the commercial pistachio industry in California in the early 1970s, there have been few midwinter freezes that significantly impacted the California pistachio industry. Overnight freezes down to 4° to 11°F have damaged 1-year-old shoot growth but not killed major scaffold or trunk wood on mature, dormant trees. The Kerman scion appears to be more tolerant of extreme cold than the rootstock species. With the introduction of the Verticillium wilt-tolerant and more vigorous and less cold-tolerant *P. integerrima*, as opposed to *P. atlantica* (parentage rootstock), late-fall and early-winter frost damage in young, nondormant trees is an increasing threat. Frost damage also appears to be increasing in young, vigorous orchards established in saline conditions. The soils in these orchards are often more saturated late in the season, delaying young, vigorous trees from entering dormancy and making them more susceptible to the late-fall freezes of recent years.

Chilling and Heat Unit Requirements

Both female and male pistachio trees have winter dormancy or chill requirements for bloom and vegetative budbreak. Generally, the most limiting climatic factor in successfully establishing pistachios is not freezing temperatures, as with most tree crops, but rather lack of the winter chill that this temperate species requires. The current method of calculating chill accumulation is a straight accumulation of hours below 45°F, with no differential weighting of higher or lower temperatures. The limited information generated suggests that the female Kerman requires approximately 800 to 900 hours and the male 900 to 1,000 hours below 45°F. In the San Joaquin Valley of California, when using the older method of chill accumulation estimation, cumulative hours below 45°F fell to 670 hours during the winter of 1977 to 1978. The resulting spring bloom and foliation of both Kerman and Peters was irregular and delayed. In the past decade, winters in California's central San Joaquin Valley have become increasingly erratic, with earlier freezes, less sustained fog cover, and more sunlight. The winters of 2013 to 2014 and 2014 to 2015 produced particularly erratic bloom, even when recorded accumulated chill hours below 45°F were sufficient. The recent weather patterns and the effect on bloom and nut set suggest that the newer method of chill accumulation estimation, the dynamic model, which differentially weights the winter temperatures above and below 45°F and incorporates early-spring heat units, would be a better method for estimating chill (Erez et al. 1990; Glozer 2008). A recent preliminary study determined that 59 chill portions will provide good bloom in the female Kerman, while approximately 69 chill portions are needed for the male Peters cultivars (Pope 2014). The typical pattern of the leaves and flowers on the cooler north side of the tree breaking and blooming before those on the warmer, more sun-exposed south side supports adoption of this method of chill calculation. A dynamic-model method of chill accumulation estimation could account for the reversal of the total chill units by the warming effect of sunlight and yield a more accurate estimate of accumulated chill. Thus far, in the San Joaquin Valley of California, total chilling accumulated during the prior winter, regardless of the current methods available for calculation, has not correlated well with subsequent harvested yields. Furthermore, recent information suggests that winter freezes may also decrease final yield by diminishing the tree's carbohydrate storage available for nut production.

The heat unit requirement for pistachio has not been well established. Based on where pistachios are grown worldwide, they require approximately 3,960 to 5,040 heat units (growing degree-days, or GDD) to mature. The following equation is used to calculate the heat unit from March through September (with temperatures in °F):

$$\text{Monthly heat unit accumulation} = [(\text{Mean monthly maximum temperature} + \text{Mean monthly minimum temperature}) \div 2) - 45 \text{ or } 50] \times \text{Number of days in the month.}$$

$$\text{Seasonal heat unit accumulation} = \text{Sum of monthly accumulations for 3/1 to 9/30}$$

A 2-year study (from 2012 to 2014) of pistachio fruit development as a function of temperature, using a base temperature of 45°F and monitoring growth by size and dry weight from nut set through nut split, demonstrated that the three currently commercial California cultivars—Kerman, Golden Hills, and Lost Hills—required between 2,500 and 3,000 GDD to mature through maximum shell hardness and split.

Elevations of 200 to 800 feet above sea level have proven to produce good yields of pistachios in the southern San Joaquin Valley of California. In 47 separate seasonal measurements at eight sites in the pistachio-growing areas in the San Joaquin Valley from 2002 through 2013, seasonal heat unit accumulation (calculated from March 1 to September 30 on an hourly basis with no maximum cutoff and 50°F as a base temperature) averaged 4,176 GDD, with a low of 3,368 and a high of 4,865. One early and economically significant symptom of borderline heat unit insufficiency is a reduction in the number of split nuts at harvest.

Even if heat unit and winter chill accumulations are sufficient, other environmental factors may make pistachio production marginal. At higher elevations, above 2,500 feet in central California, it is not uncommon for late-spring frosts to freeze new bloom. Late-spring rain and strong, desiccating winds hamper pollination by interfering with pollen dispersal and quality. Precipitation or high humidity through the growing season can greatly increase foliar fungal infections. Excessive rain at harvest may result in poor separation of the hull from the shell and increase shell staining.

Tree Growth and Reproductive Development

Seasonal Tree Growth

Once mature, the nut-bearing female tree and pollenizing male pistachio tree bear their inflorescences on 1-year-old shoot growth. Bloom in early to mid-April overlaps with vegetative shoot growth. Shoot growth begins in late April, shortly after bloom ends. Subsequent shoot growth flushes can be classified as preformed or neoformed (formed during the previous season or the current season, respectively). Preformed growth is produced from the overwintering terminal buds that received winter chilling and contain the primordia that extend and produce eight or nine leaves the following season. Preformed growth continues through late May. New vegetative, preformed shoot growth produces a shoot with lateral, alternating leaves with inflorescence buds in their axils and one to three apical vegetative buds. Sometimes there are additional flushes of vegetative shoot growth in late June and July if the trees are grown on *P. integerrima* parentage rootstock. This additional vegetative growth is called neoformed growth because the bud is formed and extends within the same season, with no intervening dormant period. Generally more prevalent in young trees, it may contribute to nonbearing juvenile trees reaching full tree size sooner. In mature trees, however, this whiplike growth is generally confined to the tree crown (often removed by dormant pruning), and it contributes little to yield the subsequent year (Spann et al. 2007, 2008).

Pistachio leaves are well adapted to hot, dry climates. They have thick, waxy cuticles to prevent water loss. A well-developed epidermis and palisade layer of cells protect and support the chloroplasts in the aqueous environment necessary for photosynthesis. The stomata, which function to let ambient air into the plant to facilitate photosynthesis, are located on both leaf surfaces. However, the density and conductance of the stomata on the lower leaf surface are greater than those of the stomata on the upper, sun-exposed surface. The stomata are also markedly unresponsive to low humidity and high heat. The result is a leaf able to maintain the turgor and temperatures conducive to photosynthesis under the hot, dry conditions of California's southern San Joaquin Valley (Al-Saghir et al. 2006).

Reproductive Relationships, Flower Bud Development, and Pollination Requirements

Pistachio trees are dioecious, meaning that female trees produce only functional female flowers (figs. 3.4 and 3.5) and male trees only functional male flowers (fig. 3.6). As nuts can only be produced from female flowers that require pollenization with the pollen produced by the male flowers, both trees are necessary for a producing orchard. In both sexes, flower buds are conspicuously larger than vegetative buds. Pistachio trees, under the intensively fertilized and irrigated production practices in the San Joaquin Valley of California, begin producing small but economically harvestable yields when 5 years old and reach full production in 10 to 12 years. Typically, the standard industry female cultivar, Kerman, begins alternate bearing when 8 or 9 years old, or the year after the first large crop is produced.

The tree simultaneously bears the current year's crop on shoots grown the previous year, as well as the buds for the following year's crop on the new apical growth of the same shoot. Nuts are borne in panicles on a rachis, referred to as a cluster (see fig. 3.2) on 1-year-old wood. The development of a cluster begins three seasons before it is harvested, as follows. At cessation of vegetative growth in season 1, usually in June, shoots terminate in a large, apical, terminal bud. In trees of reproductive age, these large terminal buds contain the primordia that will develop, next season (season 2), into a new shoot, complete with leaves and their axillary vegetative and reproductive buds. In California's San Joaquin Valley, this terminal bud begins to grow in March of season 2. As the new shoot grows, some buds in the axils of the expanding leaves on this new shoot are vegetative, and some, in late April, will rapidly begin differentiating into male or female flower buds, depending on the sex of the tree. Early flower bud differentiation includes initiation of inflorescences and their floret primordia. Differentiation continues through mid-June of season 2, and, by early July, further differentiation within these axillary flower buds ceases, and they enter a quiescent period. Prior to the off-year of the alternate-bearing cycle, many of these differentiating flower buds will not make it past this stage. They will fall from leaf axils beginning in July and continuing through October. The number of flower buds remaining in the leaf axils on shoots of a female tree in November of season 2 can provide a visual clue as to the yield potential for the following year. Alternate bearing is discussed in detail in chapter 28, "Alternate Bearing, Nut Blanking, and Shell Splitting."

In October of season 2, female florets within the axillary flower buds resume growth briefly with pistil initiation and then remain quiescent until March of season 3, when differentiation of functional female flower parts is completed. Male florets remain quiescent from July of season 2 until the following March, at which time they complete differentiation and mature by mid-spring.

In early April of season 3, both male and female trees bloom. Within the first 2 weeks of April, each male and female inflorescence, produced from a single axillary flower bud, grows into branching panicles that may consist of up to 200 individual florets. Fully formed florets are tiny, brownish green, and devoid of nectaries and petals (fig. 3.7). Each female floret has a single, superior ovary. Bloom occurs immediately before or, more commonly, during initial expansion of the first leaves. The male flowers bear pollen that is spread by wind, which pollinates the female flowers. Cross-pollination begins, with the basal florets first, when the total length of the growing inflorescence is approximately 1 inch in length and before it exhibits much panicle branching (see fig. 3.7). Bees will visit the male flowers to collect pollen but are not involved in the pistachio flower pollination. As the peduncle (stalk) of the flower bud and its attached panicles increase in size and lignify, it is referred to as a rachis. The rachis and the attached nuts (each produced from a fertilized floret) together are referred to as a nut cluster.

Figure 3.4. Female pistachio inflorescences shortly after fruit set. *Photo:* L. Ferguson.

Figure 3.5. Female inflorescences just entering bloom. Most of the florets on the expanding inflorescence have not yet opened. The florets are apetalous and have no nectaries. The flower is wind pollinated. *Photo:* L. Ferguson.

Figure 3.6. A recently expanding Randy male inflorescence. *Photo:* C. Kallsen.

Figure 3.7. A newly expanding Golden Hill female inflorescence, showing stigma opening on basal florets. *Photo:* C. Kallsen.

Bloom, Pollination, Growth, and Maturation

If dormant chill accumulation is adequate, approximately 10 percent of the 100 to 200 flowers per rachis will set fruit, averaging 14 nuts per rachis (Crane and Iwakiri 1981). Generally these are the most apical nuts on the individual stems within the rachis. Abscission of the unpollinated flowers occurs within 3 to 4 weeks after bloom. Depending on the cultivar, a percentage of the nuts (usually 5 to 30%) remaining on the trees will not contain a kernel. These fruit are generally called blanks. The primary cause of blank nuts is embryo abortion (Polito 1999). The production of blanks varies according to the cultivar. Blanks, and partially filled nuts called nonsplits, can have a significant effect on final harvested yield. As a result, they are referred to as physiological disorders. The causes of blank and nonsplit nuts are discussed in chapter 28, "Alternate Bearing, Nut Blanking, and Shell Splitting."

Pistachio nuts have an unusual growth pattern. In the first half of the season, the shell and hull develop; in the latter half of the season, the kernel grows to fill the shell. This occurs in three growth stages, identified as Stages I, II, and III. Stage I, from mid-April through mid-May in the San Joaquin Valley of California, is a period of rapid increase in the size and weight of the shell and hull. In Stage I, the pistachio shell achieves full size but is empty, since the kernel has not yet begun to grow. In Stage II, from mid-May to late August (depending on the cultivar), the shell does not increase in size but thickens and hardens by lignification, and little growth of the kernel occurs. The shell is high in hemicelluloses (38.5% in mature shells) and xylan (33.1%), around which lignin (29.4%) provides a rigid structure (Salgado et al. 2012). Initially the cells in this layer are flexible and lack these structural elements. This stage of growth is the result of lignification of only these endocarp cells. During a lignification stage, the cells begin to produce smaller amounts of the hemicellulose and lignin components, transporting these elements to the cell wall, where they are polymerized by laccases, peroxidases, and other chemical couplers (Li and Chapple 2010).

In mid- to late June, the beginning of Stage III, the kernel begins to grow, achieving nearly full size by late July or early August (fig. 3.8). The weight gains from late June through August are from kernel growth and maturation. This unusual growth pattern allows flexibility in irrigation, as there are times (during bloom, nutshell expansion, and kernel growth) when decreasing irrigation could decrease nut quality and yields through effects on size, weight, and embryo growth. Decreased irrigation during Stages I and III has been demonstrated to decrease nut size and increase nonsplit nuts, respectively, while decreased irrigation during Stage II has little effect on final yield or quality (Goldhamer et al. 2004).

The shell begins to split in late July. Full maturity (in late August or during September, depending on the heat unit accumulation and cultivar) is indicated by the easy separation of the hull from the shell. When the hull begins to separate (i.e., slip) from the shell on the majority of the nuts, the shell simultaneously begins splitting from the tip to the base, along the lateral sutures of the shell.

As the pistachio nut develops through the season, simultaneous changes occur in the chemical composition of the kernel. Pistachio kernels are initially composed of sugars, predominantly sucrose, followed by much lower levels of glucose, fructose, and inositol, a sugar alcohol. Sucrose increases sharply during the first month of kernel growth, peaking at about 35 percent dry weight in early July and decreasing rapidly to about 6 percent in mid-August, where it remains through maturity. Crude fats comprise about 4 percent of kernel dry weight in mid-July. As sugar content decreases in late July, crude fats sharply increase to 40 percent by the end of August.

Pistachio Nut Maturity and Shell Splitting

Color development of the hull also has never been well correlated with nut maturity. Nut dry weight, crude fat percentage, and percentage of nuts with fully split shells reach maximum values in late August to early September, peaking at the same time as the hulls begin to separate from the shells. Therefore, in addition to natural shell splitting, easy separation of the hull from the shell is the primary indicator of physiological maturity and dictates harvest timing. However, the actual mechanism of shell split has not been determined (Polito and Pinney 1999). Some evidence suggests that shell split is related to kernel growth and size and that the growing nut forces the pistachio shell to split (Goldhamer et al. 2004). As hull separation from the shell and shell split do not occur in blank nuts, this suggests that hull maturation and nut split require the presence of a nut kernel and a more involved process than simply an increase in kernel size.

In late July and August, leaves distal to heavily fruiting clusters often display a marked depletion of chlorophyll and early senescence. This early senescence appears to be the result of the nut fill and maturation of the current nut crop, depleting the shoot's stored carbohydrates and nitrogen in addition to the supply produced by the surrounding leaves. This senescence is normal for nut crops but is particularly pronounced in pistachio. Attempts to reverse this senescence with foliar fertilizer sprays have not proven successful. The severity and earliness of this senescence appear to be correlated with the alternate bearing that characterizes pistachios, suggesting, as have most of the recent physiological investigations, that alternate bearing is a function of the tree's fluctuating carbohydrate status, particularly during the growing season in the 1- and 2-year-old branches. Alternate bearing is discussed in more detail in chapter 28, "Alternate Bearing, Nut Blanking, and Shell Splitting."

Most leaves senesce by the end of November, and the tree remains dormant through the following March. Recent investigations suggest that pistachio leaf photosynthetic function does not decrease with age. Therefore, irrigating after harvest—up to the point where there is a danger it would prevent the trees from entering dormancy well before the first early freezes—will increase carbohydrate production for the following season.

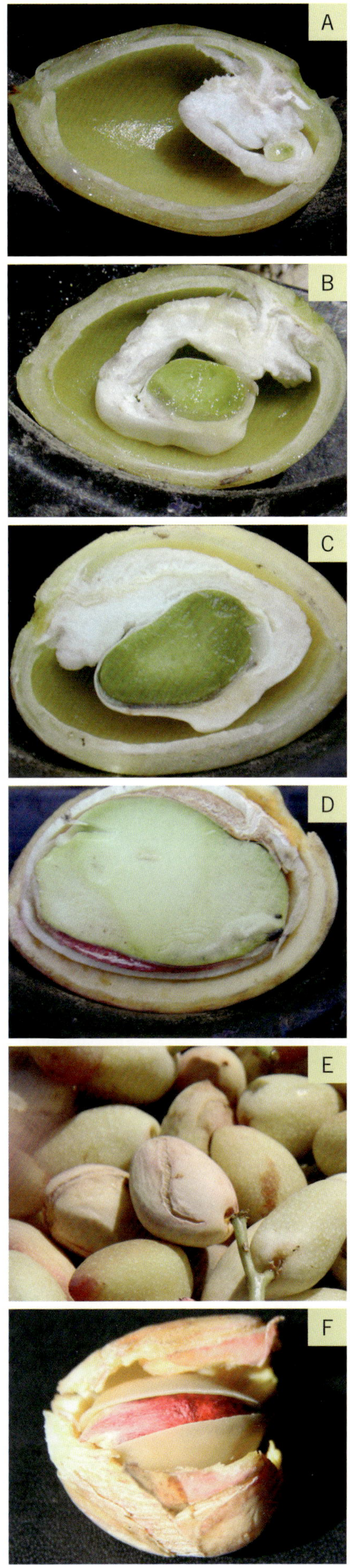

Figure 3.8. A developing pistachio embryo (A through F) within an already fully grown and hardened nutshell. Late Stage II through Stage III growth of the Kerman embryo takes place from May through late July or early August. *Photos:* L. Ferguson.

References

Al-Saghir, M. G., D. M. Porter, and E. T. Nilsen. 2006. Leaf anatomy in *Pistacia* species (Anacardiaceae). Journal of Biological Sciences 6:242–244.

Crane, J. C., and B. T. Iwakiri. 1981. Morphology and reproduction of pistachio. Horticultural Reviews 3:376–379.

Erez, A., S. Fishman, G. C. Linsley-Noakes, and P. Allan. 1990. The dynamic model for rest completion in peach buds. Acta Horticulturae 276:165–174.

Fabbri, A., L. Ferguson, and V. S. Polito. 1998. Crop load related deformity of developing *Pistacia vera* cv 'Kerman' nuts. Scientia Horticulturae 77:219–234.

Glozer, K. 2008. The dynamic model and chill accumulation. Davis: University of California, Department of Plant Sciences. UC ANR Fruit Tree website, http://ucanr.edu/sites/fruittree/files/49320.pdf.

Goldhamer, D. A., T. J. Michailides, M. Salinas, M. A. Dorster, and R. H. Beede. 2004. Using early season water stress to increase shell splitting at harvest. In California pistachio industry annual report: Crop year 2003–2004. Fresno: California Pistachio Commission. 115–117.

Hormaza, J. I., and V. S. Polito. 1996. Pistillate and staminate flower development in dioecious *Pistacia vera* (Anacardiaceae). American Journal of Botany 83:759–766.

Kallsen, C. E., and D. E. Parfitt. 2011. Comparison of scion/rootstock growth rates among U.S. pistachio cultivars. HortScience 46:1–4.

Kallsen, C. E., D. Parfitt, and B. A. Holtz. 2007. Early differences in the intensity of alternate bearing among selected pistachio genotypes. HortScience 42:1740–1743.

Li, X. and C. Chapple. 2010. Understanding lignification: Challenges beyond monolignol biosynthesis. Plant Physiology 154:449–452.

Polito, V. S. 1999. Seedlessness and parthenocarpy in *Pistacia vera* L. (Anacardiaceae): Of vascular transport to ovules. Annals of Botany 83:363–368.

Polito, V. S., and K. Pinney. 1999. Endocarp dehiscence in pistachio (*Pistacia vera* L.). International Journal of Plant Sciences 160:827–835.

Pope, K., V. Dose, D. Da Silva, P. H. Brown, and T. DeJong. 2014. Nut crop yield records show that budbreak-based chilling requirements may not reflect yield decline chill thresholds. International Journal of Biometeorology 1–9.

Salagado, J. M., B. Max, R. Rodriguez, N. Perez, and J. M. Dominguez. 2012. Extraction of ferulic acid from agro-industrial wastes and evaluation of bioconversion of ferulic acid to vanillin by *Streptomyces setonii.* Primer Congreso Iberoamericano sobre Biorrefinerias 451–457.

Spann, T. M., R. H. Beede, and T. M. DeJong. 2007. Preformation in vegetative buds of pistachio *(Pistacia vera)*: Relationship to shoot morphology, crown structure and rootstock vigor. Tree Physiology 27:1189–1196.

———. 2008. Neoformed growth responses to dormant pruning in mature and immature pistachio trees grown on different rootstocks. Journal of Horticultural Science and Biotechnology 83:137–142.

Stevenson, M. T., K. A. Shackel, and L. Ferguson. 2000. Shoot length distribution and its relation to yield of alternate-bearing pistachio trees. Journal of the American Society for Horticultural Science 125:165–168.

Part 3 Establishing the Orchard

4 Site Evaluation and Soil Physical Modification

Blake L. Sanden, Allan E. Fulton, and John P. Edstrom

Pistachios are similar to other nut crops in that high yields and nut quality are best achieved on deep, uniform, loam soils. These soils provide the optimal combination of permeability with sufficient water-holding capacity and root zone aeration. However, these deep alluvial soils are more expensive to acquire due to their limited availability, and therefore many new orchards are being developed on soils with drainage, salinity, and structural problems.

Site evaluation has two major objectives:

- determine the chemical quality of the soil and water with respect to acceptable salinity and fertility for long-term, profitable production
- determine any physical quality of the soil with respect to layering and drainage that may restrict root development and leaching

Chapter 16 in this manual addresses fertility management and chapter 15 discusses salinity management. This chapter discusses soil physical limitations, preplant site evaluation recommendations, choices among irrigation methods, land leveling, and selecting effective tillage methods. The first goal of a good physical evaluation of the site is to identify any potential limitation and determine whether modification is necessary. Avoid expensive deep-tillage operations if they are not needed. The second goal is to take corrective action before planting, when it is most feasible. Correcting a physical soil limitation after orchard establishment is usually not nearly as effective.

Evaluating Soil Physical Limitations

Evaluation and appropriate modification of orchard soils provide the following benefits:

- reduced physical barriers to drainage and root development
- increased uniformity of water infiltration and water-holding characteristics
- improved leaching of excess salts
- more uniformity and increased vigor in young orchards, requiring less time to achieve full production potential

Surface and Subsoil Variability

Orchards with soils that vary from sand to clay texture are subject to variable water stress across the orchard because of differing water-holding capacities of each soil texture. Layers of different soils in the root zone can cause waterlogging by slowing water movement through the root zone and creating temporarily saturated soil layers. This injures roots by depriving them of oxygen and enhancing conditions that favor root diseases. Some subsoil layers can form physical barriers to roots; roots cannot grow through hard, dense, or compacted layers. The result is nonuniform orchard growth and production, especially under flood or furrow irrigation, in which infiltration rates can vary considerably from one area to the next.

Soil Survey Data

The best place to begin the process of site evaluation is with the USDA Natural Resource Conservation Service (NRCS) soil surveys. California alone has 170 published surveys—from the first Fresno County survey in 1900 to the recent southwest Kern County survey in 2008. Many of these surveys are revisions of earlier surveys, changing and adding to earlier soil series descriptions. The surveys published since 1970 have the best maps (fig. 4.1) and soil series detail.

Soil maps and data are available online in these formats and locations:

- The easiest "point-and-click" access is through the University of California's SoilWeb on the UC Davis Soil Resource Laboratory website, casoilresource.lawr.ucdavis.edu/gmap/. Similar, but with more data, is the NRCS Web Soil Survey website, websoilsurvey.nrcs.usda.gov/app/.
- Full California soil surveys (published online as PDFs) are available at the USDA Natural Resources Conservation Service Soil website, nrcs.usda.gov/wps/portal/nrcs/surveylist/soils/survey/state/?stateId=CA.
- Georeferenced spatial and tabular data (more difficult to access and requiring use of GIS software) is available for California at the USDA NRCS Soil Data Access website, sdmdataaccess.nrcs.usda.gov/. Locations of NRCS offices in the United States can be found on the USDA Service Center Locator website, offices.sc.egov.usda.gov/locator/app.

For in-person help, you can also contact the local NRCS office and consult the area conservationist. Most of the full surveys are now digitized and available online. Paper copies may be available for some locations.

Identifying Physical Limitations

The soil survey, along with aerial images and personal observation of row crops previously grown in the field, is very useful for identifying zones that should be sampled and viewed separately.

Commercially available equipment (i.e., EM38 and Veris equipment) that uses electromagnetic or conductivity sensors and global positioning systems technology can also map soil variability. When properly calibrated, the sensors detect changes in soil salinity and major differences in water-holding capacity. Figure 4.2 illustrates the potential variation one might find in a possible 160-acre orchard development in western Kern County.

A thorough site evaluation uses a series of backhoe pits in the different zones. Observation pits clearly show the number and types of soil layers, the depth of the layers, and the variability of the subsoil throughout the orchard site. This information can help determine the most economical method of soil modification; how to properly set up and use deep-tillage equipment; and to what depth tillage is required. One alternative to backhoe pits is the use of a soil probe or auger to pull soil cores for evaluation. Special equipment is required for pulling undisturbed cores, which is the preferred method, but a common bucket auger can be used by an experienced agronomist or soil scientist, who can identify textural differences and potential structural problems in soil. Using your farm backhoe, or even renting one, may be cheaper in the end, as well as more revealing, since the entire profile can be viewed at once. An example pit can be seen in figure 4.3.

It is advisable to dig at least one backhoe pit per 20 acres. Where possible, locate backhoe pits in areas of the prospective orchard site that have a history of desirable as well as poor growth, for comparison (see fig. 4.2).

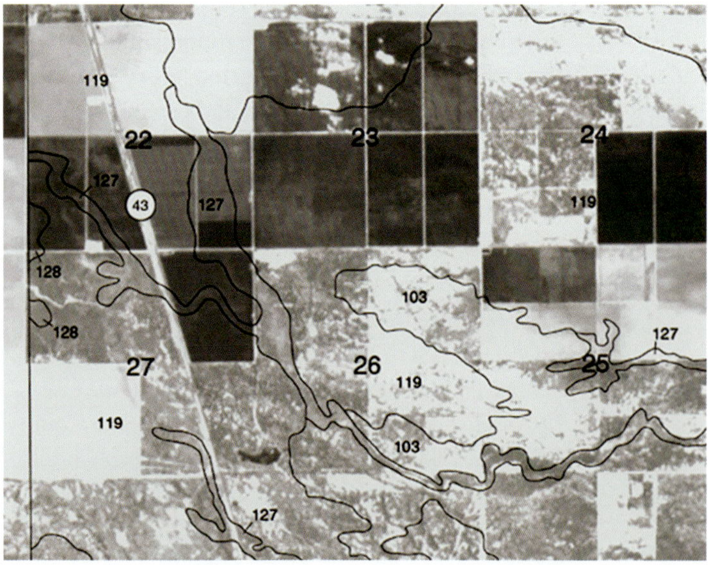

Figure 4.1. Part of Delano West Quadrangle map from western Tulare County soil survey, illustrating soil series contours. *Photo:* B. L. Sanden.

Figure 4.2. Soil type variability and required sampling areas for a 160-acre orchard development.

Figure 4.3. Backhoe pit in a Buttonwillow clay soil, with a slightly caliche silty clay layer at the 34-to-45-inch depth. Total depth is 80 inches. *Photo:* B. L. Sanden.

Figure 4.4. A marginally alkali clay loam in the top 2 feet of this almond root zone overlays a coarse, sandy loam layer. Excess moisture in the heavier layer prevented good root development in the top 2 feet. *Photo:* B. L. Sanden.

Estimating Site Evaluation Costs

The cost of evaluating the physical soil characteristics of a prospective orchard site and pulling soil samples ranges from $300 to $1,500, depending on field size and variability of soil types. This is about 0.5 percent of your eventual development cost—far less than the cost of establishing an orchard at a poor site. A backhoe service can be contracted for about $75 per hour, and the work can be completed within a day. Laboratory costs for soil and water analyses, depending on the number of tests and the lab, run from $40 to $90 per sample. Samples should be taken in a composite manner for multiple depths down to at least 4 feet. See chapter 15, "Salinity Management in Soil and Water," for a more detailed discussion on sampling and interpreting analyses.

Soil Series Evaluation Guidelines

Extensive soil modification is not always necessary to prepare an orchard site for planting. Some soils are naturally deep and relatively uniform in soil texture and structure and do not require deep tillage; however, other soils are layered and may need modification. Four types of physical soil limitations may be observed while conducting backhoe site evaluations: stratified soils, claypan soils, hardpan soils, and plowpans. Table 4.1 summarizes some common soil series in the San Joaquin Valley known to have physical soil limitations.

Stratified soils have layers with abrupt changes in soil texture beneath the surface (see figs. 4.3 and 4.4). The layers interfere with the uniform drainage of water, causing zones of poor aeration that may restrict root growth. Correcting layered soils requires mixing the soil layers.

Claypan soils have a concentrated clay layer that restricts water movement downward, thus restricting aeration and root growth in the subsoil. Clay layers can start abruptly at a depth of 12 to 24 inches below the surface, with the clay most concentrated in the upper half of the pan. The lower half gradually changes from a clay loam to a loam-textured soil. Modification designed to mix the clay layer with the rest of the soil profile will create more uniform water penetration, reduce saturation in the upper root zone, and promote extensive tree root growth.

Hardpan soils are similar to claypans, except the soil particles are glued together by hard mineral matter that will not soften, even when wet. The hardened layer can be a barrier to root growth and water percolation. Modification of hardpan soils requires fracturing and breaking rather than mixing.

Plowpans can be found in some orchard sites and are the result of tilling a soil at the same depth repeatedly. A soil with a plowpan does not necessarily have different textural layers like those of stratified, claypan, or hardpan soils. They tend to be shallow, 10 to 18 inches below the soil surface, and require only fracturing, rather than mixing, to correct the problem.

Table 4.1. San Joaquin Valley soil series with physical soil limitations but that may be considered for pistachio orchards

Stratified	Developed clay subsoil types: Moderate clay content	Claypan	Cemented hardpan
Eastside San Joaquin Valley soils			
Cajon	Borden	Cometa	Academy
Chino	Chiralar	Corning	Dinuba
Foster	Pond	Modesto	Exeter
Grangeville	Ramona	Montpellier	Fresno
Hanford	Ryer	Waukena	Lewis
Kimberlina	Snelling	Milham	San Joaquin
Nord Complex			Whitney
Visalia			
Westside San Joaquin Valley soils			
Camarillo	Avenal	Herdlyn	Dinuba
Columbia	Twisselman	Olcott	Garces
Mocho	Pleasanton	Positas	Lethent
Panoche	Rincon	Solano	
Westhaven	Rossi	Waukena	
Vernalis	Garces	Buttonwillow	

Irrigation Systems and Grade Limitations

Selecting an Irrigation Method

Selection and design of the irrigation system is influenced by site topography and soil types. The choice of irrigation system, in turn, influences final land leveling and tillage requirements. Properly designed and managed microsprinkler, surface drip, subsurface drip, and sprinkler systems help minimize physical soil limitations more than surface irrigation systems.

Cost considerations

The advantage of choosing a flood or furrow irrigation system over a microirrigation system is reduced initial capital cost. The disadvantage, especially since it takes 8 to 11 years for pistachios to reach maturity, is the 1 to 1.5 acre-foot per year of additional water required to compensate for increased evaporation and percolation losses compared with the microirrigation system. As water costs continue to increase, the cumulative cost of this additional water can be much more than the higher initial cost of the microsystem.

System uniformity

Uniform application of water with a microirrigation system depends more on hydraulic design, emitter application patterns, and system maintenance than on the uniformity of the soil. If a low-volume microsystem is selected, uniform field slope is not so critical. Microsprinklers and surface drip systems apply water at low, controlled rates that should be less than the soil infiltration rate. Also, irrigation water is applied in small quantities at high frequencies that closely match the rate of crop water use.

For furrow or border irrigation, the uniformity of applied water and subsequent pistachio water use largely depend on variable infiltration and retention characteristics of the different soils in the orchard. These factors cannot be engineered and adjusted into a set irrigation application rate as is the case with microsystems. Thus, it is virtually impossible to design a surface irrigation system with the same uniformity as a well-managed microsystem. Nevertheless, the following design and operational guidelines will help maximize the uniformity of your furrow or flood irrigation system.

Surface irrigation system guidelines:

- Irrigate in small orchard blocks that share similar physical soil characteristics to achieve more uniform water infiltration and storage.
- Provide one water discharge valve per tree row.
- Design the surface system so the field length is short enough to rapidly advance water across it, given the available water flow rates.
- Grade to uniform slopes to avoid low and high spots.
- Install a tailwater return system. Steps taken to improve irrigation uniformity of applied water by furrow or flood irrigation may result in more tailwater.

Larger irrigation sets are often more attractive to growers to minimize per-acre costs. Unfortunately, flood or furrow systems with large acreage per set usually include more soil variability. This results in lower uniformity of applied water and variable tree growth.

Leveling the Land

Orchard soils that are flood or furrow irrigated must be graded to a uniform slope. Prior to the leveling activities, heavy crop and weed residues or brush and stumps from a previous orchard must be removed. This is especially important in areas where large piles of organic debris could eventually be buried by soil from cut areas. Incorporating organic material into soil is usually beneficial; however, it can be detrimental when a large mass of material is buried deeply in compacted soil. After irrigation, the organic residue can remain wet and begin decomposing in the absence of air to produce methane gas and reduced concentrations of manganese, iron, and other organic compounds. These compounds can build up and become toxic to plant roots, killing the trees. Burning, where permitted, is effective in removing excess crop and weed residues. The remaining vegetative residue can then be incorporated into the soil by disking. Stump removal is best achieved with a backhoe or chipping machine grinder.

After the site has been cleared, some important factors need to be considered prior to leveling. These include the desired final grade, the optimal time to do the leveling, the maximum depth of cuts to be made, and the type of scraper to use.

The final orchard grade for flood or furrow irrigation depends on the soil infiltration characteristics, the length of the field, and the discharge rate of the available water supply. Generally, the final grade for medium- to fine-textured soils should be 0.08 to 0.4 feet per 100 feet (0.08 to 0.4%), depending on run length and average set times. To avoid inefficient and nonuniform irrigation, the length of surface irrigation checks should be less than 1,300 feet (¼ mile) and, if at all possible, a tailwater return system should be installed. This works very well with a standard 80-acre field that is ½ mile across and ¼ mile long.

Land leveling should be done under fairly dry soil conditions to minimize soil compaction and the breakdown of soil structure. The maximum depth of cuts depends on the initial slope of the land. If required cuts exceed 0.6 feet, the total cut should be accomplished by making two separate shallower cuts. Two types of scrapers are available for leveling: paddle type and push loading. Paddle-type scrapers create less compaction than push-loading scrapers, but the former tend to pulverize the soil to a powderlike condition that does not settle properly in fill areas.

After completing most of the land leveling, you need to determine grade stability and management strategies of cut and fill areas. Settling time is required to allow the grade to become stable. Where cuts exceed 0.6 foot, the main cuts should be made the first year and the final grade the following year, after an annual crop has been grown. A second year of annual cropping is even more preferable, ensuring that the final grade has stabilized.

After land leveling, nutritional deficiencies in the cut areas must be resolved. Nitrogen, phosphorus, potassium, zinc, and sulfur are commonly lower in exposed subsoil than in the original surface soil. This is due to lower organic matter content in the subsoil. These deficiencies can be corrected by adding fertilizer and manure and by incorporating plant residues. By planting an annual crop such as winter wheat, you will be able to better identify the amount of these materials and the field location where such amendments are needed. Furthermore, the exposed cut areas may not have as desirable a soil structure as in the original surface soil. In time, cropping, wetting and drying, and the addition of organic matter will rebuild the soil structure and establish desirable soil tilth.

Fill areas may present more difficult problems than cut areas. A given volume of soil scraped off a high spot in a field will not fill the same volume in a low area. An additional amount of soil is needed, since the soil structure has been destroyed and the soil bulk density usually increases. Clay loam to loam soils may require the volume of cut soil to be about 25 percent more soil than the intended fill space, to level the low area. In coarser, sandy soils, low areas may require 50 percent more soil volume. Fill areas will be more compacted than the original surface soils. Such compaction can cause problems with water penetration and root growth, both detrimental to establishing a productive orchard.

Filling should be done when the soil is fairly dry (to reduce compaction in the fill areas) but not so dry that the soil becomes powdery. The soil should have small clods, which indicate maintenance of some soil structure. After leveling, the fill areas may require ripping to below the depth of the original soil surface. If the fill is deep, ripping after partial leveling is desirable. Finally, if leveling and additional ripping costs are high, it is probably more cost effective in the long term to install a microirrigation system.

Crop Response to Deep Tillage

Deep tillage prior to planting is critical for orchard sites that are to be irrigated with surface systems. Thorough soil mixing improves water infiltration and drainage, promotes more extensive root growth, and creates more uniform water storage within the root zone. These factors contribute to a more manageable and productive orchard under a flood or furrow system.

Flood Irrigation

Table 4.2 illustrates the production and root development responses reported from a tillage study in which a claypan soil was modified before planting almonds to be grown under flood and border strip irrigation (Begg et al. 1998). (Similar data are not available for pistachio, but it is reasonable to expect similar responses.) In this study, production increased over 40 percent and the root count increased over 120 percent after using a 4-foot moldboard plow to thoroughly flip and mix the clay layer prior to planting. Depending on market price, the production increase would be valued at $300 to $800 per acre annually. It is unclear why slip plowing did not produce more of a response in this trial. Similar experiments have not been conducted for pistachio. Given the ability of pistachio roots to grow long distances and the longer development time to pistachio maturity, it is hard to predict the response of these trees to the different types of tillage treatments under flood irrigation.

Current economics and production potential require growers to push the yield to much higher levels for both pistachio and almond than shown in this trial. It would make sense that this would cause additional stress on the tree and result in even greater benefits from deep tillage. However, more recent trials in microirrigated orchards (one in almonds and one in pistachios, as described in the next section) showed no significant benefit to yield or tree growth from deep tillage.

Table 4.2. Responses to modifications of claypan soil in terms of yield, trunk circumference, and root count in almonds (flood and border strip irrigation)

Tillage method	Yield (lb/ac)*	Trunk circumference (in)†	Root count (per 3 ft³)*
none	1,009	14.8	78
ripper	1,120	16.6	94
slip plow	1,185	16.7	118
moldboard plow	1,433	17.0	175

Source: Adapted from Begg et al. 1998.

Notes:

*Measured in eighth year of production.

†Measured in fourth year of production.

Microirrigation

Unless some form of hardpan has been identified close to the surface, deep tillage may be unnecessary for orchards irrigated with microsystems. Field research conducted in both walnut and almond on marginal soils on the west side of Colusa County found little response to deep tillage when drip or microsprinkler irrigation was adopted (Edstrom 2008). Early trials in almonds under drip irrigation showed no positive effects from deep soil modification on a layered clay soil. As shown in table 4.3, recent experiments in microsprinkler-irrigated almonds have provided similar results (Edstrom 2008).

The soil within the research trial is a Class II Arbuckle sandy loam, 30 to 60 inches deep, underlain by a dense clay layer 10 to 15 inches thick, returning to a gravelly sandy loam below. Slip plowing to 6 feet was done on a 10-by-10-foot grid spacing to mix these layers. Despite this soil having the physical layering that might limit tree development, this field trial showed that deep slip plowing gave no improvement in almond tree size, yield, or crop quality after 8 years compared with no deep tillage (Edstrom 2008).

In another long-term study in drip-irrigated walnuts, soil excavations at a test site revealed substantial mixing of stratified soil layers and deeper walnut root development following slip plowing, but no yield or nut quality improvements were found (Edstrom 2008). The low-volume applications of water and nutrients at higher frequency with the microirrigation system appear to overcome many limitations of marginal soils without expensive deep tillage.

Table 4.3. Slip plow evaluation in microsprinkler-irrigated almonds. Nickels Soils Lab, Arbuckle, CA (2000–2008)

Year	Tree age (years)	Nut yield (lb/ac)	
		Slip plowed	Not slip plowed
2000	4	894	830
2001	5	1,070	1,243
2002	6	2,725	2,761
2003	7	2,165	2,323
2004	8	1,869	1,865
2005	9	1,548	1,841
2006	10	2,910	2,862
2007	11	2,770	2,571
2008	12	3,771	3,686
Cumulative yield		**19,722**	**19,982**

Source: Edstrom 2008.

Pistachio Tillage Trial

In January 2006, a long-term deep-tillage trial in drip-irrigated pistachios was initiated on a mixed

Buttonwillow clay overlaying and intersecting Garces and Lethent silty clay loams and Kimberlina sandy loam. A control area was included, where no tillage was done. Tillage treatments consisted of the following prior to planting:

- cotton chisels to 24 inches
- single slip plow down row to 45 inches
- triple slip plow to 45 inches, 6 feet apart
- backhoe mixing to 7-foot depth, 3 feet wide

Tillage treatments were centered on the 20-foot pistachio rows. After tillage, a 2-foot-wide band of fine soil sulfur was applied down the future tree row at 1.5 tons per total acre (15 T/ac applied) and incorporated to about 20 inches with a modified furrowing shovel–cotton chisel combination. Trees were then planted with a three-point hitch posthole auger with a 12-inch diameter. The orchard was flood irrigated the first year, as the capital was not available for a drip system. The control trees (auger only) had 18 percent smaller trunk circumference at the end of the first year but had caught up with the other tillage treatments at the end of the second year (fig. 4.5A) once the drip system was installed. After 9 years, there is no real difference in trunk circumference or yield for any of the chisel or slip plow tillage treatments. However, the intensive backhoe treatment (as a subset of the triple slip treatment) over the combined 7th, 8th, and 9th leaf harvests produced a statistically significant increase of 1,147 pounds per acre more total in-shell than the other treatments (fig. 4.5B). These trials indicate that standard deep-tillage methods may not provide significant benefits to final orchard production under microirrigation. Intensive backhoe excavation to 7 feet appeared to provide significant yield benefit, but this is a very expensive operation and was only done for a small-scale sample of 12 trees for this trial.

So, growers only have one chance (i.e., the preplanting phase) to modify the soil structure beneath the tree, and most growers consider deep tillage to be in the same category as life insurance: you hope you won't need it, but it's better to be covered.

Shallow soils located in high-rainfall environments (in excess of 15 to 20 in of average rainfall, with a high likelihood of rainfall just prior to and during leaf-out), as well as those located in floodplains, represent two conditions in which deep tillage may be beneficial even when microsprinkler or drip irrigation is used.

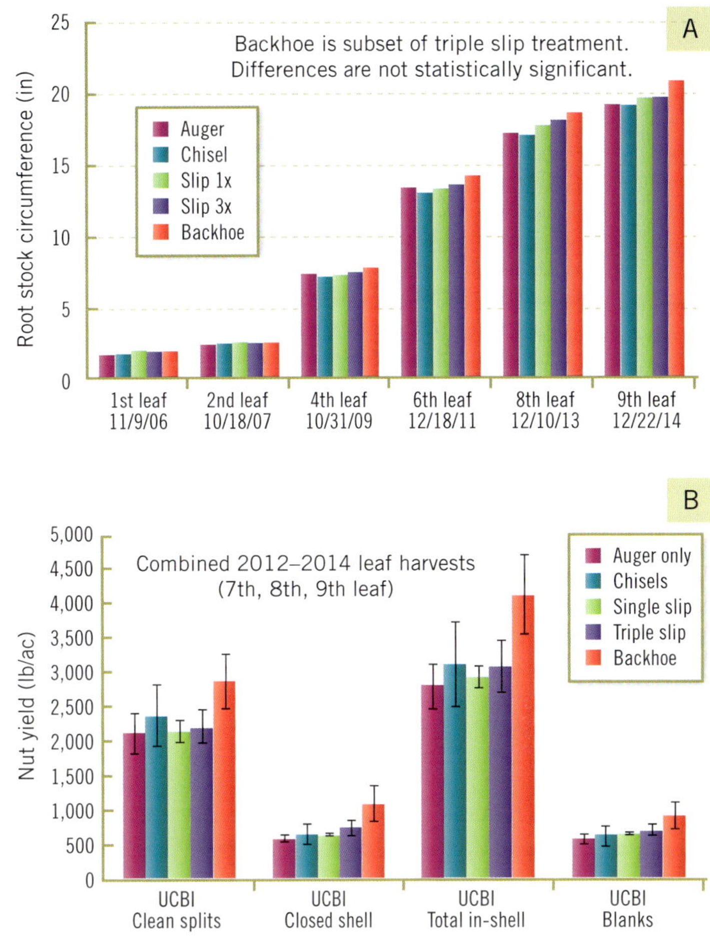

Figure 4.5. (A) Trunk circumference of UCBI rootstock under different tillage treatments. (B) Combined Kerman 3-year yield for 7th, 8th, and 9th leaf harvests. *Source:* Sanden et al. 2015.

Modifying Soil Physical Problems

Several methods are available to accomplish deep tillage prior to planting pistachios. Available equipment includes rippers, slip plows, moldboard plows, disk plows, backhoes, and trenchers. Selecting the most cost-effective method of deep tillage or soil mixing requires evaluating the type and severity of the soil problem. Remember, the purposes of modifying a soil prior to planting are to encourage uniform movement of water through the root zone, maintain adequate aeration, and increase the volume of soil available for root growth. Therefore, select the method that will most economically achieve those goals. Table 4.4 lists most options and the estimated cost.

Stratified Soils

Digging backhoe pits for each tree site is the most effective means of modifying layered soils with stratification down to 4 to 6 feet. Backhoeing is more expensive than other tillage methods because the process is slower. Typical costs range from $6 to $10

Table 4.4. Approximate costs for different methods of soil modification (based on 2008 prices)

Method	Soil depth (ft)	Cost ($/ac)
cotton ripper*	2–3	60–150
ripper*	4–6	250–500
slip plow*	4–6	400–800
moldboard plow	3–4	600–800
trencher†	3.5 (w) × 4 (d)	1,700–1,900
backhoe pits‡	4 x 4 (w) × 6 (d)	500–900
backhoe pits§	4 x 4 (w) × 6 (d)	700–1,100

Notes:

*Includes straddle pass with two ripper shanks 3 to 5 feet on either side of the initial pass down a 20-foot tree row.

†Trenching costs assume a 3.5-by-4-foot trench down a 20-foot tree row spacing and a charge of $1.50 to $1.80 per cubic yard of soil trenched.

‡Assumes farm backhoe at $60 per hour; this is less than the typical rate, due to the size of the job. Cost is based on 17-by-22-foot tree spacing.

§Cost is based on 15-by-20-foot tree spacing.

per hole, depending on the skill of the operator, farm machinery budget, and soil type. The expense of digging backhoe pits rises with increased tree density. For example, an orchard planted on a 17-by-22-foot spacing will require 116 backhoe pits per acre, while an orchard planted on a 15-by-20-foot spacing will require 145 backhoe pits per acre. The required dimensions of the backhoe pits depend on the extent of the soil layering. Generally, a pit 4 feet square and 8 feet deep is the maximum that is required and affordable. Field research using flood irrigation has indicated that the larger the pit, the more successful the results in stratified soils—but as mentioned earlier, this may not apply to microirrigation.

Trenching along the intended tree row is an alternative method of mixing abrupt soil layers. Access to equipment may be limited. Trenching is suited to soils with layering problems within 4 feet of the surface and the soil below this depth being more uniform in texture. This is because trenchers that can trench deeper than 4 feet are seldom available.

Slip plowing is the most common method of mixing stratified soils. This device uses a heavy steel beam set at a 30-to-40-degree forward angle, with a hard-faced steel shoe on the bottom of the beam that may be 12 to 18 inches wide and 24 inches tall. The forward angle of the plow pulls the shoe deep into the soil, shattering and lifting the soil in its path up along the beam, thoroughly mixing the layers along the depth of the pass. The depth of mixing is only limited by the length of the shank, the traction or stability of the field surface, and the power of the track layer (Caterpillar) pulling the implement. It usually takes a D-9 or two D-8 equivalents and two passes to reach a 6-foot depth. If the soil is extremely alkaline throughout the profile, this is an optimal time to apply sulfur to help free calcium and reduce pH. Band the sulfur along the path of the slip plow. Depending on the design of the slip plow, this band can be sucked in behind the slip plow and mixed within the top 2 to 3 feet of the profile. Other slip plows (as in the pistachio tillage trial) require another shank after plowing to accomplish this mixing. Some slip plows have a following foot attached to the point shoes and chained to the steel frame aboveground to improve mixing of layers (fig. 4.6).

Moldboard (fig. 4.7) and disk plowing provide more thorough mixing than slip plowing, but the depth of their mixing is limited to 3 or 4 feet. Ripping is less beneficial in correcting layered soils because it is designed to fracture rather than mix soils. Layered soils eventually reform after ripping.

A thorough mixing of layers is especially desirable when a heavier soil overlays a lighter texture. In this setting, the upper layer must become saturated before water percolates into the coarser layer below. This makes for worse disease problems and can cause poor root development in the upper part of the profile (see fig. 4.4).

Claypans

Since claypans typically occur near the soil surface, the moldboard plow and the disk plow are effective methods of thoroughly mixing the pan with the rest of the soil profile; they are also adaptable to larger acreage. Slip plowing is an alternative to moldboard or disk plowing for modifying claypans, but it does not mix the entire soil surface, only strips of soil that are 4 feet or more apart. Ripping a claypan soil is not advised unless the ripper shanks penetrate deeper than the depth of the restricting layer and are spaced close enough to ensure severe mixing and fracturing of the soil. Many cotton rippers used on the heavy soils of the west side of the San Joaquin Valley are spaced close enough together to accomplish this. Nevertheless, mixing rather than simple fracturing is still the best way to permanently destroy claypans. If the soil below the claypan is fairly uniform, then deep tillage below 3 feet is probably not warranted.

Hardpans

Orchards should not be planted on hardpan soils unless the layer is shallow enough that deep ripping can completely break through the hardpan into uncemented, permeable soil below. Backhoe pit evaluations are extremely helpful in determining the required ripping depth and shank spacing. It is beneficial to rip down each intended tree row. Slip plowing is generally not recommended in hardpan soils because it requires more energy than does ripping. Moldboard plowing, trenching, and digging backhoe pits are usually not advised for correcting a hardpan problem. Thoroughly mixing a hardpan layer with the rest of the profile is unrealistic and expensive if attempted, because of the cemented nature of the layer. Fracturing the hardpan by ripping is often all that is necessary to ensure adequate drainage for excess water and root penetration. If ripping cannot effectively shatter the hardpan, the site should be considered unsuitable for pistachio production, especially with flood or furrow irrigation systems.

Guidelines for Deep Tillage

Soil moisture

The soilwater content at the time of deep tillage partly determines the effectiveness of the operation. Although a moist soil requires less draft when tilling, more surface compaction occurs from the heavy equipment, so the soil does not break up and mix as readily. Dry soil conditions break up and mix more readily. Desirable soil water content for clay is 10 percent, loam 5 percent, and sandy soil 2.5 percent by weight. These are moisture levels that push most crops into permanent wilting. Soils with a shallow water table (< 6 ft from the soil surface) or excessive salts in the root zone may never dry sufficiently to effectively shatter and mix layers. Growing a crop of safflower prior to fall tillage is the best way to dry out the soil to the greatest depth, as the safflower will root more densely and deeper than most grain crops.

Effective ripping

Two factors that specifically influence the effectiveness of ripping or chiseling are the depth of the ripper shanks and the distance between them. Below are three guidelines to apply:

- The shanks should penetrate into the soil 1.5 times the depth of the targeted soil problems.
- The distance between the shanks should be equal to, or preferably less than, the desired ripping depth.
- Rip in one direction with narrowly spaced shanks rather than in two directions with a wide shank spacing.

Insert the shank 50 percent deeper than the desired ripping depth to ensure sufficient penetration through the restricting zone of soil. Setting the shank spacing to a distance narrower than the desired ripping depth is best, because ripping never breaks the soil straight across between adjacent shanks, but instead breaks it in more of an upward angle.

Figure 4.6. Slip plow with a following foot to further lift and mix layers. *Photo:* B. L. Sanden.

Figure 4.7. Moldboard plow capable of penetrating to a depth of 3 to 4 feet. *Photo:* B. L. Sanden.

A hump of undisturbed soil always remains between two ripper channels. The wider the shank spacing, the greater is the percentage of undisturbed soil. With extremely wide shank spacing, the zone of undisturbed soil may come all the way to the soil surface.

The practice of straddle ripping is probably the most common method of deep tillage when preparing to plant an orchard. It is more economical than deep tillage at 5-foot intervals across the whole field and, in most cases, just as effective.

Straddle ripping is accomplished by marking the tree rows in the field and ripping or slip plowing to the desired depth. This may require one or two passes, depending on the depth of penetration. If the soil is dry enough, fracturing will occur at an angle of about 45 degrees from the point of the chisel or plow up to the soil surface. A 6-foot-deep penetration will result in some fracturing out to 5 to 6 feet at the soil surface. Dual ripper shanks are then set up, spaced about 10 feet apart, and positioned so they straddle the initial pass with about 5 feet on either side. Since some fracturing has already occurred from the initial ripping, the straddle rip can usually be done in one pass. This ripping is usually about 1 foot shallower than the central pass.

If the central pass was to a depth of 6 feet, the straddle rip penetrated to 5 feet, and the lack of soil moisture was optimal, the final fracture zone at the soil surface should be 16 to 20 feet wide. At the 3-foot depth, fracturing should be complete between all shanks and extend to a width of 12 to 14 feet across the tree row.

References

Begg, E. L., G. L. Huntington, and W. E. Wildman. 1998. Evaluation and modification of soils. In D. E. Ramos, ed., Walnut production manual. Oakland: University of California Division of Agriculture and Natural Resources Publication 3373. 51.

Edstrom, J. P. 2008. Slip plow tillage effects in almonds. University of California Nickels Soil Lab Report.

Sanden, B., L. Ferguson, and C. Kallsen. 2015. Effect of pre-plant tillage on pistachio development under drip irrigation. California Pistachio Research Board production research reports. Fresno, CA: Administrative Committee for Pistachios website, http://www.acpistachios.org/research.php.

5 Planning and Designing the Orchard

Craig E. Kallsen, G. Steven Sibbett, and Carl Fanucchi

Pistachio trees are long-lived, and time spent properly designing an orchard can save years of farming frustrations and prevent poor economic returns. The well-designed orchard ensures the earliest economic per-acre return, the maximum per-acre production at tree maturity, and the least management cost. This chapter discusses the following orchard planning and design considerations: tree density, row length, planting designs, loading areas, pollenizer placement, and orchard floor management.

Orchard Planning

Tree Density

Pistachio culture in the United States is relatively new. The most widely planted scion-rootstock combinations in California have only been available to growers since the early 1980s. Scientific experiments designed to determine the relationship between tree density and economic returns have not been done. Planning and design suggestions included here are largely based on observation of productive orchards in the Central Valley of California consisting of trees of Kerman (the female cultivar) budded to a rootstock of *Pistacia integerrima* heritage. Suggestions for pollenizer frequency and placement are based on observation of orchards with Peters (the male cultivar). Other scion-rootstock combinations may have advantages in some environments, and spacing, densities, and male-to-female ratios may be different.

Tree and row spacing

A pistachio orchard reaches its maximum bearing potential when space over the soil is completely occupied by nut-bearing surface, with sufficient light penetration to keep lower fruitwood productive. Appropriate row spacing is particularly important because pistachio is relatively slow to begin bearing, requiring 5 to 6 years between planting and first nut harvest. The initial planting density is directly related to the orchard's onset of economic production and the time required to maximize bearing potential. The closer the trees are planted, the more likely tree thinning or earlier heavy, expensive pruning will be required for an orchard to maintain maximum bearing potential. Conversely, in a more widely spaced initial planting, the onset of economic bearing is delayed and it will take longer to reach maximum annual bearing potential. Thus, the pistachio grower has two objectives when spacing trees: ensuring earliest economic production and developing an orchard with an economically manageable maximum bearing potential. Where nuts are shaken from the tree and caught before contacting the ground, row and tree spacing is partially governed by size of commercially available catching frames. Frame sizes vary and many are adjustable, but trees spaced less than 12 feet apart within the row may make insertion of the frame difficult. Harvesting row and tree spacing greater than 24 feet may result in nuts falling outside of the frames' catchment area unless modified equipment is used.

New orchards are being planted with the assumption that tree thinning will not be necessary to ensure maximum production at maturity. However, the California pistachio industry is young, and many questions remain as to what the eventual size and bearing characteristics of increasingly older pistachio trees will be. The oldest blocks of Kerman on *P. integerrima* rootstocks were planted around 1980. Observations of well-managed trees planted in deep, well-drained soils in a rectangular design with 18 to 20 feet between rows and 15 to 18 feet between trees indicate that crowding, sufficient to

Figure 5.1. Modern, mechanized harvest equipment is large, and trees should be spaced appropriately. *Photo:* C. E. Kallsen.

Figure 5.2. A well-designed orchard allows sufficient room so equipment may be turned at the end of tree rows to avoid obstructions such as this water district reservoir. *Photo:* C. E. Kallsen.

Table 5.1. Number of trees per acre and per hectare at different row and tree spacings

Row × tree spacing (ft)	Trees per acre	Trees per hectare
20 × 17	128	316
21 × 17	122	301
22 × 17	116	288
20 × 18	121	299
21 × 18	115	285
22 × 18	110	272
20 × 19	115	283
21 × 19	109	269
22 × 19	104	257
20 × 20	109	269
21 × 20	104	257
22 × 20	99	245
21 × 21	99	245
22 × 21	94	232

reduce production, does not occur until about the 15th year. Tree crowding has been managed through the use of mechanical hedgers and toppers as well as hand-pruning. Tree thinning has been rare, and few growers consider it.

Row width. Row width should be no less than 20 feet to facilitate large equipment for harvesting and machine pruning (fig. 5.1). New pistachio orchards are being planted with row widths not exceeding 22 feet.

In-row tree spacing. New pistachio orchards are being planted with trees 15 to 20 feet apart within the row. Spacing trees at 16 feet or less leaves little room for tree growth or mechanical cross-hedging, and more expensive hand-pruning will be required to keep trees separated from one another for harvest. If trees are allowed to grow together, shaking will be transmitted to neighboring trees, resulting in nut drop outside of the harvester's catch frame. The upright tree growth characteristic of Golden Hills, a relatively new cultivar released to the pistachio industry in 2005, suggests that spacing between trees within the row can be decreased (compared with Kerman) without increasing the risk of early tree crowding. The slower growth of trees on shallow or saline soils, and thus the longer time period before crowding occurs, may provide economic justification for the closer in-row tree spacing of 15 to 16 feet in some instances. Table 5.1 lists the number of trees per acre at various row and tree spacings.

Row length. Pistachios are harvested in one direction down the row, with large equipment of limited carrying capability. Rows should not be longer than ¼ mile without a break to allow for changing harvest collection equipment or minimizing time spent traveling to and from the loading site. Additionally, growers should leave at least 35 feet from the trunk of the last tree in a row to the edge of the property to turn equipment (fig. 5.2).

Tree Placement Designs

Square or rectangle

Trees configured as a square are equally spaced within and between rows, and orchard operations can be carried out in both directions (fig. 5.3). However, to maintain the minimum of 20 feet between rows for passage of large harvesting equipment, square plantings will result in relatively low tree densities unless temporary filler trees are used for early productivity. The rectangular planting configuration is the most common tree design used for pistachio orchards. It is best adapted to the pistachio trees' slow growth habit by providing high tree densities while accommodating large equipment. Like the square design, it is also suited for mechanical cross-hedging and topping.

In most new pistachio orchards, the trees are spaced in a rectangular pattern with 20 to 22 feet between rows and 15 to 18 feet between trees. Trees spaced closer than 17 feet apart are probably too close to cross-hedge without removing excessive fruiting potential.

Offset square (diamond)

The offset system is a variant of the square or rectangular except trees in adjacent rows are offset (fig. 5.4). At any given tree spacing, the square and offset square designs have an equal number of trees per acre.

Hexagonal or equilateral triangle

Trees configured into a hexagonal or equilateral triangle design are equal distances apart when measured in any direction (fig. 5.5). This design is the most efficient in intercepting light at a given planting density as all trees are equidistant apart, which reduces mutual shading. However, this design has two disadvantages. First, trees placed in an equilateral or hexagonal design must be placed as permanent trees. This design does not allow for temporary trees because their eventual removal results in inefficient tree and row distances for optimal productivity. And second, row widths are 87 percent of the distance between trees, which must be considered with use of large equipment in the orchard.

Filler Trees

A practice that can increase early yield is to interplant temporary, or filler, trees between the trees that will be permanent. The initial expense of filler trees, continuing costs of their culture (e.g., tree training), adjustments to irrigation systems, and cost of their eventual removal must be weighed against their contribution to early productivity. In California at the present time, no pistachio orchards are being double planted with filler trees.

Filler trees are planted far enough apart so that they are harvestable with conventional equipment. They are also planted close enough together so that, when removed, the spacing of the remaining permanent trees is optimal to sustain the orchard's maximum annual bearing potential. No formula is available for calculating when filler trees should be removed. At some point, added cultural costs (such as pruning and yield reductions associated with competition for light and nutrients) become greater than any additional yield provided by a greater number of trees per acre.

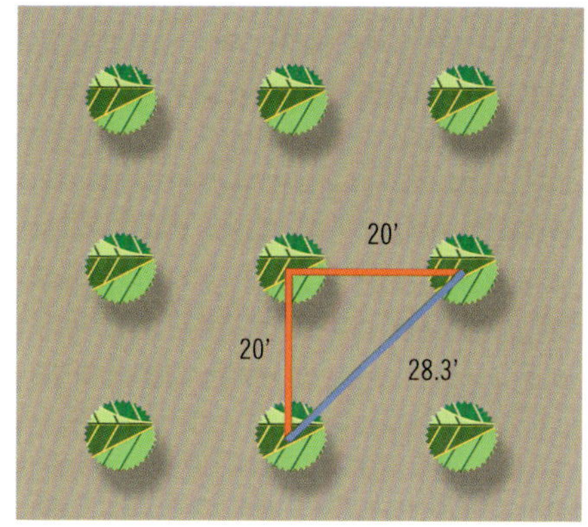

Figure 5.3. Square design (109 trees per ac at this spacing).

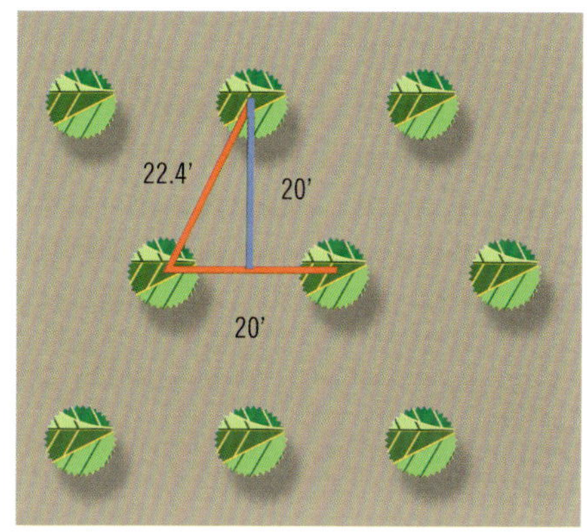

Figure 5.4. Offset square, or diamond, design (109 trees per ac at this spacing).

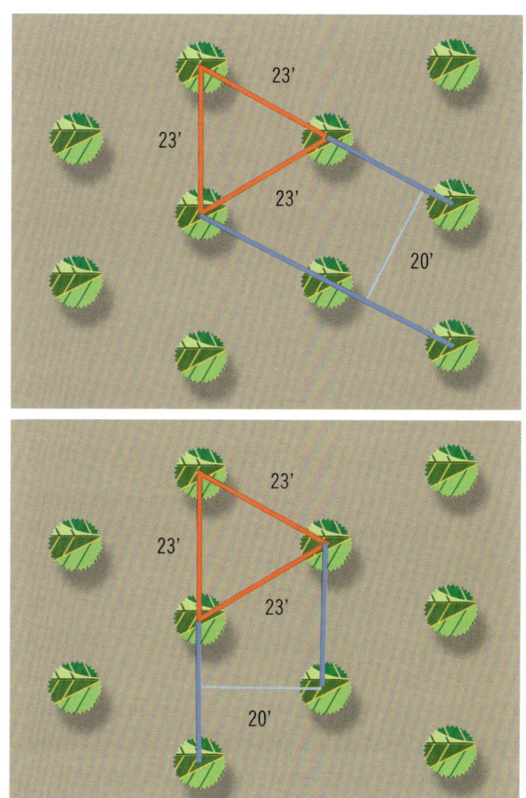

Figure 5.5. Hexagonal or equilateral triangle design (97 trees per ac at this spacing). Note that at a 23-foot tree spacing a row is 20 feet wide.

Pollenizer Placement

Pistachio trees are dioecious, with male and female flowers borne on separate trees. Because pistachio pollen is windblown, males are interspersed around the border of and within the orchard in a regular pattern. The male density within an orchard for optimal production has not been determined. Most new orchards in the Central Valley of California are now planted with males placed every fifth tree in every fifth row, a ratio of 24 female trees to 1 male. These appear to yield as well as older orchards planted with a ratio of eight Kerman to one Peters. Additional male trees are sometimes planted in border rows upwind of the orchard, but solid male rows are not needed. In some instances, in areas where few or no other pistachios are grown, the ratio of females to males has been reduced to 14 to 1 to ensure adequate pollination. Place the first pollenizer tree in the upwind corner of the orchard. Then place the remainder in the pattern described above (fig. 5.6).

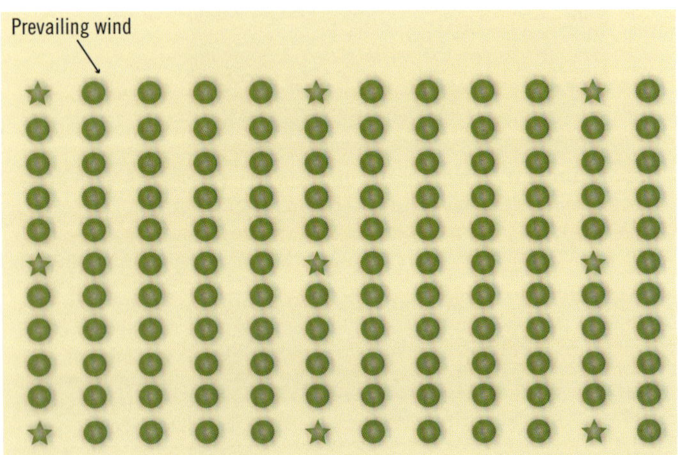

Figure 5.6. A rectangular planting system with 24 female trees (circles) to 1 male tree (stars).

Figure 5.7. A loading site facilitates a rapid and safe harvest. *Photo:* C. E. Kallsen.

Compared with many fruit and nut crops, pistachio requires a relatively long dormancy, or rest period, during the fall and winter between growing seasons for adequate production the following year. For adequate rest, the tree must experience a period of cooler temperatures, commonly referred to as chill. The chill period for pistachio production appears to be insufficient during some years in the southern San Joaquin Valley resulting in a lack of synchrony between the female cultivar and its pollenizing male. In years with insufficient chilling, the Peters male flowers later than the female and pollination may be poor. To offset this situation in orchards planted to Kerman, some growers are planting a number of earlier-flowering males, such as Randy, in addition to Peters. Also, some growers are grafting both male cultivars into the same tree to ensure that pollen is available for adequate pollination in low-chill years.

Loading Areas

Growers should include a loading area when planning a new orchard (fig. 5.7). Most California acreage is planted to a single producing cultivar (i.e., Kerman), resulting in a crop that matures at the same time in the same district, thus producing an acute demand for harvesters, especially in "on" crop years. A loading area, 50 by 500 feet, facilitates harvest and improves safety by providing space for passing equipment and efficient crop-loading operations. Enough area should exist for a typical nut elevator, truck, and passing vehicles. The loading area's length should be sufficient for trucks to turn and, if necessary, for additional trucks to get off access roads and into line for loading. A loading area should be contiguous to every 40-acre block of pistachios. If positioned properly, a loading area should serve up to 160 acres.

Service Roads

Dust generated by vehicular traffic on dirt roads that service an orchard can adversely affect pest control, air quality, and the health of workers and trees. In the United States, growers are increasingly required to implement practices to control dust particles in areas where air quality does not meet federal standards.

Various road-surfacing materials are available that can be used both to reduce dust and improve wet-weather access to the orchard. The suitability of a material for a given road application varies with the native soil, current road condition, seasonal rainfall patterns and amounts, drainage, road travel frequency, material cost, and ability to grade, blade, replace, or repair the road. Some of the materials used to control dust on farm service roads are water; lignosulfate dust binders; new or recycled asphalt; gravel washed by heavy road oil; crushed rock or decomposed granite; vinyl and acrylic emulsions and polymers; organic materials; and magnesium, calcium, or sodium salts. The advantages and disadvantages of each option should be explored in the context of cost and longevity, the needs of the specific orchard, ranch farming practices, current and future availability of materials, need for all-weather access to the orchard, and existing air quality regulations.

Harvesting Machinery

Avoid obstructing pistachio harvesting machinery. The machinery must have free access to the base of the tree to both shake the tree and catch the falling nuts. The catching frames normally extend 10 or more feet on either side of the tree. Anything that extends aboveground can be a potential obstruction during mechanical harvest (fig. 5.8). Irrigation valves and air release valves that are located within the orchard, for example, should extend less than 4 inches above ground level and should be enclosed in low vaults to prevent damage. Poles supporting electrical wires or weather-monitoring equipment should not be placed in the orchard, or they should be given sufficient clearance so that adjacent trees can be harvested easily (fig. 5.9).

Orchard Floor Management

When planning a pistachio orchard, growers must consider the orchard floor management system that will be used for both the young and mature trees. Choices include complete weed control, complete vegetative cover, strip chemical weed control, and intercropping. Further information related to orchard floor management is found in chapter 6, "Microirrigation Systems and Fertigation"; chapter 11, "Weed Management"; chapter 14, "Assessing and Improving Water Penetration"; chapter 25, "Soilborne Diseases"; and chapter 26, "Foliar, Fruit, and Branch Diseases."

Complete Weed Control

Complete cultivation

Orchards using the complete cultivation system are cultivated in both directions. The primary objectives are weed control to prevent competition with trees for water and nutrients, water penetration, and frost protection. Complete cultivation does not use herbicides, so it is a viable soil management option for organic growers. For complete cultivation to be a practical option, trees must be spaced at sufficient distance from one another to accommodate disking equipment.

Complete cultivation can be adapted to level and nonlevel land and is often used when flood or furrow irrigation systems exist. It is not well adapted to solid-set irrigation systems because cross-cultivation is difficult.

Complete cultivation is an expensive orchard floor management system because of operational costs associated with frequent use of heavy equipment. Heavy equipment compacts soils, which impedes water infiltration, and it creates dust that

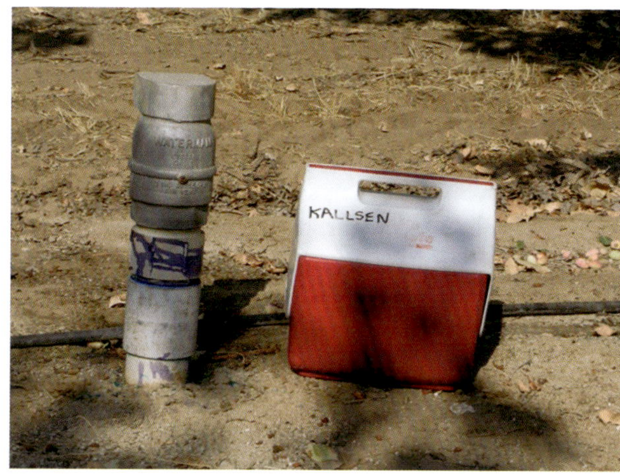

Figure 5.8. Anything that stands more than 4 inches aboveground, such as this irrigation system air release valve, can be a potential obstruction to mechanical harvest equipment and to cross-hedging pruning operations. *Photo:* C. E. Kallsen.

Figure 5.9. Orchards should be designed so that equipment access to the trees and orchard floor is not limited by poles and other obstructions. *Photo:* C. E. Kallsen.

may encourage mite pests. Complete cultivation may increase the potential for damage to the trees or roots. On rolling land, cultivation increases potential for soil erosion—and in areas of high rainfall, it may not always be done in a timely fashion, thus further exacerbating soil erosion.

Complete chemical weed control

The orchard floor can be kept free of weeds with the use of chemical herbicides. This orchard floor management system can use both low-volume sprinklers and flood systems. If a flood system is used, the land should be dead level, as the absence of disking usually results in reduced percolation, increased runoff, and ponding of water on the lower elevations of the orchard.

Where weeds are completely controlled with chemicals, land preparation for irrigation or harvest is not required. Dusty conditions are reduced, as cultivation equipment is not used.

Orchards using this system tend to be warmer than those with cover crops, which is an advantage in areas where late frosts occur or in colder areas during the first winter following planting.

Water penetration can be a problem on a clean orchard floor. This is particularly evident where the impact of droplets from high-volume sprinklers destroys the surface soil structure.

Complete Vegetative Cover

With this system, a complete cover of resident or planted vegetation is maintained on the orchard floor throughout the year. The plant cover is mowed regularly to prevent interference with irrigation or harvest. This system is adapted to both flat and rolling land.

Resident or planted vegetation minimizes soil compaction and has been shown to improve water penetration. Vegetative ground covers allow quicker reentry of equipment, including harvesting equipment, after a rain or irrigation. They also allow better traction, which is especially important in rolling terrain. Ground covers reduce dust and erosion compared with complete cultivation, but they may increase rodent populations, especially gophers. Mowing is less costly than disking, as it can be done faster with lighter and less expensive equipment. Vegetation in the orchard results in a colder orchard than one where vegetation is eliminated; but since pistachio trees bloom in April, increased frost hazard associated with complete covers is of less concern than with other crops. Actively growing

Figure 5.10. In the southern San Joaquin Valley of California, the middles between the tree rows are commonly disked and the berm within the tree row treated with pre- and postemergent herbicides. *Photo:* C. E. Kallsen.

vegetation increases water use by as much as 25 percent, and fertilizer use may increase as well. Irrigation emitters may also interfere with mowing. Flood or high-impact sprinklers, often necessary for maintaining vegetation that covers 100 percent of the orchard floor, can result in high orchard humidity. This increases the incidence of fungal disease, such as Alternaria late blight and Botryosphaeria panicle and shoot blight.

Strip Chemical Weed Control

With cultivation of the row middle

Weeds in row middles are controlled by cultivation, while weeds in the tree row are controlled with herbicides (fig. 5.10). This system of orchard floor management is commonly used in conjunction with low-volume irrigation systems located in tree rows but also where furrow or flood irrigation systems are used.

Continued disking in row middles may result in a raised berm within the tree row that will interfere with cross-hedging. On rolling ground, use of a disk can result in terracing of the orchard. An elevation difference of 12 inches between adjacent terraces can create problems during nut harvesting, as uneven ground complicates shaking of trees and rapid movement of equipment.

With vegetation in the row middle

This orchard floor management system combines chemical weed control in the tree row with mowed planted sod or resident vegetation in the row middle (fig. 5.11). It is the most widely adopted system.

Strip chemical weed control with vegetation is also often used in conjunction with low-volume irrigation systems that do not irrigate the row middles. In the drier parts of California, resident vegetation is allowed to grow in the row middles. This vegetative cover is not irrigated, and it dies or becomes dormant in late spring, greatly reducing management requirements. Where late precipitation is greater or irrigation water is more available and cheaper, the middles may be planted to sod or other cover crop and mowed to control growth throughout the season.

Little information exists on how resident or other planted ground covers either positively or negatively influence insect, mite, nematode, and plant diseases. Effects may also vary depending on species of plants, pests, and weather. Planted ground covers and resident vegetation may increase gopher populations.

Figure 5.11. A winter cover crop of planted or natural vegetation requires little summer maintenance in areas of low rainfall, where the orchard is irrigated by low-volume systems and the berms within the tree row are treated with pre- and postemergent herbicides. *Photo:* C. E. Kallsen.

Intercropping

The practice of intercropping is sometimes used in the case of pistachios. This is because the tree's relatively slow early growth and the required heavy wood removal associated with early tree training extends the length of time before middles are shaded and before economic bearing occurs. To provide income before the pistachios begin producing, some growers plant a row crop, such as watermelons, beans, or wheat, between the rows of young trees. Care must be taken in choosing the intercrop to ensure that chemicals used on both the intercrop and the pistachios are compatible and registered for both crops. Intercrops, like weeds, may compete with the pistachios for water and nutrients, and they may provide habitat for harmful insect and mite pests. Managers must constantly monitor the cultural and harvest practices required for producing an intercrop to ensure that the first pistachio harvest is not delayed as a result of reduced growth or damage to trees.

6 Microirrigation Systems and Fertigation

Lawrence J. Schwankl

Low-volume irrigation systems, often referred to as microirrigation systems, allow the operator to carefully manage both the quantity and timing of irrigation applications, and they are frequently used for pistachios. High-volume sprinkler irrigation systems are infrequently used and have been associated with increased incidence of fungal diseases, such as Alternaria late blight and Botryosphaeria panicle and shoot blight, that attack the leaves and the nut clusters.

Microirrigation Systems

Microirrigation systems (surface drip, subsurface drip, and microsprinklers) are increasingly used in orchards because of their many advantages. These are partial-coverage irrigation systems, which wet only a portion of the orchard root zone. Generally, it is recommended that 30 to 60 percent of the potential orchard root zone volume be wetted for good tree performance.

Advantages of Microirrigation Systems

- A high degree of water application uniformity, often the highest of all irrigation systems in use.
- Excellent control of the amount and timing of irrigation. Small, frequent irrigations (often daily with drip irrigation) can be applied to match the trees' water needs. Runoff is minimized because of the low application rates, and deep percolation losses can also be minimized if the correct amount of water is applied. The irrigation frequency can be fine-tuned to provide the right amount of soil moisture for optimal tree performance.
- Easy irrigation of irregular terrain.
- Weed growth is minimized, since only a portion of the orchard floor is wetted.
- Fertilizers and other chemicals can be injected through the microirrigation system, targeting the active root zone and increasing the application safety as well as the chemical use efficiency.

Disadvantages of Microirrigation Systems

- High initial cost.
- Frequent maintenance and timely management is needed to maintain the system and avoid clogging of the emitters by physical particles, organic materials, or chemical precipitates.
- Irrigation water must be pressurized, resulting in energy costs. The required pressures are generally less than those needed for sprinkler systems, but they are higher than those of flood irrigation systems.
- Cover crops cannot be grown year-round due to incomplete wetting of the orchard floor.

Components and Considerations

The various types of microirrigation systems—microsprinkler, surface drip, and subsurface drip systems—are made up of the same basic components. Figure 6.1 shows the components of a typical microirrigation system. Generally, a microirrigation system consists of

- pump
- flow meter
- mainlines and submains
- drip or microsprinkler lateral lines
- valves
- filter(s)
- injection equipment

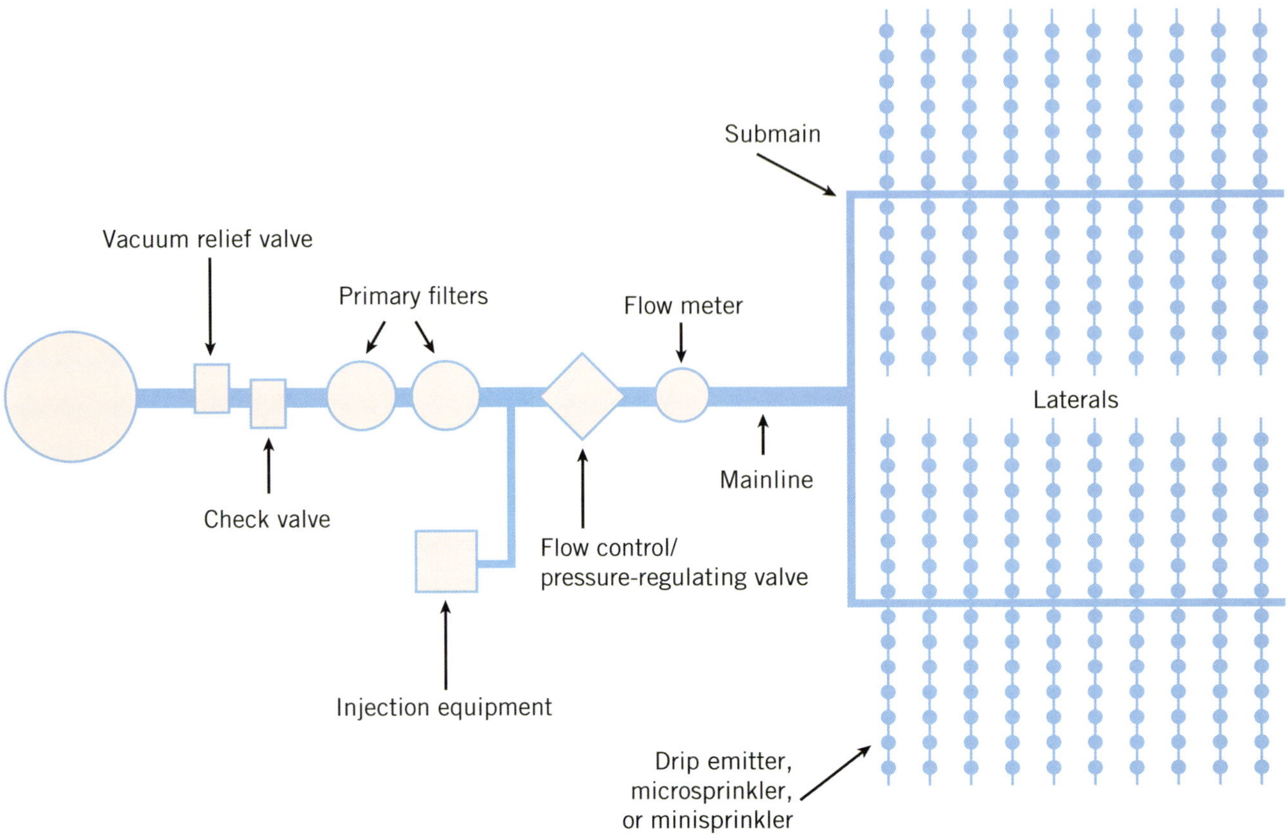

Figure 6.1. Components of a microirrigation system.

For the most part, microirrigation systems differ only in the emitter spacing, the type of emission device used, and the size of the components. The type of emitter used affects the size of the other components. Microsprinklers, for instance, generally require larger filters, mainlines, and submains compared with drip systems.

In a microirrigation system for trees, the emitter spacing and discharge rate depend primarily on the tree spacing and water needs of the trees. The emission devices must be capable of supplying each tree with enough water during the peak water use periods to satisfy the evapotranspiration (ET) requirement.

Following is a brief overview of the components and operation of a microirrigation system for trees (see fig. 6.1).

Pumping plant

It is important to select a pump and motor (or engine) that will deliver the correct pressure and flow rate at the highest possible efficiency. The microirrigation system designer determines the flow rate and pressure to be delivered by the pump, and the pump supplier uses this information to select the most efficient pump for a given system.

Flow meters

It is very important that a flow meter be part of the irrigation system. Knowing the flow rate is necessary for determining the amount of water being applied, which, in turn, is critical to efficient irrigation and scheduling. A propeller meter, which displays the flow rate or total water applied or both, gives an accurate measurement. A flow meter can also be a valuable maintenance tool in determining whether clogging or leaks are occurring.

Injection equipment

Microirrigation is well suited to injecting chemicals such as soluble fertilizer. Various types of injection equipment, differential pressure tanks (batch tanks), venturi devices, electrical or water-driven pumps, and solutionizer machines can be used, depending on the chemical applied, the accuracy level needed, and the injection rate required.

Valves

Valves are the control mechanisms of microirrigation systems. Several types are common: control valves to turn the irrigation system on and off, air or vacuum relief valves (which allow air to escape when the system is turned on and to enter when the system is shut down), and check valves (which prevent undesirable flow reversal).

Pressure-regulating valves are important for maintaining a constant operating pressure in the system. A virtue of microirrigation systems is their ability to deliver a uniform amount of water to each location they serve so that water is applied evenly over the field. Because of pressure differences throughout the system and variability in emitter manufacture, even new systems may not apply water completely evenly. A carefully designed system can use pressure regulators in mains, submains, laterals, or hoses—as well as using pressure-compensating emitters—to overcome pressure differences and variation in emitter discharge rates.

Filters

Selecting the appropriate filter requires consideration of water quality factors. Particulate matter (such as sand) in the water can be removed with vortex filters (frequently referred to as sand separators). Screen, disk, or sand media filters are also effective in removing particulate matter.

Organic matter such as algae or slime can be removed using screen, disk, or sand media filters. But, since organic matter can quickly clog a screen or disk filter and is difficult to flush from the screen or disk elements, sand media filters are the usual choice for filtering surface waters containing algae and slime. If a screen or disk filter is going to be used with water containing organic contaminants, filters with automatic backwash capability are often a good choice.

All filters must be periodically backwashed. The pressure drop across the filter indicates when backwashing is required. Backwashing can be accomplished either manually or automatically, with automatic backwashing taking place either on a defined schedule or when the filter senses a predetermined pressure drop. The water used to backwash is frequently discharged out of the system.

Mainlines and submains

Main and submain pipes, usually made of PVC, deliver water to the lateral lines and emitters. The mains and submains must be sized carefully, with the cost of the pipe balanced against pressure losses caused by elevation differences across the orchard and friction as water moves through the pipe. A qualified microirrigation system designer should design the system.

Lateral lines

Emitters are attached to tubing, or lateral lines, usually made of polyethylene. The length and diameter of the lateral lines to be used also depend on economics, balancing the tubing size and cost against pressure loss along the lateral. If the lateral lines are too long or the wrong diameter, then emitters may discharge water at different rates, resulting in nonuniform irrigation.

Emitters

The many different types of microirrigation emitters available can be grouped generally into punch-in drippers, in-line drippers, and microsprinklers. These three groups are further split into pressure-compensating (PC) emitters and emitters without pressure compensation. Choosing which microirrigation emitter to use depends not only on physical issues such as field topography, soil type, and water quality but also on the personal preferences of the manager.

Surface drip. Surface drip irrigation, along with microsprinkler irrigation, is one of the most commonly used microirrigation systems in tree crops. The drip emitters can either be punched-in to the drip tubing or formed integrally in the drip hose (in-line) at a specific spacing by the manufacturer. Both single and double drip lines per tree row can be used. Double drip lines are usually chosen to achieve a greater wetted area in the orchard or to increase the application rate of the drip system. Increasing the wetted area is frequently an issue in orchards with sandy soils, where water does not move well laterally (poor subbing).

Double-line drip systems are initially more expensive than single-line systems, but they may offer other advantages. By increasing the application rate of a drip system, double-line systems require fewer operating hours to satisfy tree water needs. This gives the manager more time to accommodate other orchard cultural practices, such as spraying or mowing, when the drip system is not operating. A higher application rate may also allow the manager to more easily use off-peak power rates.

For pistachios, which take 7 to 10 years to reach full canopy cover and become full bearing, using punch-in emitters allows growers to increase irrigation efficiency in the first 2 to 3 years of growth. This is done by placing a drip emitter on either side of the young tree and then adding the full number of emitters starting at the 3rd or 4th leaf.

Subsurface drip. Orchard subsurface drip irrigation (SDI) systems are most often internal in-line drippers in hard drip tubing, due to the product durability and ease of installation. Drip tape products are seldom used in orchard applications. Both single-line and double-line subsurface drip systems have been used successfully. Subsurface drip systems are usually installed to a depth of 10 to 24 inches. The deeper depths are chosen to minimize surface wetting, but they make installation and repairs more difficult. Advantages of subsurface drip systems include reduced weed growth since the soil surface usually stays dry, reduced irrigation system damage during harvest or other cultural practices, and the ability to irrigate almost anytime, even right up to harvest. The major disadvantages of subsurface drip are the difficulty in detecting clogging, the hazard of root intrusion into the emitters, and the potential pinching off of buried hoses by expanding tree roots. Root intrusion can be mitigated by using drip products with herbicide-impregnated emitters, but these products are more expensive and the efficacy of the herbicide declines with time. Some herbicides have a chemigation label that can be used to help control root intrusion.

Microsprinklers. The advantages of microsprinklers compared with drip emitters include a larger wetted area, often a higher application rate, less susceptibility to particulate and chemical clogging since the orifices are larger, and easier visual inspection for clogging. While wetting a larger area may be a benefit for tree growth, it is a disadvantage due to increased weed growth. The larger pattern also means that microsprinklers may use more water than drip systems to achieve the same growth in young trees, due to the higher rate of evaporation. Another disadvantage of microsprinklers is that insects can enter or lay eggs in the microsprinkler orifice, causing them to clog. A number of microsprinkler products are now available with bug caps to prevent this from occurring.

Operation and Maintenance

Microirrigation systems can apply irrigation water quite efficiently, but only if they are operated and maintained properly. Irrigation scheduling—determining when to irrigate and how much water to apply—is critical to operating the system efficiently. Effective irrigation scheduling requires knowing how much water the tree is using or has used since the last irrigation (the evapotranspiration, or ET, of the tree) and how much water the irrigation system applies in a given period of time (the application rate). Determining the application rate of a microirrigation system can be confusing at times, so a procedure is discussed at the end of this section. That information can be used to calculate how long the microirrigation system should be run.

While a well-designed system can deliver water with a high degree of uniformity, the system must be properly maintained to keep the application uniform. The principal cause of nonuniformity in microirrigation systems is emitter clogging by particulate or organic matter, lime precipitates, or iron precipitates.

Particulate (silt and sand particles) clogging can occur with both groundwater and surface water sources, but it can often be prevented by good filtration. Microirrigation emitter manufacturers frequently specify a degree of filtration for their emitters.

Biological clogging is caused by bacterial slimes, algae, and other materials that are either present in the irrigation water or growing in the irrigation system. Biological clogging is more frequent with surface water supplies than with groundwater, but groundwater can have biological clogging associated with iron bacterial slimes. Maintenance practices to minimize biological clogging include injection of a biocide, such as chlorine, and good filtration. Serious biological clogging may require continuous chlorine injection at 2 to 5 parts per million (ppm) chlorine, measured at the end of the lateral line. Less severe problems may require only periodic chlorine treatments at 10 to 20 ppm every few weeks.

Chemical precipitate clogging can often be anticipated by an irrigation water analysis. For example, irrigation water with a pH of 7.5 or higher and a bicarbonate level of 2 milliequivalents per liter (meq/l) or greater has an increased risk for calcium carbonate precipitate clogging. Waters with iron levels of 0.5 to 1.0 ppm or higher are at risk for iron precipitate clogging. Calcium carbonate precipitate clogging can be prevented or remediated by injecting acid to lower the water pH to 7.0 or less. Iron precipitate clogging is usually handled by storing the water in a reservoir in which the iron precipitate is allowed to settle prior to withdrawing the irrigation water.

More information on maintaining microirrigation systems is available in *Microirrigation of Trees and Vines* (UC ANR Publication 3378) and *Maintaining Microirrigation Systems* (UC ANR Publication 21637).

Selecting an Irrigation Application Rate

When designing an irrigation system, it is often difficult to choose an application rate. A major objective is to minimize runoff from the orchard. While this is often not a problem with microirrigation systems, it can occur. Ideally, water should soak in where it is applied. The infiltration rate of the soil is not easy to determine since it changes during an irrigation and may change across the season. During irrigation, the infiltration rate begins at its highest rate and decreases during the irrigation until a relatively constant, final infiltration rate is reached (fig. 6.2).

For microsprinklers, this final infiltration rate should be equal to or greater than the water application rate so that no runoff occurs. This does not hold true for point source drippers, where the application rate to the small wetted area under the emitter will almost always exceed this constant, final infiltration rate. Often the best guidance for choosing a water application rate can be made by observing what has worked on other orchards with similar soil conditions, slope, and so on.

When designing an irrigation system, it is preferable to choose the correct application rate at the design stage, but emitter sizes can be retrofitted. It is better to overestimate the application rate during design and later decrease emitter flow rates if needed, because underground piping, pump sizing, and so on are based on a maximum system flow rate. If emitters with greater discharge rates are later retrofitted to the system, it is likely that the piping and pump capacity will be too small, resulting in inadequate pressure.

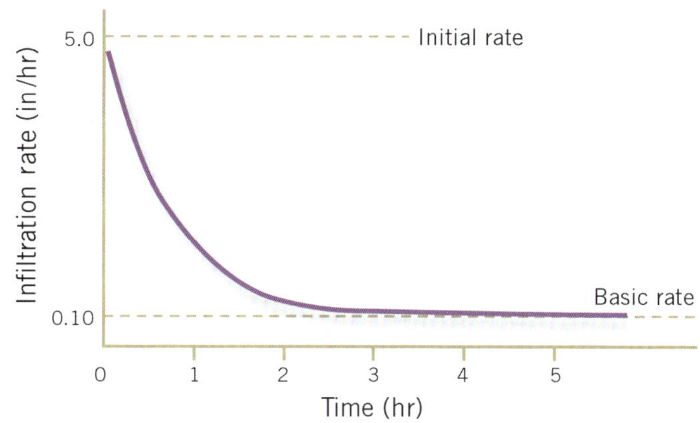

Figure 6.2. Typical infiltration rate curve showing a high infiltration rate at the beginning of the irrigation, decreasing to a lower, constant final infiltration rate.

Irrigation Uniformity

How uniformly or evenly water is applied is also important to good irrigation water management. An irrigation system with uniform water application means that each tree will receive nearly the same amount of water during irrigation.

Application Rate

Determining the application rate of microirrigation systems can be confusing because information on irrigation scheduling and tree water use is usually presented in inches per day (in/day), while discharge from microirrigation emitters is measured in gallons per hour (gph). The water use of the tree and the application rate of the emission device(s) determine how long the drip and microsprinklers should be operated.

Calculating the operating time

- **Step 1:** Convert the tree water use (ET) value from inches per day to gallons per day, using the following formula (or see table 6.1):

  Tree water use (gal/day) = Tree spacing (ft^2) × Tree water use (in/day) × 0.623

  For example, if tree spacing is 20 feet by 20 feet and tree water use is 0.3 inches per day:

  Tree water use (gal/day) = (20 ft × 20 ft) × 0.3 (in/day) × 0.623 = 75 gal/day

- **Step 2:** Determine the application rate per tree of the irrigation system in gallons per hour (gal/hr). For both drip emitters and microsprinklers, this requires determining the number of emission devices per tree and the discharge rate per emission device (gal/hr/emitter):

  Application rate (gal/hr) = Number of emission devices per tree × Discharge rate per device (gal/hr)

  Drip emitters: For example, if there are four drip emitters per tree and the discharge rate per emitter is 1 gal/hr:

  Application rate per tree (gal/hr) = 4 (emitters/tree) × 1 (gal/hr) = 4 gal/hr

  Microsprinklers: For example, if there is 1 microsprinkler per tree and its discharge rate is 12 gal/hr:

  Application rate per tree (gal/hr) = 1 (microsprinkler/tree) × 12 (gal/hr) = 12 gal/hr

- **Step 3:** Determine the irrigation system operation time in hours per day. This requires using the tree water use (determined in Step 1) and the application rate (determined in Step 2) in the following formula (or see table 6.2):

  Net hours of operation per day = Tree water use (gal/day) ÷ Application rate (gal/hr)

Table 6.1. Pistachio water use (gal/day) for various tree spacings and tree water use (in/day)

Tree spacing (ft^2)*	Tree water use (in/day)							
	0.05	0.1	0.15	0.2	0.25	0.3	0.35	0.4
50	2	3	5	6	8	9	11	12
100	3	6	9	12	16	19	22	25
150	5	9	14	19	23	28	33	37
200	6	12	19	25	31	37	44	50
250	8	16	23	31	39	47	55	62
300	9	19	28	37	47	56	65	75
350	11	22	33	44	55	65	76	87
400	12	25	37	50	62	75	87	100
450	14	28	42	56	70	84	98	112
500	16	31	47	62	78	94	109	125

Note: *Tree spacing (ft^2) = row spacing (ft) × tree spacing in the row (ft).

Drip emitters:

Net hours of operation per day = 75 (gal/day) ÷ 4 (gal/hr) = 18.8 hr/day

Microsprinklers:

Net hours of operation per day = 75 (gal/day) ÷ 12 (gal/hr) = 6.3 hr/day

Note that table 6.2 gives the same hours of operation for these examples.

Calculating the depth of water applied

An alternative method of determining the application rate is to calculate the depth of applied water (in inches) by monitoring a flow meter (if available) at the pump. The following formula can be used:

Applied depth of water (in) = Pump discharge (gpm) × Irrigation time per set (hr) ÷ [449 × Acres irrigated per set]

Frequency of Irrigation

Determining how often to irrigate requires consideration of microirrigation system capacity as well as soil and tree parameters. While the goal is to supply just the right amount of water to each tree at the right time, it is also necessary to maintain aeration in the wetted zone to supply oxygen to the roots. An additional requirement is that water penetrate deep enough to provide for some leaching and sufficient depth of rooting.

Historically, many drip irrigation systems were designed to operate many hours daily to meet ET demands. This minimized the total system flow rate and kept the cost to a minimum. However, this required systems to run nearly continually during July, resulting in constant saturation under drippers and increased potential for Phytophthora and other root rots. A more prudent design is to require the drip system to operate only about half to two-thirds of the time during the summer to meet the peak tree water demands.

Microsprinkler systems have a higher application rate than do drip systems. Depending on the amount of water applied daily and the maturity of the orchard, 3 to 7 days between irrigations is common for microsprinkler systems during peak ET periods. Microsprinkler irrigation duration should be more than just a few hours. Short irrigation events with microsprinklers result in only the surface soil being wetted. Particularly in young orchards, this shallow water may be lost to evaporation and not be available for tree uptake. Deeper penetration of irrigation water is desirable, and this is often achieved by longer irrigation events such as 24- to 48-hour sets. To determine the most efficient irrigation duration and frequency in a new orchard development, soil moisture monitoring with an auger or other device is helpful both before and after irrigation. Concerns about salinity in the soil or irrigation water, as well as the possibility of shallow (perched) groundwater, need also to be considered when setting irrigation frequency and duration.

Table 6.2. Hours of operation per day for various application rates (gal/hr) and tree water use (gal/day)

Tree water use (gal/day)	Application rate (gal/hr)										
	1	2	4	6	8	10	12	14	16	18	20
5	5.0	2.5	1.3								
10	10.0	5.0	2.5	1.7	1.3	1.0					
15	15.0	7.5	3.8	2.5	1.9	1.5	1.3	1.1			
25		12.5	6.3	4.2	3.1	2.5	2.1	1.8	1.6	1.4	1.3
50			12.5	8.3	6.3	5.0	4.2	3.6	3.1	2.8	2.5
75			18.8	12.5	9.4	7.5	6.3	5.4	4.7	4.2	3.8
100				16.7	12.5	10.0	8.3	7.1	6.3	5.6	5.0
125				20.8	15.6	12.5	10.4	8.9	7.8	6.9	6.3

Fertigation

Fertigation is the injection of fertilizers through an irrigation system. Microirrigation systems are well suited to fertigation because of their frequency of operation and because water application can be easily controlled. Applying fertilizers through a microirrigation system

- allows fertilizer distribution to be as uniform as the water application
- allows flexibility in timing fertilizer application
- reduces the labor required for applying fertilizer
- allows less fertilizer to be applied
- can lower costs

Fertilizer Solubility

To be injected, fertilizers must be soluble. Fertilizers delivered as a solution can be injected directly into the irrigation system, while those in a dry granular or crystalline form must be mixed with water to form a solution. Fertilizer materials differ widely in water solubility, depending on the physical properties of the fertilizer as well as on irrigation water temperature and pH. Dry fertilizers are mixed into a tank containing water until the granules or crystals are dissolved and the desired concentration is reached. The solution is then injected into the irrigation system. With solutionizer injection machines, the injected material may be in a slurry form, which goes into solution once it is mixed with the irrigation water.

Nitrogen Sources

The fertilizer most commonly injected is nitrogen, as many soluble nitrogen sources work well in fertigation. Following is a list of common nitrogen sources, with information on their use in fertigation:

- **Anhydrous ammonia** and **aqua ammonia** are nitrogen sources that increase the water pH, which may result in a precipitate if calcium or magnesium, along with comparable levels of bicarbonate, are present in the irrigation water. Volatilization of nitrogen (loss to the atmosphere) may also occur when anhydrous or aqua ammonia are used.
- **Urea** is relatively soluble in the irrigation water and is not strongly held by soil particles, so it moves deeper into the soil than ammonia products. Urea is transformed by hydrolysis into ammonium, which is then fixed to soil particles.
- **Ammonium sulfate, ammonium nitrate,** and **potassium nitrate** are relatively soluble in water and cause only a slight shift in the pH of the soil or water.
- **Calcium nitrate** is relatively soluble in water and causes only a slight shift in the soil or water pH. If the water is high in bicarbonate, however, the calcium content may lead to precipitation of calcium carbonate (lime).
- **Ammonium phosphate** can cause soil acidification. If calcium or magnesium levels are high enough in the irrigation water, precipitates may form and clog the drip emitters. (For precautions in using ammonium phosphate, see "Phosphate Sources," below.)

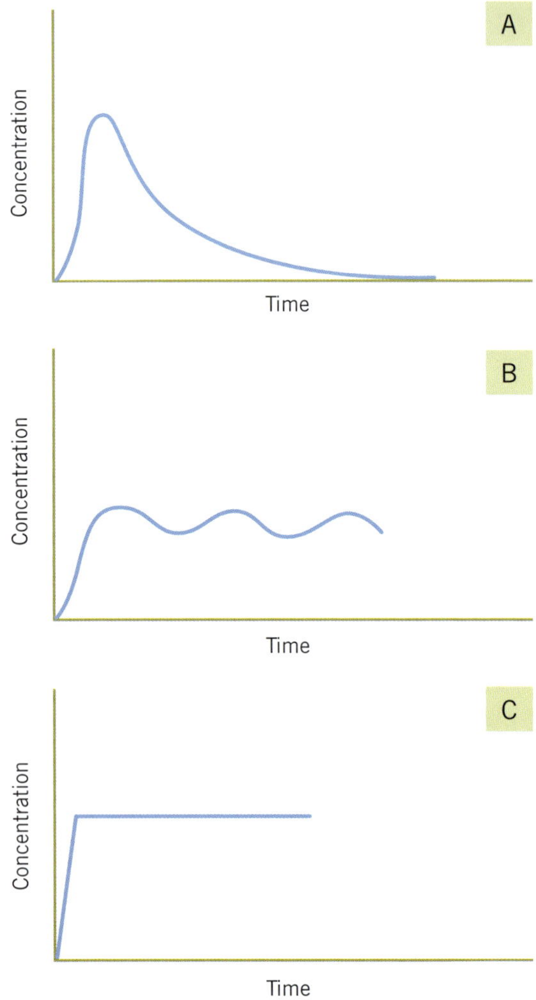

Figure 6.3. Chemical concentration levels during injection using a batch tank (A), venturi injector (B), and positive displacement pump (C). *Source:* Adapted from Bosconer 1987.

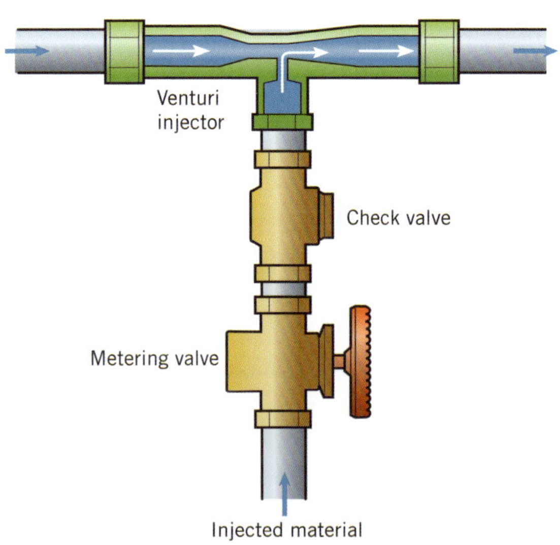

Figure 6.4. Venturi device.

Phosphate Sources

Using phosphate fertilizers may cause chemical or physical precipitate clogging. The calcium and magnesium content and the pH of the irrigation water should be considered, since calcium phosphate and magnesium phosphate precipitates may form when the water pH is higher than 7.5. Acidifying the water with sulfuric acid or phosphoric acid keeps the irrigation water pH low and minimizes precipitation problems. Phosphorus is quickly fixed to soil particles and does not move readily into the soil profile, but it has been found to move more easily under microirrigation than under conventional irrigation methods.

Potassium Sources

Injecting potassium fertilizers usually causes few problems, but caution should be used if potassium fertilizers are mixed with other fertilizers. Potassium, like phosphorus, is fixed by soil particles and does not move readily through the soil profile. Potassium is usually applied in the form of potassium chloride, but for crops sensitive to chloride, potassium sulfate or potassium nitrate may be more appropriate. Potassium sulfate is not very soluble and may not dissolve well in irrigation water.

Injection Devices

Many different substances can be injected through irrigation systems, including chlorine, acid, fertilizers, herbicides, micronutrients, nematicides, and fungicides. Of these, fertilizers are most commonly injected. Chlorine or acid injection is used in microirrigation systems to prevent clogging caused by biological growths (algae and bacterial slimes) and chemical precipitation (particularly calcium carbonate).

Irrigators wishing to inject chemicals can choose from a variety of injection equipment, including differential pressure tanks, venturi devices, positive displacement pumps, small centrifugal pumps, and solutionizer machines.

Differential pressure tanks

Differential pressure tanks, often referred to as batch tanks, are the simplest injection devices. The inlet of a batch tank is connected to the irrigation system at a point of pressure higher than that of the outlet connection. This pressure differential causes irrigation water to flow through the batch tank containing the chemical to be injected. As the irrigation water flows through the batch tank, some of the chemical goes into solution and passes out of the tank and into the downstream irrigation system. Because the batch tank is connected to the irrigation system, it must be capable of withstanding the operating pressure of the irrigation system.

While relatively inexpensive and simple to use, batch tanks have the disadvantage that, as irrigation continues, the chemical mixture in the tank becomes increasingly dilute, thus decreasing the concentration in the irrigation water (fig. 6.3A). If a set amount of a chemical, such as fertilizer, is to be injected and concentration during the injection is not critical, using batch tanks may be appropriate. If the chemical concentration must be kept relatively constant during injection, batch tanks are not appropriate.

Venturi devices

Venturi devices (fig. 6.4), often referred to as Mazzei injectors, consist of a constriction in a pipe's flow area that causes a negative pressure, or suction, at the throat of the constriction. Mazzei is a trade name for a particular brand of venturi injector; venturi injectors are also available from other manufacturers.

The venturi injector is frequently installed across a valve or other point where between 10 and 30 percent of the pressure is lost because of friction in the venturi. This means that the inlet of the venturi injector must be at a pressure 10 to 30 percent higher than the outlet port. Because of this significant pressure loss, the injector should be installed parallel to the pipeline so that flow through the injector can be turned off with a valve when injection is not occurring. The injection rate of a venturi device is determined by the size of the venturi and the pressure differential between inlet and outlet ports. Injection rates as high as 700 gallons per hour are possible with large venturi devices.

Venturi injectors can also be installed with a small centrifugal pump that draws water from the irrigation system, increases its pressure while moving the water through the venturi, and then returns the water and chemical back into the irrigation system.

Venturi devices (fig. 6.3B) are inexpensive and relatively simple to operate, but they do not inject chemicals at as constant a rate as positive displacement pumps. Injecting with venturi devices may be sufficiently accurate for applications such as fertilizer injection, however.

Positive displacement pumps

Positive displacement pumps are piston or diaphragm pumps that inject at precise rates. The pumps are powered by electricity or gasoline, or they are driven by water. The water-driven pumps can be installed in locations that lack electrical power. When a constant and precise injection concentration is needed, positive displacement pumps are preferable (fig. 6.3C). Positive displacement pumps are the most expensive of the injection devices, with costs for electric pumps running $750 or more.

Centrifugal pumps

Centrifugal pumps are frequently used for injecting fertilizers. These pumps have a greater flow rate than do the positive displacement pumps or most venturi injectors, which makes them appropriate for higher injection rate applications. Centrifugal pumps can be driven by either electricity or gas engines. Using a centrifugal pump in conjunction with a flow meter can be helpful in controlling injection rates.

Solutionizer machines

Solutionizer machines were developed to inject materials that are not readily soluble. Their most common use is for injecting finely ground gypsum through the irrigation system, but they are also used for injecting fertilizer products such as potassium sulfate. Solutionizer machines inject a slurry of material into the irrigation line, where it then mixes and goes into solution. In microirrigation systems, it is important that these materials be injected upstream of the system filters to ensure that insoluble materials are filtered out and do not clog the emitters. For example, gypsum materials, which are 95 percent pure, may still contain up to 5 percent insoluble materials. This would mean that for every 100 pounds of gypsum material injected, 5 pounds of insoluble material might be present. Dry fertilizer materials may also contain significant insoluble material that could contribute to emitter clogging.

Injection Point

The injection point should be located so that the injected fertilizer and the irrigation water can become thoroughly mixed well upstream of any branching of the flow. Because of concerns over fertilizers being flushed out when the microirrigation system filters are backwashed, the injection point should be downstream of the filters in all except the solutionizer injection systems. To ensure that no contaminants are injected into the microirrigation system, a good-quality screen or disk filter should be installed on the line between the chemical tank and the injector. The system should be allowed to fill and come up to full pressure before injection begins. Following injection, the system should be operated to flush the fertilizer from the lines. Leaving residual fertilizer in the line may encourage clogging from chemical precipitates or organic sources such as bacterial slimes.

Chemigating Uniformly

Once injection begins, the injected material does not immediately reach the emitters. There is a travel time for water and injected chemical to move through a microirrigation system. Measurements made in commercial orchards indicate that this travel time may range from 30 minutes to well over an hour, depending on the microirrigation system design. To ensure that application of any injected material is as uniform as the water applications, take the following steps:

- **Step 1:** Determine the travel time of chemicals to the farthest hydraulic point in the microirrigation system. This is a one-time determination and can be done by injecting chlorine into the system (also a good maintenance procedure) and tracing its movement through the system by testing the water for chlorine with a pool or spa test kit.
- **Step 2:** The injection period should be at least as long as it takes the injected material to reach the end of the last lateral line (determined in Step 1). A longer injection period is usually preferable.
- **Step 3:** Once injection is stopped, the irrigation should continue for as long as it took the injected material to reach the end of the farthest lateral (Step 1). A longer, postinjection irrigation period is usually preferable.

Make sure, especially with injected materials that easily travel with the water (e.g., nitrate materials), that there is no overirrigation that moves water and injected material through the root zone. Such overirrigation could waste the injected material and lead to groundwater contamination.

Chemigation Safety

Appropriate care should be exercised when handling all fertilizer materials, and the safety of personnel should be of highest priority. Environmental safety associated with fertigation should also be a priority. Fertigation regulations vary from state to state. In California, chemigation safety regulations apply only to labeled chemicals, not to fertilizers. While there are no California state regulations concerning fertigation, there may be local regulations—and these should be checked and followed. The same safety equipment required for injection of labeled chemicals is also useful for environmental protection when fertigating.

Figure 6.5 shows a sample chemigation layout with safety devices for preventing environmental contamination. Numerous other approved layout configurations are possible that incorporate different injectors and other safety devices. The safety devices in figure 6.5 include the following:

- **Chemigation check valve,** located between the water source and the injection point, prevents chemical from moving back to the water source. The check valve has a one-way, spring-loaded flap inside that allows water to pass only downstream. It also has an air vent-vacuum relief valve and a low-pressure drain upstream of the one-way flap closure (fig. 6.6). The vacuum relief valve prevents a vacuum from forming, which could draw chemical through the closed check valve. If some chemical does leak past the closed check valve, it will drain out of the low-pressure drain, which is open when the irrigation system is shut down but closes when the irrigation system is pressurized.

  Even if there is no expectation that chemicals other than fertilizer will be injected through the irrigation system, installation of a chemigation valve is prudent when fertigation is planned. Backflow of fertilizer to a well or other water source can result in surface and groundwater contamination. At a minimum, leaving room to install a chemigation valve at a later date is wise.
- **Electronic interlock** between the water pump and the fertilizer injector pump prevents operation of the injector if water is not being pumped.
- **Check valve** in the line from the injector to the irrigation system prevents water from flowing back through the injector and overflowing the chemical storage tank.
- **Normally closed solenoid valve** (or a normally closed, hydraulically operated valve) between the chemical tank and the injector keeps chemical in the tank from flowing into the irrigation system when the system is not operating.
- **Pressure switch** in the irrigation system, interlocked to the pump, shuts down the irrigation and injection systems if there is a break in a pipeline or any other cause for a drop in operating pressure.

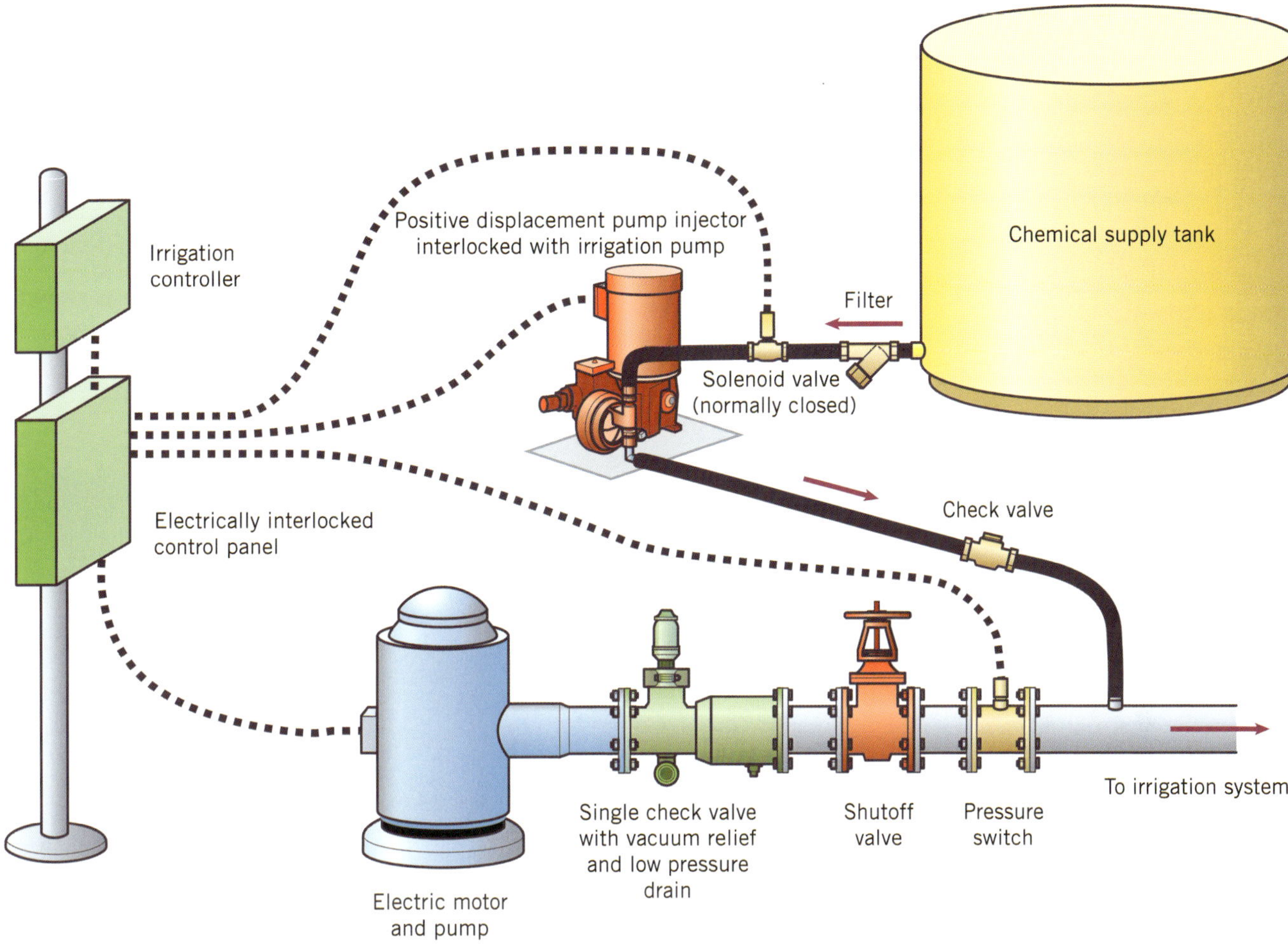

Figure 6.5. Typical layout of chemigation injection system with safety devices installed.

Summary

Irrigating pistachios using a microirrigation system promotes efficient irrigation, as it applies the water the trees need at the time they need it. To do this, however, the manager must know the tree water demands and the microirrigation system application rate. How to determine tree water needs and microirrigation application rates is discussed in chapter 13, "Irrigation Management." The system also must be maintained properly, since emitter clogging can reduce the application uniformity. In addition, microirrigation systems are well suited to applying fertilizers (fertigation) and other chemicals (chemigation) through the irrigation system, often resulting in more efficient and safer chemical applications.

Figure 6.6. Double chemigation check valve protecting a well from backflow contamination. A double check valve is required in some states and is installed for safety redundancy. *Photo:* L. Schwankl.

References

Bosconer, I. How irrigation lines can serve double duty. 1987. Agricultural Engineering May/June 1987.

Schwankl, L. J., and T. Prichard. 2001. Chemigation in tree and vine microirrigation systems. Oakland: University of California Division of Agriculture and Natural Resources Publication 21599.

Schwankl, L. J., B. R. Hanson, and T. L. Prichard. 1998. Micro-irrigation of trees and vines. Oakland: University of California Division of Agriculture and Natural Resources Publication 3378.

———. 2008. Maintaining microirrigation systems. Oakland: University of California Division of Agriculture and Natural Resources Publication 21637.

7 Pistachio Cultivars

Dan E. Parfitt, Craig E. Kallsen, and Joseph Maranto

The edible pistachio of commerce is the species *Pistacia vera* L. In addition to many named cultivars, significant populations of wild germplasm exist, primarily in central Asia from Turkey to Afghanistan (Parfitt et al. 2012). The species has long been propagated for nuts throughout the Mediterranean and Middle East. Several reports suggest that the Romans were responsible for the spread of *Pistacia vera* L. within the Mediterranean basin. Wild pistachio nuts are typically much smaller than the cultivated pistachio, usually about 1 centimeter long and nonsplit. There is considerable variation in size and other characteristics among wild pistachio nuts (fig. 7.1), especially those from the Bagtiz region of Turkmenistan and adjacent areas of Iran, considered the center of diversity for pistachio (Whitehouse 1957).

Europeans have recognized the differences in the flavor of pistachios produced in different parts of the world. Since the 1800s, Sicilian pistachios have been mentioned among European consumers as some of the best-tasting nuts. The nuts are normally small, and a high percentage are unsplit and filled with dark green kernels. The pistachio nut produced in Middle Eastern countries is larger, has a yellowish kernel, and is considered by some to have less flavor. Pistachios grown in Tunisia and neighboring countries are generally small and are considered to have a good but not exceptional flavor. Many pistachio cultivars (cultivated varieties) are grown in more than one country and have different characteristics in each country, especially for fruit shape and color of the kernel, which varies from light yellow to deep green. A survey made in the United States showed that pistachio consumers do not have a preference

Figure 7.1. Nuts from two different trees in the Bagtiz region of Turkmenistan show the range of size and shape in wild pistachio populations. *Photo:* D. E. Parfitt.

for nut shape, which could vary from oblong (almond shaped) to round (filbert shaped). European consumers formerly preferred the elliptical to the round shape. Some scientists (e.g., J. Maranto) believe that the Italian cultivars have a superior flavor, although D. E. Parfitt has identified excellent flavor in Persian cultivars. Flavor is a very subjective parameter to evaluate, and ultimately what emerges as the "correct" pistachio flavor will be an average of the total public perception. A study by Kader et al. (1982) using organized taste panels as well as measurement of various fat and sugar concentrations showed no significant differences in overall flavor among Kerman, Red Aleppo, Trabonella, and Bronte cultivars. Kerman had more linoleic and less oleic acid than the other cultivars and thus could be considered most healthy. However, the differences were only about 4 percent. Kerman also had more sugars than the other cultivars, including reducing sugars, which can be a problem when roasting. (Reducing sugars can cause browning or discoloration when heated.) Trabonella was somewhat more oily and bitter than the other cultivars, while Red Aleppo was most crisp or crunchy. However, Kerman had a substantially higher firmness score than the other cultivars. Recent publications by Tsantili et al. (2010, 2011) identified significant differences in shell and kernel color, protein content, oil composition, phenolics, and antioxidants among Aegina, Pontikis, Bronte, Cerasola, Joley, Kerman, Mumtz, and Sirora.

European Cultivars

Napoletana is the predominant variety in Sicily. Less often planted are Agostana, Girasola, Notaloro, Cappuccia, and Femminello. Trabonella and Bronte are Sicilian cultivars with similar characteristics. The nut color is greenish and the nut shape is longer and thinner than Kerman or Iranian cultivars. The nut size is considerably smaller than Kerman, and the nut quality under California conditions is poor, with a high level of nonsplits in some seasons and significant disease and pest problems. Sfax, Mateur, and El Guettar are grown in Tunisia. Mateur may be the best of these and is being tested in Spain. Sfax produces large, tight nut clusters, but nut size, yield, and percent splits are quite inferior to Kerman nuts. Aegina (Aegenes) and the more recently released Pontikis are grown in Greece. Aegina appears to be very susceptible to *Botryosphaeria dothidea*, perhaps due to its very early flowering in the spring. Pontikis has a moderately large fruit, with kernel weight at 55 percent of fruit weight; the shape of both fruit and kernel is oblong-ovate. It splits much better than Aegenes (same as Aegina), with 90 to 98 percent splits. Blank nut percentage is about 5 to 10 percent, the same as Aegenes, and yields are similar to Aegenes. However, the adaptation of Pontikis to other growing areas outside of the Athens region has not been determined. Lamarka is the main cultivar in Cyprus. Sirora was developed in Australia (Maggs 1990). The industry has remained small in Australia, so this cultivar has not been widely planted. Kastel and Rashti, grown in Israel, are similar in some aspects to our Kerman variety. Neither Kastel nor Rashti has been tested directly against Kerman in California yield trials. Rashti has large nuts, high split percentage, and a good flavor. The tree structure is similar to Kerman, as is its alternate-bearing characteristic. It is a late-maturing cultivar, probably several weeks after Kerman. This has been an issue during years when late summer and fall have been very cool. Under these conditions it may not ripen before the start of winter rains. Kastel seems to be very similar to Kerman in most characteristics.

Turkish and Syrian Cultivars

Turks prefer Uzun and Kirmizi; less favored is Turkish Red Aleppo. In Syria, the Red Aleppo is more common. Achoury, Alemi, El Bataury, Obiad, and Ayimi are also grown in Turkey. Red Aleppo was used as a cultivar during the early development of the California industry. It splits and yields well but may not produce as well as Kerman. The nut size tends to be somewhat smaller than Kerman.

Iranian Cultivars

Major pistachio cultivars grown in Iran are Momtaz, Ohadi, and Kalehghouchi. Other cultivars are Ghermeza, Tbeahimi, Ogah, and Wahidi. Another Iranian cultivar, Rafsanjani, is considered to be a promising cultivar in Iran, and it is being tested for adaptation in Azerbaijan. In addition to these cultivars, many other named cultivars are grown in Iran, either locally or across the country. Some of these items may be the same as the aforementioned cultivars but with different local names. Ohadi produces attractive nuts that are slightly smaller than Kerman nuts. Kalehghouchi has very large nuts as well as a good yield. It has attracted some interest in California because of its nut size and good split percentage, which may be better than Kerman under California conditions. In replicated trials budded in 1998 and located on the west side of Kern County, Kalehghouchi and another Iranian variety, Aria, were tested against Kerman.

Table 7.1. Average characteristics of UC (6th through 11th leaf) and Iranian (5th through 7th leaf) pistachio varieties on the west side of the southern San Joaquin Valley of California

Characteristic	University of California cultivars			Iranian cultivars		
	Kerman	Golden Hills	Lost Hills	Kerman	Kalehghouchi	Aria
nut yield (lb/ac)*	3,036 a[†]	3,849 b	2,895 a	2,470 b	2,539 b	2,162 a
payable yield (lb/ac)	2,712 b	3,655 b	3,460 b	2,276 b	2,276 b	2,026 a
edible, split, in-shell yield (lb/ac)	2,389 a	3,460 b	2,631 a	2,080 a	2,334 b	1,880 a
payable yield (%)[‡]	30.2 a	34.8 c	32.6 b	29.4 a	33.3 b	30.2 a
edible, split, in-shell (%)[†]	67.8 a	84.6 b	85.5 b	74.4 a	84.8 c	77.2 b
total insect damage (%)[†]	1.8 b	0.0 a	0.2 a	1.3 a	1.1 a	0.4 a
loose shells and kernels (%)[†]	0.5 a	0.6 a	3.20 b	0.5 a	1.4 b	5.2 c
individual nut weight (g)	1.24 a	1.26 a	1.47 b	1.23 a	1.44 c	1.35 b
average full bloom date[#]	Apr 5	Mar 31	Mar 31	Apr 5	Mar 31	Mar 31
harvest readiness date	Sep 12	Aug 29	Aug 30	Sep 16	Sep 14	Sep 1

Notes:

*lb/ac x 1.1208 = kg/ha.

[†]Values within the same row for each trial followed by different letters are significantly different at $P \leq 0.05$ by Fisher's protected LSD test.

[‡]Based on percentage of the total dried green weight of the sample measured at harvest.

[§]Based on percentage of hulled, total dry sample adjusted to 5% moisture.

[#]Estimated from 2003 to 2007 observations.

Kalehghouchi trees in the 7th to 10th leaf yielded similarly to Kerman, while Aria yields were lower (table 7.1). Nut split percentages and weights were higher for both of these cultivars than for Kerman. Both Kalehghouchi and Aria flower 5 to 10 days earlier than Kerman. Kalehghouchi matures at about the same time as Kerman, while Aria matures about 2 weeks earlier than Kerman. Kalehghouchi produces excessive vegetative growth on mature trees under California management conditions. It has a tendency to produce many long "whips" and "hanger" branches on which a considerable fraction of the fruit is borne. This makes shaking difficult or requires considerable additional pruning to maintain tree structure. Aria, when tested in the same trials, produced nuts with poor shell-hinge strength, resulting in excessive loss of shells and kernels during hulling. The location of flower buds on new branches has made training this cultivar difficult and has resulted in sunburned nut clusters.

These limited replicated trials have demonstrated that a particular cultivar's success in a foreign country does not guarantee successful performance in the United States. In California, very few experimental comparisons of horticultural performance have been conducted among Kerman and cultivars from other countries.

Cultivars from the Former Soviet Union

Pistachio cultivar collections exist at several of the Vavilov Institute germplasm collections in the independent republics of the former Soviet Union. With the separation of these republics from Russia, funds to maintain and curate these collections were lost. The collections are in poor condition and records for some of them are no longer available (Parfitt, personal observation). Some of the selections maintained at these sites are of commercial quality, and some are not.

California Cultivars

Less than 20 named cultivars have been imported into the United States. Female cultivars that were introduced into California by the USDA in the early 1900s included Red Aleppo from Syria, Bronte and Trabonella from Sicily, Sfax from Tunisia, Kastel and Rashti from Israel, and some from other countries. Additional unnamed *Pistacia vera* germplasm was introduced through the USDA Plant Introduction Station at Chico, which was closed in 1967. However, very little *Pistacia vera* germplasm has been imported into the United States since the station was closed. Several importations of *Pistacia* species germplasm have been made by the USDA and are maintained at the National Clonal Germplasm Repository (NCGR), Davis, California.

Kerman, a California-developed cultivar, was selected in 1936 (after being collected by W. E. Whitehouse in 1929) and released for trial in 1957 by the USDA Plant Introduction Station in Chico. The release of Kerman occurred just prior to the major growth of the industry in California. Kerman is now the primary female cultivar commercially grown in all regions in California. Kerman is a very good cultivar, producing large yields of large and attractive nuts, especially in the primary growing areas of the southern San Joaquin Valley. It is not a perfect cultivar, however. It has a strong alternate-bearing tendency, a high percentage of blank nuts in some years, a relatively high level of nonsplit nuts, and a light greenish-yellow kernel with almost minimal flavor when dried at commercial temperatures. Kernel color and flavor have not been an issue with American consumers, who have not been exposed to European cultivars with different flavor characteristics. Kerman is a relatively late-maturing cultivar, which is not usually a problem in the San Joaquin Valley with its large number of heat units. However, Kerman's later maturity means that it can be exposed to a third flight of navel orangeworm. Beyond the direct losses from damaged nuts, navel orangeworm infestations have been implicated in the development of higher levels of *Aspergillus flavus* contamination and consequent aflatoxin contamination. In the Sacramento Valley during years with low heat unit accumulation, Kerman may not mature until after the first fall rains, with a concomitant increase in disease problems.

In 1980, another open-pollinated seedling introduced as a seed from Damghan, Iran, was selected at the University of California, Davis, by J. Crane and named Joley, in honor of the former director of the Chico USDA Station. Joley has been planted in a few orchards in California and in the state of New Mexico, where a limited acreage of pistachios is grown. The pistachios from this cultivar are considered by some to be among the best tasting of those developed in California. The tree has moderate vigor, and it blooms and matures about 10 days earlier than Kerman. It has almond-shaped nuts similar to Trabonella or Bronte (Sicilian varieties). The kernel color is greener than that of Kerman, but the nut size is significantly smaller, and the nonsplit percentage can be high in some years. There are few blanks; however, in some instances shell removal by consumers is not as easy as for Kerman. On a few young trees grown in Kern County, the yield has never been high. It does not appear to be a commercially viable cultivar.

Another cultivar, Lassen, was developed by Whitehouse from the same seed lot as Kerman and released in 1962 from the USDA Plant Introduction Station in Chico. It is very similar to Kerman with respect to nut characteristics. It has never been tested in replicated trials, but it has good potential as a cultivar. Data from yield trials with Kerman is not available. Damghan was also developed from this seed collection. It has very large nuts but appears to be very low yielding under California conditions.

Two new female cultivars were released in 2005 from the pistachio breeding program of Parfitt et al. (Chao et al. 1998; Kallsen et al. 2009) Five years (6th through 11th leaf) of production and phenological data have been taken on these cultivars (see table 7.1). Golden Hills (Parfitt et al. 2007) is a new female cultivar that flowers 5 to 7 days before Kerman and matures about 2 weeks earlier than Kerman. Early maturity permits efficient use of harvesting and processing equipment in much the same way that cultivar maturity series for peaches facilitates use of limited resources for harvesting and marketing. Earlier harvest also allows this cultivar to avoid infestation by the most damaging September flight of navel orangeworm. This cultivar produced 35 percent higher grower-paid yield than Kerman during the first 5 years that harvestable yields were collected from a replicated trial in the southern San Joaquin Valley. Nut size and weight were similar to Kerman's (fig. 7.2), but percentage of blank nuts was lower than for Kerman during 3 years of data collection for this characteristic. Golden Hills has a balanced shell-hinge strength that provided for a split nut percentage about 25 percent greater than Kerman's during 5 years of data collection. Golden Hills has more but smaller scaffold branches than Kerman does, producing a smaller, bushier tree after 3 to 4 years of training. Nut size is similar to Kerman's and insect damage was less, as the result of an earlier harvest.

Lost Hills is a new female cultivar from the breeding program of Parfitt et al. (2008). It flowers 4 to 7 days before Kerman and matures about 2 weeks earlier than Kerman. As with Golden Hills, earlier maturity results in much less navel orangeworm damage at harvest (< 0.2 percent). This cultivar produced 28 percent higher grower-paid yield than Kerman over the first 5 bearing years. Nut size and weight were larger than for Kerman (fig. 7.3). The percentage of splits was about 26 percent higher than for Kerman. Lost Hills produced more lost and loose kernels than did Kerman during hulling as well (3.2%). This is a potentially undesirable characteristic from a processing perspective. Lost Hills appears to be a less alternate-bearing cultivar than do Kerman or Golden Hills, but additional evaluation data in later years will be needed to confirm this observation. Flowering was more uniform for both this cultivar and Golden Hills than for Kerman during low-chilling seasons. This resulted in a more uniform maturity and less difficulty in determining the correct time to harvest for maximum splits and minimum staining. Evaluation of Lost Hills and Golden Hills is continuing.

Several additional female selections from the California breeding program are being tested with replicated trials in the southern San Joaquin Valley. In addition to yield and the usual nut quality characteristics, these selections are being evaluated for flowering date and harvest date. The development of a harvest date series, extending the harvest period to more than a month prior to Kerman harvest, is a major objective, since expanding pistachio acreage has created capacity issues for the processors. Earlier flowering cultivars may be less susceptible to the effects of low chill years, which may become more common in the future.

Male Cultivars

The California pistachio industry is based on one male cultivar, Peters. Peters is a good pollinator and was found in the early 1900s by A. B. Peters at Fresno, California. However, the parentage is unknown. It produces abundant pollen that is shed over a relatively long period (about 2 weeks), and its pollen viability over time (durability) is very good. In recent years, under low-chill conditions it has shed pollen at the end of the receptive period for Kerman, resulting in poor and irregular pollination. Other male cultivars, the Chico selections 02-16 and 02-18 imported from Russia, are early and late blooming compared with Peters. Pollen from 02-18 is somewhat less durable than pollen from Peters (Polito et al. 1988). Nazareth, Ask, and Chico males are also grown sporadically in some locations in California. Chico was introduced from the Chico USDA Plant Introduction Station in 1962 as PI150646. It is reported to be a male seedling of *Pistacia vera*, selected from seed introduced under PI73396 from Aleppo, Syria, probably of species hybrid origin and tested as Chico 23. Observations of leaf characters and bloom period (Parfitt, personal observation) suggest that it is probably an interspecific hybrid between *Pistacia vera* L. and *Pistacia integerrima* L. It is a prolific pollen producer and blooms early, courtesy of its *P. integerrima* parentage. Chico sheds pollen prior to and during the earliest part of the Kerman bloom period. If xenia (male contribution) effects are present in pistachio for nut size, then this cultivar should not be used. (Xenia effects have not been systematically researched in pistachio.) Ask and Gazvin were introduced from Israel a few years ago and may have some value as pollenizers. They flower earlier than Peters but have poor pollen durability.

Figure 7.2. A comparison of Golden Hills harvested and dried nuts with Kerman nuts. *Photo:* D. E. Parfitt.

Figure 7.3. A comparison of Lost Hills harvested and dried nuts with Kerman nuts. *Photo:* D. E. Parfitt.

A number of new pollen sources were evaluated as part of the pistachio breeding program conducted by Parfitt et al. They were evaluated for quantity of pollen produced, the period over which pollen was shed, pollen viability percentage at pollen shed, and pollen durability (the length of time during which pollen viability remained high). Two selections were made to complement Peters, flowering earlier and later than Peters, respectively. The early-flowering selection, Randy, was released in 2007 (Parfitt et al. 2010). Randy flowers 1 to 3 weeks earlier than Peters. It is characterized by a relatively long bloom period (in excess of 2 weeks), a characteristic that Peters shares. (Most male pistachios have a flowering period of about 1 week.) Peak flowering is 1 to 2 weeks earlier than peak flowering for Kerman. The pollen is more durable than Peters pollen (75% viable, declining to 35% viable after 29 days of storage, compared with Peters initial viability of 45%, declining to 15 to 5%

after 27 days). Although it was selected as a pollenizer for early-flowering cultivars from the breeding program of Parfitt et al., Randy may be useful as a Kerman pollenizer during low-chill seasons, when there is less overlap in flowering period between Kerman and the later-flowering Peters. It flowers too early to serve as the primary pollenizer for Kerman, however. The bloom period for Randy closely matches that of Kalehghouchi and would be a better pollenizer choice for this cultivar than Peters. (Some reports of poor yield in Kalehghouchi in California may be attributable to poor pollenization with Peters.)

Several male selections have been identified in the California breeding program to provide flowering period overlap with early female selections and to serve as improved pollenizers for Kerman. These selections are being evaluated for flowering date, length of flowering period, and quantity and quality of pollen.

Future Cultivars

Pistachio breeding programs are located in Spain, Turkey, and Israel. Research is probably being conducted in Iran as well, but it is less well documented. Australia had a large breeding program in the past, but it was discontinued due to loss of funding. Potential cultivar materials from the California program continue to be tested and some new seedlings are produced for evaluation under California conditions. Several superior cultivars may be selected from these populations. Over the longer term, superior cultivars may be derived from European or Middle Eastern breeding programs. As noted previously, these materials may not be adapted to California conditions.

References

Chao, C. T., D. E. Parfitt, L. Ferguson, C. Kallsen, and J. Maranto. 1998. Breeding and genetics of pistachio: The California program. Proceedings of the 2nd International Symposium on Pistachios and Almonds. Acta Horticulturae 470:152–161.

Joley, L. E. 1979. Pistachios. In R. A. Jaynes, ed., Nut tree culture in North America. Hamden, CT: Northern Nut Growers Association. 163–174.

Kader, A. A., C. M. Heintz, J. M. Labavitch, and H. L. Rae. 1982. Studies related to the description and evaluation of pistachio nut quality. Journal of the American Society for Horticultural Science 107:812–816.

Kallsen, C. E., D. E. Parfitt, J. Maranto, and B. Holtz. 2009. New pistachio varieties show promise for California cultivation. California Agriculture 63(1): 18–23.

Maggs, D. H. 1990. The Australian Pistachio Sirora. Fruit Varieties Journal 44:178–179.

Parfitt, D. E. 1997. Pistachio. In R. M. Brooks and H. P. Olmo, eds., The Brooks and Olmo register of fruit and nut varieties. Alexandria, VA: ASHS Press. 581–582.

Parfitt, D. E., S. Kafkas, I. Batlle, F. J. Vargas, and C. E. Kallsen. 2012. Pistachio. In M. L. Badenes and D. H. Byrne, eds., Fruit breeding. NY: Springer. 803–826.

Parfitt, D. E., C. E. Kallsen, B. Holtz, and J. Maranto. 2008. 'Lost Hills': A new pistachio cultivar for California. HortScience 43(1): 247.

———. 2010. 'Randy' male pistachio. HortScience 45(7): 1113–1115.

Parfitt, D. E., C. E. Kallsen, J. Maranto, and B. Holtz. 2007. 'Golden Hills' pistachio. HortScience 42(3): 694–696.

Polito, V. S., J. G. Luza, and S. A. Weinbaum. 1988. Differential low-temperature germination responses by pollen of *Pistacia vera* clones with different bloom dates. Scientia Hort 35:269–274.

Pontikis, C. A. 1986. 'Pontikis' pistachio. HortScience 21:1074.

Tsantili, E., K. Konstantinidis, M. V. Chirsopoulos, and P. A. Rousssos. 2011. Total phenolics and flavonoids and total antioxidant capacity in pistachio (*Pistachia vera* L.) nuts in relation to cultivars and storage conditions. Scientia Hort 129:694–701.

Tsantili, E., C. Takidelli, M. V. Christopoulos, E. Lambrinea, D. Rouskas, and P. A. Roussos. 2010. Physical, compositional and sensory differences in nuts among pistachio (*Pistachia vera* L.) varieties. Scientia Hort 125:562–568.

Whitehouse, W. E. 1957. The pistachio nut: A new crop for the western United States. Economic Botany 11:281–321.

8 Pistachio Rootstocks

Louise Ferguson, Robert H. Beede, Heraclio Reyes, Blake L. Sanden, Stephen R. Grattan, and Lynn Epstein

A rootstock is defined as a rooted plant or stem into which a bud is inserted or a graft is made. Rootstocks are plants grown specifically to provide the roots for the fruit-producing portion (called the scion) of the plant. Rootstocks are used to adapt both the female fruit-bearing and male pollen-producing scions to climate, soilborne diseases and nematodes, insect pests, and soil and water quality factors, as well as to enhance tree yield or nut quality. For many fruit trees—for example, citrus—rootstocks were developed to combat pathogens in the soil or in (or transmitted by) scion buds. The initial pistachio cultivar evaluations demonstrated that the *Pistacia vera* seedling tree that produces the edible pistachio nut was susceptible to the soilborne pathogens *Verticillium dahliae* and *Phytophthora* spp. and to nematodes (Woodruff 1979). As a result, California's pistachio crop is now produced entirely on rootstocks.

When the California pistachio industry started in the late 1960s, all rootstocks were produced from germinated seed, since the *Pistacia* spp. used for rootstocks could not be easily propagated as rooted cuttings. The *Pistacia* spp. used for rootstocks, like the nut-bearing pistachio trees, are dioecious. The female flowers (producing rootstock seed) are wind pollinated by pollen from the flowers of the male trees. The resulting rootstock seed are sexual products, as they have a male and female parent. In the early 2000s, asexual (clonal) micropropagation of pistachio rootstocks was introduced. Seedling rootstocks produce better root systems, but, like all sexual products, they have diversity within the rootstock population, since they have genes from both the female and male parents. Clonally propagated rootstocks are genetically identical to the plant from which they were generated and are therefore uniform. However, they lack diversity within the population produced from that plant.

Genetic diversity or the lack of it, depending on how the plant population reacts to pathogens and conditions, can have both advantages and disadvantages. A recent development in the California pistachio industry illustrates this. In the early 2000s, specific individual rootstocks were cloned from a population of selected hybrid seedling rootstocks. No experimental trial data has been collected on the performance of these specific rootstock populations. For the first decade, however, multiple orchards produced on these clonal rootstocks were strikingly uniform and commercially productive. In 2010, severe morphological characteristics began manifesting increasingly in young orchards on these clonally propagated rootstocks. Subsequent investigations demonstrated that the rootstocks were infected with the bacterium *Rhodoccus* sp. Researchers are now investigating the way *Rhodoccus* sp. initially infected these clonal rootstocks, as well as the bacterium's potential for infecting other rootstocks, both seedling and clonal, in the nursery and in the field. Developing this data and methods to control—or resist—this bacterium will require years of research. This sequence of events has demonstrated that while clonal rootstocks can produce superior yields, they can also have limiting vulnerabilities (Stamler et al. 2015).

Thus far, little research has been done on the pest and pathogen susceptibilities, and even less research exists on the genetic composition or variability of the single species or interspecific hybrids of the current commercial pistachio rootstocks. Rootstock quality control in the nursery consists of visually evaluating each seedling for trueness to type (roguing) and discarding morphologically different rootstocks from germination through sale at 6 to 18 months. However, as the industry expands, it is inevitable that susceptibilities, of both rootstocks and cultivars, to pests and pathogens will develop. For this reason the California pistachio industry should consider developing a rootstock and cultivar clonal protection program such as the one developed by the California citrus industry (Vidalakis et al. 2014).

The process of producing rootstocks is covered in detail in chapter 9, "Rootstock Production and Propagation." Seedling and clonal rootstocks are initially grown in the greenhouse and, when larger, are transferred outside. Currently, commercial rootstocks are generally not sold with the scion bud inserted, or budded. They are transplanted in the orchard as 6-to-18-month-old rootstocks and budded with the female or male scion within the first 6 months in the ground. Two-year-old budded rootstocks are currently used as transplants when individual trees die after initial orchard establishment. However, this practice of planting unbudded rootstocks and later field budding could change in the future, due to the potential for disease transmission via budding. Also, as more patented cultivars are developed, it is easier to document a scion bud's origin, as is now done in the citrus industry (Vidalakis et al. 2014), when budded in a greenhouse.

In California and other pistachio-growing areas in the United States, the commercial nut-producing species *Pistacia vera* is budded into seedling and clonally propagated rootstocks of other inedible (due to small kernel size) *Pistacia* species and their interspecific hybrids. Five commercial rootstocks have been used in California: three *Pistacia* species and two interspecific hybrids. The species are *P. terebinthus, P. atlantica, P. integerrima*, and the hybrids are *P. atlantica* pollinated by *P. integerrima* and the reciprocal cross. The first two, no longer commercially available, are referred to by their

Figure 8.1. Leaves of California's currently most popular rootstocks. (A) The hybrid of *Pistacia atlantica x Pistacia integerrima,* sold commercially as UCBI, as both a seedling and cloned rootstock. (B) *Pistacia atlantica,* Atlantica, an early rootstock no longer sold commercially. It was the rootstock of choice in the first California orchards but lost popularity when it was demonstrated to be susceptible to *Verticillium dahliae.* (C) *Pistacia terebinthus,* Terebinthus, never a commercial rootstock in California but still existing in some of the earliest orchards. (D) *Pistacia integerrima,* sold commercially as Integerrima, PGI, or Pioneer Gold I, as both a seedling and cloned rootstock. All of California's rootstocks are derived from two species, *P. atlantica* and *P. integerrima. Photo:* L. Ferguson.

species names, Terebinthus and Atlantica, and the latter two are generally known by their trade names or abbreviations. The species *P. integerrima* is represented by the seedling population from a *P. integerrima* × *P. integerrima* cross referred to as Integerrima, Pioneer Gold I, or PGI. The hybrid of a specific *P. atlantica* female pollenized by a specific *P. integerrima* male is known as University of California, Berkeley I, or UCBI. The reciprocal cross, *P. integerrima* pollenized by *P. atlantica*, known as Pioneer Gold II, or PGII, is also no longer commercially available under this name. The same cross is now marketed as Platinum. The parents for this cross are genetically different from those used for UCBI. Since the early 2000s, multiple commercial nurseries have been introducing new seedling and clonal rootstocks. However, all these newer rootstocks are still progeny of the same *P. atlantica* and *P. integerrima* parents. The experimental data presented in this chapter is from trials on Atlantica, PGI, UCBI, and PGII rootstocks completed by 2005. All of these rootstocks have characteristics that make them suitable for specific planting situations. Understanding these differences will allow better selection of a rootstock for specific climates, soils, water quality, and soilborne pests and diseases. Figure 8.1 shows the leaf shape differences in the rootstock leaves of the four *Pistacia* species used for California's commercial rootstocks.

During the early years of the pistachio industry in California, as noted above, the *P. vera* cv. Kerman female cultivar was found to have no resistance to nematodes (Woodruff 1979) or to the five species of the soilborne fungal pathogen *Phytophthora* spp. and *Verticillium dahliae*, also known as Verticillium wilt. Therefore, the other available species were used as rootstocks. Atlantica quickly became the primary rootstock, as Terebinthus was more difficult to bud and produced less uniform trees. The first pistachio orchards were planted on these two rootstocks in the southern San Joaquin Valley, where both rootstock species proved susceptible to Verticillium wilt. A small number of Integerrima seedlings selected from an Iranian seedling tree planted at the USDA Plant Introduction Station in Chico were tested in a *Verticillium*-infested soil test plot in the San Joaquin Valley and found to be tolerant of the disease. This Integerrima seedling rootstock, quickly commercialized as Pioneer Gold I (PGI), rapidly became the dominant rootstock in California's rapidly growing industry. As hundreds of thousands of trees on Atlantica rootstock declined or died of Verticillium wilt, they were replaced with trees on Integerrima rootstocks. New plantings were exclusively on Integerrima, until the first hybrids of Atlantica and Integerrima were introduced in the late 1980s. Lee J. Ashworth, of the University of California, Berkeley, pollenized a female Atlantica with pollen from a male Integerrima to produce the UCBI (University of California Berkeley I). Shortly thereafter, the reciprocal cross, a female Integerrima pollenized by a male Atlantica, was produced by a commercial nursery and named PGII (Pioneer Gold II). These two species and their hybrids were the four rootstocks tested from 1989 through 2005 in multiple California locations. The four rootstocks—Atlantica, Integerrima, UCBI, and PGII—were evaluated for effects on scion vigor, early production, yield, and alternate bearing, as well as their tolerance of Verticillium wilt, cold, and salinity. The results of these trials led to Integerrima and both seedling and cloned UCBIs becoming the two major rootstocks in California. By 2000, Atlantica and PGII were no longer commercially available. The following sections highlight these trial results in discussions of the individual rootstocks (Ferguson, Beede, et al. 2002; Ferguson, Poss et al. 2002).

Pistacia terebinthus

While *P. terebinthus* was commonly planted in the early days of the California pistachio industry, it has fallen out of favor for the following reasons: it is susceptible to Verticillium wilt; it was generally perceived to be more difficult to bud than *P. atlantica* or *P. integerrima*; its use often resulted in less uniformity compared with the other rootstocks; and it is less vigorous than the other commonly used rootstocks. Currently, it is seldom used. However, *P. terebinthus* does have some characteristics that may make it useful in specific situations, and it may be valuable in future breeding programs. Among the five rootstocks, *P. terebinthus* is the most cold hardy. Pistachios grown on this rootstock have been reported to withstand temperatures as low as 15°F with no damage. In nutritional studies where nutrient levels of Kerman cultivar leaves on different rootstocks were compared at seven locations throughout the state, trees on *P. terebinthus* rootstock often had the highest nutrient levels. In these studies, *P. terebinthus* was more efficient than other rootstocks in absorption of zinc and copper, micronutrients that are known to be deficient in soils of some California pistachio orchards. Results of a rootstock trial in soil infested with Armillaria root rot in Yolo County have indicated that *P. terebinthus* is more resistant to this disease than the other rootstocks.

Pistacia atlantica

Until *P. integerrima* was discovered to be tolerant of Verticillium wilt, *P. atlantica* was the most common rootstock. *Pistacia atlantica* is more cold hardy than *P. integerrima* and the two interspecific hybrids but less cold hardy than *P. terebinthus*. It is susceptible to Verticillium wilt. It is less vigorous than *P. integerrima* and the two interspecific hybrids, and it is slightly slower to come into production. *Pistacia atlantica* is more efficient at uptake of zinc than *P. integerrima* and UCBI, but it absorbs copper less efficiently than *P. terebinthus* or PGII.

Because of greater cold tolerance than *P. integerrima* and the low incidence of Verticillium wilt in the Sacramento Valley, *P. atlantica* was initially the most common rootstock in this region. Since 1981, however, the desire for greater vigor has led to increased plantings of *P. integerrima* and, in the 1990s, the interspecific hybrid UCBI (*P. atlantica* × *P. integerrima*).

Pistacia integerrima

After *P. integerrima* was discovered to be tolerant of Verticillium wilt, it became the most widely planted rootstock in California. It is vigorous, buds easily, and results in a uniform orchard. However, it is the least cold tolerant of the commonly used rootstocks and can be damaged by winter freezes and early frosts, particularly if they occur when the tree is young and not dormant. Results of the statewide rootstock trials, which were terminated in 2002, indicated that it has growth rates and early production superior to *P. atlantica* but inferior to the two interspecific hybrids. Nutrient studies have shown *P. integerrima* to be less efficient at boron, zinc, and copper uptake than *P. atlantica* and *P. terebinthus*.

The rootstocks with *P. integerrima* parentage, both the single species rootstock and its hybrids discussed below, produce Kerman scion growth not exhibited by scions growing on *P. atlantica* rootstocks. A Kerman scion on a *P. atlantica* rootstock produces only preformed axial fruit-bearing inflorescence buds (i.e., buds preformed within the axial buds the preceding year that bear fruit the following year). A study of the dormant terminal and lateral buds of immature and mature Kerman female trees demonstrated that trees on rootstocks with *P. integerrima* parentage, particularly when young, produce neoformed buds, buds that are both formed and grown within the same season. The result is shoots with multiple sequential flushes of growth within a single season. The potential advantage of neoformed growth is that it may allow young trees to reach bearing sooner, as once the leaves mature and become net exporters of carbohydrates, they support tree growth. Neoformed growth has little advantage in mature producing trees, as there are generally less than 60 neoformed shoots per tree, primarily in the tree crown, where these longer shoots are more likely to be removed through annual dormant pruning. Thus, neoformed growth is not the factor that generates the higher yields of trees on *P. integerrima* parentage rootstocks, other than possibly producing a larger tree. It has been hypothesized that the neoformed growth produced by *P. integerrima* parentage rootstocks is associated with greater resource uptake or transport by these rootstocks (Spann et al. 2007).

Interspecific Hybrids

Of the two original interspecific hybrids, PGII and UCBI, only the latter is currently commercially available. UCBI is the result of the reciprocal cross, *P. atlantica* pollenized by *P. integerrima*, on single trees and using controlled pollination in which all foreign pollen is excluded. All current commercial UCBI seedling parent trees were initially generated from a single male and female tree. The pollination is controlled, as *P. integerrima* pollen is harvested 2 to 4 weeks before the *P. atlantica* female tree flowers are receptive. The pollen is introduced into a closed greenhouse in which the *P. atlantica* tree is grown. Thus, both hybrid rootstocks are interspecific hybrids and both exhibit seedling variability. Both of the hybrid rootstocks have increased vigor compared with *P. atlantica* and generally equal or greater vigor compared with *P. integerrima*. Statewide rootstock trials have demonstrated that trees on UCBI rootstocks produce the best yields through year 13, followed by trees grown on PGII, PGI, and *P. atlantica* (Ferguson, Beede et al. 2002; Ferguson, Poss et al. 2002).

Verticillium wilt is a fungal pathogen that persists in the soil for years as microsclerotia and has multiple alternate cultivated (tomatoes and cotton) and wild (the Solanaceae, or nightshades) hosts. It could easily become a major problem again, as there is no control beyond rootstock resistance and tolerance. Chapter 25, "Soilborne Diseases," discusses this pathogen in detail. Laboratory screening tests and early field studies indicate that the two hybrids differ in their mechanisms for surviving Verticillium wilt (Epstein et al. 2004). PGI is tolerant; it can be heavily infested but not manifest symptoms. UCBI is moderately resistant; it does not easily become infected but is mildly symptomatic when infected. Neither rootstock succumbs to Verticillium wilt. Both *P. atlantica* and PGII become heavily infected (though the latter is more susceptible), and they exhibit symptoms and suffer high mortality rates (Epstein et al. 2004). The introduction of *P. integerrima* parentage rootstocks has rendered this once devastating pathogen a minor concern within the industry. Cotton is a known host of Verticillium wilt. When the cotton industry declined in the 1990s due to international competition, pistachio orchards were planted on *P. integerrima* rootstocks in former cotton fields.

Nutritional studies indicate that PGII is relatively efficient in zinc uptake compared with *P. terebinthus*; *P. atlantica* is intermediate; and UCBI and *P. integerrima* are the least efficient. For uptake of boron, PGII was found to be slightly less efficient than *P. integerrima* and slightly more efficient than UCBI. UCBI is similar to *P. atlantica* and *P. terebinthus* in terms of boron uptake. Preliminary results indicate that PGII and *P. atlantica* are superior to *P. integerrima* and UCBI for absorption of copper (Brown et al. 1991).

Some evidence of incompatibility between Kerman and UCBI was observed in the late 1980s in some locations in California and Arizona. Approximately 25 to 35 percent of the trees demonstrated rootstock scion incompatibility. It is believed that this incompatibility was related to a single maternal Atlantica tree used in Lee J. Ashworth's breeding program. After elimination of this tree as a parent for UCBI, there have been few reports of true rootstock scion incompatibility. Variability in vigor within seedling UCBI rootstock populations has also been observed. However, as nurseries continue to improve their seedbed rogueing, this problem has decreased. Thus far, there appear to be no rootstock scion incompatibilities between the currently commercial rootstocks, the new female cultivars Golden Hills and Lost Hills, and their pollenizing male, Randy. Golden Hills does display some overgrowth at 6 to 10 years, but the graft unions appear strong.

California Pistachio Rootstock Trials

Rootstock selection for most tree crops is generally a history of the most limiting factors in given climates, with different qualities and quantities of soil, water, and disease pressures. Pistachios are no exception. The *Pistacia vera* cv. Kerman female cultivar and the Peters male, which until recently were the sole nut-producing and pollenizing cultivars in California, are not germinated or cloned and grown on their own roots because they are susceptible to nematodes and the soilborne fungi *Verticillium* and *Phytophthora*. The new female cultivars, Golden Hills, Lost Hills, and Gumdrop have always been grown on rootstocks, so their sensitivities to these pathogens are unknown. However, as *P. vera* cultivars, there is a good possibility that they are also susceptible to these pathogens. The *P. atlantica* and *P. integerrima* hybrids were developed for greater yield and vigor as well as *Verticillium* tolerance, but they are now also demonstrating a salinity tolerance that would allow orchard establishment in marginal ground and the use of marginal-quality water.

In four long-term field trials from 1989 through 2015 and in a single greenhouse trial in 1999 we investigated the cold and salinity tolerance, *Verticillium* resistance, and yield and alternate-bearing behavior of trees grown on *P. atlantica, P. integerrima*, and the two interspecific hybrids UCBI and PGII seedling rootstocks (Ferguson et al. 2004; Sanden et al. 2008, 2010). These 24 years of data collection and unexpected freezes have delineated the relative merits of the four seedling rootstocks now used in California. The following tables detail these results and incorporate results from earlier trials and other researchers. The information is presented in the order of most limiting factors: first, the site potential for cold damage, soilborne diseases, and soil salinity, followed by the effects of micronutrient uptake and the horticultural effects on scion yield, nut quality, and alternate bearing.

Cold Tolerance

Table 8.1 gives the relative cold tolerance of budded rootstocks less than 2 years of age. This data was obtained in the winter of 1989 to 1990, when the newly planted rootstock trials experienced freezing temperatures ranging from 4° to 12°F over 11 nights. In our trials, 41 percent of the trees on *P. integerrima* rootstock died. Trees on the other four rootstocks demonstrated the relative cold tolerances given in table 8.1. Trees on *P. terebinthus* are more cold tolerant than trees grown on *P. atlantica*, followed by trees on UCBI and PGII. Scions on all rootstocks suffered freeze damage but recovered readily. Similar temperatures are not as damaging to a mature orchard, as the rootstock is less exposed and the scion cultivar Kerman is much more cold tolerant than trees on young rootstocks, particularly if they are not dormant.

Soilborne Fungal Disease Tolerance

Table 8.2 contains the relative soilborne fungal disease tolerance rankings of the current pistachio rootstocks. A recent, long-term Verticillium wilt research trial demonstrated that while trees on both *P. integerrima* and UCBI survived well in a heavily infested plot, the two rootstocks had different defense mechanisms (Epstein et al. 2004). As discussed above, trees on *P. integerrima* rootstock, while heavily infected with the pathogen, tolerated it well. Trees on UCBI rootstock were much less likely to become infected but, when infected, displayed slightly decreased vigor. The fact that trees on *P. integerrima* rootstock are easily infected with *Verticillium* suggests that, as they age, they might display decreased vigor. Trees on both *P. atlantica* and PGII rootstocks were easily infected by *Verticillium* and suffered marked declines in vigor and yield as well as high mortality. Surprisingly, trees on PGII rootstock were more susceptible than trees on *P. atlantica* rootstock. *Verticillium* infection was probably the reason for the poor performance of trees on PGII rootstock when they were commercially released in the early 1990s.

Phytophthora root and crown rot is currently not a problem except in saturated situations. However, as more rootstocks are planted in marginal ground and irrigated with saline water, or as nematode populations develop, Phytophthora could become a problem. The relative Phytophthora-susceptibility ranking given in table 8.2 is from Iranian data.

Armillaria root rot has not proven to be a problem in pistachio production. However, if declining almond orchards are replaced with pistachios, Armillaria could possibly become a problem.

Salinity Tolerance

Two trials, an 8-year field trial started in 1994 in an established orchard (Sanden et al. 2004) and a 1-year greenhouse trial with budded seedling rootstocks in 1999 (Ferguson, Beede et al. 2002; Ferguson, Poss et al. 2002), were done to determine the relative and absolute salinity tolerance of the three commercial rootstocks. PGII was eliminated from the greenhouse trial as by 1999 it was no longer commercially available. A third long-term trial, from 2005 to 2014, was done to demonstrate what the previous two trials had not addressed: the ability to establish an orchard under saline conditions. At the time this trial was established, the seedling PGI and UCBI rootstocks were the major commercial rootstocks.

Table 8.3 integrates the results of these three trials, showing relative salinity tolerance of the five pistachio rootstocks. The greenhouse trial used 2-year-old budded trees, and the percentage decrease in scion growth relative to the control treatment was the measurement of salinity tolerance. In the 8-year field trial, yield decrease (as a function of the salinity of irrigation water and root zone soil water) was the indicator of salinity tolerance. In the field trial, trees on all three rootstocks tolerated eight sequential seasons of irrigation with water adjusted to 8 decisiemens per meter (dS/m). Above this, at 12 dS/m, all four rootstocks had yield decreases, but UCBI appeared the most sensitive. However, in the companion greenhouse trial, trees on *P. integerrima* rootstock demonstrated the most sensitivity to salinity, losing a full 50 percent of their growth increase when root zone salinities reached 8 dS/m.

Salinity tolerance trials done in a greenhouse sand tank have three advantages over those done in an orchard. The well-drained sand tanks eliminate the soil saturation that results as salinity increases; the salinity levels are constant; and roots in the sand tank cannot avoid the salinity by proliferating outside the saline treatment zone. For these reasons, the results from the greenhouse trial are more reliable than the data from the field trial. Nevertheless, results from both trials supported the conclusion that all four pistachio rootstocks are tolerant of soil root zone salinity up to 8 dS/m.

However, neither of these first two salinity trials investigated the primary production question: what are the salinity limits for establishing an economically productive orchard? The second (8-year) field trial, using the then four commercial rootstocks discussed above, suggested that pistachios can tolerate

Table 8.1. Relative cold tolerance of budded pistachio rootstocks less than 2 years old

Rootstock	Cold tolerance
P. atlantica	2
P. integerrima	4
PGII	3
P. terebinthus	1
UCBI	3

Note: Ranked from 1 to 4 with 1 indicating the most cold tolerance and 4 the least.

Table 8.2. Relative disease tolerance of budded pistachio rootstocks

Rootstock	Verticillium	Armillaria	Phytophthora
P. atlantica	3	3	1
P. integerrima	1	3	3
PGII	4	3	—
P. terebinthus	—	1	—
UCBI	1	1	1

Note: Ranked from 1 to 4 with 1 indicating the most disease tolerance and 4 the least.

Table 8.3. Relative salinity tolerance of pistachio rootstocks

Rootstock	Salinity tolerance
P. atlantica	1
P. integerrima	4
PGII	—
P. terebinthus	—
UCBI	2

Note: Ranked from 1 to 4 with 1 indicating the most salinity tolerance and 4 the least.

Table 8.4. Relative micronutrient uptake efficiency of pistachio rootstocks

Rootstock	Cl	Na	Zn	B	Cu
P. atlantica	2	2	3	3	2
P. integerrima	1	1	4	1	3
PGII	—	—	1	2	1
P. terebinthus	—	—	2	2	1
UCBI	3	4	4	4	3

Note: Ranked from 1 to 4 with 1 indicating the most nutrient uptake efficiency and 4 the least.

a soil water extract up to 8 dS/m without significant yield reductions (Sanden et al. 2004, 2008). However, these trees were 5 years old when the trial was established—under good conditions and with good-quality irrigation water. Though attempts were made to prevent the possibility, trees in the highly saline plots may have had roots extending into freshwater supplies in adjacent nontreatment plots.

The third and largest trial, done from 2005 to 2014, was designed to evaluate establishing a pistachio orchard under moderately saline conditions, 3.2 and 5.2 dS/m. The cumulative in-shell yield over those 4 years of bearing was the final integrator of salinity tolerance. In this trial, Kerman trees on both PGI and UCBI rootstock demonstrated significant yield declines above 5 to 6 dS/m. However, the yield declines were only 3.1 percent for Kerman on PGI and 1.4 percent on UCBI, respectively. Most importantly, the results of this trial supported the results of the 1-year greenhouse trial that demonstrated that trees on UCBI rootstocks were significantly more tolerant of salinity than trees on PGI rootstocks. This third trial is discussed in detail in chapter 15, "Salinity Management in Soil and Water."

Micronutrient Uptake Efficiency

Table 8.4 gives the relative micronutrient uptake efficiency of the five rootstocks. This table summarizes results from several trials and gives only relative guidelines. Generally, it demonstrates that trees on the two most commercially important rootstocks, *P. integerrima* and UCBI, are prone to zinc and copper deficiencies and that the latter rootstock is also prone to boron deficiencies (Brown et al. 1991). In older orchards established with trees on *P. atlantica* rootstock, replants with *P. integerrima* or UCBI rootstocks can often be distinguished in the early spring by the delayed and uneven leafing out of foliage that characterizes a zinc deficiency.

Nutrient uptake efficiencies of sodium and chloride given in table 8.4 were generated in a sand tank trial. They demonstrated that trees on *P. integerrima* rootstock had much higher uptake (and translocation to the scion) of sodium and chloride. Interestingly, in this trial only boron produced specific ion toxicity in the form of an even marginal leaf burn. Both sodium and chloride partitioned equally into green and necrotic scion leaf tissue. This tendency for trees on *P. integerrima* rootstock to take up and translocate sodium and chloride to the scion could be deleterious in saline conditions. In the trial that produced the sodium and chloride data shown in table 8.4, trees on *P. integerrima* rootstock had a significantly greater decrease in yield compared with trees on *P. atlantica* and UCBI rootstocks at the same salinities. At 8 dS/m, trees on *P. integerrima* rootstock had a 50 percent decrease in growth, whereas trees on *P. atlantica* and UCBI rootstocks had no decrease in growth.

Effect of Rootstock on Marketable Yield

Table 8.5 demonstrates the effect of rootstock on final marketable yield of Kerman grown on the four different rootstocks. The data in this table was produced in three identical rootstock trials grown in three different microclimates from 1989 through 2001 (Ferguson, Beede et al. 2002; Ferguson, Poss et al. 2002). To reduce a potential cultivar effect, all the trees in these three trials were budded with buds from the same male and female tree. The data in this table is based on the first 5 producing years. The trees did not begin producing marketable crops until 1997. Alternate bearing commenced with a very low crop in 1999, the third bearing year.

Trees on UCBI rootstocks produced significantly better yields than trees grown on the other three rootstocks in all locations and in all years (table 8.6). Averaged over all 5 years of crop production and all three trial locations, trees on UCBI rootstock produced an average of 45.3 percent more marketable crop than trees on *P. atlantica* rootstock, 19.1 percent more than trees on *P. integerrima* rootstock,

Table 8.5. Effect of pistachio rootstock on first 5 bearing years of cumulative yield of marketable dry, in-shell, split nuts

	California county plot location	*P. atlantica*	*P. integerrima*	PGII	UCBI
average % decrease in yield produced by trees on other rootstocks using UCBI rootstock as the 100% control	—	45.3	19.1	15.1	—
alternate-bearing index	—	0.67	0.68	0.66	0.67
cumulative yield, 1989 through 2001*	Fresno	7,394 c	9,442 b	9,539 b	11,095 a
cumulative lb/ac at 112 ♀ trees/ac	Kern	11,418 c	13,052 b	13,301 b	15,258 a
	Madera	4,258 d	5,643 c	6,285 b	7,170 a

Note: *Values within a row followed by different letters are significantly different. Absence of letters indicates no significant differences within a row.

and 15.1 percent more than trees grown on PGII rootstock. It must be remembered that this was only the first 5 years of economic production. Thus, these results may only reflect the early-bearing years. The only definitive way to demonstrate if this difference in effect of rootstock on Kerman scion productivity persists through the trees' bearing life is to again measure yield and quality in these ongoing orchard trials, in which all trees are pruned to an equivalent canopy size. There was no significant difference in the alternate-bearing indexes among trees on the different rootstocks. On a scale of 0 to 1 with total alternate bearing being 1 and no alternate bearing 0, the alternate-bearing indexes on the four rootstocks ranged from 0.66 to 0.68.

Within single crop years there were no consistent significant differences in nut size or weight among nuts produced from trees on the four different rootstocks. Similarly, within single crop years there were few consistent significant differences in the percentage of blank and nonsplit nuts produced by trees on the four different rootstocks. Differences in nut weights and the percentage of blanks and splits did differ among years, but they did so in a similar fashion for Kerman on all the rootstocks. For example, in heavier crop years the nonsplit percentages of nuts rose in trees on all four rootstocks.

An analysis of the components of yield, clusters per tree, nuts per cluster, and nut size demonstrates that trees on UCBI rootstock yield better because this rootstock produces a larger tree. This, in turn, results in more clusters per tree, not a higher density of clusters or more or bigger nuts per cluster. This is reflected in the average relative trunk circumference in trees grown on the different rootstocks. The more productive trees are the larger ones by both trunk girth and canopy size. This also suggests that trees on different rootstocks, when pruned to the same canopy size, may be equally productive.

The data demonstrates that rootstock affects the vigor of the Kerman scion, producing a larger tree with more clusters. Rootstock obviously does not affect alternate bearing, as the calculated alternate-bearing indexes were approximately equal and

Table 8.6. Relative yield ranking of Kerman female scion on pistachio rootstocks

Rootstock	Yield
P. atlantica	4
P. integerrima	3
PGII	2
P. terebinthus	—
UCBI	1

Note: Ranked from 1 to 4 with 1 indicating the most yield and 4 the least.

Table 8.7. Relative yield and tolerance ranking of pistachio rootstocks in San Joaquin Valley, 1989–2002

Rootstock	Yield	Limiting factor		
		Frost	Salinity	Verticillium wilt
P. atlantica	4	1	1	3
P. integerrima	3	4	3	1
PGII	2	3	—	4
UCBI	1	2	2	2

Note: Ranked from best (1) to worst (4).

independent of absolute crop load among trees on the different rootstocks. This supports the view that alternate bearing is a function of the scion cultivar alone, and that mitigation or elimination of this characteristic will not come from rootstock breeding.

In summary, the results of these rootstock trials indicate the following major conclusions. *Pistacia atlantica* is the most cold tolerant of the historically commercial rootstocks, followed by the currently commercial UCBI and *P. integerrima* rootstocks. Trees on *P. integerrima* and UCBI rootstocks are equivalent in their performance under Verticillium wilt pressure but appear to have different mechanisms for coping with the disease. Both are significantly more tolerant of Verticillium wilt than trees on *P. atlantica* or PGII rootstocks, in that order. All four rootstocks can be established with soil water salinity of up to 6 dS/m in the root zone, but they will sustain yield declines if established under higher salinities. Trees on *P. integerrima* rootstock are the most saline sensitive, as they have a tendency to take up and translocate more sodium and chloride into the Kerman scion more readily than trees on UCBI or *P. atlantica* rootstocks, in that order. Trees on UCBI produce the best yields among trees on all the rootstocks when planted at relatively low density, followed by those on PGII and *P. integerrima* rootstocks. All three result in better yields than *P. atlantica* rootstock. However, these differences in yield appear to be a function of tree size. Therefore, trees on less vigorous rootstocks planted at closer densities could potentially be equally productive. Table 8.7 summarizes the relative tolerance ranking of these four pistachio rootstocks.

As the pistachio industry continues to grow, a systematic rootstock and cultivar quality control and tracking program (like the Citrus Clonal Protection Program maintained by the California citrus industry) would be beneficial (Vidalakis 2014). The production, monitoring, and distribution of disease-free, true-to-type budded rootstocks would greatly facilitate investigation of the problem that the California pistachio industry experienced with clonally produced UCBI and the bacterial pathogen *Rhodoccus facians* (Stamler et al. 2015). Also, the information discussed in this chapter was largely developed before 2005. With the proliferation of new rootstocks (developed with different propagation, production, and orchard establishment methods) and the molecular information currently being developed about the California pistachio rootstocks, a new generation of rootstock investigation is needed.

References

Brown, P., Q. Zhang, and L. Ferguson. 1991. Nutrient uptake efficiencies of various pistachio rootstocks. California pistachio industry annual report 1991, 158–162. Fresno: California Pistachio Commission.

Epstein, L., R. H. Beede, S. Kaur, and L. Ferguson. 2004. Rootstock effects on pistachio trees grown in *Verticillium dahliae*-infested soil. Phytopathology 94(4): 388–395.

Ferguson, L., R. H. Beede, H. Reyes, and P. Metheney. 2002. California pistachio rootstock evaluations. International Society of Horticultural Science, 3rd International Symposium on Pistachios and Almonds, 2001. Acta Horticulturae 591:63–66.

Ferguson, L., J. A. Poss, S. R. Grattan, G. M. Grieve, D. Wang, C. Wilson, T. J. Donovan, and C.-T. Chao. 2002. Pistachio rootstocks influence scion growth and ion relations under salinity and boron stress. Journal of the American Society for Horticultural Science 127(2): 194–199.

Sanden, B. L., L. Ferguson, D. Corwin, and C. E. Kallsen. 2010. Using saline groundwater for large-scale development and irrigation of pistachio interplanted with cotton. XIV GREMPA Meeting on Pistachios and Almonds. Options Mediterraneennes Series A: Mediterranean Seminars 201 No. 94. 31–36.

Sanden, B. L., L. Ferguson, C. Kallsen, D. Corwin, B. Marsh, and B. Hutmacher. 2008. Large-scale utilization of saline groundwater for irrigation of pistachios interplanted with cotton. Western Pistachio Association 2007 production research reports. Fresno, CA: Western Pistachio Association.

Sanden, B. L., L. Ferguson, H. C. Reyes, and S. C. Grattan. 2004. Effect of salinity on evapotranspiration and yield of San Joaquin Valley pistachios. Proceedings of the 4th International Symposium on Irrigation of Horticultural Crops. Acta Horticulturae 664:583–589.

Spann, T. M., R. H. Beede, and T. M. DeJong. 2007. Preformation in vegetative buds of pistachio *(Pistacia vera)*: Relationship to shoot morphology, crown structure and rootstock vigor. Tree Physiology 27:1189–1193.

Stamler, R. A., J. Kilcrease, C. Kallsen, E. J. Fichtner, P. Cooke, M. L. Putnam, R. J. Heerema and J. J. Randall. 2015. First report of *Rhodococcus* isolates causing pistachio bushy top syndrome on 'UCB-1' rootstock in California and Arizona. Plant Disease (in press).

Vidalakis, G., D. J. Gumpf, M. Polek, and J. A. Bash. 2014. The California Citrus Clonal Protection Program. In L. Ferguson and B. Grafton-Cardwell, eds., Citrus production manual. Oakland: University of California Division of Agriculture and Natural Resources Publication 3539. 117–130.

Woodruff, J. G. 1979. Tree nuts: Production, processing, products. Westport, CT: AVI Publishing.

9 Rootstock Production and Propagation

Brent A. Holtz, Dan E. Parfitt, and Louise Ferguson

California pistachios are grown primarily on three rootstocks (two species and one interspecific hybrid), all members of the genus *Pistacia*. They are named Atlantica *(P. atlantica)*; Pioneer Gold I, or PGI *(P. integerrima)*; and University of California, Berkeley I, or UCBI, which is a hybrid between the *P. atlantica* female and the *P. integerrima* male. Two other rootstocks have occasionally been grown in California, namely *P. terebinthus* and Pioneer Gold II, or PGII (the *P. atlantica* female crossed with the *P. integerrima* male).

Pistachio rootstocks are produced from seed. Originally, the seed was produced from open-pollinated trees, with control of the genetic composition of the seed being limited to the source tree. More recently, closed crosses (control of both male pollen sources and female source trees) have become standard. UCBI is produced from closed crosses. PGI was originally produced from open-pollinated trees but more recently from selected male and female trees. (Note that PGI is a trademark and does not denote the specific genetic composition of the progeny, unlike UCBI, which is produced from an individual female clone and male clone. Thus, there is more potential for variability among the non-UCBI hybrids.) In the San Joaquin Valley, PGI and UCBI are the most commonly used rootstocks. In colder areas outside the San Joaquin Valley, *P. atlantica* and UCBI are the most commonly planted rootstocks. Only two cultivars (both of them *P. vera*) are widely used as scions in California. They consist of the female cultivar Kerman and the male pollenizer Peters. Two additional female cultivars, Golden Hills and Lost Hills, and a male cultivar, Randy, have been recently released by the University of California. Small acreages of Kalehghouchi and Red Aleppo also exist.

Rootstock Seed Germination

The requirements for seed germination vary little among the rootstock species. Rootstock seed are relatively small, with a hard, unsplit endocarp, or shell. Seed coats can be removed by rubbing the seed across screens. Seed are usually stratified in a refrigerator (36° to 40°F) for 1 to 3 months before they are sown, usually in February.

After stratification, the seed usually are soaked in tap water for 2 to 48 hours at room temperature. Some nurseries dip their rootstock seed in 1 to 5 percent muriatic or hydrochloric acid (HCl) combined with sodium hypochlorite (NaOCl), which is available as household bleach (at a rate of 3 to 5% bleach in a water solution), to wear away the seed coat (scarification). This is followed by a thorough wash. Germination rates of *P. terebinthus* have been doubled and the incubation time reduced by treating seed with concentrated sulfuric acid (H_2SO_4) to scarify the seed. Thorough rinsing after the acid treatment is essential to prevent complete digestion of the seed coat and damage to the embryo.

After the acid treatment and wash, seed are usually soaked for another 1 to 2 hours before sowing. After soaking, water is drained away and the seed are placed in a single layer between moistened, clean (sometimes bleached for disease control) layers of white cotton Turkish towels. The excess towel is then folded over the seed and rolled up. Another moistened towel is often placed over the rolled-up towel containing the seed. The seed are kept moist (sprayed with distilled water) and incubated at 70° to 80°F for 3 to 6 days. Seed are checked daily (sometimes twice) and removed with sterile tweezers when they have germinated and their hypocotyls have emerged.

Planting

After seed germinate, they are planted, usually in Jiffy liner pots (Jiffy 7 or 757), flats, fumigated nursery beds, or directly into plastic planting cones. Some nurseries lightly sterilize their germinated seed with low concentrations of bleach (3%) in water. (Note that high levels of bleach used for an extended period of time can kill the seed.) Most commonly, germinated seed are planted into Jiffy liner pots and placed in greenhouses kept at 70° to 90°F. The germinated seed should be watered regularly but not overwatered, because overwatering can result in damping-off diseases. The benches on which these liner pots or flats are placed should be clean and sterile, because several plant pathogens can survive on dirty benches and infect young pistachio rootstock seedlings.

Transplanting

The seedlings should be 3 to 4 inches tall by mid-April. At this point, they can be transplanted into pots and placed outside where they will receive full sun. Seedlings transplanted from flats may initially suffer transplant shock, but they will eventually develop a more fibrous root system if broken and damaged roots are pruned and remaining roots uniformly spread out. Seedlings transplanted from the Jiffy peat pots will not suffer transplant shock, but care should be taken not to break down the edge of the peat pot when transplanting. Irregular water distribution may delay seedling root development in the surrounding soil.

Seedlings are usually transplanted into various potting mixtures depending on the nursery. Standard seedling mixes usually contain soil, humus, perlite, and peat moss in various concentrations, which are usually amended with sand to improve drainage and texture. Various commercially available mycorrhizae mixes are often added to the potting mixes to improve the uptake of nutrients (especially nitrogen and phosphorus) through the symbiotic relationship between the rootstock and fungi. After transplanting, the seedlings are watered and fertilized regularly and are suckered every 10 days to keep their terminal buds growing. The rootstocks are tied to 3-foot bamboo stakes when they are 10 to 16 inches tall.

Seedlings may be transplanted into various pots, typically 6 inches in diameter and 14 to 16 inches deep. Some pots, such as those made of tar paper, felt, or paper fiber, can be used and planted directly into the orchard, avoiding transplant shock due to root disruption. Tall plastic pots of similar size also are frequently used, especially by nurseries better suited for the extended maintenance of rootstocks in lath houses and outdoor nurseries. The plastic pots are more durable and reduce evaporation through the pot itself. Plastic pots also provide more protection for the seedlings during shipping and handling, but they must be removed at planting. Pistachio rootstocks are highly susceptible to injury and care should be taken when handling them.

Transplanting into the Orchard

Pistachio rootstocks, whether budded or not, are usually planted from potting containers directly into the orchard. Potted plants are delivered in plastic or fiber pots 14 to 16 inches tall. It is possible to transplant bare-root nursery stock, but because bare-rooted trees are very sensitive to drying, this is not a common practice. Proper tree care after planting is more crucial to survival than the type of container in which the trees were grown. Trees should be well watered immediately after planting, and soil moisture should extend well below the 3-foot root zone. A second irrigation should follow within 2 weeks, followed by a regular fertilization and irrigation program. Rootstocks are staked to ensure straight tree growth for budding and to protect them from excessive wind.

Transplanting 3–4-Inch Seedlings Directly into the Orchard

Some growers have had experience with direct planting of very small seedlings into the field and have found that, with appropriate care after planting, high levels of survival and excellent growth can be obtained. In an effort to reduce the amount of time that seedlings spend in nurseries (and thus reduce the expense of the rootstock), 4-month-old UCBI seedlings are sometimes transplanted directly into the orchard. These seedlings are typically germinated in January and planted directly into 2-by-10-inch plastic planting cones. The germinated seed are planted at approximately ¼ inch deep and hand-watered daily with a fine mist. The cones are packed into plastic fresh-fruit boxes (15 by 16 by 5 inches) and placed in greenhouses, where they are given bottom heat at 70° to 80°F. The seedlings are between 6 to 10 inches high when they are planted in orchards in April and May. At planting time, the tip of the planting cone is cut off, and the rootstock is easily pushed out of the tube into a small hole created with a 2-inch-diameter bulb planter. The planting cones appear to enhance early root growth by providing a wider and deeper space for root growth. Planting germinated seed directly into cones, as opposed to using Jiffy pots, also provides an earlier opportunity to inoculate the planting mix with mycorrhizae. Direct planting of *P. vera* seedlings into the field using Jiffy 7s has been done with a high level of success by one of the authors. This is an approach that may be useful as transportation and planting costs increase.

Since the seedlings are very small and vulnerable, they can be protected in the field with plastic grow tubes that are open at each end and can be attached to planting stakes (fig. 9.1). The ventilated plastic grow tubes protect young rootstock seedlings from high winds, rodents, sunburn, and herbicides while providing an ideal microclimate for growth. The ventilated plastic tubes (Treessentials/Snap N' Grower) can actually snap apart and back together, which facilitates removal and replacement when budding. These tubes have also been used successfully to start grape vineyards. The bottom end of the tube should be buried at least ½ inch under the soil surface to avoid creating a chimney effect, which could cause the seedlings to rapidly dry out. Various fertilizers are often added to enhance rootstock growth.

Budding should be done in July as described below. By July the seedlings should be between 20 to 26 inches tall. It may be necessary to snap or pinch off some of the growth above the bud to decrease competition with the bud. After budding, the grow tubes can again be placed around the seedlings to push the Kerman or Peters buds.

UCBI appears to be better suited for early orchard planting, compared with *P. integerrima*, because of strong early growth and improved cold tolerance. With appropriate post-transplant care, successful planting of any rootstock should be possible with these techniques. If the seedlings are not sold the first year, the nursery can transplant them into typical potting containers and sell them as 14-month-old rootstocks the following year.

Figure 9.1. Grow tube. *Photo:* B. A. Holtz.

Budding

Scion cultivars Kerman and Peters are propagated via standard budding and grafting techniques. Budding and grafting are similar in that, with both techniques, a portion of one plant is joined to another to form a compound genetic system of one scion and one rootstock cultivar. With budding, a single bud is placed on a stock plant; with grafting, a larger portion of the stem is used as the scion. When planting a new pistachio orchard, rootstocks are planted in February or March and budded in late June, July, or early August. Budding is usually done in early summer, when the scion buds are mature and dormant and the bark on the rootstock is slipping and active. Red leaves on the new rootstock growth are a good indicator that the bark is slipping. If bud take is unsuccessful, rootstocks can be budded again until September. Unlike many tree crops, best budding success is observed when the weather is hot. Some rootstocks are budded in the nursery and sold as replants 20 to 26 months later, either in October/November or the following February/March (fig. 9.2). When *P. atlantica* was the rootstock of choice, budding was usually performed in the second season or very late in the first season, since its growth is slower compared with UCBI or *P. integerrima*.

T-budding is the most efficient method in commercial pistachio production, and it is used 99 percent of the time. T-budding is performed when the rootstock bark is slipping during the early summer and the scion buds have matured and hardened (fig. 9.3). Chip buds and patch buds have also been used successfully to bud pistachio scion onto rootstock (fig. 9.4). Patch budding, however, requires a larger-caliper rootstock and is therefore more likely to be done in the field in the spring or fall. Chip and patch budding are not common procedures in California. Rootstocks sometimes produce excess sap,

Figure 9.2. Nursery-budded tree. *Photo:* B. A. Holtz.

called gumming, which can be a problem when budding in early summer.

Although rarely done, two types of grafts have been used with pistachios. The standard cleft graph (fig. 9.5) is the more common of the two and may be done either in a greenhouse or field. The saddle, or inverted cleft graft, has also been used to graft pistachios. Both are usually more successful if performed in late summer. Saddle grafts may be superior for grafting difficult materials. Cambium matching is critical for success with either type of graft.

Healthy rootstocks are essential for successful grafting or budding of pistachios. Trees should be the correct size for the procedure to be used; scion and rootstock diameters should be similar in size. Generally, trees with a diameter of ¼ to ½ inch are useful for T-budding, chip budding, and saddle grafts. Larger trees, between ½ and ¾ inch in diameter, are needed for patch buds. Trees from ¼ inch to more than ¾ inch in diameter can be cleft grafted as long as a good cambial match between scion and rootstock is maintained. Accurate matching of cambia is also required for chip buds.

Careful selection of ideal scion buds is essential for successful budding. Selection of scion wood from semihardened water sprouts or nonfruiting shoots is generally the best. Since most dormant buds on mature pistachio trees are flower buds, propagators select vegetative buds based on their size and location. Vegetative buds are almost always smaller than flower buds and tend to occur at the base of current-year growth on flowering shoots. Vegetative buds tend to be more pointed than flower or fruit buds, which are more rounded. Young, rapidly growing shoots with greater than 20 inches of current growth tend to have more vegetative buds. Generally, trees less than 6 years old are the best source of budwood.

For successful budding and grafting, a razor-sharp budding or grafting knife should be used to make smooth cuts (fig. 9.6). The bud or scion should be dormant, and the rootstock should be growing in a nursery or planted in an orchard (fig. 9.7). The cambial zones of the scion and rootstock should make intimate contact at least at one point (fig. 9.8). With cleft grafting, do not line up the cambia precisely, but slant the scion into the cleft so that the cambia cross at one point. Wrap or secure the union to prevent movement (fig. 9.9). Budding and grafting wraps can be made of various materials, but the budding rubber is the most commonly used. It is secured by looping the end under the last wrap (fig. 9.10). Waxes are available to help prevent drying of the scion and cambial zones, but they are not typically used when T-budding.

Good systems for production of clonal scions on seedling rootstock are currently available and widely used commercially. The exact procedures by which pistachio nurseries produce their products are proprietary.

Micropropagation of Pistachio Rootstocks

Clonal micropropagation of pistachio (*Pistacia vera* L.) using tissue culture techniques has been reported in the scientific literature since the early 1980s (Barazi and Schwabe 1982). Since then, numerous researchers have reported successful propagation of *P. vera* using various medium modifications and culture conditions, including Onay (2000) and Tilkat et al. (2008). Propagation protocols for *P. atlantica*,

P. khinjuk, P. terebinthus, and *P. integerrima* have also been reported (Pontikis 1984; Picchioni and Davies 1990; Parfitt and Almehdi 1994; Tilkat et al. 2005). Propagation of *P. vera* is relatively straightforward (Barghchi and Alderson 1983a, 1983b). At least one nursery has developed an effective large-scale program of clonal micropropagation using combinations of published and customized protocols.

The general approach involves sterilizing actively growing vegetative shoot tips and placing them in an artificial sterile medium combined with a support matrix. Shoot tips are usually prewashed in running water for a number of hours. Sterilization is accomplished with a mild sodium hypochlorite solution (10 to 15% bleach for 10 to 30 minutes) followed by repeated washes with sterile water. A surfactant or wetting agent such as Tween 20 is often added to the sodium hypochlorite solution. The sterilization step is critical, since insufficient sterilization results in contaminated shoot tips, while excessive sterilization results in dead shoot tips. Exact concentrations of the sterilization solution and length of sterilization time vary among investigators, depending partly on where and when the tissue was collected. Generally, field-collected tissue is more contaminated than greenhouse-grown shoot tips. Levels of exogenous and endogenous fungi and bacteria increase later in the season. Internal bacterial contamination is especially problematic (Cassells 1991) and can be partially avoided by collecting shoots from greenhouse-grown plants in the spring. Parfitt and Almehdi (1994) avoided many of these problems by using a carbon-deficient medium supplemented with carbon dioxide (CO_2) to propagate the shoot tips.

After sterilization, shoot tips measuring up to 1 centimeter are placed on a solid medium. This could be Difco agar, Gelrite, or other sterile medium into which are incorporated nitrogen, phosphorus, potassium, minerals, and combinations of auxin (IAA and IBA) and cytokinin (TDZ, BA, and kinetin) growth regulators. The ratio of auxin to cytokinin is critical to successful micropropagation and is usually determined empirically by the investigator. A high auxin to cytokinin ratio will favor rooting while a high cytokinin to auxin ratio will favor shoot formation and multiplication. A carbon source (CO_2 or a sugar) is also included. Vitamins are sometimes added to specific media. Most media have been developed from Murashige and Skoog's medium (1962) or the Driver Kunyuki Walnut Medium (DKW) (McGranahan et al. 1987). Parfitt and Almehdi (1994) developed a modified medium for superior shoot multiplication of both *P. vera* and *P. integerrima*

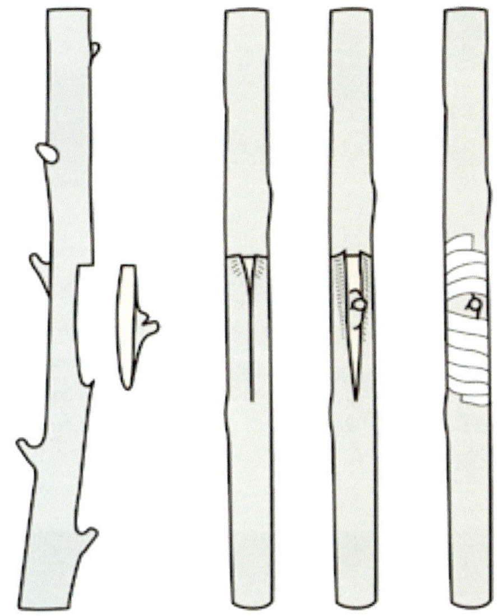

Figure 9.3. T-budding is the most common method used to bud pistachio scion onto rootstock. The shield is cut from the budstick and inserted into a T-cut on the stock.

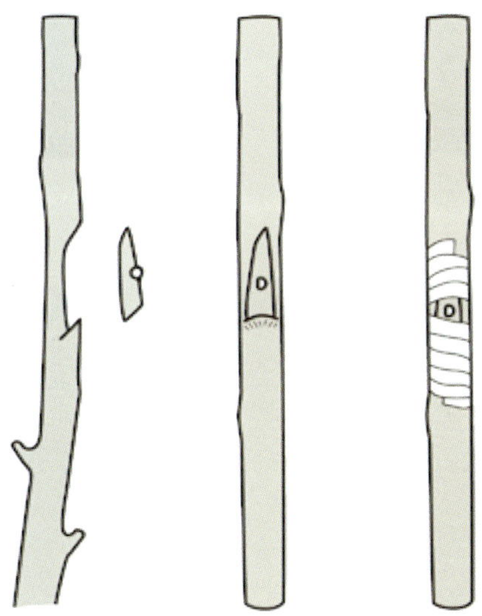

Figure 9.4. Chip and patch buds.

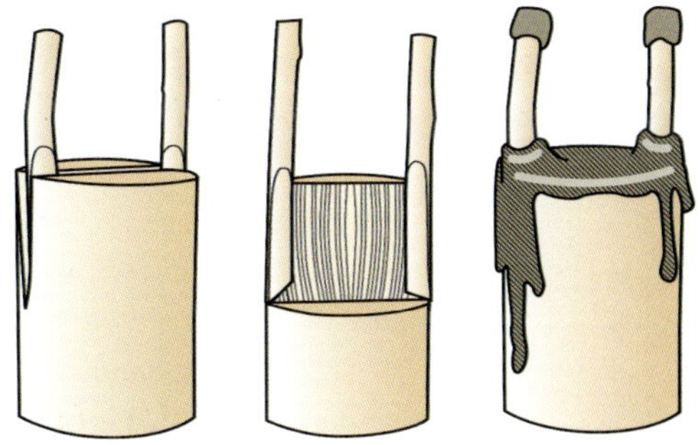

Figure 9.5. Cleft graft.

Figure 9.6. Shield with bud is cut from budstick. *Photo:* B. A. Holtz.

Figure 9.8. Shield is inserted into the T-cut. *Photo:* B. A. Holtz.

Figure 9.7. T-cut is made on the rootstock. *Photo:* B. A. Holtz.

Figure 9.9. Inserted bud is wrapped with rubber band. *Photo:* B. A. Holtz.

Figure 9.10. Finished and wrapped T-bud. *Photo:* B. A. Holtz.

through manipulation of micronutrient levels and replacement of the sucrose carbon source with CO_2 and high light intensity. After a multiplication phase, explants were separated and moved to a Gelrite-based rooting medium without a sugar carbon source, followed by acclimatization in a perlite-based solid substrate (fig. 9.11).

The final and perhaps most critical step in the micropropagation process is acclimatization of the rooted explants. A variety of approaches can be used, few of which are emphasized or described in the literature. The general concept is to gradually expose the rooted plantlets to lower levels of relative humidity (RH). Because the cultured plantlets are developed in an atmosphere that is maintained at 100 percent RH, the leaf surfaces do not develop a wax layer to restrict transpiration. Further, in some of the literature, researchers speculate that the stomata function at a low level in vitro and require a period of acclimatization to begin to function normally. If cultured plants are taken from medium directly to a pot in the greenhouse or to the field, they will die. A frequently used protocol is to move the plantlets to a perlite-solid medium that is kept moist with a balanced N-P-K nutrient solution supplemented with micronutrients, such as Hoagland's solution or a customized solution similar to that used for rooting but without growth regulators. High humidity conditions can be maintained by using clear covers or a fog or mist system in the greenhouse. Light levels should be increased and humidity reduced over a period of 4 or more weeks.

Some general micropropagation references are Bonga and Durzan (1987), Conger (1981), Debergh and Zimmerman (1991), Ahuja (1993), and Hartmann et al. (1997).

Figure 9.11. UCBI explants. *Photo:* D. E. Parfitt.

References

Ahuja, M. R. 1993. Micropropagation of woody plants. Dordrecht; Boston: Kluwer Academic Publishers.

Barazi, Z., and W. W. Schwabe. 1982. Rooting softwood cuttings of adult *Pistacia vera*. Journal of Horticultural Science 57(2): 247–252.

Barghchi, M., and P. G. Alderson. 1983a. In vitro propagation of *Pistacia* species. Acta Horticulturae 131:49–60.

———. 1983b. In vitro propagation of *Pistacia vera* L. from seedling tissues. Journal of Horticultural Science 58(3): 435–445.

Bonga, J. M., and D. J. Durzan, eds. 1987. Cell and tissue culture in forestry. Vol. 1. Dordrecht; Boston: Martinus Nijhoff Publishers.

Cassells, A. C. 1991. Problems in tissue culture: Culture contamination. In P. C. Debergh and R. H. Zimmerman, eds., Micropropagation: Technology and application. Dordrecht; Boston: Kluwer Academic Publishers. 447–469.

Conger, B. V., ed. 1981. Cloning agricultural plants. Boca Raton, FL: CRC Press.

Debergh, P. C., and R. H. Zimmerman, eds. 1991. Micropropagation: Technology and application. Dordrecht; Boston: Kluwer Academic Publishers.

Hartmann, H. T., D. E. Kester, F. T. Davies, Jr., and R. L. Geneve. 1997. Plant propagation: Principles and practices. 6th ed. Upper Saddle River, NJ: Prentice Hall.

McGranahan, G. H., J. A. Driver, and W. Tulecke. 1987. Tissue culture of Juglans. In J. B. Bong and D. J. Durzan, eds., Cell and tissue culture in forestry. Vol. 1. Dordrecht; Boston: Martinus Nijhoff Publishers. 261–271.

Murashige, T., and F. Skoog. 1962. A revised medium for rapid growth and bioassays with tobacco tissue cultures. Physiologia Plantarum 15:473–479.

Onay, A. 2000. Micropropagation of pistachio from mature trees. Plant Cell Tissue and Organ Culture 60(2): 159–162.

Parfitt, D. E., and A. A. Almehdi. 1994. Use of high CO_2 atmosphere and medium modifications for the successful micropropagation of pistachio. Scientia Horticulturae 56(4): 321–329.

Picchioni, G. A., and F. T. Davies, Jr. 1990. Micropropagation of *Pistacia atlantica* shoots from axillary buds. Plant Propagator Newsletter. 14–15.

Pontikis, E. A. 1984. In vitro propagation of *Pistacia terebinthus* L. Plant Propagator 30(3): 14–15.

Tilkat, E., C. Isikalan, and A. Onay. 2005. In vitro propagation of khinjuk pistachio (*Pistacia khinjuk* stocks) through seedling apical shoot tip culture. Propagation of Ornamental Plants 5:124–128.

Tilkat, E., A. Onay, H. Yildirim, and H. C. Ozen. 2008. Micropropagation of mature male pistachio *Pistacia vera* L. Journal of Horticultural Science and Biotechnology 83:328–333.

10 Planting and Training Young Trees

Robert H. Beede, Louise Ferguson, Chris Wylie, and Carl Fanucchi

Skillful training of young pistachios (primarily the Kerman cultivar in California) is essential to produce a canopy that facilitates harvesting and other cultural operations. Unlike other nut crops, pistachios require repeated in-season tipping of the shoots destined to become the major structural limbs, and this is performed primarily during the second and third growing seasons. This labor-intensive practice is necessitated by two pistachio growth characteristics: the vigorously growing young shoots do not develop sufficient girth to remain upright, and their apical dominance suppresses the natural formation of lateral branches. Therefore, timely in-season heading cuts (tipping) prevent the future structural branches from bending beyond 45 degrees. They also force lateral branches that create a compact canopy. As many as six passes are performed during the second growing season to tip small quantities of new growth off the shoots destined to form the main structural framework of the tree. Tipping divides the shoots into lateral branches at a desired location, and it further assists in maintaining upright main branches. Repeated in-season tipping of current growth is a standard practice for the Kerman cultivar, which represents 90 percent of the California acreage. The more rapid girth development in the new cultivars, Golden Hills and Lost Hills, results in a more naturally upright tree. However, these cultivars also express apical dominance similar to Kerman, and thus frequent tipping is required for compact canopy development. Pistachio trees in California are trained to open vases, leaving much of the interior wood for early fruiting. Depending on the rootstock, pistachio trees have low to moderate vigor and require 5 to 7 years of growth before producing a significant crop. Although applicable to other cultivars (also referred to as scions to distinguish them from the rootstock), the training and pruning guidelines suggested in this chapter are based on Kerman, the main female cultivar grown in California at this date.

The Principles of Tree Training

A discussion of tree training can be lengthy. There are four important concepts to remember:

- All pruning, and therefore training, is dwarfing. A pruned tree may have the desired shape, but it will always have less total growth than an unpruned tree.
- Dormant training (winter) is the most invigorating time to prune. Removing portions of a tree at this time results in more resources (carbohydrates stored in the wood) available to the remaining growing points. New shoots therefore grow faster in the spring and ultimately attain greater length than those of an unpruned tree. Pistachio trees grow slowly and thus require dormant pruning to create the growth rate required for structural branch development.
- In-season training (tipping) has the greatest dwarfing effect on trees. Commonly called summer pruning, in-season tipping reduces the amount of foliage and thus the amount of carbohydrates produced by photosynthesis. Growers should not delay in-season tipping when training young trees, because it wastes shoot growth and excessively reduces photosynthetic area. In-season tipping performed for branching should therefore only remove 2 to 3 inches of shoot growth beyond the length desired for the structural branches. Avoid removing more than 6 inches from any single shoot.
- Heading cuts (removing only part of a limb) stimulate more vigor (faster growth) than thinning

cuts (removing an entire limb at its point of origin). Heading cuts are most often used on young trees to force branching and produce long shoots with rapid growth. Thinning cuts are used to control tree shape, remove unwanted limbs, and maintain tree vigor at a desired level.

The First Growing Season

Planting and Budding

Pistachio plants are marketed in three forms: 1-year-old seedling rootstock whose seed is derived from specific, isolated mother trees; 1-year-old seedling rootstock budded to the desired *P. vera* cultivar with growth that will become the trunk; and micropropagated (clonal) rootstocks derived from seedling selections. One-year-old rootstock is sold as a large container plant 6 inches in diameter, rather than the bare-root tree common in other tree crops. Clonal rootstock, often only a few months old upon delivery, is sufficiently small to be nursery grown in pots small enough for delivery as multiple plants within a single container.

Planting of unbudded, 1-year-old rootstock seedlings begins in late January or February, while they are still dormant or in the early leafing stage. Planting dormant trees allows for a much greater margin of error than planting after leafing, but the potential for tree-killing, freezing temperatures exists. Growers must also be careful not to overwater the dormant plants and damage the limited root system. Planting continues into May. Early summer plantings have also been successful but are not desirable due to the potential for excessive tree loss and insufficient growth for field budding in August or September.

Before leaving the nursery, dormant 1-year-old seedlings are typically topped while dormant to approximately 24 inches, irrigated, and placed in standard fruit bins for shipment. Since the rootstocks are usually leafing out, it is essential that they be covered during transport to the orchard site. This is often done by capping the tree-filled bin with an empty one or transporting on flatbed trucks with tarped sides. Shortly after arrival and unloading by forklift, provisions must be made to water the seedlings if necessary due to weather conditions or planting delays. Under no circumstances should nursery trees become dry and stressed. The seedling root balls must be kept moist to remain intact during transfer to the field and planting. Bin trailers are also needed to transport the bins filled with trees to the field.

The rootstocks are grown in a variety of containers, which often become the trademark of an individual nursery. The traditional container is a

Figure 10.1. Removal of plastic liner base prior to planting. *Photo:* R. H. Beede.

5-mil plastic liner that is 6 inches in diameter and 14 to 16 inches deep. Although cumbersome to handle and transport, it provides sufficient rooting volume to minimize root compaction, and it holds sufficient water to reduce plant stress during transport and planting. The liner is removed by cutting off approximately 1 inch of compacted root mass from the bottom of the stock (fig. 10.1). This not only facilitates removal of the liner once the tree is placed in the planting hole, but it also eliminates roots that have grown in a circle at the bottom of the liner and can become pot-bound in the field. Once the tree is placed at the proper depth in the hole, the liner is then pulled up and over the rootstock. A reusable, hard plastic pot, similar in volume to the liner but with open mesh sides, has recently been developed for 1-year-old seedlings. The open mesh appears to greatly enhance root volume and change the root architecture by allowing the roots to grow radially in the pot and then be "air pruned" once they reach the outer pot edge. The hard plastic mesh pot is hinged on one side and pinned on the other. It is removed by pulling the pin and spreading the two sides like a clamshell. With other mesh pots, the tree is removed by pushing up the base of the pot.

Clonally propagated rootstocks are typically much smaller (4 to 12 in tall) than seedlings when planted. They are commonly delivered in hard plastic pots (solid or mesh sided) 3 to 4 inches square and 7 to 9 inches deep inches deep. The smaller trees are less costly to transport and plant but require careful handling and water management due to their limited root volume. The smaller volume also allows plant removal from the solid-side container by gently pulling at the base of the tree stem above the soil line. Clonal and seedling plants are also now produced in biodegradable paper pots, which are not removed when planted. They range from 2 to 4 inches in diameter, depending on desired plant size, and are 5 to 9 inches deep. Eliminating tree removal and pot recycling greatly facilitates planting.

The planting holes are either hand-dug or machine augered, and the stock is inserted with the top of the root ball slightly above the soil level. This compensates for settling after watering and reduces the potential for excessive moisture around the crown (soil-tree interface) of the seedling (fig. 10.2). Tamping the backfilled soil by foot is done sparingly to avoid damage to the small roots extending from the crown. The bamboo stick supporting the 1-year-old plants is also removed at planting, and the resulting hole is covered with soil to prevent drying of the crown roots (fig. 10.3).

Figure 10.2. Proper planting depth of 1-year-old seedling nursery stock. (left) *Photo:* R. H. Beede.

Figure 10.3. Hole left by bamboo support stick must be filled to avoid drying of root crown. *Photo:* R. H. Beede.

A stake is placed at the tree site before or after planting so that a shoot, which will become the tree trunk, can be trained up the stake. It is preferable that tree staking occurs immediately on the large plants so that all the stress factors associated with planting are consolidated prior to the required postplant irrigation. The trees are typically staked with a 6-foot-tall, 2-inch-square grape stake driven approximately 18 inches into the soil. Steel stakes are gaining popularity due to their ease of installation, rot resistance, and elimination of breakage and warping. Most California growers place the stake at the edge of the root ball of the large container trees, about 3 inches away from the tree on the downwind side (fig. 10.4). Growers planting small clonal plants usually install the stake prior to planting, since it does not interfere with digging the much smaller hole, and it eliminates potential damage to the limited root system.

Rootstocks budded in the nursery are handled like field-budded trees. They can be planted any time, providing that irrigation is immediately available. Fall planting dates must allow sufficient time to initiate adequate dormancy to prevent injury to the scion (i.e., the selected cultivar above the rootstock) from mid-November freezes. If planted in the fall, the scion (also the future trunk of the tree) is tipped at the nursery just enough to facilitate handling, but no more than 48 inches. This allows for reheading of the tree in the winter to the desired 42 inches. Budded trees can also be planted prior to leaf-out, between late January and early March. It is preferred to have them established an estimated 4 weeks prior to leaf-out to allow for root development into the native soil. Nursery budded trees are then trained as second-year trees during their first year in the field.

Figure 10.4. Proper orientation of the stake, young tree, and drip emitter. Place the stake on the downwind side of the young rootstock at the edge of the root ball, not in it. *Photo:* R. H. Beede.

Clonal rootstocks range from 4 to 12 inches tall when delivered to the grower. Delivery date presently appears more dependent on when the plants reach commercial size in the nursery rather than when it is optimal to plant them. Planting in late August to early September provides adequate time for the clonal rootstocks to establish and be capable of withstanding November freezes. However, there are risks associated with planting this late. A late March through early May planting allows the smaller clonal rootstocks to establish under more favorable growing conditions with less risk of stress, which stunts plant development. Planting clonal rootstocks in June and July exposes them to high temperatures, which can slow root development and greatly reduce growth. Midsummer planting dates also make it very difficult to develop sufficient rootstock growth and girth for successful budding later that season. Most clonal rootstocks planted at this time require special budding procedures in the spring with dormant buds, a practice at which only a few budders are successful. September budding is possible, providing that a sufficient percentage of the trees have reached ⅜ of an inch in size, but the risk of bud injury from early frost is greater. Once planted, clonal rootstocks require the installation of an 18-to-24-inch-tall, plastic, corrugated growing tube to enhance early growth, reduce stress from direct sun and wind, and prevent herbicide damage to the lower foliage and tender trunk.

After planting, preemergent herbicides can be applied down the tree row, providing there is rainfall or irrigation water to incorporate them. Hand-weeding by contract labor is also done around the base of the trees in the first year. Most growers place 18-to-24-inch-tall wraps around the rootstocks to minimize trunk injury from contact herbicides, a practice that is now required by most herbicide labels for nonbearing trees. Weed control is critical for minimizing tree loss from gopher damage.

After planting, the rootstock is initially allowed to grow with little manipulation. In April (3 to 4 weeks after planting), a suckering crew makes its first pass and removes any growth from the rootstock that is within 8 inches of the ground to facilitate weed management. One vigorous shoot is loosely tied to the stake to develop a straight trunk, reduce wind breakage, and facilitate budding.

Budding can begin in mid-July, as long as the size of the stem to be budded is ⅜ of an inch in diameter. However, experience has shown that delaying budding until late July or early August provides greater vigor and more uniform scion development. How early one buds depends on tree size, uniformity, and the availability of good budwood. Spring dormant chip budding is possible, but it requires special skills to achieve an acceptable percent take. Three to 5 days after summer budding, at least 50 percent of the growth above the bud and on other branches is removed. The crew also partially girdles (notches) the rootstock ½ an inch directly above the Kerman bud, using a 24-tooth-per-inch hacksaw blade. As soon as a sufficient number of buds begin pushing, the remaining rootstock growth is suppressed by terminal pinching to prevent competition with growth of the Kerman bud. If the bud does not push but appears to be alive, the rootstock is only lightly tipped to prevent excessive loss of leaf area. Test the viability of inactive buds by touching the leaf petiole subtending it. If it resists abscission from light force, the bud is most likely dead.

Pistachios can be successfully budded until late September if the bark still slips. Fall buds are not encouraged to grow, due to the risk of loss from sudden cold temperatures. Excessive late-season growth also predisposes the Kerman scion to cold injury should night temperatures reach about 27°F for an hour or more. Tender, late-season scion growth is also susceptible to infection by *Botrytis cinerea* if exposed to wet weather. Fall buds pushing voluntarily should be tied up loosely to prevent loss from mechanical operations during the winter. Fall tying also aids in developing a straight trunk (Peterson 1985).

Trunk Development

Successful July budding of optimally managed, vigorous rootstocks makes it possible to achieve 6 feet of scion growth in the first growing season. The bottom 42 inches of this scion growth will become the budded tree's trunk. Once the bud has grown 4 to 6 inches, training up the stake begins. First, the rootstock is pulled to the stake and firmly secured, using 1-inch, 8-mil plastic tape or some other strong fastener. This reduces the distance between the developing Kerman shoot and the stake. Subsequent ties are made on the Kerman shoot using green ½-inch, 4-mil plastic tape or 6-to-8-inch-long, ¼-inch-wide black soft ties. The initial growth of the developing shoot is tied to the rootstock it emerged from. Subsequent growth is tied to the stake after every 6 to 8 inches of new growth. The frequency of ties is dependent on growth rate, orchard uniformity, and wind conditions. The last tie should be at the top of the stake to prevent a bow in the future trunk. Well-managed trees can grow 1 inch daily, requiring crews to make field passes every every 10 to 14 days. Low and flat laterals developing from the trained Kerman shoot are pinched rather than removed to provide the greatest possible leaf area for greater trunk girth. Strong, upright laterals are removed

Figure 10.5. Pistachio tree near the end of the first season. *Photo:* R. H. Beede.

to prevent competition with the main shoot. The remaining rootstock leaves below the Kerman shoot are similarly pinched for the same reason (fig. 10.5).

Heading of the Kerman shoot to force primary scaffold branches is not performed during the first season, due to the resulting loss of carbohydrates critically needed for increasing shoot girth and winter cold hardiness. Also, the remaining growing period is often so short that the lateral shoots created by the heading cut fail to acquire sufficient length and girth to be valuable as primary branches. Limited lateral branching is also common, making it difficult to achieve proper tree shape. Therefore, this is not recommended.

The First Dormant Pruning

The first winter pruning consists primarily of removing rootstock growth on successfully budded trees, along with the removal of dormant shoots on unbudded trees judged unsuitable for future budding. Trees with sufficient Kerman growth are headed at 42 to 44 inches to force primary scaffold development in the spring. Dormant heading has been consistently observed to force more laterals on pistachios than in-season heading (three to five versus two to three). Any rootstock remaining above the bud union is headed 3 to 4 inches above the union. This stub may be used to secure the base of the tree to the stake for straightening purposes without placing tension on the still-tender bud union. It is best to have young tree pruning completed by mid-February.

Fall-budded trees with little or no Kerman scion growth are suckered, and the rootstock above the bud is headed at 3 to 4 inches. Kerman buds having grown only a few inches are left untied until spring, when their greater flexibility allows tying without damaging the bud union. Dormant Kerman shoots less than 42 inches high are left untipped to allow the terminal bud to continue growth in the spring.

Unbudded trees are assessed for their vigor and pruned accordingly. Poorly growing rootstocks are headed at one-third to one-half their height. Seedlings with less than 6 to 8 inches of growth require replacement, especially if gophers have been active in the area. Selective thinning cuts are made to the lower third of vigorous seedlings that are the height of the stake or greater. The main shoot tied up the stake the previous season is headed at 28 to 32 inches to force laterals in the spring. One or more of these can then be scion budded in May. The most vigorous scion shoot is then trained up the stake. Unlike the scion (Kerman) shoot developed from August budding in the first growing season, tipping of the scion at 42 inches is permitted on second-year, spring-budded rootstock, because the plant is now well established and it has the entire season for canopy development.

The Second Growing Season

Training

Fall-budded trees are trained up the stake and then tipped at 42 inches when they have reached a height of 44 to 46 inches. Use ½-inch, 4-mil plastic tie tape for training young shoots, not the heavier 1-inch tape. The heavier tape contracts during cold nights and strains the bud union. Rootstock growth below the bud and lateral growth from the Kerman shoot are again suppressed by pinching. This provides additional leaf area but decreases competition with the growing terminal. Growth close to the ground is suckered to reduce possible herbicide damage; this should be performed early to prevent excessive competition and significant loss of leaf area.

Trees whose Kerman bud reached sufficient height the previous season to allow dormant heading at 42 inches will push several laterals near the cut. These will become the primary scaffolds. To ensure that the growth angle of the laterals is not adversely affected by the stake, remove the two uppermost ties during the dormant pruning. Three or four laterals are selected, based on their uniform distribution around the trunk and a growth angle of 30 to 45 degrees. If possible, select laterals from buds that are slightly staggered vertically on the newly developed trunk. Primary branches originating from the same location on the trunk can result in mature scaffold limb breakage due to included bark and weaker attachment. Primary shoots can grow vigorously and may require tying into a vase-shaped position in May or June. Failure to do so results in flat scaffolds and an excessively open canopy.

Moderately vigorous laterals can also develop lower on the trunk. Experience suggests that these must be tipped early (at the 3-to-6-node growth stage) to avoid competition with the developing primary shoots. Single leaves or short spurs along the developing trunk can be left, since they aid in trunk diameter growth.

Primary shoots are allowed to grow to about 15 inches before they are tipped to 12 to 14 inches. This usually occurs by early June (fig. 10.6). Experience indicates that the removal of small quantities of tender shoot tissue results in better lateral bud break than the removal of large amounts of growth associated with delayed tipping. Heading into more mature growth often results in only one new shoot, and thus, no division. Lateral branching can supposedly be improved by clipping the leaf blade off its petiole at the second node position from the terminal cut. Two lateral shoots are then more likely to grow, rather than just the one nearest the pinched tip. Lateral shoots arising from primary branches are referred to as secondary shoots. The secondary shoots are allowed to grow to 15 inches before they are tipped to the same length as the primaries. The lateral shoots arising from the secondary branches are referred to as tertiary shoots. Shoots are called branches when they transition from green succulent tissue to brown woody tissue.

Young pistachio tree scions, especially Kerman, tend to be of insufficient girth, and primary limbs are often unable to support the weight of new secondary growth. Tying of limbs is therefore often practiced during this development stage (fig. 10.7). It is generally not advisable to attempt development of

Figure 10.6. Primary shoots from the trunk ready for tipping to 12 to 14 inches. *Photo:* R. H. Beede.

Figure 10.7. Cross-tying primary branches to support and maintain uprightness. *Photo:* R. H. Beede.

secondary and tertiary branching in the same year, because summer pruning often occurs too late in the season. Secondary branches would require tipping at 12 to 14 inches by mid-July to have sufficient vigor to produce tertiary shoots 30 inches in length by mid-September. Excessive summer pruning during the second season leaves the tree weak and spindly, and it predisposes trees to cold injury in the winter. Achieving primary, secondary, and tertiary branching all in the second growing season requires the heat units and excellent soils typical of Kern County.

Unbudded rootstocks are handled in much the same manner as developing Kerman trees; this facilitates instructions given to field crews. Budding should take place as soon as possible to avoid trees becoming too large for a high bud take.

The Second Dormant Pruning

Trees with secondary branches are thinned to two or three branches arising from each primary limb. When selecting these branches, one should avoid retaining limbs directly opposite one another if possible. It is preferable to have branches vertically spaced 3 to 4 inches apart down the primary limb. Following selection, each is headed to 12 to 14 inches and tied up, if necessary (fig. 10.8).

Trees headed in season to force primary branching are thinned to three or four well-distributed scaffolds. Each primary is then headed to 12 to 14 inches and tied to form an upright vase shape. In vigorous orchards, it is common for these trees to achieve both primary and secondary scaffold development during the second season if the primaries attained 14 inches of growth by mid-July. Pruning guidelines for trees at earlier training stages have been previously discussed.

Figure 10.8. Young pistachio tree after the second dormant pruning. *Photo:* R. H. Beede.

The Third, Fourth, and Fifth Growing Seasons

Training activities during the third season are very similar to those of previous years. Trees with secondary branching in the winter will produce tertiary branches during the third growing season. Tertiary branches are typically not tipped during the third growing season. Leaving them untipped provides needed photosynthetic area to increase limb diameter and structural strength. In-season tipping of the tertiaries should only be performed in orchards with unusually high vigor. Third-season training efforts should therefore focus on rootstock sucker removal and training of trees that are behind in development. This is also true for the fourth and fifth growing seasons.

The Third, Fourth, and Fifth Dormant Seasons

Pruning trees with developed tertiary branches is very similar to pruning those with only developed secondary branches. Two or three shoots per secondary limb are selected for location and strength and headed at 18 (Golden Hills) to 22 (Kerman) inches in length (fig. 10.9). Some tying of trees is still necessary to ensure maintenance of the desired shape. Long branches with good girth beyond the tertiary branches are typically cut in half, and some minor limb removal (thinning) is performed in the center of dense trees. Limbs that have grown in a better location than those previously selected can now be incorporated into the branching system. Avoid heading structural branches at the location along the branch where multiple lateral buds are compressed within a short distance (less than 1 in). Heading directly above this whorl of buds forces the growth of multiple short shoots (spurs), which can enhance early fruitfulness but prevent continued terminal shoot growth and use as a structural branch.

It is unwise to open the center of the young pistachio canopy early in its development. Removing the centers of trees 3 to 8 years in age reduces their early-bearing potential and forces outside shoots to grow excessively long. Interior branches headed lightly will develop lateral spurs farther from the terminal. These spurs will then differentiate flower buds at most of their node positions and bear fruit the following season. This is especially important for male Peters trees, as they tend to produce insufficient numbers of flower buds for pollination in the early-bearing years. Hence, once the basic tree shape has been secured with sound training during the first 4 years, early bearing is promoted by reducing the number of thinning cuts made per tree. This distributes the total growth over a greater number of growing points, resulting in shorter shoots with higher fruitfulness. Excessive pruning of the centers to prevent shading and fruitwood loss during the early years is unnecessary, as light is not yet limiting in these canopies. The removal of interior branches should focus on branches crossing over the center and those competing excessively with more desirable structural limbs. Delayed pruning of young vigorous trees until shortly before budbreak should also be avoided. Field observations of delayed leafing and poor lateral bud development suggest that it upsets the hormonal balance within the tree.

Figure 10.9. Young pistachio tree before the third dormant pruning. The horizontal red marks on the third-year shoot growth indicate where the heading cuts should be made during the dormant season. *Photo:* R. H. Beede.

Compared with training other nut crops, pistachio tree training is labor intensive and expensive. However, executing this critical production component well will pay huge dividends as the orchard matures. Pruning of mature canopies will be more systematic and less corrective. It will also require fewer difficult decisions by contract laborers. Harvesting will be more efficient due to increased nut removal and reduced catch frame damage from low limbs. Proper training is a worthy investment.

References

Peterson, B. 1985. New pistachio training technique. In California pistachio industry annual report 1985, 127–129. Fresno: California Pistachio Commission.

Part 4 Production

11 Weed Management

Kurt J. Hembree

Weed management plays an important role in the production of pistachios in California. Weeds must be controlled because they compete for water and nutrients. Weeds can harbor damaging insect pests, and they can be good habitat for gophers and other rodents. Weeds can also interfere with the distribution of water in low-volume irrigation systems.

A weed management plan should be in place before planting trees and continue through the life of the orchard. Weed control is especially critical during the first years of establishment, since competition from weeds can reduce tree vigor and productivity (fig. 11.1). Weed management decisions are influenced by market economics, scale of production, labor and equipment costs, herbicides available, weeds present, herbicide resistance, terrain, irrigation and soil type, and grower philosophy. Several chemical, mechanical, and cultural tools are available to aid in weed management efforts. Specific strategies and tools used may vary between fields and years. Weeds are commonly managed with herbicides in a strip 4 to 6 feet wide within the tree row (in-row), while the area between the rows (middles) is repeatedly tilled, mowed, or sometimes planted to cover crops.

Weed Identification

Identifying weeds correctly is necessary for effective management. It saves money by helping eliminate ineffective treatments and by reducing spray trips. Several important weed characteristics should be understood before choosing management options: weed life cycle (annual, biennial, or perennial) (fig. 11.2), weed type (broadleaf, grass, or sedge), weed growth habit (prostrate or erect), reproduction by seed or other structures (rhizomes, tubers, stolons,

Figure 11.1. Nutsedge competing for moisture with a new pistachio tree. *Photo:* K. J. Hembree.

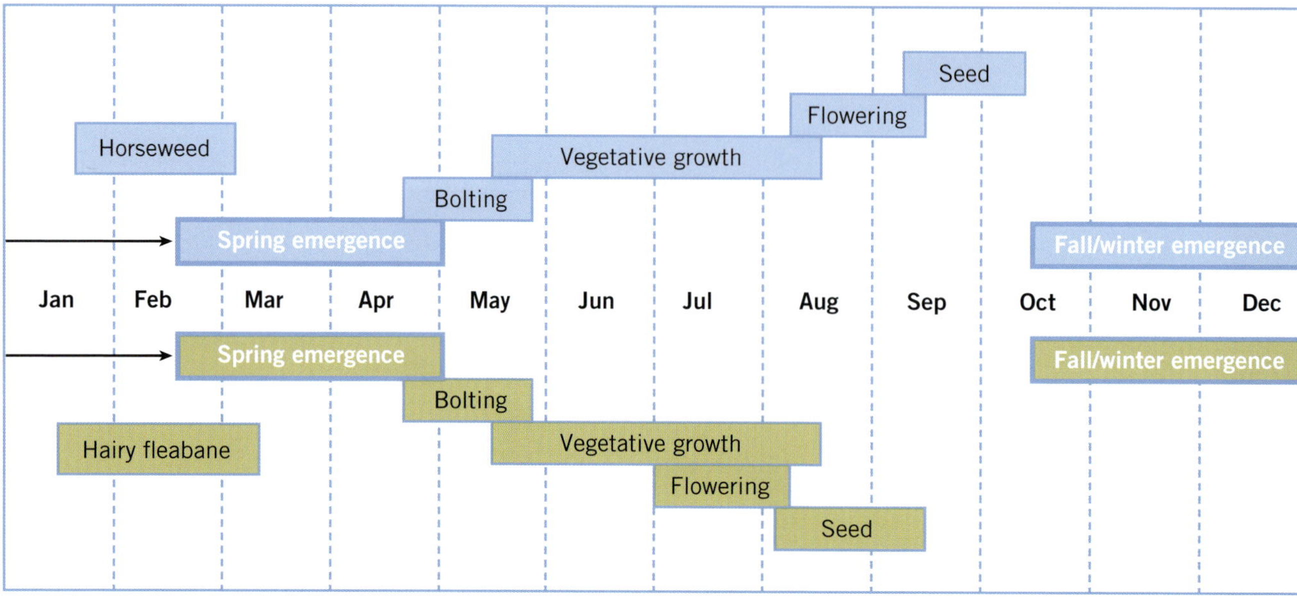

Figure 11.2. Life cycle of horseweed and hairy fleabane in the central San Joaquin Valley of California. *Source:* K. J. Hembree.

etc.), and susceptibility to herbicide and mechanical control. Examples of weeds common to pistachio orchards are shown in table 11.1. Resources helpful for identifying weeds and their growth characteristics include *Weeds of California and Other Western States, Growers Weed Identification Handbook, Weeds of the West* (see the References for more information), and the UC IPM website, ucipm.ucanr.edu. UC Farm Advisors, specialists, and chemical company representatives can also help identify weeds. To have weeds identified by another person, collect all portions of the plant (alive, if possible)—including aboveground stems and leaves, roots, and flower or seed heads—and transport them in a sealed plastic bag under cool conditions. Weed populations can be reduced if weeds are properly identified and appropriate strategies employed. Weed identification is particularly important when selecting herbicides and rates.

Soil Considerations

Soil characteristics are important because soil texture and organic matter can influence timing of cultivations and residual activity of herbicides. In light-textured soils, annual species like puncturevine *(Tribulus terrestis)*, large crabgrass *(Digitaria sanguinalis)*, and witchgrass *(Panicum capillare)* or perennial species like nutsedge (*Cyperus* spp.) and bermudagrass tend to be prevalent. In heavier-textured soils, perennials like curly dock *(Rumex crispus)* and field bindweed *(Convolvulus arvensis)* are common.

Soil texture also influences the rate of soil-residual (preemergent) herbicides used. Lower rates are usually needed on sandy, light soils or where organic matter is low, and higher rates are needed on clay or clay loam soils. Most of the preemergent herbicides registered in pistachio bind fairly tightly to soil particles and are not likely to leach into the tree root zone (compared with other more soluble herbicides like simazine, diuron, and norflurazon). However, some people have reported seeing injury to trees treated with flumioxazin in sandy soils watered with low-volume irrigation.

Mechanical cultivation can also be impacted by soil texture. Light-textured soils dry faster than heavy-textured soils, so more frequent cultivation is normally needed on lighter soils where irrigation occurs more frequently. In heavier soils, weeds should be cultivated when they are in their seedling stage of growth and the top few inches of soil is slightly moist. This maximizes control and equipment efficiency.

Irrigation and Rainfall Considerations

Irrigation method, amount of water applied, and rainfall affect weed growth, timing and frequency of cultivation, residual activities of preemergent herbicides, and timing of foliar (postemergent) herbicides. Annual weeds are difficult to control once they develop extensive root systems, especially in heavier, wet soils. Cultivating row middles should be performed prior to an irrigation or rainfall event under slightly moist conditions.

Pistachio orchards are irrigated primarily with drip, microsprinklers, microjets, or misters. Low-volume systems like these improve irrigation uniformity and efficiency compared with other systems, so weed control is needed to enhance emitter performance and facilitate system monitoring. Overwatering can increase the rate of breakdown of preemergent herbicides, which are degraded faster in moist, warm soils than in dry, cold soils. However, they tend to be more stable on heavier soils than lighter soils, persisting longer under periodic wetting. As these herbicides break down, weed escapes will occur within the wetted patterns of the emitters. Here, weeds grow vigorously and require close monitoring and timely treatment with postemergent materials.

Preemergent herbicides used in pistachios require incorporation or activation by water. The time between application and rainfall or irrigation must be relatively short (table 11.2), before they significantly degrade. The first rainfall or irrigation following treatment affects herbicide performance and determines how far it moves into the soil; subsequent irrigations and rainfall are less important to movement. Additionally, products injected into irrigation water (chemigation) degrade relatively fast through hydrolysis.

Since the majority of preemergent herbicides used in pistachios are not very mobile in the soil, losses through soil leaching or runoff are rare. Runoff can occur, however, where land is severely sloping and soil is physically moved following a severe rain event or a break in the irrigation system.

Postemergent herbicide performance is indirectly related to the method and duration of irrigation. These herbicides only control emerged weeds, so increasing the frequency or duration of irrigation increases the vigor of emerged weeds. In the case of spotted spurge, growth is rapid and viable seed is produced in less than 30 days. Therefore, treating soon after weed emergence is critical. The potential for multiple periods of weed emergence, or weed flushes, also increases in the wetted area. In comparison, weeds growing on the perimeter of wetted patterns can be drought tolerant and more difficult to control. Cudweed (*Gnaphalium* spp.), hairy fleabane *(Conyza bonariensis)*, horseweed *(C. canadensis)*, and turkey mullein *(Eremocarpus setigerus)* are examples. In any case, weeds escaping preemergent treatments should be treated with postemergent materials shortly after emergence.

Field Selection

Many considerations go into deciding where to plant pistachio trees, including weed management. Select fields free of perennial weeds, like nutsedge, field bindweed, bermudagrass, and johnsongrass, because

Table 11.1. Annual and perennial weeds commonly found in pistachio orchards

Winter annuals	Summer annuals
annual bluegrass *(Poa annua)*	annual morningglory (*Ipomoea* spp.)
burning nettle *(Urtica urens)*	annual sowthistle *(Sonchus oleraceus)*
cheeseweed *(Malva parviflora)*	barnyardgrass *(Echinochloa crus-galli)*
chickweed *(Stellaria media)*	cupgrass (*Eriochloa spp.*)
filaree (*Erodium* spp.)	cutleaf eveningprimrose *(Oenothera laciniata)*
foxtail (*Setaria* spp.)	hairy fleabane *(Conyza bonariensis)**
henbit *(Laminum amplexicaule)*	horseweed *(Conyza canadensis)**
mustard (*Brassica* spp.)	lambsquarters *(Chenopodium album)*
shepherd's-purse *(Capsella bursa-pastoris)*	large crabgrass *(Digitaria sanguinalis)*
wild radish (*Raphanus sativus*)	prickly lettuce *(Lactuca serriola)*
Biennials	**Perennials**
Hooker's eveningprimrose *(Oenothera hookeri)*	bermudagrass *(Cynodon dactylon)*
wild carrot *(Daucus carota)*	dallisgrass *(Paspalum dilatatum)*
	field bindweed *(Convolvulus arvensis)*
	johnsongrass *(Sorghum halepense)*
	purple nutsedge *(Cyperus rotundus)*
	yellow nutsedge *(Cyperus esculentus)*

Note: *May germinate in fall and grow like a biennial (see fig. 11.2).

Table 11.2. Incorporation and activation requirements of some preemergent herbicides

Herbicide	Time to incorporation	Moisture needed for activation (rainfall or irrigation water)
flumioxazin (Chateau)	< 28 days	¼"
isoxaben (Gallery T&V)	< 28 days	¼"
mesotrione (Broadworks)	7–10 days	¼"
napropamide (Devrinol)	< 10 days	¼"
oryzalin (Surflan, etc.)	< 21 days	½"
oxyfluorfen (Goal 2XL, etc.)	< 28 days	¼"
pendimethalin (Prowl H_2O)	< 21 days	½"
rimsulfuron (Matrix FNV)	< 21 days	½"
thiazopyr (Visor)	< 21 days	½"

Note: Refer to a chemical representative for specific products available but not listed. Always read and follow label directions carefully before using any pesticide.

they are difficult to eradicate once established. If problematic annual weeds are present, dedicate efforts to control them as well.

Unfortunately, no inexpensive or easy solutions exist to control established perennial weeds ahead of planting. Many growers use repeated applications of glyphosate or other postemergent herbicides to accomplish this. Burying vegetative reproductive structures deep (12 to 20 in) in the soil profile with a Kverneland plow or other similar equipment can be helpful. Fumigating the soil with methyl bromide or metam sodium can also be useful for reducing the weed population and controlling nematodes and other soilborne pests.

Weed Scouting

Scouting for weeds is an important but often overlooked activity. While there is no one method of scouting that will be effective for every person, programs can be tailored to meet individual needs. An example of factors that go into scouting for weeds is shown in table 11.3. Early detection of weeds that have recently been introduced or escaped control efforts is useful. It helps in the effort to prevent new weeds from establishing, spreading, or shifting to species that are more difficult to eradicate, like hairy fleabane. Weed species and populations vary by field and year. By scouting for weeds, you can evaluate the current strategy and make adjustments as the weed flora changes. Routinely scouting fields is cost effective because it helps eliminate ineffective treatments (see table 11.3).

To maximize control and reduce costs, conduct weed surveys following each treatment. Pay close attention to low-lying areas, where water accumulates, and around low-volume drip emitters or microsprinklers. Scouting the field at least twice a year (midwinter and midsummer) will give an indication of the weeds present. Watching for weeds escaping treatment can show if the current program is working or if different tactics need to be employed.

It is desirable to become familiar with all the weed species in the orchard and to monitor each of them. If certain weeds are not tracked, they can soon dominate, which is referred to as a weed shift. In other cases, weeds may develop resistance to an herbicide and require a new method of control. To help avoid these issues, scout fields frequently and rotate with effective herbicidal, mechanical, or other control tools whenever possible. Recorded observations help point out areas in the field where weeds such as field bindweed, mullein, and nutsedge may pose a problem. By scouting for weeds, you can alter the program based on a change in the weed pattern. Some time spent scouting for weeds will help in the selection of control options and the reduction of long-term costs. Maintain written records of scouting results for future reference.

Weed Management Methods

It would be ideal if one method of weed management could be used in all pistachio orchards, but this is obviously not possible. Many local conditions influence management decisions, including topography, soil structure and texture, soil permeability, type of irrigation, availability of equipment and labor, herbicide cost, accessibility to the orchard during wet conditions, and the particular weeds present. The specific strategy for managing weeds will vary from site to site, but it usually includes the combination of mechanical, chemical, and cultural options.

Mechanical Options

The goal of using mechanical control is to destroy weeds (primarily annual weeds) by uprooting or burying them, or by maintaining them at a low vegetative height. Weed burial works best on small weeds less than 4 inches tall or wide. Larger weeds are controlled by destroying the root-shoot connection or by slicing, cutting, or turning the soil to separate the root system from contact with the soil. Perennial weeds have extensive root systems and are difficult to kill with a single tillage operation. In most cases, 80 to 90 percent of perennial plant structures are found below the soil surface, and only 10 to 20 percent are seen aboveground. The top growth of perennials is destroyed with tillage, but new shoots are generated from underground reserves. Disking 3 to 4 inches deep depletes the root system and forces the weeds to use up food reserves. Over time, the reserves are depleted and the plants die.

Mechanical weed control is almost exclusively limited to the row middles, while herbicides involve in-row application. However, weeds often escape control where the treated tree row interfaces with tillage in the row middles (fig. 11.3), and they need to be controlled with postemergent herbicides. In some instances, cross-disking is used to kill weeds in newly planted orchards before irrigation tubing is installed. Care should be exercised when cross-disking to prevent mechanical damage and invasion by crown-rotting organisms. (Disadvantages to disking include increased soil erosion on sloping land, soil compaction, dust, reduced water infiltration, and a general decline in productivity.) To avoid damage, leave a 1-to-2-foot strip of undisturbed soil next to the trees.

Weeds escaping control can be removed by hand or treated with postemergent herbicides.

In some cases, resident weeds are repeatedly mowed (four to six times per year) in the middles and maintained in a low-growing state. This helps reduce soil erosion, especially on sloping soils. This also allows for easier field access during wet conditions and reduces soil compaction. Mowing too close to the soil surface will create dust and should be avoided.

Weeds can also be suppressed in the row middles by using cover crops. Cover crop species and management techniques vary. Select a cover crop that will not become competitive with the trees. Some clovers can spread to tree rows and are difficult to control with herbicides, so they should be selected carefully. To preserve surface cover, mow the cover crop at a recommended height. For information about cover crop varieties and management, refer to the UC SAREP Cover Crop Resources website, sarep.ucdavis.edu/covercrop/res. Some growers also control weeds in the middles with repeated applications of postemergent herbicides using shielded boom sprayers.

Chemical Options

Pre- and postemergent herbicides are the most common means of controlling weeds in pistachio. In some cases, herbicides have both pre- and postemergent activity (like oxyfluorfen). In other cases, combinations of products (tank-mixed) or sequential applications of products will be needed to provide broader, economic control. Identify the weeds, and read and follow label directions before using herbicide products. Table 11.4 shows herbicides registered in pistachio orchards in California and their application timing based on tree age.

It is important to know which herbicides control which weeds before spraying. Refer to tables 11.5 to 11.7 and appropriate product labels for weeds controlled, use rates, and application timings. The tables presented here serve as guidelines for weed management, so consult the product label and pest consultants for written recommendations. (For information on current registrations, consult your local agricultural commissioner, pest control adviser, or UC Cooperative Extension Farm Advisor.)

Preemergent herbicides

Preemergent herbicides are applied to the soil surface to control germinating weeds. They do not control dormant weed seed or emerged weeds. To be effective, they must be moved or incorporated into the top 2 to 3 inches of soil, where most weed seed germinate. If they are not applied or incorporated uniformly, weed control will be inconsistent and trees may be injured. Using rainfall during the dormant period is the most common means of soil incorporation and activation. Preemergent herbicides must be incorporated or activated within a certain period to be effective (see table 11.2). Where rainfall is limited, solid-set sprinkler irrigation can be used in lieu of rainfall.

Table 11.3. Weed scouting criteria and general guidelines

Criteria	Examples of questions for developing scouting guidelines
1. Know your farm.	Which weeds are present? When are specific weeds a problem? How do soil conditions and terrain vary? Are there drainage problems in certain areas? Are weeds present on field edges, fence lines, and irrigation canals or ditches? What control methods have been successful in the past and which have failed?
2. Develop a strategy.	What chemical or mechanical options are needed for the weed flora? Are ample equipment and labor available for timely control? Is weed resistance an issue? Are areas surrounding the field being considered? Is it economically affordable?
3. Implement a strategy.	Does it meet label and timing requirements? Are weather and field conditions favorable? Is equipment being observed during operation?
4. Review results and modify the approach.	Were the efforts effective? Which weeds still persist? Did unanticipated events affect results? Were there any successes? Were there any side effects (erosion, e.g.)? What were the costs versus benefits? How should the program be modified?

Figure 11.3. Weed growth between the treated strip and disked middles. *Photo:* K. J. Hembree.

Table 11.4. Herbicides registered for pistachio orchards in California and application timing based on tree age

Herbicide	Application timing (in relation to planting, bearing, or tree age)
Preplant herbicides	
metam sodium* (Vapam HL,* etc.)	before planting
pendimethalin (Prowl EC or Prowl H_2O)	before planting
Postplant, preemergent herbicides	
flumioxazin (Chateau)	at least 1 year old
indaziflam (Alion)	at least 3 years old
isoxaben (Trellis)	nonbearing only
mesotrione (Broadworks)	at least 1 year old
napropamide (Devrinol)	anytime after planting
oryzalin (Surflan, etc.)	anytime after planting
oxyfluorfen (Goal 2XL, GoalTender, etc.)	anytime after planting
pendimethalin (Prowl EC)	nonbearing only
pendimethalin (Prowl H_2O)	anytime after planting
penoxsulam + oxyfluorfen (Pindar GT)	at least 9 months old
rimsulfuron (Matrix FNV)	at least 1 year old
thiazopyr (Visor)	nonbearing only
Postplant, postemergent herbicides	
carfentrazone (Shark)	anytime after planting
clethodim (SelectMax)	nonbearing only
fluazifop-p-butyl (Fusilade DX)	nonbearing only
glufosinate (Rely 280)	anytime after planting
glyphosate (Roundup, Glyphomax, etc.)	anytime after planting
oxyfluorfen (Goal 2XL, GoalTender, etc.)	anytime after planting
paraquat* (Gramoxone Inteon*)	anytime before or after planting
pyraflufen (Venue)	anytime after planting
saflufenacil (Treevix)	at least 1 year old
sethoxydim (Poast)	anytime after planting
2,4-D amine* (Dri Clean,* etc.)	at least 1 year old

Notes: Refer to your chemical representative for specific products available but not listed. Always read and follow label directions carefully before using any pesticide.

*Requires a restricted use permit.

Some preemergent products break down rapidly once sprayed on the soil surface. For example, napropamide requires rainfall or irrigation within 5 to 10 days for activation. The emulsifiable concentrate (EC) formulation of pendimethalin is so volatile on moist soil that mechanical incorporation is required within hours of treatment. Other herbicides, like oryzalin and flumioxazin, can remain on the soil surface for longer periods (3 to 4 weeks) before incorporation is needed. It should be noted that breakdown of herbicides by volatilization and UV light begins soon after herbicides are applied. Therefore, the closer the treatment is to rainfall, the better the efficacy.

Preemergent herbicides are applied in-row, directed toward the base of the trees in a narrow (4-to-6-ft) strip of soil (or strip spray). When applying a strip spray, the amount of water and product used needs to be calculated based on the actual area treated. This is necessary in order to stay within recommended label rates, preserve efficacy, and limit potential crop injury.

It is also important to treat soil that is free from leaves, dried weeds, and other debris. Some herbicides, like oryzalin and napropamide, can adhere to dried weeds and leaves, in which case they may not be moved to the soil surface. Good soil contact is especially important for some herbicides, like oxyfluorfen and flumioxazin, which work by creating a weed barrier at the soil surface. While some preemergent herbicides, like thiazopyr, can be applied at a lower spray volume (10 to 20 gal/ac), most preemergent products should be applied at about 30 gallons per acre (GPA) to ensure adequate soil coverage. To improve soil contact, leaves and other debris should be mechanically blown off the tree rows before treatment (fig. 11.4). The length of residual activity and the effectiveness of some products can also be reduced if the soil is disturbed after treatment. Walking across the treated area to prune or perform other functions disturbs the soil and reduces weed control where products like flumioxazin are used. Therefore, apply these kinds of herbicides after most foot traffic has occurred.

Some preemergent herbicides can also be applied through low-volume drip microsprinkler irrigation systems during spring and summer, when rainfall is unlikely. Weed control will be limited to the soil area being wetted by the irrigation. Materials that can be injected through the irrigation system include oryzalin, pendimethalin, and rimsulfuron. When using this procedure, calculate the herbicide dose based on the actual area being wetted. Refer to the appropriate herbicide labels for treatment recommendations and application procedures.

Preemergent herbicides can persist from a few months to a year or more, depending on application rate, rate of degradation, and whether leaching or runoff occurs. Generally, preemergent herbicides used in pistachio irrigated with drip or microsprinkler irrigation are degraded faster than where furrow or flood irrigation is used. Tank-mixing preemergent products often broadens and lengthens the spectrum of weeds controlled, compared with using each product alone. For example, oxyfluorfen controls a wide array of broadleaf weeds but is weaker on grassy species. Therefore, adding another herbicide to the spray tank, like oryzalin, will increase the control of grass species and other weeds not controlled by

Table 11.5. Susceptibility of annual broadleaf weeds to herbicides in pistachio orchards in California

	Preemergent herbicides										Postemergent herbicides									
	flumioxazin (Chateau)	indaziflam (Alion)	isoxaben (Trellis)	mesotrione (Broadworks)	oryzalin (Surflan, etc.)	oxyfluorfen (Goal, etc.)	pendimethalin (Prowl H_2O)	penoxsulam + oxyfluorfen (Pindar GT)	rimsulfuron (Matrix SG)	thiazopyr (Visor)	carfentrazone (Shark)	clethodim (Select Max)	fluazifop (Fusilade DX)	glufosinate (Rely 280)	glyphosate (Roundup, etc.)	paraquat (Gramoxone, etc.)	pyraflufen (Venue)	saflufenacil (Treevix)	sethoxydim (Poast)	2,4-D amine (Orchard Master, etc.)
Annual broadleaves																				
cheeseweed	C	C	C	C	P	C	P	C	C	P	C	N	N	C	P	P	P	C	N	P
chickweed	C	C	C	C	C	P	C	C	C	P	C	N	N	C	C	C	C	—	N	N
clovers	—	P	P	P	N	P	N	C	—	—	P	N	N	P	P	P	P	—	N	N
cocklebur	—	—	—	C	N	P	N	—	P	N	C	N	N	P	C	C	C	C	N	C
cudweed	—	C	C	C	N	N	N	C	—	C	P	N	N	P	C	N	C	—	N	P
fiddleneck	—	C	C	C	C	C	C	C	C	C	C	N	N	P	C	P	P	—	N	P
filarees	C	C	C	C	P	C	N	C	C	C	C	N	N	C	P	P	P	C	N	C
goosefoot	C	C	C	C	C	C	C	C	P	C	C	N	N	P	C	C	C	C	N	C
groundcherry	C	—	C	C	N	C	N	P	C	P	C	N	N	C	C	C	C	C	N	C
groundsel, common	C	C	C	C	P	C	N	C	C	C	C	N	N	P	C	C	C	C	N	P
hairy fleabane	P	C	C	C	N	P	N	C	C	P	P	N	N	C	P	P	P	C	N	C
henbit	C	P	C	C	C	C	C	C	C	P	C	N	N	C	C	C	C	—	N	P
horseweed	C	C	C	C	N	P	N	C	C	P	P	N	N	C	P	P	P	C	N	C
knotweed, common	—	P	C	—	C	P	C	P	C	C	P	N	N	P	P	P	C	—	N	P
lambsquarters	C	C	C	C	C	C	C	C	C	P	C	N	N	P	C	C	C	C	N	C
London rocket	C	C	C	C	P	C	P	C	C	P	C	N	N	C	C	C	C	—	N	C
morningglory	C	P	C	P	P	C	N	—	N	—	C	N	N	C	C	P	C	C	N	P
mullein, turkey	—	—	C	P	N	P	N	—	—	C	P	N	N	C	P	P	P	—	N	P
mustard	C	C	C	C	N	C	P	C	C	P	C	N	N	C	C	C	C	C	N	P
nettle	C	C	C	C	P	C	N	C	C	C	C	N	N	C	N	P	C	C	N	P
nightshade	C	C	C	C	N	C	N	C	P	P	C	N	N	C	C	C	C	C	N	C
pigweeds	C	C	C	C	C	C	C	C	C	P	C	N	N	C	C	C	C	C	N	P
prickly lettuce	P	P	C	C	N	C	N	C	P	C	C	N	N	C	C	P	C	C	N	C
primrose, evening	—	P	C	C	P	P	P	C	—	C	P	N	N	C	C	C	C	—	N	—
puncturevine	C	—	C	C	C	P	P	P	C	P	P	N	N	P	C	C	P	C	N	P
purslane	C	C	C	C	C	C	C	C	C	C	N	N	N	C	C	C	C	C	N	P
Russian thistle	C	C	C	—	P	P	P	C	P	P	P	N	N	C	C	C	C	C	N	P
shepherd's-purse	C	C	C	C	N	C	P	C	C	C	P	N	N	C	C	P	C	C	N	C
sowthistle	P	C	C	C	P	C	N	C	P	C	N	N	N	C	C	P	C	C	N	P
spotted spurge	C	C	C	N	P	P	P	P	C	P	P	N	N	C	C	C	C	C	N	P
wild radish	C	—	C	C	P	C	N	C	C	C	P	N	N	C	C	C	C	C	N	C
willowherb	C	C	P	—	P	P	—	C	—	—	P	N	N	C	P	N	P	C	N	P

KEY: N = no control; C = control; P = partial control; and — = no information.

Note: Refer to your local dealer or chemical representative for specific herbicide products available but not listed. Always read and follow the label directions carefully before using any pesticide. Ratings reflect appropriate timing and dose according to label recommendations. Consult the manufacturer's product label for specific treatment recommendations.

Table 11.6. Susceptibility of annual grass and perennial weeds to herbicides in pistachio orchards in California

	Preemergent herbicides										Postemergent herbicides									
	flumioxazin (Chateau)	indaziflam (Alion)	isoxaben (Trellis)	mesotrione (Broadworks)	oryzalin (Surflan, etc.)	oxyfluorfen (Goal, etc.)	pendimethalin (Prowl H_2O)	penoxsulam + oxyfluorfen (Pindar GT)	rimsulfuron (Matrix SG)	thiazopyr (Visor)	carfentrazone (Shark)	clethodim (Select Max)	fluazifop (Fusilade DX)	glufosinate (Rely 280)	glyphosate (Roundup, etc.)	paraquat (Gramoxone, etc.)	pyraflufen (Venue)	saflufenacil (Treevix)	sethoxydim (Poast)	2,4-D amine (Orchard Master, etc.)
Annual grasses																				
annual bluegrass	C	C	N	N	C	P	C	C	C	C	N	C	N	C	C	P	N	N	N	N
barnyardgrass	C	C	N	N	C	P	C	P	C	C	N	C	C	C	C	P	N	N	C	N
bromegrasses	P	C	N	N	C	N	C	—	C	C	N	P	P	C	C	P	N	N	P	N
canarygrass	P	—	N	N	C	P	C	—	—	C	N	C	C	C	C	P	N	N	C	N
crabgrass, large	C	C	N	N	C	N	C	P	C	C	N	C	C	C	C	C	N	N	C	N
fescues	P	C	N	N	C	N	C	—	C	P	N	P	P	P	C	P	N	N	P	N
foxtails	C	C	N	N	C	N	C	—	C	C	N	C	C	P	C	C	N	N	C	N
junglerice	C	C	N	N	C	P	C	P	C	C	N	C	C	P	C	P	N	N	C	N
lovegrass	C	C	N	N	C	P	C	—	P	P	N	C	C	C	C	P	N	N	C	N
ryegrass, Italian	P	C	N	N	C	N	C	P	C	C	N	C	C	C	C	P	N	N	C	N
sandbur	C	C	N	N	P	N	C	—	—	C	N	C	C	C	C	P	N	N	C	N
sprangletop	P	C	N	N	C	N	C	—	—	C	N	C	C	P	C	N	N	N	C	N
wild barley	P	C	N	N	C	P	C	—	P	C	N	C	C	C	C	P	N	N	C	N
wild oat	C	C	N	N	C	P	P	C	P	—	N	C	C	C	C	P	N	N	C	N
witchgrass	P	—	N	N	C	P	C	C	—	—	N	C	C	P	C	P	N	N	C	N
Perennials (from seed or seedlings)																				
bermudagrass	N	—	N	N	C	N	C	N	N	C	N	C	C	C	C	P	N	N	C	N
dallisgrass	—	—	N	N	C	N	C	N	N	C	N	C	C	C	C	N	N	N	C	N
johnsongrass	C	—	N	N	C	N	C	N	P	C	N	C	C	C	C	C	N	N	C	N
field bindweed	—	—	C	P	P	N	P	P	P	C	C	N	N	C	C	P	P	C	N	P
Perennials (established)																				
bermudagrass	N	N	N	N	N	N	N	N	N	N	N	P	P	P	P	N	N	N	P	N
dallisgrass	N	N	N	N	N	N	N	N	N	N	N	P	P	P	P	N	N	N	P	N
johnsongrass	N	N	N	N	N	N	N	N	N	N	N	P	P	P	P	N	N	N	P	N
field bindweed	N	N	P	N	N	N	N	N	P	P	P	N	N	P	P	N	N	N	N	N
nutsedge, purple	N	N	N	N	N	N	N	N	P	P	N	N	N	P	P	P	N	N	N	N
nutsedge, yellow	N	N	N	N	N	N	N	N	P	C	N	N	N	P	P	P	N	N	N	N

KEY: N = no control; C = control; P = partial control; and — = no information.

Note: Refer to your local dealer or chemical representative for specific herbicide products available but not listed. Always read and follow the label directions carefully before using any pesticide. Ratings reflect appropriate timing and dose according to label recommendations. Consult the manufacturer's product label for specific treatment recommendations.

oxyfluorfen. If weeds are present at the time of application, a postemergent herbicide should be added to the tank.

Postemergent herbicides

Postemergent herbicides have either contact or translocated (systemic) activities. Contact herbicides (like paraquat and glufosinate) kill only the green parts of the plant they actually contact. Therefore, a higher spray volume (30 to 50 gal/ac) is usually required to provide adequate coverage of the weeds. Smaller weeds (< 4 in tall or wide) are easier to control than larger ones. The spray mix should not come into contact with any green foliage, immature bark, developing buds, flowers, or nuts; otherwise, crop injury can occur.

Systemic herbicides (like glyphosate and sethoxydim) move within the plant phloem or xylem from one part of the plant to another. Herbicide movement is typically toward growing points, like newly developing shoots, roots, and leaves. However, some products, like glyphosate, can move to aboveground tissues and also to the roots when carbohydrates are transported downward. For field bindweed control, it is more effective to treat with glyphosate in the fall, as the plant begins to transport sugars to the roots before going dormant. Systemic herbicides work better on young, vigorous weeds than on mature or moisture-stressed ones. Since these herbicides must move through the weed to kill it, control is usually slow (2 to 4 weeks).

Postemergent herbicides are applied as a directed in-row spray or as a spot treatment. Like preemergent herbicides, the amount of herbicide and water needs to be adjusted for the area within the tree row actually being sprayed. Postemergent herbicides can be applied alone or tank-mixed with other postemergent materials to broaden the control of weeds, or they can be mixed with preemergent herbicides during the dormant period to kill actively growing weeds.

Pistachio trees are sensitive to spray drift from postemergent herbicides. Sprayers equipped with spray shields should be used whenever possible to prevent the herbicide from contacting the tree foliage or other sensitive plants. Plastic-coated grow tubes should be used on new trees to help prevent damage by vertebrate pests, wind, and contact with the herbicide spray. Damage can still occur if equipment or excessively windy conditions do not favor uniform application.

Postemergent herbicide activity is influenced in part by the degree of weed growth in the wetted area near emitters or microsprinklers. Weeds emerging near emitters grow vigorously, making early timing of application important and usually requiring multiple treatments for season-long control. Weeds should be no taller than 4 inches and have fewer than 12 leaves at the time of treatment. Weeds like cudweed, hairy fleabane, horseweed, and mullein, when growing along the perimeter of the wetted area, may be drought tolerant and difficult to control, so they should be treated shortly after emergence before they become moisture stressed.

Figure 11.4. Leaves blown off tree row before preemergent herbicide application. *Photo:* K. J. Hembree.

Postemergent herbicides also require the addition of a surfactant or adjuvant to help improve absorption through the leaves. In most cases, a crop oil concentrate (COC), methylated seed oil (MSO), or nonionic surfactant (NIS) is used, depending on manufacturer recommendations. Additionally, adding sprayable-grade ammonium sulfate to the water in the spray tank before the herbicide is added will assist in plant uptake. A rate of 10 to 17 pounds per 100 gallons of water is generally needed. Most herbicides work best at a spray pH of 6 to 7. When the water pH is above 7 or 8, citric acid or other acidifying agent should be added to help buffer the water. Refer to the herbicide label for specific herbicide spray tank pH recommendations.

Cultural Options

Cultural options are generally limited to irrigation and row middle management. Since the majority of pistachios are irrigated with low-volume drip or microsprinklers, breakdown of preemergent herbicides can be expected, especially under high soil temperatures during summer. Consequently, repeated spot treatments with postemergent herbicides will likely be needed. However, some

Table 11.7. Herbicide use rates, modes of action, PHI, and other label recommendations for use in pistachio orchards in California

	Use rate per acre	Mode of action	Preharvest interval	Application information
Preemergent herbicide				
flumioxazin (Chateau WDG)	6–12 oz/app 24 oz/season	Group 14 (WSSA)	60 days	bearing/nonbearing; ¼–½" rain in 21–28 days; spray volume 20–40 gpa; can tank-mix; will not disturb treatment
indaziflam (Alion)	5–6.5 fl oz/app 10.3 fl oz/season	Group 29 (WSSA)	14 days	bearing/nonbearing at least 3 years old; ¼" rain within 21 days; 10–30 gpa; can tank-mix
isoxaben (Trellis)	0.66–1.33 lb/app 1.33 lb/season	Group 21 (WSSA)	60 days	bearing/nonbearing; ½" rain in 21 days; spray volume at least 10 gpa; can tank-mix
mesotrione (Broadworks)	3–6 fl oz/app 12.0 fl oz/season (not to exceed 3 applications per year)	Group 27 (WSSA)	30 days	bearing/nonbearing at least 12 months old; ¼v rain within 7–10 days; spray volume 10–60 gpa; can tank-mix; provides postemergent burndown on established weeds
oryzalin (Surflan A.S., etc.)	2–6 qt/app 12 qt/season 2.4–7.1 lb/app 14 lb/season	Group 3 (WSSA)	0 days	bearing/nonbearing; ½–1" rain in 21 days; spray volume 20–40 gpa; can chemigate; can tank-mix
oxyfluorfen (Goal, GoalTender, etc.)	5–6 pt/app (2 lb/gal) 6 pt/season (2 lb/gal) 2.5–3 pt/app (4 lb/gal) 3 pt/season (4 lb/gal)	Group 14 (WSSA)	7 days	bearing/nonbearing; ¼" rain in 28 days; spray volume 20–40 gpa; can chemigate; can tank-mix
pendimethalin (Prowl H_2O)	2–6.3 qt/app (H_2O) 6.3 qt/season (H_2O) 2.4–4.8 qt/app (3.3EC) 4.8 qt/season (3.3EC)	Group 3 (WSSA)	60 days (H_2O) 365 days (3.3EC)	(H_2O) — bearing/nonbearing; ½" rain in 21 days; spray volume 20 gpa; can tank-mix; (3.3EC) — nonbearing; preplant
penoxsulam + oxyfluorfen (Pindar GT)	1.5–3 pt/app 4.5 pt/season	Group 2 + 14 (WSSA)	60 days	bearing/nonbearing at least 9 months old; applied up to green leaf tissue; ½" rain within 14 days; spray volume 10–30 gpa; can tank-mix; do not disturb after treatment
rimsulfuron (Matrix SG)	2–4 oz/app 4 oz/season	Group 2 (WSSA)	14 days	bearing/nonbearing > 1 year old; ½" rain in 21 days; spray volume 20–40 gpa; spray pH > 5; can chemigate; can tank-mix
thiazopyr (Visor 2E)	2–4 pt/app 4 pt/season	Group 3 (WSSA)	365 days	nonbearing; ½" rain in 28 days; spray volume 20–40 gpa; split treatments of 2 pt/ac for nutsedge control; can tank-mix
Postemergent herbicide				
carfentrazone (Shark EW)	2 fl oz/app 7.9 fl oz/season	Group 14 (WSSA)	3 days	bearing/nonbearing; weeds < 4" tall; adjuvant used; spray volume > 30 gpa; tank-mix; spray pH 5–8
clethodim (Select Max)	9–16 fl oz/app 64 fl oz/season	Group 1 (WSSA)	365 days	nonbearing; grasses; retreat perennial grasses; adjuvant used; spray volume 30–40 gpa
fluazifop (Fusilade DX)	16–24 fl oz/app 72 fl oz/season	Group 1 (WSSA)	365 days	nonbearing; grasses; retreat perennial grasses; adjuvant used; spray volume 30–40 gpa
flumioxazin (Chateau WDG)	4–6 oz/app 24 oz/season	Group 14 (WSSA)	60 days	bearing/nonbearing; weeds < 4" tall; adjuvant is used; spray volume > 30 gpa; can tank-mix
glufosinate (Rely 280)	3.6–6 pt/app 21.5 pt/season	Group 10 (WSSA)	14 days	bearing/nonbearing; weeds < 6" tall; bright, humid, warm conditions; spray volume > 30 gpa; can tank-mix
glyphosate (Roundup, etc.)	0.5–1.5 lb ai/app 10.5 lb ai/season	Group 9 (WSSA)	3 days	bearing/nonbearing; weeds < 6" tall; AMS or buffering agents to bring pH to 5–6; spray volume 5–40 gpa; tank-mix
oxyfluorfen (Goal, GoalTender, etc.)	1–6 pt/app (2 lb/gal) 6 pt/season (2 lb/gal) 0.5–3 pt/app (4 lb/gal) 3 pt/season (4 lb/gal)	Group 14 (WSSA)	7 days	bearing/nonbearing; weeds < 4" tall; adjuvant used; > 30 gpa; tank-mix
paraquat (Gramoxone Inteon, etc.)	2.5–4 pt/app 20 pt/season	Group 22 (WSSA)	7 days	bearing/nonbearing; weeds < 6" tall; restricted use permit; adjuvant used; spray volume > 30 gpa; can tank-mix
penoxsulam + oxyfluorfen (Pindar GT)	0.5–1 pt/app 4.5 pt/season	Group 2 + 14 (WSSA)	60 days	bearing/nonbearing at least 9 month old; applied after nut set; broadleaf weeds < 4" tall; NIS or COC adjuvant is used; spray volume 20–30 gpa; can tank-mix
pyraflufen (Venue)	0.7–4 fl oz/app 6.8 fl oz/season	Group 14 (WSSA)	prebloom	bearing/nonbearing; broadleaf weeds < 4" tall; adjuvant used; spray volume > 30 gpa; can tank-mix
saflufenacil (Treevix)	1 oz/app 3 oz/season	Group 14 (WSSA)	7 days	bearing/nonbearing at least 12 months old; broadleaf weeds < 4"; MSO must be used; spray volume 20–40 gpa; can tank-mix
sethoxydim (Poast)	2.5 pt/app 10 pt/season	Group 1 (WSSA)	15 days	bearing/nonbearing; grasses; retreat perennial grasses; adjuvant used; spray volume 30–40 gpa
2,4-D amine (Orchard Master CA, etc.)	2–3 pt/app 6 pt/season	Group 4 (WSSA)	60 days	bearing/nonbearing; broadleaf weeds < 6" tall; restricted use permit; spray volume 20 gpa; can tank-mix

Note: Refer to your local dealer or chemical representative for specific herbicide products available but not listed. Always read and follow the label directions carefully before using any pesticide. Ratings reflect appropriate timing and dose according to label recommendations. Consult tables 11.5 and 11.6 along with the manufacturer's product label for specific treatment recommendations.

preemergent herbicides (oryzalin, oxyfluorfen, and rimsulfuron) can be applied through low-volume irrigation during summer to extend preemergent control and reduce postemergent spray trips. In most cases, combinations of preemergent strip sprays during dormancy followed by chemigation of preemergents during summer can reduce or eliminate the need for postemergent materials. Cover crops are also used to help improve soil nutrition and reduce soil compaction while suppressing weeds.

Herbicide Application Equipment

Herbicide sprayers must be accurately calibrated and maintained to deliver the proper amount of herbicide and water. Several types of shielded sprayers are available, including boom and boomless sprayers and low-volume applicators (like the Enviromist or Bubco) (fig. 11.5). Most boom or boomless sprayers deliver 15 to 60 gallons per acre, at an operating pressure of 15 to 60 pounds per square inch (psi), whereas controlled droplet applicators (Bubco, Enviromist, etc.) deliver less than 5 gallons per acre of a concentrated spray mix. Also available is a type of specialized equipment that sprays postemergent herbicides (primarily glyphosate) only when it detects green weeds. For example, the WeedSeeker (smart sprayer) is equipped with light sensors and solenoids to spray green weeds and not bare soil. It is designed to be most efficient when weeds appear in patches, reducing the amount of herbicide needed by more than 50 percent.

Spray nozzle selection is also important to consider for maintaining efficacy and reducing risk of crop injury (discussed later). Some herbicides require special application requirements, so read and follow label recommendations carefully. Clean out the sprayer and nozzles following each application to prevent sprayer contamination and potential crop injury from leftover residues. Some herbicides, like 2,4-D amine, flumioxazin, and others, are active at very low doses and can adhere to hoses and other sprayer parts, making cleaning especially important. To avoid sprayer contamination problems, follow label directions carefully when rinsing out spray equipment after use, regardless of the herbicides used.

Proper Application of Herbicide Sprays

Keep Sprays on Target

When it comes to spraying herbicides, cutting corners, even under difficult economic conditions, defeats the purpose of spraying. Improper application procedures and worn nozzles can lead to reduced weed control and crop injury. Newly planted trees are more sensitive to spray damage than mature trees, although damage to the latter can still occur.

Since herbicides are designed to kill plants, pistachio trees are also susceptible to damage (fig. 11.6). Therefore, it is very important to apply herbicides to the weeds and prevent off-target herbicide movement (spray drift) as much as possible. Apply treatments at the appropriate time and under optimal environmental conditions. Be especially cautious when spraying outside rows or bordering areas. The degree of herbicide-related injury varies when trees are exposed to herbicide spray drift. It can depend on type of herbicide, herbicide dose and age, tree health, timing of exposure to the herbicide, and other factors. Herbicide drift symptoms may range from slight yellowing of the foliage to malformation of the leaves, spotting of the fruit, or complete crop loss and tree death.

Trees are less susceptible to damage during winter, when they are dormant. It is important to take advantage of pre- and postemergent herbicides during dormancy to help reduce weed population as much as possible when trees are less prone to damage. Regardless of when spraying occurs, be

Figure 11.5. Shielded sprayer applying a tank mix of herbicides. *Photo:* K. J. Hembree.

Figure 11.6. Young pistachio tree injured by herbicide spray drift. *Photo:* K. J. Hembree.

aware of the surroundings and keep the spray mix on the target site.

The goal of applying herbicides is to maximize the amount of herbicide reaching the target weeds and minimize the amount reaching off-target areas. Proper application procedures will maximize herbicide performance and limit crop injury. Spray drift occurs as a result of the physical movement of an herbicide from the target area to areas where the herbicide is not intended. Spray drift injury primarily occurs when spray droplets are carried from the application site downwind and deposited on a sensitive plant surface (buds, leaves, shoots, etc.). Although less likely to occur, spray drift can also be the result of vapor drift (evaporation of an herbicide from a plant surface) and particle drift (evaporation of spray carrier droplets, which leaves concentrated herbicide droplets).

Spray drift is influenced mainly by droplet size (fig. 11.7) and environmental conditions during application. High wind, low humidity, and high temperature favor drift. Spray droplets with a mean diameter of less than 200 microns (μm) are more susceptible to drift than larger droplets. The longer they remain airborne, the greater is the chance that they will be carried by wind away from the treatment site. In addition to droplet size, spray booms and nozzles raised higher than the manufacturer's recommended height increase the risk of drift. The influences of spray boom and nozzle height, environmental conditions, and other factors on drift potential are shown in table 11.8.

Selecting the appropriate spray equipment and using good application techniques will reduce the risk of herbicide spray drift and crop injury. Before treating weeds, consider the following important factors to help reduce spray drift:

- **Label instructions.** Read and follow all label directions before spraying. Information printed on the label describes ways to maximize herbicide performance and reduce spray drift and crop injury.
- **Environmental conditions.** Spray when winds are minimal (3 to 7 mph). Since many herbicides are volatile at temperatures above 80°F, spray toward the end of the day as it begins to cool. Use amine formulations of 2,4-D, rather than ester formulations, because they are less volatile. Do not spray when it is foggy, because very small droplets (or fines) can remain suspended in the air a long time.
- **Buffer zones.** Use adequate buffer zones surrounding the application site to protect other crops, gardens, waterways, or sensitive areas that are downwind.
- **The sprayer.** Use spray shields whenever possible because they reduce drift from 35 to 75 percent, depending on the model and wind conditions. Practice routine sprayer maintenance and calibration.
- **Application timing.** Apply preemergents before weeds emerge and postemergents when weeds are less than 4 inches tall. Be cautious during bud swell or in the presence of foliage.
- **Nozzle selection and spray pressure.** Use spray nozzles with an orifice size larger than 02, unless otherwise specified by the product label (fig. 11.8). Keep the spray pressure below 60 psi and use at least 10 gallons per acre, unless otherwise specified. To make large changes in spray volume, change the travel speed or select a different nozzle size rather than adjusting the spray pressure. Changes in spray pressure should only be made for fine adjustments in flow rate. Spray nozzles designed to produce coarse droplets reduce drift from 50 to 95 percent. Refer to table 11.9 and the spray nozzle owner's manual for appropriate nozzle selection.
- **Spray boom and nozzle height.** Operate the spray boom and nozzles at their lowest recommended height. For 80-degree tips, nozzle height is 30 inches, and for 110-degree tips, nozzle height is 20 inches. Orient nozzles to spray in a forward direction to allow for height reductions. Be cautious when orienting the nozzles across the row, as this increases the risk of spray drift across the field.
- **Travel speed.** A travel speed of 3 to 6 mph is generally recommended; do not spray above 7 mph.
- **Spray additives.** Drift retardants can reduce spray drift by up to 95 percent. Other additives, like

ammonium sulfate, improve efficacy of postemergent sprays but also thicken or alter the spray's viscosity. Additions like this to the tank may alter droplet size and reduce flow rate. In this case, check the spray droplet pattern on water-sensitive paper before adding the herbicide(s) to the tank.

Weed Management and Drought

Prolonged periods of dry conditions affect weed growth, alter herbicide performance, and complicate weed management strategies. Competition for water increases as soil moisture becomes limited, so weed control is even more important when water is scarce. Fewer weeds are usually observed during dry conditions, and they tend to be less vigorous. However, some weeds, like Russian thistle *(Salsola iberica)* and field bindweed, take advantage of limited surface moisture and develop extensive root systems early, making them more competitive and difficult to control. Other weeds, like hairy fleabane and horseweed, can adapt to both wet and dry conditions.

When soil moisture is adequate, a properly timed shallow cultivation can be used to control weeds in the middles. During dry conditions, however, tillage can reduce soil moisture and potentially limit crop growth. It is important to understand how the performance of pre- and postemergent herbicides changes under dry or drought conditions. UV light degrades preemergent herbicides from the soil surface if they are not activated by water and moved into the soil soon after treatment. This is not a concern in annual crops, where irrigation water or mechanical methods are used for incorporating herbicides. However, in perennial crops like pistachios, growers rely on rainfall for herbicide incorporation and activation. Although each product label specifies how long the herbicide can remain on the soil surface before activation by moisture (see table 11.2), some degradation by sunlight can be expected once applied. During dry years, preemergent treatments should be applied as close to rainfall as possible to ensure adequate activation and residual control. Irrigation water can also be used to activate herbicides in lieu of rainfall.

Performance of postemergent herbicides also changes during dry conditions. Systemic herbicides are especially impacted under drought conditions. Examples of herbicides commonly affected include 2,4-D amine, glyphosate, and sethoxydim. Water-stressed weeds have thicker waxy leaves (cuticle), reducing herbicide absorption. Plant growth processes are also altered, reducing movement of

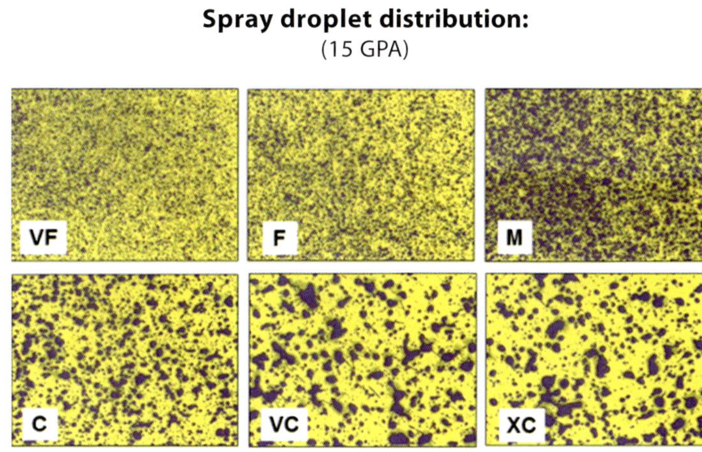

Figure 11.7. Spray droplet deposition based on droplet size category. (VF = very fine; F = fine; M = medium; C = coarse; VC = very coarse; XC = extremely coarse.) *Source:* K. J. Hembree.

Figure 11.8. Drift from two nozzle sizes: (A) size 01 and (B) size 04. *Photo:* K. J. Hembree.

Table 11.8. Influence of environmental conditions and other factors on spray drift

Spray factor	Leads to more drift	Leads to less drift
spray particle size	smaller	larger
wind speed	higher	lower
air temperature	higher	lower
relative humidity	lower	higher
air stability	no vertical movement	vertical movement
herbicide volatility	volatile	nonvolatile
nozzle orifice size	smaller	larger
spray droplet size	very fine to fine	medium to extremely coarse
spray pressure	higher	lower
spray release height	higher	lower

Table 11.9. Nozzle types and their operational features

Nozzle type	Spray angle (°)	Nozzle orifice size*	PSI	Spray height (in)	Contact herbicide	Systemic herbicide	Drift control
Air Induction (AI)	80 110	015–06 015–08	30–100	30 20	G	E	E
Air Induction XR	110	015–06	15–90	20	G	E	E
Drift Guard TwinJet (DGTJ60)	110	015–08	30–60	20	VG	E	VG
Extended Range (XR)	80 110	01–15 01–15	15–60	30 20	E (> 30 psi) G (< 30 psi)	E (> 30 psi) G (< 30 psi)	E (> 30 psi) G (< 30 psi)
Turbo TeeJet Induction (TTI)	110	015–06	15–100	20	NI	E	E
Turbo TwinJet (TTJ60)	110	02–05	20–90	20	E (> 30 psi) VG (< 30 psi)	E (> 30 psi) E (< 30 psi)	VG (> 30 psi) VG (< 30 psi)
TwinJet (TJ60)	65 80 110	01–08 01–10 02–10	30–60	35 30 20	E	NI	NI

KEY: E = excellent rating; VG = very good rating; G = good rating; NI = no rating indicated.

Notes: Refer to your chemical representative for specific products available but not listed.

*These are the numbers stamped on spray nozzles to indicate a given size. The numbers refer to the spray nozzle flow rate when it is operated at a pressure of 40 psi. (For example, a number of 01 = 0.1 gpm at 40 psi; 015 = 0.15 gpm at 40 psi; 04 = 0.4 gpm at 40 psi, etc.)

herbicides within the weeds to sites of action, which can reduce control.

Contact herbicides do not require movement within weeds, so they are less likely to be affected by drought. However, drought-stressed weeds can have spindly shoots or leaves, making them harder to kill. Examples of herbicides commonly affected include carfentrazone, paraquat, and oxyfluorfen. Using a higher label rate and spray adjuvant can help compensate for drought-stressed conditions. Using a nitrogen-based additive (like ammonium sulfate) or a crop oil adjuvant, or both, may also help increase absorption into the leaves of stressed weeds. Apply a spray mixture volume high enough to adequately cover the weeds and increase the chance of absorption. In most cases, a spray volume of 30 to 50 gallons per acre is needed. Applying treatments toward the end of day also increases efficacy.

A Word of Caution When Using Herbicides

Read and follow label instructions carefully before using any herbicide. Information presented here is designed to supplement herbicide label recommendations. Information concerning limitations on use is also available from regulatory officials, agricultural chemical trade channels, and University of California Cooperative Extension offices in each county. Observe all precautions and restrictions on the herbicide label regarding use of protective clothing and equipment, handling and storage, and protection of pets, domestic animals, wildlife, fish, and desired vegetation. A permit is required from the county agricultural commissioner for purchase and use of restricted herbicides like paraquat and 2,4-D amine.

Since preemergent herbicides can persist in the soil for a relatively long time after application, discontinue their use at least 1 year before tree removal. Plan ahead and cultivate or use postemergent sprays for weed control before tree removal, and follow label recommendations regarding plant-back restrictions. If it is necessary to replant trees in the field and preemergent herbicides have been applied, use clean, untreated soil to backfill around the tree's roots. Shut the sprayer off when stopping or turning at the ends of rows and apply at a constant speed, since slowing down increases the amount of herbicide applied.

Always adjust the amount of water and herbicide needed for broadcast, strip, or spot treatments. The herbicides and rates discussed here are within current label recommendations at the time this publication was written. They also represent those generally found to be safe and effective by University of California weed scientists. Adjust rates to meet specific regions, soil types, and conditions.

Designing a Weed Management Program

Weed management decisions should be designed around three crop phases: (1) before planting, (2) after planting or nonbearing, and (3) established planting or bearing. In some cases, herbicide choices are limited based on the growth stage of the trees.

Before Planting

The primary goals of managing weeds before planting are to eradicate or reduce perennial weeds, practice good land preparation, and control annual weeds to reduce tree-weed competition. Perennial weeds (like bermudagrass, field bindweed, and nutsedge) have deep rooting systems and are very competitive for soil water and nutrients. If they are allowed to become established, it may take several years of control efforts to reduce the population sufficiently. If possible, select fields free of perennial weeds. Several options can be used to help reduce the level of infestation of perennial weeds, including mechanical tillage, burial, and herbicides.

If perennial weeds are present, repeatedly disk the field during summer when the soil is dry to destroy the aboveground vegetation and dehydrate the vegetative propagules (rhizomes, roots, tubers, etc.). A Kverneland plow (or other similar equipment) can be used before final land preparation to bury perennial propagules and seed of annual weeds more than 12 inches deep into the soil profile (fig. 11.9).

Repeated applications of glyphosate can also be used to suppress perennials. To be most effective, they should be applied in the fall when carbohydrates move down into the roots. Field bindweed growth can be reduced for 1 to 2 years by spraying with high rates of glyphosate, waiting a week, and then undercutting the roots with a reclamation blade 16 to 18 inches deep. For nutsedge control, treat before the plants have four leaves.

Select fields relatively free of annual weed species to help reduce costs after the trees are planted. Pre- and postemergent herbicides are commonly used before planting to reduce the weed population. Pendimethalin is commonly applied in a narrow strip down the center of the tree rows and disk incorporated 2 to 4 inches deep prior to planting to control germinating grasses and some small-seeded broadleaf species. At planting time, backfill the holes using clean, untreated soil to prevent injury to the tree roots.

Postemergent herbicides have little to no soil residual activity, so they are often used to kill small, emerged annual weeds before planting. If water is available, border up the field, irrigate, and then treat with a postemergent material to kill the weeds.

Soil solarization is a nonpesticidal method of controlling some soilborne pathogens, nematodes, and weeds. With this method, weed seed can be killed by placing clear plastic sheets directly on moist soil for 4 to 8 weeks during summer to superheat the soil and kill the seed. It leaves no toxic residues and can be easily used on small or large acreage. Soil solarization also improves soil structure and increases the availability of nitrogen and other essential plant nutrients. Disposing of the plastic film properly is a necessity to consider when using this approach. For additional information on this technique, see *Soil Solarization* (Elmore 1997). Fumigating with metam sodium, Telone, or other products can also be used in conjunction with tarping the field.

After Planting

The primary goals of managing weeds during the nonbearing years are to reduce the weed population as much as possible and prevent crop injury. The nonbearing stage of tree growth lasts from 5 to

Figure 11.9. Modified moldboard plow burying weed propagules 16 inches deep. *Photo:* K. J. Hembree.

7 years, depending primarily on the rootstock and general growing conditions. After planting, rooting is limited and competition from weeds can be high, leading to reduced growth and possibly a delay in production. As a result, it may take several years to overcome these early competitive effects. Extreme care should be exercised when trees are very young to protect them from chemical or physical injury. While plastic-coated grow tubes can help reduce the likelihood of herbicide-related injury, they are not a guarantee for complete protection. Be cautious when using chemical and nonchemical control measures in newly planted orchards to ensure that tree health and vigor are maximized.

Mechanical and Chemical Strategies

Mechanical or chemical techniques, or a combination of these, are used to manage weeds during the nonbearing years. In most situations, weed control is managed differently within the rows and in the row middles. Since trees are planted on a grid, mechanical methods can be practiced on terrain easily traversed by equipment. Here, cross-cultivating may be used after planting to destroy about 95 percent of the weeds. Hand-weeding or spot treating with postemergent herbicides will be needed to control escaped weeds around individual trees. To prevent disease-related issues, avoid physically damaging the trees. Cross-disking cannot be used once irrigation tubing is installed. Weeds should be small and the soil moist enough to allow the equipment to flow smoothly without pulling up large clods. Hand-held weed eaters may also be used within the row—but their nylon cords can injure the trunk, and they are expensive to operate on large acreages. Generally, three or four cultivations will be needed yearly for control.

Managing weeds in the row middles is usually done with disks, tillers, or mowers. It takes three or four trips a year to manage vegetation effectively. The equipment should be set to cut shallowly to readily destroy small weeds. As weeds mature, they can clog equipment and are difficult to control. Cultivation in preparation for planting a winter annual cover crop also helps suppress weed growth.

Herbicides play an important role in reducing weed populations and contributing to early growth. The plastic-coated grow tubes used to protect trees from wind damage are also valuable for decreasing exposure to herbicides. In all cases, direct herbicides toward the base of the trees to reduce the risk of tree injury. Several herbicides are registered for use in nonbearing pistachio plantings in California (see table 11.4). Herbicides labeled for use in nonbearing orchards can only be applied up to 12 months before nuts are harvested. In other cases, herbicides can only be applied after orchards have been planted for a given period of time, which may or may not coincide with the nonbearing cycle. Therefore, it is important to plan ahead for herbicide selection.

Preemergent herbicides are applied to the in-row areas once the trees have settled into the soil after planting. Tank mixes of preemergent products are commonly used to broaden the spectrum of weed control. All the preemergent herbicides used in pistachio orchards, with the exception of the emulsifiable concentrate (EC) formulation of pendimethalin, require water (usually rainfall) for activation. Pendimethalin EC requires mechanical mixing into the soil, so it is normally used before planting. The flowable (F) formulation of pendimethalin (sold as Prowl H_2O) can be applied to the soil surface and incorporated or activated with moisture.

It is also important to mechanically blow leaves and other debris off the soil prior to treatment to increase herbicide-soil contact. Also, burning down emerged weeds with postemergent herbicides—or mulching them in mechanically before preemergent herbicides are used—allows for better distribution of the soil treatment, especially where napropamide, oryzalin, or flumioxazin are used.

The criteria for selecting preemergent herbicides should include knowledge of the weeds present, soil texture, and irrigation method and frequency. These factors can influence the herbicide activity and crop safety. Refer to the appropriate product label for specific recommendations regarding these and other factors. If weeds are present at the time of treatment, postemergent products should be added to kill the existing weeds.

A common herbicide program in newly planted pistachios might include applying a strip spray of preemergent products (like flumioxazin plus pendimethalin) during late fall to early winter, just ahead of rainfall, and spraying again before bud swell. If horseweed or hairy fleabane is present, substituting rimsulfuron for flumioxazin will improve control. This is also a good time to add 2,4-D amine or glyphosate to the mix (or both) to burn down existing weeds. Oxyfluorfen, used at high label rates, complements napropamide, oryzalin, and pendimethalin. It controls a wide spectrum of summer and winter annuals, including cheeseweed *(Malva parviflora)* and chickweed *(Stellaria media)*. Oxyfluorfen should not be applied near the time of bud swell and early fruit set, or tree injury may result. Oxyfluorfen can be applied to bearing pistachio orchards from May 1 to 7 days prior to harvest. This gives greater flexibility for controlling late-germinating winter annuals like filaree (*Erodium* spp.) and cheeseweed. It is also effective on summer annuals, including lambsquarters (*Chenopodium album*) and nightshade (*Solanum* spp.). If yellow nutsedge is a problem in nonbearing fields, apply thiazopyr as a split treatment in the fall and winter to move the herbicide down into the top few inches of the soil surface, where nutsedge sprouts. A single, reduced rate of thiazopyr is used when susceptible annual weeds are targeted.

Preemergent products can also be injected directly into low-volume drip and microsprinkler irrigation, including rimsulfuron, oryzalin, oxyfluorfen, and napropamide. This is typically done during late spring and summer to control weeds in the tree rows as winter preemergent treatments begin to break down and summer weeds proliferate. Herbicides injected through the irrigation system should meet label guidelines and should be applied for the duration of the irrigation, leaving only enough time to flush the system at the end of the application. It is also very important to have adequate mixing of the herbicide(s) in the area between where the product is injected at the pump and where the individual irrigation lines "T" off the mainline. Again, the amount of herbicide used should be based on the actual area being wetted by the emitters.

Some growers prefer only to use postemergent herbicides in nonbearing trees during the winter to control a wide variety of weeds. This could mean using paraquat, for example, or tank-mixing glyphosate and 2,4-D amine. If 2,4-D amine is used alone or mixed with another product, trees must be at least 1 year old. Many crops, especially grapes, are very sensitive to 2,4-D amine, so use caution when spraying. A permit from the agricultural commissioner is required to apply 2,4-D amine, and special restrictions may exist, depending on location and time of year. Regardless of the specific postemergent products used, they should be applied to young weeds at a sufficient spray volume to adequately cover the weeds. This is particularly important when contact herbicides are used. Apply contact-type products in a spray volume of 30 to 50 gallons per acre. Before mixing herbicides, read the appropriate product labels to determine if product mixing is allowed.

If annual and perennial grasses are the primary problem, a selective postemergent grass herbicide (like clethodim, fluazifop, or sethoxydim) might be used. These materials do not control nutsedge or broadleaf weeds, and clethodim is the only one of the three that controls annual bluegrass. Perennial grasses should be sprayed when the shoots are 8 to 12 inches long in the spring and summer. Several treatments may be required before their storage reserves are depleted and the plants are killed. The grass herbicides are generally safer to use around young trees than are nonselective contact materials (like glufosinate) or nonselective systemic materials (like glyphosate). The grass herbicides do, however, require the addition of an adjuvant (either a nonionic or nonphytotoxic oil), which increases the risk of injury to green tissue on the trees, particularly under hot conditions. Use spray shields where possible to reduce the risk of tree damage. Row middles are sometimes sprayed with postemergent herbicides, in lieu of tillage. Here, it is also very important that shielded sprayers are used to reduce the likelihood of spray drift.

Never will a single treatment with an herbicide or combination of herbicides control all the weeds for the season in pistachio orchards. In most cases, one or two preemergent treatments will be required from October through February (along with postemergent products if weeds have emerged), followed by two or three postemergent treatments for season-long control. Refer to tables 11.5 to 11.7 for a list of herbicides currently registered for pistachios, the weeds they control, their rates, timings, and general application guidelines. Always read and follow all label recommendations before making any herbicide application to control weeds.

Established Trees

Established, or bearing, pistachio trees are ones that have begun bearing marketable nuts, which usually takes 5 to 7 years from planting. Time to tree maturity and production status depend on numerous factors, including rootstock, growing region, and degree of weed competition. Shading of the soil surface by the tree canopy is limited and offers little in the way of weed suppression. Weeds will still grow and can interfere with normal cultural practices. For example, weeds disrupt the application pattern of low-volume microsprinklers. They do not generally hinder harvest, since nuts are shaken into catch frames, unlike almonds, which are harvested off the ground. The primary strategies for weed management during this period should be aimed at preventing additional weed seed production, preventing weed resistance, and fine-tuning strategies to reduce management costs.

Weed Seed Prevention

Weeds should be destroyed before they produce seed, a practice that will help deplete soil seed reserves. Table 11.10 shows the seed production of some commonly found weeds. Allowing a few weeds to set seed can significantly increase the seed reserves in the soil for years. Common purslane *(Portulaca oleracea)* seed, for example, can remain viable in soil for more than 20 years. Killing young weeds also conserves soil moisture, important where water is limited. A clean soil surface also reduces the risk of frost damage. Winter weeds also serve as a good overwintering site for destructive insects (e.g., lambsquarters for lygus bug, black mustard for green soldier bug, London rocket for false chinch bug), so they must be controlled.

Herbicide-Resistant Weeds and Management

The development of herbicide-resistant weeds is not a new phenomenon. It is a naturally occurring evolutionary process in response to a selection pressure (i.e., herbicides). Currently, 393 different herbicide-resistant weed biotypes have developed globally, of which 23 have been identified in California, including hairy fleabane and horseweed (table 11.11). Other important weeds that have developed resistance to herbicides elsewhere (but are common to pistachio orchards in California) include lambsquarters, redroot pigweed *(Amaranthus retroflexus)*, and prickly lettuce *(Lactuca serriola)*.

Herbicide resistance is defined by the Weed Science Society of America (WSSA) as "the inherited ability of a plant to survive and reproduce following exposure to a dose of herbicide normally lethal to the wild type." In other words, the weed was once controlled by the herbicide at a labeled rate but is no longer controlled. This should not be confused with herbicide tolerance, which is defined as "the inherent ability of a plant to withstand the specific effects of a particular herbicide" (WSSA). It is also important to note that herbicides do not cause mutations that lead to resistance. Rather, they select for biotypes of the plant (or weed) that already have genes that confer insensitivity to the herbicide.

The three primary factors that lead to or accelerate the development of herbicide resistance are herbicide chemical properties, weed characteristics, and cultural practices used. The chemical properties of some herbicides can lead to more rapid development of resistance than those of other herbicides. Herbicides with a single site of action are more prone to developing resistance among weeds than herbicides with multiple paths of activity. An example in pistachios would be rimsulfuron (an ALS-inhibiting herbicide). In addition, herbicides that are detoxified by a common metabolic pathway in weeds are more likely to develop resistance among weed species.

Characteristics in weed growth also contribute to resistance. For example, the top 10 most common weeds in the world that have developed herbicide resistance have an annual growth habit, are widely distributed, and proliferate in a variety of climatic and environmental conditions. This suggests that genetically diverse species are most likely to contain a gene that confers resistance to a particular herbicide. For example, glyphosate-resistant horseweed (fig. 11.10) is an annual species, produces at least 250,000 seed per plant, is wind disseminated, and grows under a variety of habitats (including wet or dry soils, deep or shallow soils, gravelly roadways, and shade or full sunlight).

Cultural practices that increase the selection pressure on a weed population increase the likelihood of resistance. For example, relying solely on herbicides for weed control increases the risk of resistance. Also, repeated use of the same herbicide active ingredient (like glyphosate) or products with the same mode of activity greatly increases the risk of developing weed resistance. Once a weed develops resistance to glyphosate, it will be resistant to all products containing the same active ingredient (found in products like Roundup, Glyphomax, and numerous others). Furthermore, the likelihood of developing resistance is increased by using increasingly higher rates of the same herbicide over time.

It is important to recognize resistant weed populations early so that strategies can be implemented to stop them. In most cases, growers observe about 30 percent failure in the control of a particular species before suspecting resistance. Herbicide-resistant weeds are not always easy to identify. They occur naturally within a given population but cannot be

Table 11.10. Weeds and their seed production

Common name	Species	Seed per plant
horseweed	*Conyza canadensis*	250,000
large crabgrass	*Digitaria sanguinalis*	150,000
redroot pigweed	*Amaranthus retroflexus*	117,400
common purslane	*Portulaca oleracea*	52,300
barnyardgrass	*Echinochloa crus-galli*	40,000
hairy fleabane	*Conyza bonariensis*	40,000
puncturevine	*Tribulus terrestris*	10,000
wild oat	*Avena fatua*	250

Table 11.11. Herbicide-resistant weeds in California

Weed	Situation	Herbicide mode of action	Year of resistance
common groundsel *(Senecio vulgaris)*	asparagus	PS II inhibitors	1981
perennial ryegrass *(Lolium perenne)*	railways, roadsides	ALS inhibitors	1989
smallflower umbrella sedge *(Cyperus difformis)*	rice	ALS inhibitors	1993
California arrowhead *(Sagittaria montevidensis)*	rice	ALS inhibitors	1993
Russian thistle *(Salsola iberica)*	roadsides	ALS inhibitors	1994
wild oat *(Avena fatua)*	barley, wheat	pyrazoliums	1996
redstem *(Ammania auriculata)*	rice	ALS inhibitors	1997
ricefield bulrush *(Scirpus mucronatus)*	rice	ALS inhibitors	1997
late watergrass *(Echinochloa phyllopogon)*	rice	ACCase inhibitors	1998
late watergrass *(Echinochloa phyllopogon)*	rice	thiocarbamates	1998
rigid ryegrass *(Lolium rigidum)*	almonds	glycines	1998
long-leaved loosestrife *(Ammania coccinea)*	rice	ALS inhibitors	2000
barnyardgrass *(Echinochloa crus-galli)*	rice	thiocarbamates	2000
barnyardgrass *(Echinochloa crus-galli)*	rice	ACCase inhibitors	2000
early watergrass *(Echinochloa oryzicola)*	rice	ACCase inhibitors	2000
early watergrass *(Echinochloa oryzicola)*	rice	thiocarbamates	2000
littleseed canarygrass *(Phalaris minor)*	onions	synthetic auxins	2001
smooth crabgrass *(Digitaria ischaemum)*	rice	synthetic auxins	2002
horseweed *(Conyza canadensis)*	roadsides, trees/vines	glycines	2005
hairy fleabane *(Conyza bonariensis)*	roadsides, trees/vines	glycines	2007
hairy fleabane *(Conyza bonariensis)*	roadsides, trees/vines	glycines and bipyridiliums	2009
junglerice *(Echinochloa colona)*	corn, roadsides, trees/vines	glycines	2011
Palmer amaranth *(Amaranthus palmeri)*	alfalfa, corn, cotton, roadsides, trees/vines	glycines	2012

visually distinguished from susceptible biotypes until the population increases over time. Resistant weeds generally begin to appear in clumps or patches, unless their seed is windborne, which makes this appearance unlikely. Thus, scouting for the initial evolution of resistance patterns of weed escapes is a critical aspect of management and prevention.

Rarely does weed resistance develop because of a complete herbicide failure from a single application. Even isolated escapes are seldom associated with resistant weeds, although it is certainly possible. Weed escapes appearing in single-line patterns in a field are usually associated with a plugged spray nozzle or some other application error. Other causes for weeds escaping herbicide treatments include misidentification of the weeds, weed tolerance to the herbicides and the rates used, weed emergence following application, degradation of preemergent

Figure 11.10. Herbicide-resistant horseweed (bottom) and susceptible horseweed (top) exposed to four times the label rate of glyphosate. *Photo:* K. J. Hembree.

herbicides, and application at an insensitive stage of weed development (treated too late).

As soon as weed resistance is suspected, alternative management efforts need to be employed in order to achieve effective weed control and preserve the herbicide's effectiveness. Preventing weed resistance from occurring is best achieved when an integrated weed management approach is used. Killing weeds before they produce seed has the best long-term effect. To prevent and reduce the level of resistance, do not use the same herbicide (or herbicides with a similar mode of action) year after year. Rather, rotate or tank-mix herbicides with differing modes of action. Mechanical cultivators give excellent control of these weeds. Finally, rotating chemical and mechanical options whenever possible reduces the risk of developing weed resistance. It is also important that field edges be maintained clean to help prevent the introduction of new weed seed into the orchard.

Mechanical and Chemical Strategies

Weed management with mechanical and chemical options during the bearing years is not too different from that of the nonbearing years, although there are a few more herbicides to select from. During the bearing years, most growers continue to apply pre- and postemergent strip sprays and mechanically control weeds between the tree rows. Some growers, however, elect to use shielded applications of glyphosate or other postemergent materials to control weeds in the row middles.

As discussed earlier, cultivation is usually practiced with a disk and is limited to the row middles. Annual weeds should be cultivated at the seedling stage of development, and repeated applications need to be made as emergence occurs. For perennial species, like field bindweed and bermudagrass, cultivate seedling plants before they are 3 weeks old or before they begin forming perennial reproductive structures. Cultivating for nutsedge control is not generally recommended, since this can increase the spread of underground nutlets through the field. Equipment must be set properly and monitored closely during operation to prevent damage to the trees. Mechanical tools should not be used during wet conditions, as this promotes soil compaction and erosion problems.

Weeds growing in the row middles can also be mowed close to the soil surface to suppress weeds, which also allows for access of equipment during wet conditions, and this can be done repeatedly. Interseeding with low-growing cover crops is also an option, as discussed earlier.

Several herbicides are available that can be applied alone or tank-mixed for control of most annual weed infestations in bearing pistachio orchards (see tables 11.5 to 11.7). Once the population of weeds is reduced, preemergent materials can be used every 2 to 3 years to keep the weeds at a manageable level. Periodic, timely postemergent treatments are needed to maintain control and prevent new seed production. In the case of most annual weeds, four or five treatments are usually required. Field accessibility and ability to treat when weeds are small is important for this to be effective, as large weeds are not easily controlled. Consider weed species present, soil type, irrigation method, tree age, local restrictions, and other factors when making weed management decisions.

Over time, a change (shift) in the types of weeds present may be observed. These weeds may be annuals that are tolerant to the herbicides being used or perennials that flourish without annual weed competition. Refer to notes obtained during field scouting to help adjust herbicide selection before annual and perennial weed species shifts occur and become difficult to eradicate. Rotating herbicides or using combinations of herbicides helps eliminate this problem and reduces the likelihood of developing weed resistance. Preemergent herbicide rates should be adjusted to the lowest rate that provides effective control. Always adjust the herbicide rate to match the soil with the coarsest texture if there is soil variability within the field.

If perennial weeds become established, control options are generally limited to postemergent materials. The stage of growth of perennials at the time of treatment is important to consider. For example, to control established johnsongrass, postemergent herbicides (like glyphosate or sethoxydim) should be applied when regrowth is 8 to 12 inches long. Follow-up treatments are required to deplete the food reserves and kill the weed. If perennial weeds occur in patches, spot treating can be effective and reduce the amount of herbicide needed, compared with strip treatment. Nutsedge can be effectively controlled with postemergent herbicides when plants have four leaves or less, after which new underground nutlets, or tubers, are formed. For nutsedge, continue to spray as new flushes of nutsedge emerge (about every 28 days) during the season. It often takes diligence and 2 to 3 years of properly timed sprays to eliminate this weed. If annual weeds are present and appear patchy throughout the field, consider using smart sprayer technology, such as the WeedSeeker sprayer or other similar equipment. This cuts herbicide costs by 50 percent or more in most cases.

Specific herbicide properties (like solubility, volatility, half-life, etc.) can influence the performance of herbicides. For example, under certain conditions the EC formulation of oxyfluorfen can volatilize off wet soil, move upward in the air, and damage green tissue in the tree canopy. To resolve this issue, consider using the flowable (F) formulation of oxyfluorfen, which tends to volatilize less. The soil activity is similar for both formulations when used at the same active ingredient rate, or active ingredient per acre (ai/ac). However, the postemergent activity of the F formulation is about one third less than that of the EC formulation, so rates may need to be increased for controlling emerged weeds. Factors like these and others must be considered before selecting herbicides.

Summary

Effective weed control can be achieved in California pistachio orchards with both chemical and nonchemical measures. Such measures can help resolve weed control issues, based on individual needs. When herbicides are used, however, it cannot be overemphasized that proper weed identification and accurate application is imperative for cost-effective control. By having knowledge of the weed infestation and becoming familiar with the available weed control options, growers can produce pistachios relatively free of weeds in a manner that is economically feasible.

References

Ball, D. A., D. W. Cudney, S. A. Dewey, C. L. Elmore, et al, eds. 2003. Weeds of the West. 9th ed. Laramie: Western Society of Weed Science, in cooperation with the Western United States Land Grant Universities Cooperative Extension Services and the University of Wyoming.

DiTomaso, J. M., and E. A. Healey. 2007. Weeds of California and other western states. Oakland: University of California Division of Agriculture and Natural Resources Publication 3488.

Elmore, C. 1997. Soil solarization: A nonpesticidal method for controlling diseases, nematodes, and weeds. Oakland: University of California Division of Agriculture and Natural Resources Publication 21377.

Fischer, W. 1996. Grower's weed identification handbook. Oakland: University of California Division of Agriculture and Natural Resources Publication 4030.

Heap, I. 2010. International survey of herbicide-resistant weeds. Weed Science website, http://www.weedscience.org.

Ingels, C., R. Bugg, G. McGourty, and P. Christensen. 1998. Cover cropping in vineyards: A grower's handbook. Oakland: University of California Division of Agriculture and Natural Resources Publication 3338.

Nasso, V., et al. 1983–2009. Crop data management systems. Crop Data Management Systems, Inc. website, http://www.cdms.net.

Shrestha, A., B. Hanson, and K. Hembree. 2008. Glyphosate-resistant hairy fleabane documented in the Central Valley. California Agriculture 62(3): 116–119.

Shrestha, A., K. Hembree, and N. Va. 2007. Growth stage influences level of resistance in glyphosate-resistant horseweed. California Agriculture 61(2): 67–70.

Shrestha, A., K. Hembree, and S. Wright. 2008. Biology and management of horseweed and hairy fleabane in California. Oakland: University of California Division of Agriculture and Natural Resources Publication 8314.

UC SAREP (Sustainable Agriculture Research and Education Program). Cover Crop Resources. UC SAREP website, http://www.sarep.ucdavis.edu/covercrop/res.

UC Statewide Integrated Pest Management Program, UC IPM website, http://ucipm.ucanr.edu.

12 Pruning Mature, Bearing Trees

Robert H. Beede and Louise Ferguson

Three factors must be considered when developing a pruning program for mature pistachio trees. First, pistachios are alternate-bearing trees. A large crop, borne on 1-year-old wood, reduces the fruiting potential for the following year by triggering the abscission of flower buds produced at the base of the current season's growth. Pruning has been shown to mitigate this phenomenon by regulating crop load and increasing the ratio of nonbearing to bearing shoots in the following years. (For further information on alternate bearing, see chapter 3, "The Pistachio Tree: Physiology and Botany," and chapter 28, "Alternate Bearing, Nut Blanking, and Shell Splitting.")

Second, pistachios express strong apical dominance in their growth habit. Explained in detail within this chapter, this characteristic becomes more pronounced with tree age, exacerbating the problem of bearing the crop largely on the canopy periphery. This results in structural branches becoming more horizontal, increasingly shaded, and less fruitful than upright limbs, which have better light exposure.

Third, the current orchard design of 16 to 17 feet within the row and 18 to 20 feet between rows strives to achieve early bearing without excessive crowding at orchard maturity. Failure to limit trees to their allotted space reduces crop capture by existing harvest equipment and further reduces orchard fruitfulness because of shading (figs. 12.1 and 12.2).

Proper pruning of mature trees seeks to minimize the effects of these factors and maintain the canopy at its optimal yield potential. The productivity benefits from proper annual pruning are optimized only when complemented by proper irrigation, fertilization, and pest control. This chapter discusses the physiological principles and specific considerations for pruning pistachios as they relate to growth and fruiting habits. It also describes methods for implementing mechanical pruning.

Figure 12.1. Unpruned mature pistachio. Its size no longer allows commercial harvest, and the majority of the crop is borne on the canopy periphery. *Photo:* R. H. Beede.

Figure 12.2. A properly managed, mature pistachio canopy has fruitwood uniformly distributed throughout the canopy, thus making it sufficiently stiff for efficient mechanical harvesting. *Photo:* R. H. Beede.

The Growth Habit of Pistachio

Pistachios express strong apical dominance in their vegetative growth habit, meaning that vigorous current-season shoots develop from the terminal bud of the previous season's growth with little or no lateral branch development. Vegetative buds lower on the branch are capable of producing long, vigorous shoots, but usually they develop only short spur growth due to the hormonal suppression exerted by the terminal bud. This characteristic becomes more pronounced with tree age. Thus, branches may elongate for years without substantial lateral branching, causing the tree to bear fruit farther and farther from its central axis. Early pistachio pruning studies indicated that apical dominance could be overcome in older trees by cutting the entire canopy back into 3- and 4-year-old wood (Crane and Iwakiri 1980) This treatment produced new shoots on 14- to 17-year-old wood. Another study used light and severe pruning cuts to overcome apical dominance, but reinvigoration of older wood was not observed (Wolpert 1986).

The absence of copious lateral branching also slows development of pistachio branch girth, due to less photosynthetic area. This increases the tendency for main fruiting limbs to become less upright with each successive crop. In almond, a tree much less affected by apical dominance, vigorous upright lateral shoots (water sprouts) develop where the main limb bends downward from crop load. Providing there is adequate light, these shoots serve as natural replacements for main bearing limbs. In contrast, apical dominance in pistachio promotes the loss of upright tree shape by the absence of vigorous water sprouts on the main bearing branches. It also limits the growth and leaf area of fruiting spurs in the lower canopy to the degree that they eventually die or fail to develop flower buds. Less growth from flat limbs on the sides of the canopy also promotes added growth on upright shoots in the top of the tree, where light is abundant. Although highly fruitful, the upper canopy eventually grows beyond the height manageable by hand-pruning crews, and it may be difficult to remove and catch the nuts with trunk-shaking harvesters. Therefore, with bearing pistachios, the major objectives of pruning are to confine the trees to their allotted space, facilitate mechanical harvesting, and promote light penetration for viable fruitwood and nut production throughout the tree.

Pistachios, like other deciduous trees, have latent vegetative buds. Latent buds are subterminal buds that remain undeveloped for a season or longer. However, they remain viable in older wood and replace broken or older bearing branches, as they grow with characteristically high vigor (fig. 12.3).

Recent studies of pistachio growth and development (Spann et al. 2009) have provided greater understanding of the biology of short spurs versus long, whiplike shoots. Both originate from vegetative meristems that contain a given number of node positions differentiated during winter dormancy. These preformed nodes range in number from seven to nine. Shoots produced from lateral buds behind the terminal meristematic bud usually grow only until their predetermined nodes have developed. Their terminal meristem then "hardens off" rather than differentiating new shoot tissue for additional current-season growth. The resulting short growth is a preformed shoot. Preformed shoots usually become highly fruitful spurs the following season (fig. 12.4).

Figure 12.3. A highly vigorous neoformed shoot originating from a latent bud forced into growth by severe pruning. (left) *Photo:* R. H. Beede.

Figure 12.4. A nine-node preformed shoot commonly referred to as a spur. Note that flower buds are differentiated at every node. The dead branch resulted from not having left a vegetative bud above the fruit buds the previous season. *Photo:* R. H. Beede.

The long, whiplike growth produced from the terminal buds on 1-year-old wood also originates as preformed shoots. However, when these shoots are in an upright and well-lighted position, apical dominance enhances their growth rate. Consequently, they experience two and sometimes three flushes of growth in one season. The first flush, occurring in early spring, is preformed growth. The subsequent flushes originate from tissue developed in-season by the terminal meristem. The final shoot length depends on tree vigor and the growing conditions. Shoots with this growth characteristic are called neoformed. Dormant heading cuts of vigorous 1-year-old shoots increase the likelihood of more neoformed growth, since pruning reduces the number of vegetative buds and provides more resources for those remaining (fig. 12.5).

Figure 12.5. Unpruned (A) and pruned (B) branches on a young pistachio tree. Note that the unpruned branch has drooped below the desired upright position, and the lateral nodes have only produced preformed spurs. This contrasts with the pruned branch, which produced two upright neoformed shoots without lateral spurs but, it is still useful for canopy development. *Photo:* R. H. Beede.

The Fruiting Habit of Pistachio

Pistachios bear their crop laterally on 1-year-old wood. An individual inflorescence bud produces as many as 200 individual flowers on a structure known as a panicle. During bloom, the panicle resembles a 1-to-2-inch-long pine cone in shape, with the individual flowers surrounding its periphery. Once pollinated and fully developed, the panicle is similar to a grape cluster, consisting of an elongated central axis with lateral branches, laden with fruit. Only about 10 percent of the potential flowers set. Growers refer to the structure supporting the developing fruit as the rachis, and the term *cluster* applies collectively to the rachis and fruit.

Dormant 1-year-old fruitwood is categorized as either short spur-type growth or long whiplike growth. Typical for all spur-bearing fruit trees, spur growth is responsible for most of the fruit production. The long neoformed growth is essential for canopy development, and it is dominant in young trees. Neoformed growth is also found in the upper canopy of older trees, but it represents only a small percentage of the total seasonal growth. Neoformed growth increases in older trees with pruning severity. Although fruitful on the basal seven to nine nodes, neoformed growth in older trees is used primarily for replacing older branches that have become excessively flat from repeated cropping. Often more spur development occurs along the length of neoformed growth in young trees, where apical dominance is less pronounced. However, greater apical dominance in older trees usually limits spur growth to the buds closest to the terminal. This necessitates dormant heading of the neoformed growth to approximately 24 inches to maintain the compressed fruiting canopy that facilitates efficient mechanical harvesting (figs. 12.6 and 12.7).

Pruning Effects on Growth and Yield

Pruning of mature, deciduous trees has several objectives. One is to confine the tree to its allotted space. This, for reasons already described, is of major importance in pistachio. Secondly pruning is performed to renew fruitwood and distribute light throughout the canopy. The invigorating effect of pruning also helps maintain tree health and longevity. Pruning can also improve nut removal at harvest and reduce equipment damage from low branches. A 7-year trial involving severe mechanical hedging and topping, combined with normal hand-pruning, showed that cumulative split nut yield for the severely pruned trees was not different from the hand-pruned control. Further, severe hedging and topping, followed by normal hand-pruning in subsequent years, strongly and persistently mitigated alternate bearing (Ferguson et al. 1991, 1995).

As with all deciduous fruit tree species, pruning has a dwarfing effect on total tree growth. Unpruned trees have a greater total leaf area and overall biomass than pruned trees. Pruned trees appear more vigorous, as relatively more nitrogen and carbohydrates are available to fewer growing points, but they still produce less total growth than an unpruned tree (despite the length of their rapidly growing shoots).

Dormant pruning removes stored carbohydrates in the branches while simultaneously removing sites

Figure 12.6. Neoformed branches in the upper canopy left untipped the previous winter. Most of the preformed spurs are near the branch terminal. Insufficient branch girth prevents efficient fruit removal due to insufficient energy transfer from the shaker. *Photo:* R. H. Beede.

Figure 12.7. Abundant neoformed branches created in the upper canopy from winter topping. These must be cut back to 24 inches to control tree height and allow harvest of the fruit borne on their basal buds. *Photo:* R. H. Beede.

for new leaves capable of carbohydrate production. Reduction in leaf number also results in less root growth. Summer pruning is especially dwarfing. It removes shoots on which the tree has already expended resources producing, and it removes leaves capable of carbohydrate production.

Heading and Thinning Cuts

Deciduous fruit and nut trees are pruned using heading and thinning cuts. The different response to both types of cuts can be used effectively to manipulate pistachio tree growth.

Heading cuts remove the terminal ends of shoots and branches. They are extensively used in tree training to direct lateral branching and maintain high vigor for canopy development (figs. 12.8 and 12.9). Heading cuts are more invigorating than thinning cuts because they remove more buds per unit length as internodal distance is less near the branch terminal. Also, fewer stored carbohydrates are removed due to the taper of the branch. Heading cuts are performed on mature trees to force lateral branching in 3- to 5-year-old wood, and the cuts are also done to reduce the length of long 3-year-old shoots in the top of the canopy. Heading of 1-year-old shoots is commonly referred to as tipping (fig. 12.10). Heading cuts need to be distributed over the canopy for uniform invigoration, since the pruning effect is localized. The resulting neoformed growth (whips) requires tipping the following winter for the purposes of shortening the limb and inducing lateral branching in the desired location. It is important to leave at least one vegetative bud above the flower buds produced near the base of the shoot. Failure to do so results in the branch dying back to the next growing lateral after fruiting (see fig. 12.4). Heading cuts can also be used to control neoformed growth and increase fruiting by tipping whips immediately

Figure 12.8. A heading cut removes a portion of a branch and stimulates vigorous shoot growth close to the pruning cut. *Photo:* R. H. Beede.

Figure 12.9. Heading cuts in the upper canopy reduce branch length and stimulate new growth. Note also how this cut oriented the branch into a more upright position. *Photo:* R. H. Beede.

Figure 12.10. Tipping 1-year-old neoformed growth in the upper canopy. Tipping becomes difficult in canopies taller than 16 feet. *Photo:* R. H. Beede.

Figure 12.11. One-year-old branch with multiple vegetative buds where the shoot terminal temporarily stopped growth before initiating another flush. Avoid pruning directly above this zone on branches needed for canopy expansion. *Photo:* R. H. Beede.

Figure 12.12. Pruning directly above the zone of vegetative bud compression results in multiple spur growth in the form of wagon wheel spokes. This prevents the branch from growing vigorously and serving as a structural limb. This cut is purposely performed on flat branches to reduce further terminal growth and enhance fruitfulness. *Photo:* R. H. Beede.

above a growth flush (fig. 12.11). This causes the numerous vegetative buds, which typically differentiate at these points, to produce preformed shoots (spurs) that assume the appearance of wagon wheel spokes (fig. 12.12). This becomes a highly fruitful unit the following season and prevents further terminal growth unless it is reheaded below the spur whorl. These cuts should not be performed on structural branches.

Thinning cuts remove entire branches where they originate (fig. 12.13). They are less invigorating because, in general, they remove fewer buds but more carbohydrates than heading cuts. Thinning cuts are used to reduce the density of branches within the canopy and thus improve light and potential fruitfulness for the remaining branches (fig. 12.14). Thinning cuts are also used to manage the outside of the canopy by removing branches that have become flat or that extend beyond the harvest catch frame. Large selective thinning cuts are a good alternative to mechanical topping for reducing canopy height. Looking up from the interior

Figure 12.13. Thinning cuts remove an entire branch at its point of origin. They are intended to reduce limb number and enhance light penetration within the canopy. *Photo:* R. H. Beede.

Figure 12.14. The result of the thinning cut in figure 12.13 is reduced weight and better light for the remaining upper branch. *Photo:* R. H. Beede.

of a canopy, the branch producing the most shading within the lower canopy can be removed at its insertion. Generally not more than one large canopy-lowering thinning cut per tree should be done per dormant-season pruning.

Hand- versus Machine Pruning

Pistachio growers have the option of using hand crews or mechanical hedging and topping equipment, or a combination of both for pruning. Each method has advantages and disadvantages. Designing the optimal pruning program should consider both the sustainability of the tree canopy's fruit-bearing capacity and net return.

Hand-Pruning

Hand-pruning allows each tree canopy to be modified individually. Heading and thinning cuts can be applied as needed to optimize light interception, stimulate new growth in weak areas, and control tree size.

Yield data collected in one pruning trial in western Kings County (Beede et al. 2002) showed that properly hand-pruned pistachio trees produced more pounds of split nuts over a 6-year period than trees that were single- or double-side hedged by machine. The difference between hand-pruned trees and those single-side hedged annually was 1,348 split nut pounds over 6 years. Double-sided hedging every other year differed significantly from hand-pruning, by 1,923 pounds of split nuts. However, considering that hand-pruning large, mature trees can easily cost $400 per acre compared with $50 per acre for mechanical pruning, the complications and costs associated with hand-pruning a large acreage may not be justifiable.

Hand-pruning crews or their individual members vary in skill from novice to expert. Therefore, it is essential that the grower has a clear understanding of what type of pruning is desired and then is capable of communicating the objectives clearly and simply. One method of accomplishing this goal is to divide the instructions into four steps. The first is to remove branches that are broken, too low, crossing over the center, or overlapping one another. These are the relatively obvious cuts that can be performed by less skilled crew members. Second, thinning cuts must be made to the outside of the canopy with the objective of pushing the bearing branches upright. This is accomplished by thinning off branches that are at less than 45 degrees from horizontal. Branches must be cut back sufficiently so that the remaining ones are pointing upward and not outward. Few thinning cuts are needed in the center of pistachios, since apical dominance causes most of the new growth to occur on the canopy periphery. The third step is to perform selected heading cuts in areas closer to the tree axis that are weak or too long to effectively shake. These will hopefully stimulate latent buds to develop into new shoots that can be used to renew fruitwood. The final step is to use pole pruners to tip the 1-year-old whips produced primarily in the top of the canopy. Shorter growth can also be tipped elsewhere with the goal of creating as much new spur growth as possible. It is essential that the tipping crew know the difference between flower and vegetative buds so that at least one of the latter is retained on the tipped 1-year-old wood. Periodic review of the crew's work with the foreman avoids disappointment and arguments later.

Mechanical Pruning

Mechanical pruning is fast and inexpensive, and large acreages can be pruned with a single machine. Mechanical pruning uses equipment with a gang of rotating circular saws set at predetermined heights, angles, or widths (fig. 12.15). All wood in the machine's path is removed by indiscriminate heading cuts (fig. 12.16). Depending on the severity of the cuts and the orchard vigor, the shoot growth produced may require hand-pruning the following season (fig. 12.17). Generally, mechanical pruning needs complementary hand-thinning and -tipping cuts to increase light interception in the lower canopy.

Side hedging

As with hand-pruning, the objectives of mechanical pruning must be clearly understood to avoid surprises. Research conducted over 4 years in two locations prior to both on- and off-years showed no difference in yield between moderate single- and double-sided hedging (cuts made into 1- and 2-year-old wood) and hand-pruned controls (Ferguson et al. 1991). Unlike in an earlier trial employing severe machine pruning (Ferguson et al. 1991, 1995), moderate side hedging had no effect on mitigating alternate bearing. The absence of yield differences between hand-pruning and machine side hedging was supported by additional research showing that pistachio trees compensate for the flower buds removed during pruning by increasing the number of nuts set per cluster (Beede et al. 1991). Pruned trees also had greater nut size than unpruned trees.

A 6-year mechanical pruning trial conducted in a dense and vigorous 15-year-old Kerman pistachio

Figure 12.15. A double-boom, stationary-blade mechanical pruner performing side hedging. Ground conditions must be smooth to avoid large swings in cut severity. *Photo:* R. H. Beede.

Figure 12.16. The size and number of indiscriminate heading cuts performed by side hedging depend on the severity of canopy overgrowth in the row middles and the distance of the cutting plane from the tree axis. *Photo:* R. H. Beede.

Figure 12.17. Growth resulting from mechanical side hedging can be very vigorous and usually requires hand-tipping and -thinning the following winter. *Photo:* R. H. Beede.

orchard on *P. integerrima* rootstock suggested that there is less yield fluctuation by hedging one side every year compared with hedging both sides every other year. This fluctuation also resulted in significantly fewer split nuts from the mechanically pruned trees (11.7 %) than the hand-pruned trees over the 6-year trial period. Cumulative yield data also suggested that the fluctuation might lessen with repeated side hedging. This observation may be associated with the larger amount of growth removed when hedging was initiated before the 1997 growing season. Due to the density of the test orchard, the large amount of growth removed may have been more than the tree could compensate for (Beede et al. 2002).

The yield difference between moderate single- and double-sided hedging in any given year is also affected by the potential crop. Experiments indicate that double-sided hedging should not be performed prior to a low-production year. Support for this recommendation was found in four of the six seasons studied. In 1999, when split nut production was only 1,900 pounds per acre in the hand-pruned trees, double-sided hedging significantly reduced split nut yield by 316 pounds (19 %) per acre compared with single-sided hedging. In 2000, split nut production in the hand-pruned trees was above 4,000 pounds, and no significant yield difference was recorded between single- and double-sided hedging. In fact, the double-sided hedge treatment (pruned the previous year) produced significantly more split nuts (363 lb) in 2000 than the trees hedged annually on one side. In 2001, the second off-year, double-sided hedging reduced split nut yield by only 227 pounds relative to single-sided hedging. Finally, in 2002, when split nut yield was 5,308 pounds, no statistical differences were recorded between trees hedged on one or two sides (Beede et al. 2002).

Mechanical topping

It is still unclear whether mature pistachio canopy heights can be managed by repeated mechanical topping at the same height without suffering significant yield loss (Beede et al. 2002). However, the use of topping to replace manual tipping in young trees appears to be an acceptable, cost-saving practice that has already been implemented by growers (fig. 12.18). Repeated mechanical topping of 1-year-old growth by 50 percent allows additional gains in tree height that eventually cause shading out of the lower canopy. A significant portion of the crop in the uppermost canopy can also be flung beyond the harvester catch frames during shaking. Correction

Figure 12.18. Mechanical topping is sometimes practiced in younger orchards to reduce hand-labor costs. *Photo:* R. H. Beede.

requires a remedial topping to reduce tree height to 14 to 15 feet (fig. 12.19). Data collected thus far suggests that the yield loss from topping is correlated to the potential crop rather than physiological conditions associated with on-and-off-year bearing cycles. Results also suggest that topping for tree size control prior to an off-year promotes low yield the following season and accentuates alternate bearing.

Experimentation with in-season retopping to control vigorous growth appears promising (Beede et al. 2002). Trees retopped in early August grew very little, and nut quality was not adversely affected that year.

The mechanical pruning experiments conducted thus far strongly suggest that side hedging should be implemented before crowding occurs so that yield loss the first year can be minimized. Growers also must ensure that the orchard floor is smooth and free from depressions. Small holes within the orchard floor can cause saw blades to dip deeply into the canopy. Growers wishing to perform tipping of 1-year-old wood can easily do

Figure 12.19. Topping of mature, tall trees requires the removal of 4 or more feet of growth in order to return the canopy height to a level that can be managed from the ground. Insufficient topping severity also minimizes improved light penetration within the canopy. *Photo:* R. H. Beede.

Figure 12.20. A three-blade, rotating cutter head is commonly used by orchardists performing their own mechanical pruning. *Photo:* R. H. Beede.

so with a three-bladed, rotating star machine that can be mounted on a large tractor (fig. 12.20). The five-bladed, rotating star machine makes cleaner cuts than the three-bladed machine when performing moderate side hedging and topping. Travel speed must be monitored to prevent the rotating head from breaking large structural branches before they are cut (fig. 12.21). Dull saw blades aggravate this problem. When performing orchard reinvigoration, where cuts are often 2 inches or greater in diameter, breakage appears to be much less with a hedger employing a stationary boom with five or six saws mounted on it. This provides a cutting plane that is much less likely to cause unwanted branch breakage.

Figure 12.21. Saw blade sharpness and speed of travel are important factors in minimizing limb splitting from hedgers using a rotating head rather than a stationary boom design. *Photo:* R. H. Beede.

In summary, mature pistachio trees grow relatively slowly and produce unbranched shoots due to their apically dominant growth habit. Annual pruning is therefore necessary to maintain the canopy of each tree within its allotted space, using heading cuts to shorten the fruitwood and keep it upright. Thinning cuts are used to provide adequate light distribution throughout the canopy and reduce lower limb loss from shading. Tipping of the long 1-year-old shoots in the uppermost part of the tree is necessary for both height control and maintenance of a compressed fruiting area for efficient mechanical shaking. Hand-pruning has the potential of maximizing production, provided that it is performed by skilled individuals. However, the cost and time required may not be compensated by the yield increase compared with properly performed mechanical pruning. Research shows that pistachios compensate for moderate side hedging by increasing the fruit set and nut size within the remaining inflorescences. Potential crop load also influences the yield lost from the multiple, indiscriminate heading cuts resulting from hedging one or both sides. These factors have made mechanical pruning a routine practice in many commercial orchards. Mechanical topping can substitute for hand-tipping crews in young orchards, provided that the trees are relatively uniform in height. Topping of mature trees must be approached with caution, since current research shows its effect on yield reduction is correlated with the severity of the topping. Aggressive hedging and topping can both be employed for canopy reconstruction of older, mismanaged orchards or orchards suffering from severe alternate bearing. Complete yield recovery under these circumstances may require 3 or more years.

References

Beede, R. H., L. Ferguson, K. Munger, D. Munger, and D. Rose. 2002. Pistachio canopy management by moderate mechanical hedging and topping: Sixth year progress report. In California pistachio industry annual report, 109–112. Fresno: California Pistachio Commission.

Beede, R. H., L. Ferguson, D. Rose, and J. Padilla. 1991. Effect of pruning on the yield components of pistachio. In California pistachio industry annual report, 90–92. Fresno: California Pistachio Commission.

Beede, R. H., D. Rose, J. Padilla, and D. Thomas. 1992. Growth responses from on-year pruning and its effect on pistachio yield and nut quality for the following off-year. In California pistachio industry annual report 1992, 127–131. Fresno: California Pistachio Commission.

Crane, J., and B. T. Iwakiri. 1980. Preliminary notes on pruning mature pistachio trees. In California pistachio industry annual report 1980, 118–119. Fresno: California Pistachio Commission.

Ferguson, L., J. Maranto, and R. H. Beede. 1995. Mechanical pruning mitigates alternate bearing of 'Kerman' pistachios (*Pistacia vera* L.). HortScience 30(7): 1369–1372.

Ferguson, L., J. Maranto, and B. Blackwell. 1991. Mechanical pruning of pistachios: Final progress report. In California pistachio industry annual report 1991, 140–143. Fresno: California Pistachio Commission.

Ferguson, L., J. Maranto, R. H. Beede, S. Gaona, D. Castle, A. Garza, and R. Fanucchi. 1991. Mechanical pruning of pistachios: Final progress report. In California pistachio industry annual report, 105–107. Fresno: California Pistachio Commission.

Spann, T. M., R. H. Beede, and T. M. DeJong. 2009. Contributions of short- and long-shoots to yield of 'Kerman' pistachio (*Pistacia vera* L.). Scientia Horticulturae 121:495–500.

Wolpert, J. 1986. Nitrogen and potassium nutrition and preliminary results on pruning of 13-year-old trees. In California pistachio industry annual report, 80–85. Fresno: California Pistachio Commission.

13 Irrigation Management

David A. Goldhamer and Robert H. Beede

The goal of irrigation scheduling is to supply the proper amount of water to the orchard on a timely basis. Good irrigation management ensures that an adequate supply of soil moisture is maintained throughout the season. Informed irrigation decision making requires knowledge of the water use requirements of the trees and the irrigation system performance, including the uniformity of application and delivery rates. Information on how productivity is affected by suboptimal irrigation practices also can be used to guide irrigation decision making. Most of the U.S. pistachio production is in California, and virtually all orchards are irrigated. This differs markedly from the rest of the world, where much of the pistachio production is under dryland conditions. A huge difference in productivity exists between irrigated and dryland trees. For example, the average dryland yield in Turkey is about 3 pounds per tree compared with 37 to 40 pounds per tree in California (Kanber et al. 1993). Much of the irrigation research during the last 30 years has taken place in California, including work on tree water use, response to water stress, regulated deficit irrigation, and drought irrigation strategies. California pistachio growers have adopted much of this research due, in part, to the fact that a high percentage of the acreage is located in areas where water cost is high or availability is low.

Pistachio trees have a reputation of being drought tolerant, that is, able to survive and even produce modest crops with very little water. However, drought tolerance does not mean that pistachio trees require little water for optimal performance. Indeed, research has shown that mature pistachio trees can utilize water at a fast rate—much faster than other fruit and nut trees (Goldhamer et al. 1985; Walker et al. 1988). If pistachio trees are underirrigated during stress-sensitive periods, important tree processes can be adversely affected. Thus, one must recognize that the drought tolerance of the pistachio refers to its ability to survive under severe water-stressed conditions; it does not mean that the trees will grow rapidly or have high productivity if they are poorly irrigated.

Pistachio is distinguished from other nut crops by having more quality components. In addition to a high level of alternate bearing, blanking and kernel abortion (failure of the kernel to develop) can render up to 60 percent of the fruit load worthless. Also, endocarp dehiscence (shell splitting), which is required to produce the highest-value nuts, can reduce crop value if it does not occur (Polito and Pinney 1999). Closed-shell nuts at harvest cannot be marketed as snack food, which is the largest market for pistachios. Another type of nut that is not commercially viable is early splits (nuts that split well before the onset of normal shell splitting). Not only are these nuts worthless but also they are prone to fungal disease infection that can lead to the formation of aflatoxin (Doster and Michailides 1995). Finally, the percentage removal of filled nuts by mechanical shaking also impacts harvest yields. All of these yield components can be affected by irrigation management.

Three fundamentally different approaches are used today for scientific irrigation scheduling: monitoring soil moisture levels by hand or with various instruments and replenishing soil water in a timely fashion; monitoring tree water status directly with a pressure chamber or indirectly with canopy temperature measurements and applying water so that desired tree stress is not exceeded; and estimating how much water the crop is using and applying water to match this crop water requirement. The latter technique is known as the water budget and involves knowledge of the soil, plant, and climate.

This chapter will focus on the water budget approach, which we believe is the most cost-effective and comprehensive water management technique currently in use. Successful implementation of the water budget enables us to answer the two questions necessary for effective irrigation: (1) when to irrigate and (2) how much water to apply. Since the vast majority of California's pistachio orchards are drip or microsprinkler irrigated, we will focus on these irrigation methods. However, we will also briefly cover issues pertaining to the water budget approach that relate primarily to surface irrigation (border strip and furrow). These include soil water-holding capacity, allowable depletion, and rooting depth.

The Water Budget

Since the orchard water budget is essentially an exercise in balancing water additions and losses from the orchard, the irrigation requirement is the difference between the water losses and effective rainfall. Thus, the irrigation requirement can be calculated by the following formula:

$$\text{Irrigation requirement} = ET_c - \text{Effective rainfall} + \text{Irrigation system losses}$$

Crop evapotranspiration (ET_c) is the sum of evaporation from the soil (E) and transpiration from leaves (T). Effective rainfall is the total rainfall amount stored in the profile, an amount that can be difficult to estimate. While the amount of rainfall stored in the soil depends on its intensity and duration, experience has shown that this value usually ranges between 50 and 70 percent of total winter rainfall. Rather than use estimates, the best way to evaluate effective rainfall is to measure the depth of wetted soil at the beginning of the season. As for losses other than ET_c, deep percolation of water below the root zone and surface runoff can occur when using flood, furrow, basin, or border irrigation, even with the best management. While this waste can be minimized, it is usually not economically feasible to eliminate these losses entirely.

For drip- and microsprinkler-irrigated orchards, the only pieces of information needed for the water budget are ET_c, rainfall, and irrigation system losses. For surface-irrigated orchards, you need additional information on soil water-holding capacity, root zone depth, and allowable depletion levels so that irrigations can be scheduled. These data allow you to estimate the size of the soil water reservoir—and by knowing the rate that water is being used, you can determine when to irrigate next and how much water will be needed to refill the reservoir. Figure 13.1 illustrates this situation by comparing it to that of replacing water in a tank. It should be emphasized, however, that storing soil water is not as simple as holding water in a reservoir. The problems of replenishing the root zone profile and the ability of trees to extract stored water will be addressed in the infiltration discussion below.

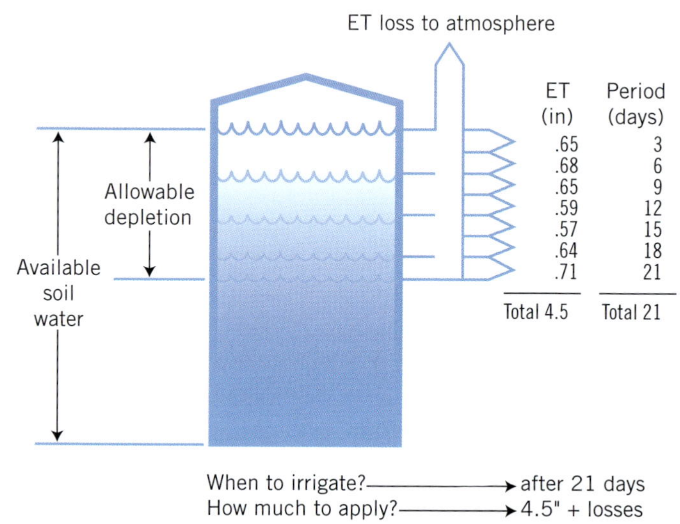

Figure 13.1. Irrigation scheduling using the water budget approach is analogous to replacing water in a tank.

Infiltration

Water applied to a field infiltrates into the soil quickly at first and then slows down as irrigation continues. With surface irrigation methods (flood, furrow, border), the soil infiltration rate controls the amount of water that infiltrates and therefore may dictate the length of time you should irrigate. This complicates irrigation management. With drip, microsprinklers, and traditional impact sprinklers, the system application rate determines the amount of water infiltration, assuming the minimum soil intake rate is not exceeded.

After irrigation, soil water drains rapidly initially. As the large soil pores empty, the soil conducts water much less readily. After 3 or 4 days, the rate of water movement slows to very little and can be neglected for our purposes. At this point, the remainder of the soil water can be considered stored. The water content of the soil at this stage is called field capacity (FC) and is the upper limit of water storage. A practical lower limit of soil water content below which crop growth is severely reduced by water stress has been defined as the permanent wilting point (PWP). Although pistachio trees do not visibly wilt, the PWP is as significant for them as for plants that show wilting.

The difference between FC and PWP is termed the available water content (AWC). Table 13.1 shows the range and average AWC of various soil types in units of inches of water per foot of soil. While several terms are used to express soil water contents, inches of water per foot of soil is a practical unit and can be visualized as the depth of water obtained if all available water were extracted from a 1-foot depth of soil. As shown in table 13.1, sands, with their relatively small total pore space, do not store large amounts of water. What water is held is easily removed by plant roots. Clay soils, because of their large total pore space, have a large AWC. However, their small water-filled pores have attractive forces that tend to resist water extraction by plants. Intermediate-textured soils, the loams, have good water-holding properties—and because of their wide range of particle sizes, they are readily able to release their water for plant use. Once the AWC is known, the total water-holding capacity of the profile is easily determined by multiplying the AWC by the depth of the root zone.

Precise determination of AWC is generally not necessary for irrigation scheduling. Indeed, the expressions FC, PWP, and AWC are actually concepts and should not be thought of as absolute, fixed amounts; they are practical estimates used to represent what are actually continuously changing water contents in the profile.

Table 13.1. Estimates of available water content for different soil types

Soil type	Available water content	
	Range (in/ft)	Average (in/ft)
coarse-textured sand	0.50–1.25	0.90
sandy loam	1.25–1.75	1.50
silty clay loam	1.50–2.30	1.90
clay	1.60–2.50	2.10

Allowable Depletion

While it is useful to compare soil water storage to a reservoir, it is not entirely accurate. As soil water content decreases, it becomes more difficult for roots to extract the remaining water, even though it is well above the PWP. This is because after the large soil pores give up their water, the smaller pores must assume two important functions: (1) to store water and (2) to conduct water moving between the soil and plant roots. Small pores not only hold water tightly, but they require water to travel exceedingly slowly through them. These factors combine to limit water uptake as soils dry out. Thus, crop growth decreases before the entire root zone reaches the PWP. For this reason, you should usually irrigate before the root zone water content reaches a level that restricts growth below its maximum potential.

Unfortunately, no single soil water level can be recommended for all situations. The safe amount of depletion (called allowable depletion [AD] or yield threshold depletion [YTD] and usually referred to as a percentage of the total available water in the root zone) depends on numerous factors, including rooting density, soil texture, and weather. Figure 13.2 illustrates the relationship between AD and plant growth in two extreme situations. One example is a low-root-density crop on a heavy (clay) soil, grown under hot, windy conditions. Here a depletion of more than 30 to 40 percent of the available root zone moisture may affect crop growth. On the other hand, a densely rooted crop grown on a light (sandy) soil under mild conditions may be able to tolerate an AD of 70 to 80 percent before growth rate drops. Allowable depletion determination is therefore not easily made, but precise estimates are not required. The particular sensitivity to water stress during bloom, rapid shell and kernel development, and shoot initiation and growth periods suggests that irrigating pistachio trees at relatively small depletion levels during these periods is advisable. At other times, 50 to 75 percent depletion has been used successfully in California orchards. Remember that the AD concept is important only for surface irrigation and sprinklers; it is irrelevant for drip and microsprinklers. Indeed, the goal of these high-frequency systems is to maintain soil moisture at relatively constant levels over the season as opposed to conventional methods that involve wetting and drying cycles.

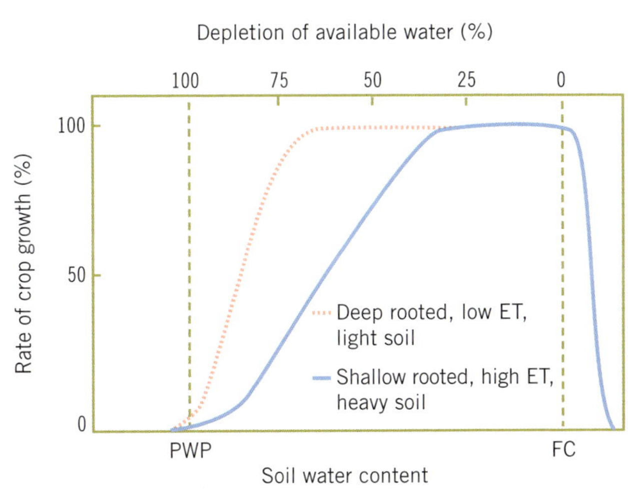

Figure 13.2. Relationship between allowable depletion of soil water and plant growth in two extreme situations.

Water Requirements and Methods of Estimating ET_c

The water budget procedure can be used successfully only if ET_c is known. ET_c depends on climatic, plant, soil, and orchard management factors, each of which is briefly discussed below.

Weather Conditions

Weather conditions largely determine ET_c rates. Because both processes that make up ET_c involve vaporizing water, the energy status of the atmosphere is of primary importance. Components of the energy balance include solar radiation (sunlight intensity), air temperature, humidity, and wind speed. Additionally, if an orchard is bordered upwind by bare ground, advective energy can cause ET_c rates to increase drastically and must be taken into account in estimating crop water use.

Evaporation

Evaporation is important only when the soil surface is wet. After irrigation, water evaporates from the soil at the same rate that trees transpire. As the soil surface dries out, upward water flow is drastically reduced, and surface evaporation decreases rapidly. Thus, the total amount of water evaporated depends on the area of the orchard floor that is wetted and the number of irrigations.

Plant Factors

Pistachio trees have stomata, the small orifices that allow for both transpiration and carbon dioxide assimilation, located on both sides of the leaves (unlike other deciduous species, whose stomata are limited to the underside of leaves). This enhances the potential transpiration for pistachio. The most significant plant factor affecting ET_c is the size of the total leaf area intercepting solar radiation. Thus, the size of the tree canopy, the tree spacing, and the stage of leaf development during the season all influence crop water use. Rather than trying to measure leaf area, research indicates that the degree of plant cover (shade) of the orchard floor correlates well with sunlit leaf area. The relationship between plant cover and ET_c is important in determining irrigation schedules for young orchards. Field studies with other deciduous tree crops show that ET_c reaches its maximum when about 60 percent of the ground is shaded by tree canopies at midday (Fereres et al. 1982) (fig. 13.3). The relationship for pistachio trees has not been established, but since the canopy architectures are similar, the almond research is most likely applicable to pistachio trees. Note that in figure

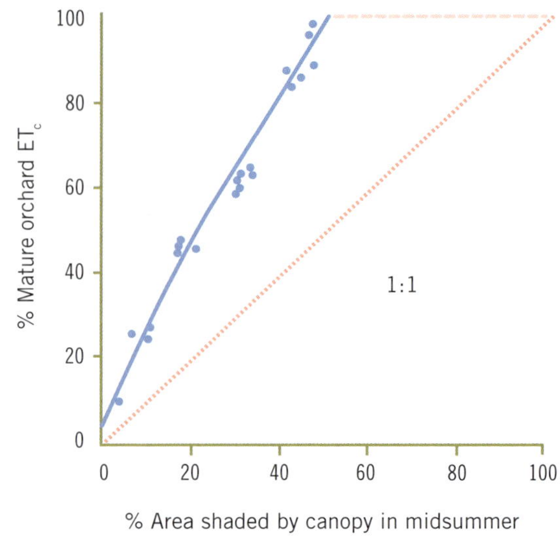

Figure 13.3. Relationship between ET_c and percentage of shaded area of the orchard floor. This was developed using almond; it is assumed to be similar for pistachio, although the latter has not been developed. *Source:* Fereres et al. 1982.

13.3 the relationship between percentage shade and ET_c deviates considerably from a one-to-one basis. This is presumably because the area of the orchard floor that receives direct sunlight transfers energy to the tree canopies by advection, thereby increasing the ET_c rate.

While most factors affecting ET_c cannot be manipulated by growers, irrigation system and orchard floor management can influence water usage. As noted earlier, the frequency of irrigation and size of the wetted surface area influence evaporation from the soil (E). With good management, the irrigator can control both factors and limit E loss. With furrow irrigation of young orchards, one furrow can be used on either side of the tree row rather than completely wetting the ground with multiple furrows. Studies have shown that drip and microsprinklers can significantly decrease the rate (in/day) of E relative to a full coverage system because of a limited wetted surface. However, since this surface is almost constantly wet, there is little seasonal E savings compared with conventional irrigation systems.

Although cover crops can have many benefits, an undesirable consequence is that they use a considerable amount of water. Cover crops or uncontrolled weeds can increase seasonal orchard ET_c by 20 to 25 percent in deciduous orchards and by much more in young trees. Therefore, the cost and availability of water should be taken into account when considering cover crops.

Since climatic conditions have the greatest influence on ET_c, many mathematical formulas have been developed to estimate ET_c based on meteorological measurements. Values for reference crop evapotranspiration (ET_o) are derived from these empirical equations. This value approximates ET_c from a well-watered pasture grass. Another index of evaporative demand is evaporation from a free water surface (Epan). But since Epan is strictly a physical mechanism and T is biologically controlled by the leaf stomata, ET_o has been shown to correlate better with actual crop ET_c than Epan. Therefore, crop water use estimates made with ET_o values are potentially more accurate. Long-term historical average daily ET_o has been compiled for all locations in California and is available in UC ANR Publication 21454 (*Irrigation Scheduling: A Guide for Efficient On-Farm Water Management*). Daily ET_o values typical for the southern San Joaquin Valley are shown in table 13.2. It should be noted that seasonal ET_o values within this area can vary by around 10 percent from the values shown, depending on location.

The relationship between ET_c and ET_o, when expressed as a ratio (ET_c / ET_o) is called the crop coefficient (K_c). The K_c varies with the crop and its stage of growth, but it is assumed to be independent of location in the pistachio-growing areas of California.

Table 13.2 also gives ET_c estimates on a bimonthly basis for pistachio trees grown in the southern San Joaquin Valley under clean cultivated conditions during a normal weather year. Long-term mean ET_o estimates and research-generated K_c values (also shown in table 13.2) were used to calculate ET_c as inches/period, inches/day, and gallons/tree/day assuming a 17-by-17-foot tree spacing. Note that the peak pistachio ET_c rate (late June through early July) is approximately 0.35 inches/day, which is significantly greater than previously published values for other deciduous trees. This high pistachio ET_c is reflected by a peak K_c of 1.19; in other words, pistachio trees can use water at a rate that is 19 percent greater than a well-watered grass crop. Seasonal pistachio ET_c is about 44 inches in the southern San Joaquin Valley.

Long-term average ET_c data can be successfully used for irrigation scheduling even though a normal year seldom occurs. Common sense should be used to modify irrigation schedules based on long-term averages if the season has drastically higher or lower temperatures or winds than normal. If you want more accurate data, current (real-time) ET_o estimates are available for more than 100 locations in

Table 13.2. Long-term, historical average ET_c for mature pistachio trees under clean cultivation in the southern San Joaquin Valley

Date	ET_o (in/day)	K_c*	ET_c (in/period)	ET_c (in/day)	ET_c (gal/tree/day)†
Apr 16–30	0.20	0.43	1.3	0.09	15
May 1–15	0.23	0.68	2.4	0.16	28
May 16–31	0.26	0.93	3.9	0.25	44
Jun 1–15	0.28	1.09	4.5	0.30	54
Jun 16–30	0.39	1.17	5.0	0.33	60
Jul 1–15	0.30	1.19	5.3	0.35	64
Jul 16–31	0.29	1.19	5.4	0.34	61
Aug 1–15	0.26	1.19	4.7	0.31	56
Aug 16–31	0.24	1.12	4.3	0.27	49
Sep 1–15	0.21	0.99	3.1	0.21	37
Sep 16–30	0.18	0.87	2.3	0.15	28
Oct 1–15	0.13	0.67	1.3	0.09	16
Oct 16–31	0.10	0.50	0.8	0.05	9
Nov 1–15	0.07	0.35	0.3	0.02	4
TOTAL			**44.7**		

Notes:

*Crop coefficient for grass reference crop (ET_o): $ET_c = (K_c) \times (ET_o)$.

†Based on 17-by-17-foot tree spacing. The following equation can be used to calculate individual tree ET_c for other spacings: gal/tree/day = ET_c (in/day) × spacing (ft^2) × 0.622 (gal/in-ft^2).

California from the CIMIS network of automated weather stations. To illustrate how years vary, daily ET_o data from the Madera CIMIS weather station for the coolest (2006) and hottest (2008) years from 2002 to 2010 is shown in figure 13.4. The California Department of Water Resources (DWR) makes ET_o and other weather data available free of charge on their CIMIS website, cimis.water.ca.gov/cimis/. Also, several newspapers and radio stations in California report daily reference crop water use data. You must be careful to recognize whether the data are estimates of Epan, ET_o, or some other reference value, in that they each represent different measurements, and the appropriate set of crop coefficients must be used to avoid errors.

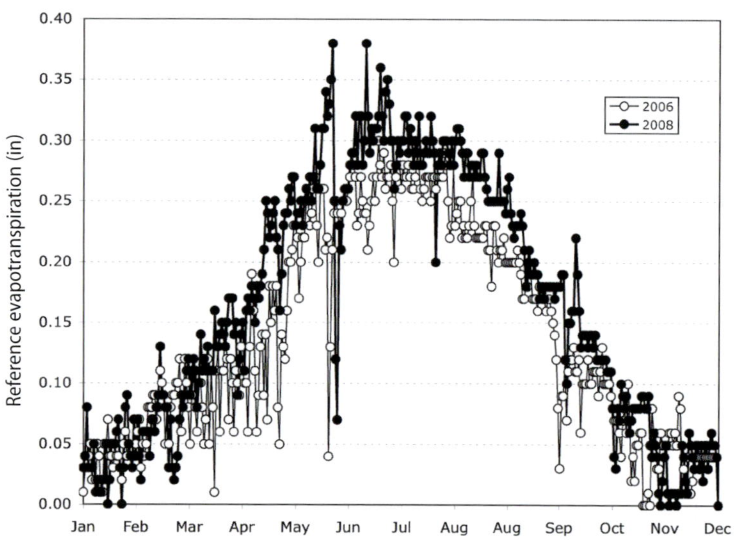

Figure 13.4. Daily reference evapotranspiration (ET_o) for Madera, CA, for the coolest (2006) and hottest (2008) years from 2001 to 2010. Data from the CIMIS weather station. August appears twice on the lower axis due to a spacing requirement of the software used to generate this graph. However, it denotes the same 30- to 31-day interval as that for the other 11 months.

Irrigation Efficiency

The information discussed so far can be used to estimate when to irrigate and how much water to supply in order to satisfy ET_c (net irrigation requirement). When water is applied to an orchard, however, some losses are unavoidable, especially with surface irrigation methods. Such losses must be considered in calculating the actual amount of water to be applied (gross irrigation requirement). The type of irrigation used, soil and climatic conditions, and water management practices largely determine irrigation efficiency.

Water applied to a field can be lost by runoff, percolation below the root zone, evaporation, and, with sprinklers, spray evaporation and drift. The goal of good on-farm water management is to minimize these losses. For instance, runoff can be minimized by using an irrigation system design that prevents it or reuses the collected tailwater. Application efficiency (Ea) is a term commonly used to describe how well growers irrigate. It is defined as the percentage of applied irrigation water that is available for crop use and can be calculated by the following formula:

$$Ea = \text{Applied water available for plant use} \div \text{Water applied}$$

In general, Ea is directly related to how uniformly water can be applied over the surface. Therefore, the method of irrigation is of prime importance.

Each irrigation method differs in how uniformly water can be applied. With surface irrigation, the intake properties of the soil and the rate at which water moves over the field determine the uniformity of infiltration. The faster the water moves to the bottom of a basin or run, the smaller the difference will be in the opportunity time for infiltration between the top and bottom of the field. The distribution of water under impact sprinkler irrigation depends mostly on system design, including spacing, nozzle type and size, riser height, and operating pressure. With drip and microsprinklers, design and maintenance determine the uniformity of application. In general, sprinkler and drip or microsprinkler systems can be operated with higher efficiencies than surface methods, since soil intake properties are of minor importance. Moreover, estimating application efficiency with these systems is much easier than with surface methods.

Because application efficiencies vary, each situation must be evaluated for Ea. In practice, it is nearly impossible to measure the ET_c from individual areas within a field, and thus we usually measure the uniformity of water applied as an indication of the maximum Ea for an irrigation system. This measurement is called the distribution uniformity (DU), which is simply the ratio of depth of water applied to the driest 25 percent of the field divided by the average field application. UC Cooperative Extension, the Natural Resources Conservation Service, DWR

Table 13.3. Gross estimates of distribution uniformity (DU) associated with different irrigation methods

Irrigation system	Ea (%)
basin	70–80
border strip	70–80
furrow	65–75
sprinkler	75–85
drip, microsprinkler	85–95

Mobile Irrigation Labs, and private consultants may offer assistance in evaluating systems. Table 13.3 shows gross estimates of DU associated with different irrigation methods.

Developing Irrigation Schedules

Once you understand each component of the water budget approach, it is a simple procedure to determine an actual irrigation schedule. For orchards irrigated with drip or microsprinkler systems, you must decide on an irrigation frequency and then apply water to meet ET_c and system losses due to application nonuniformity. For surface-irrigated orchards, ET_c is totaled until it exceeds a predetermined percentage of the total available water in the tree root zone. At that time, you need to irrigate and should apply the amount equal to the ET_c loss plus the unavoidable loss associated with each irrigation. The two following examples use long-term historical ET_o to show actual development of irrigation schedules for a mature pistachio orchard: first for a microsprinkler system and then for a border strip system.

General information for examples 1 and 2

Location: San Joaquin Valley, California
Tree spacing: 17 by 17 feet
Soil: sandy loam
Available water storage capacity (AWC): 1.5 in/ft
Rooting depth: 6 feet
Allowable depletion: 50 percent of total AWC

Example 1. Microsprinkler irrigation system

Assumptions:
Distribution uniformity ÷ application efficiency: 90 percent
Application rate: 11 gal/sprinkler/hr (one sprinkler per tree)
Desired irrigation frequency: every 3 days

Step 1. Calculate the water use rate (7/1 to 7/15, for example).

Orchard ET_c
= $(K_c) \times (ET_o)$
= (1.19) × (0.27 in/day)
= 0.32 in/day

Individual tree ET_c
= (orchard ET_c) × (tree spacing) × (conversion factor)
= (0.32 in/day) × (17 × 17 ft) × (.622 gal/in-ft²)
= 58 gal/tree/day

Step 2. Calculate the irrigation amount.

Amount to apply
= ET_c ÷ application efficiency
= 58 gal/day ÷ 0.90
= 64 gal/tree/day
= 192 gal every 3 days

Step 3. Calculate the set time (duration of water application).

Set time
= Amount to apply ÷ application rate
= 192 gal ÷ 11 gal/hr
= 17.5 hr every 3 days

Step 4. Calculate the irrigation frequency (days between irrigations) when irrigation district deliveries require you to take a set amount of water for a full 24 hours.

Irrigation frequency
= Irrigation system application/day ÷ orchard ET/day
= 11 gal/tree/hr × 24 hr/day ÷ 64 gal/tree/day
= 4.1 days

Example 2. Border strip irrigation system

Assumptions:
System delivery: 500 gal/min/ac
Application efficiency: 80 percent

Step 1. Estimate available moisture in the root zone.

Total available water (AW)
= AW × rooting depth
= (1.5 in/ft) × (6 ft)
= 9.0 in

Step 2. Calculate the amount of depletion allowed between irrigations.

Allowable depletion (AD)
= (total AW) × (depletion percent)
= (9.0 in) × (0.50)
= 4.5 in

Step 3. Estimate normal orchard water use.

Historical ET_c data from table 13.2 for a clean-cultivated orchard are plotted as cumulative ET_c versus time in figure 13.5.

Step 4. Decide when to irrigate.

Assuming that the soil water profile is full as the season begins (from a combination of winter rainfall and postharvest irrigation), deciding when to irrigate is simply a matter of periodically determining when the cumulative ET_c equals the amount of depletion allowed (from Step 2). This procedure is illustrated in figure 13.5 and results in a total of nine irrigations, taking place from late May through mid-September.

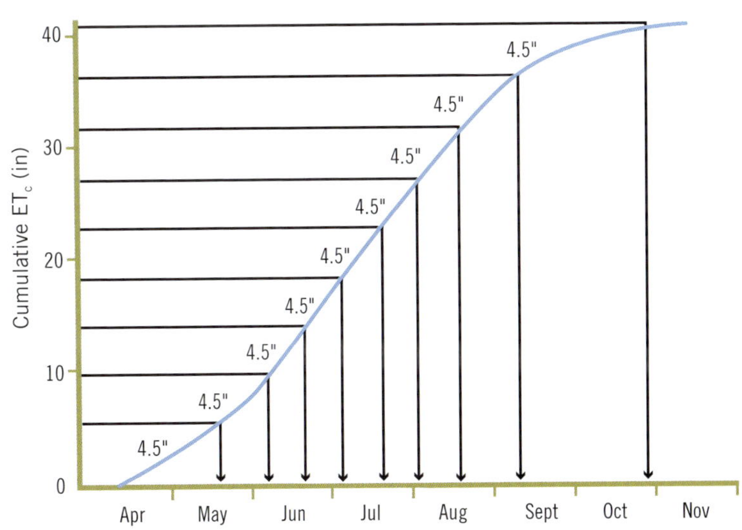

Figure 13.5. Graphical means of estimating irrigation dates using the water budget approach and assuming long-term historical ET.

(Note that if the root zone is only partially wet at the beginning of the season, the initial total available stored water can be estimated by soil probing.)

Step 5. Calculate the irrigation amount.

Amount to apply

= Depletion amount ÷ application efficiency

= 4.5 in ÷ 0.80

= 5.6 in × 27,100 gal/ac-in

= 152,000 gal/ac

Step 6. Calculate the set time.

Set time

= Amount to apply ÷ delivery rate

= 152,000 ÷ 500 gal/min

= 300 min/ac

= 5 hr/ac

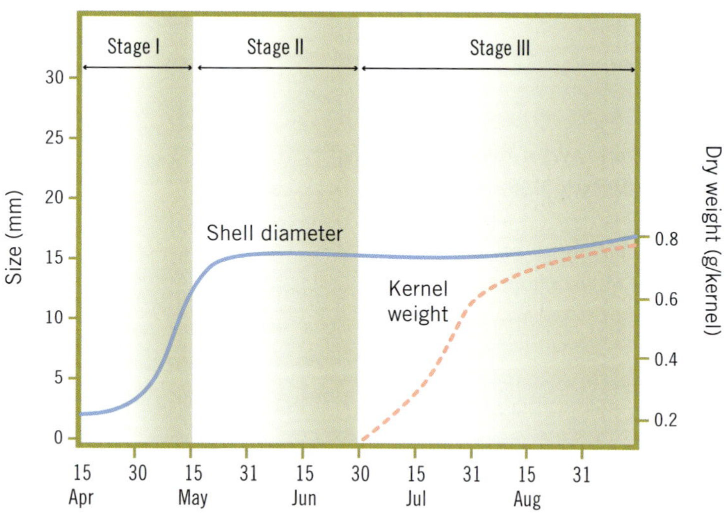

Figure 13.6. Generalized time course development of shell and kernel growth for San Joaquin Valley pistachios.

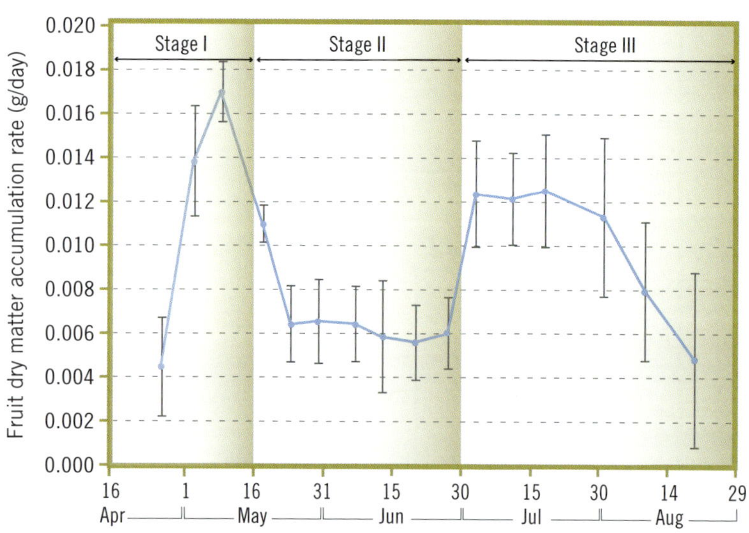

Figure 13.7. Time course development of dry matter accumulation in pistachio nuts, illustrating the three growth stages. Vertical bars are plus and minus one standard error of the mean.

Verification of the Water Budget

The water budget procedure is based on sound agronomic principles. However, even if the most accurate information is used, it is a good idea to check soil moisture levels periodically. This can be done with a soil probe or auger, examining how soil samples behave when squeezed in the hand. Alternatively, soil or plant-based instruments can be used.

The reproductive growth in pistachio trees can be divided into three stages based on the development of the nut component parts: the hull plus shell and the kernel. The development patterns of the nut components are shown in figure 13.6 and the impact on dry matter accumulation in the nuts in figure 13.7 (Goldhamer and Kanber 2012). While the dates associated with each of the growth stages vary between years (due primarily to differences in weather and location within the state), we use average dates for the southern San Joaquin Valley in our discussion. Be aware that these dates will be delayed as one moves northward in California.

The hull plus shell grows rapidly from late April through mid-May, after which full size is attained. This period is referred to as Stage I. However, the kernel does not normally begin to grow until early July. From mid-May through early July, the primary growth activity in the nut is the thickening of the shell. This period, characterized by a relatively low rate of dry matter accumulation in the nut, is known as Stage II (see fig. 13.7). Rapid growth of the kernel begins in early July and continues as harvest is approached. The biofix for this period, which is known as Stage III, is the appearance of a distinct green color in the small kernel. Research has identified the concomitant vegetative growth associated with these stages and their eventual importance as locations for fruiting positions.

Growth Stage Responses to Water Stress

Effects during Stage I

Two types of shoot growth occur: preformed and neoformed (Spann et al. 2009). The preformed shoots are generally short and are finished growing in Stage I, while the neoformed shoots are mostly long, with growth that can occur later in the season. Most of the reproductive buds, which were differentiated the previous season, are found on preformed growth.

There is generally no reduction in preformed shoot length due to Stage I water stress. However, stress during Stage I can significantly reduce neoformed growth. In essence, Stage I stress can prevent the short shoots from growing into long shoots. However, this does not decrease the number of fruiting positions, since they are located mostly on the short shoots.

Water stress during Stage I can reduce nut size by about 6 percent relative to fully irrigated trees, and it can also increase shell splitting at harvest by up to about 15 percent (Goldhamer and Kanber 2012). One theory is that Stage I stress impacts shell growth more than kernel growth, resulting in a greater splitting percentage. Closed-shell nuts can be as low as 5 percent of the harvested nut load and as high as 60 percent. On the downside, Stage I stress can increase the size of the shell opening (the distance between shell halves at the distal end of the nut). This can result in the shell detaching from the kernel during commercial nut processing, and the loose kernels can decrease the harvest value. The decision to use a Stage I irrigation stress strategy to enhance the production of more valuable split nuts would depend on whether the grower had a sufficiently severe problem with the production of closed-shell nuts.

Effects during Stage II

Water stress during Stage II has relatively little impact on fruit yield and quality. Individual nut weight might be reduced, but this requires severe water stress. Possibly the most important tree process that is negatively impacted by Stage II stress is early shell splitting. Early splits are nuts that split well before the onset of normal shell splitting, and thus they are not commercially viable. Moreover, they are susceptible to fungal diseases that can eventually result in aflatoxin contamination. The most sensitive stress period appears to be mid-May through early June.

Effects during Stage III

Stress imposed during Stage III can have dramatic negative impacts on virtually all the yield components of pistachio, with the notable exception of fruit load. These include a reduction in nut size, shell splitting, and harvestability. Remarkably, even with no irrigation during Stage III, researchers have found no effect on total tree nut load. However, Stage III is considered to be the most stress-sensitive period of the season in pistachio.

Effects during Postharvest

From harvest to the onset of defoliation, there is very little outward appearance of tree activity. Following the removal of fruit, reproductive bud differentiation resumes and continues through October. Trees generally defoliate in mid- to late November due to leaf senescence, which is accelerated by low temperatures. Research has shown little, if any, negative impact of water stress during postharvest.

The amazing ability of pistachio trees to maintain fruit loads even when stressed for much of the season highlights the importance of preformed shoot growth on eventual fruit yield. It suggests that there is enough preformed shoot growth during the first month of the season (mid-April to mid-May), when the trees are relying on stored winter rainfall rather than irrigation to achieve full fruit loads.

Irrigation Impacts on Alternate Bearing

It has been speculated that a long duration of water stress aggravates alternate bearing, suggesting that irrigation could alter periodicity by making more carbohydrates available during peak periods of carbon demand. However, California research found that Stage I stress during an alternate-bearing on-year resulted in more than a threefold increase in fruit load the following season (the subsequent off-year) relative to fully irrigated trees (Goldhamer and Kanber 2012). The mechanisms responsible for this are unknown—and it should be emphasized that, in this particular study, early-season stress was possible only because winter rainfall was abnormally low. The following season, the winter rainfall eliminated any Stage I stress, but the fruit loads of this irrigation regime were 25 percent lower than the fully irrigated trees. This pattern continued in the succeeding season, when this irrigation regime had a fruit load 25 percent higher than trees under full irrigation. It appears that regardless of why higher yields occur in a normally off-year, the one-time higher yields can change the alternate-bearing pattern for the following years.

Crop load also influences the impact of deficit irrigation on the various yield components of pistachio. All the yield components, with the exception of harvestability, are generally more negatively impacted in the on-year. Thus, growers can anticipate greater negative impacts from serious droughts during on versus off alternate-bearing years. In fact, a possible management strategy with very limited water supplies under microirrigation would be to cut off irrigation on the trees with low fruit loads (those in the off-year), making that water available for the trees in the orchard with the high fruit loads.

Regulated Deficit Irrigation

Regulated deficit irrigation (RDI) is a technique that purposely stresses trees during certain stages of tree and fruit growth in order to reduce ET_c (i.e., reduce consumptive use of water, a true water savings) while minimizing or eliminating negative impacts of stress on fruit yield or quality. This approach differs from season-long stress in that the deficit irrigation is restricted to stress-tolerant periods. First applied on stone fruit in Australia and New Zealand, successful RDI depends primarily on inducing stress during periods of slow vegetative and reproductive growth. Based on the previously mentioned research on the stress sensitivity of the different growth stages outlined above, various RDI strategies that imposed water deficits in Stage I, Stage II, and postharvest were tested in California at many sites and over many years (Goldhamer and Kanber 2012).

It was concluded that irrigating at 50 percent of potential ET_c during Stage II and 25 percent ET_c during postharvest did not affect production. Based on these results, an optimal RDI regime (table 13.4) was established that reduces ET_c by about 20 percent (8 to 10 in). In some cases, significantly higher shell splitting occurred with deficit irrigation during Stage I (leaf-out through mid-May). Closed shell at harvest was reduced by about 50 percent, with a concomitant 4 to 8 percent reduction in individual nut weight. Thus, if growers have a relatively high percentage of closed shell at harvest, Stage I stress could be a tool to improve the yield of marketable product. On the other hand, growers with no closed-shell problem may not want to sacrifice fruit size for a small increase in split nuts or the risk of producing early-split nuts.

Reducing the ET_c with RDI is a true water savings as opposed to most other irrigation management practices, where losses are only temporary. Consider what occurs when a surface irrigation system, such as flood, is replaced by drip or microsprinklers. The latter apply water more uniformly to the orchard, and this reduces the loss of water to deep percolation below the root zone. However, if that water eventually moves into a groundwater table where it can be pumped and used again for irrigation, the original lost water is not a true loss, and no water is saved by the higher application uniformity achieved with the drip or microsprinkler system. However, if the percolated water moves into a saline sink, it would be rendered useless for irrigation and then would be considered a true loss. Other undesirable consequences of deep percolation are extra pumping costs and water quality degradation.

Table 13.4. Mature pistachio orchard water use (ET_c) estimated for Madera, CA, for full irrigation and a suggested RDI regime

Growth stage	Approximate phenology	Period	ET_o* (in)	K_c†	Full ET_c in period (in)	Proposed RDI level (%)	Proposed RDI ET_c (in)
Stage I							
	leaf-out	Apr 16–30	3.0	0.43	1.3	100	1.3
	shell expansion	May 1–15	3.4	0.68	2.3	100	2.3
Stage II							
	shell hardening	May 16–31	3.9	0.93	3.6	50	1.8
	shell hardening	Jun 1–15	4.4	1.09	4.8	50	2.4
	shell hardening	Jun 16–30	4.6	1.17	5.4	50	2.7
	nut filling	Jul 1–15	4.2	1.19	5.0	100	5.0
	nut filling	Jul 16–31	4.9	1.19	4.9	100	4.9
Stage III							
	nut fill/shell split	Aug 1–15	4.8	1.19	4.8	100	4.8
	shell splitting	Aug 16–31	3.9	1.12	3.9	100	3.9
	hull slip	Sep 1–15	3.3	0.99	3.3	100	3.3
	harvest	Sep 16–30	2.2	0.87	2.2	25	0.6
	postharvest	Oct 1–15	1.4	0.67	1.4	25	0.4
Postharvest							
	postharvest	Oct 1–31	0.7	0.50	0.7	25	0.2
	postharvest	Nov 1–15	0.4	0.35	0.4	25	0.1
Totals				**44.0**		**33.6**	

Notes: Irrigation schedule early in the season must take into account stored winter rainfall.
*ET_o = reference water use.
†K_c = crop coefficient.

RDI appears to be a promising technique to save water while maintaining or even enhancing the production of marketable product. The key is being able to impose the desired stress at the correct time. The deeper the soil and the greater the soil moisture reserves going into Stage II, the longer the lag between cutting back on irrigation and the onset of tree stress. RDI is much easier to impose with drip or microsprinkler irrigation due to the greater control of applied water.

Drought Irrigation Strategies

With RDI, the issue is whether to impose stress, such that ET_c is reduced while top yields of high-quality nuts are achieved. During droughts, the issue is not whether stress occurs, it is when it will be imposed. Growers should attempt to restrict the bulk of water stress to the stress-tolerant Stage II and postharvest periods while maintaining, as much as possible, minimal stress during Stage III. Suggested drought strategies for seasonal water availabilities of 18, 24, 30, and 36 inches are shown in table 13.5.

Pistachio Production Function

Production functions relate crop yield to water use. With most field and row crops, there is a linear, one-to-one, functional relationship between production and water use. For example, if crop water use is reduced by 25 percent, the yield of marketable product is reduced by 25 percent. This is because the consumptive use of water use by plants occurs mainly by transpiration (T). The uptakes of carbon dioxide (the molecule that provides the carbon that is the building block for the entire plant, including the fruit) and T are directly related since both use the same orifices (stomates) for leaving or entering the plant. Indeed, the plant is often said to trade water for carbon.

Data from numerous RDI trials in California, where primarily Stage II and postharvest stresses were imposed, as well as from some limited European studies, were used to produce a pistachio production function (fig. 13.8). The data were developed for Kerman with a variety of rootstocks, and there were no identifiable differences in behavior among rootstocks. While it is much easier to simply measure applied irrigation rather than estimate total

Table 13.5. Suggested drought strategies for seasonal water availabilities of 36, 30, 24, 18, and 12 inches

Date	Applied water										
	Potential ET_c in period (in)	36 in		30 in		24 in		18 in		12 in	
		(% ET_c)	(in/period)	(% ET_c)	(in/period)	(% ET_c)	(in/period)	(% ET_c)	(in/period)	(% ET_c)	(in/period)
Apr 16–30	1.3	100	1.3	50	0.6	50	0.6	25	0.3	10	0.1
May 1–15	2.3	100	2.3	50	1.2	50	1.2	25	0.6	10	0.2
May 16–31	3.6	50	1.8	25	0.9	25	0.9	10	0.4	10	0.4
Jun 1–15	4.8	50	2.4	25	1.2	25	1.2	10	0.5	10	0.5
Jun 16–30	5.4	50	2.7	25	1.3	25	1.3	10	0.5	10	0.5
Jul 1–15	5.0	100	5.0	100	5.0	75	3.8	75	3.8	10	0.5
Jul 16–31	4.9	100	4.9	100	4.9	75	3.6	75	3.6	50	2.4
Aug 1–15	4.9	100	4.9	100	4.9	75	3.7	75	3.7	50	2.5
Aug 16–31	3.9	100	3.9	100	3.9	75	3.0	75	3.0	50	2.0
Sep 1–15	3.3	100	3.3	100	3.3	75	2.5	75	2.5	50	1.6
Sep 16–30	2.2	100	2.2	25	0.6	25	0.6	10	0.2	10	0.2
Oct 1–15	1.4	100	1.4	25	0.3	25	0.3	10	0.1	10	0.1
Oct 16–31	0.7	100	0.7	25	0.2	25	0.2	10	0.1	10	0.1
Nov 1–15	0.4	100	0.4	25	0.1	25	0.1	10	0.0	10	0.0
Total	**44.1**		**37.2**		**28.4**		**22.9**		**19.3**		**11.2**

Note: Potential ET_c at Madera, CA, is used; growers should use potential ET_c for their locations and deficit ET_c percentages in this table to determine their drought irrigation schedules.

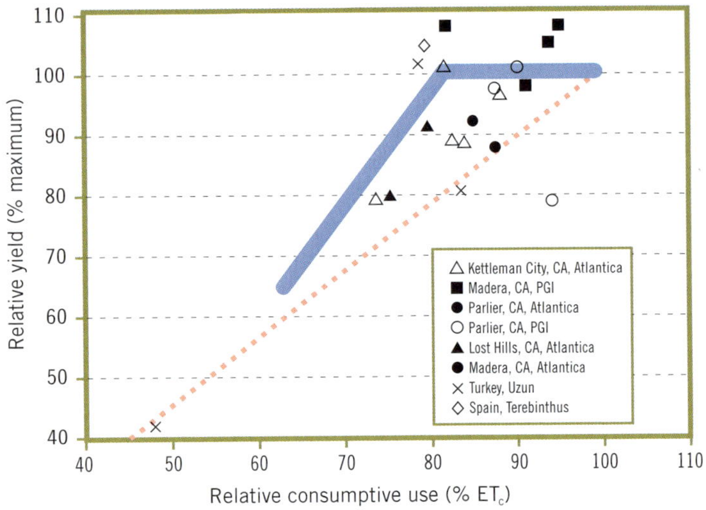

Figure 13.8. Production function developed for female pistachio scion cultivars on a variety of rootstocks, using RDI strategies that imposed stress only during Stage I, Stage II, and postharvest for at least a 4-year period. Eight studies from the United States and Europe met this criterion and are presented. Dashed line shows linear regression from full yield, full ET_c through zero yield, and 7% ET_c (the level that might be necessary for tree survival).

water use (ET_c), it is impossible to prepare a meaningful production function from multiple trials using only applied water. This is because the contribution of stored rainfall to seasonal ET_c varies so much between experimental locations. Thus, ET_c must be used in the production function.

Since RDI can successfully be used with pistachio trees, the pistachio production function is an improvement over the one-to-one relationship between yield and ET_c found with most crops (see fig. 13.8). Our production function shows that ET_c can be reduced by about 20 percent while still achieving maximum yield. With the linear, one-to-one production function found in most crops, one would expect a 20 percent yield loss. We must emphasize that a 20 percent reduction in ET_c will normally translate into a higher percentage reduction in applied water, since rainfall must be taken into account when considering ET_c. For example, if ET_c is 44 inches, of which 8 inches came from the trees' use of stored winter rainfall, the applied water would have been 36 inches. A 20 percent reduction in ET_c would reduce consumptive use by about 7 inches, a 19 percent reduction in applied water.

Additional Considerations

Using Pressure Chamber Measurements

Scientists have long known that the plant is the best indicator of its well-being relative to water. Thus, monitoring plant water status is useful to help guide irrigation management. The pressure chamber is a tool that allows growers to measure tree water status with measurements known collectively as water potential. As trees become water stressed, the water potential in the leaves decreases as it becomes more and more difficult for the trees to extract soil water. This is reflected in the pressure chamber measurements; the higher the reading, the lower the tree water potential.

Using the pressure chamber is simple, but it requires a consistent technique and observance of certain rules. In general, a leaf or small spur is removed from the tree, placed in the pressure chamber, and then subjected to compressed gas that is added to the chamber until xylem sap appears at the cut end of the petiole. At this point, the pressure is read on the gauge, and this value is considered the leaf or spur water potential. The higher the reading, the more stressed is the tree. Readings are generally taken from about noon to 2:00 p.m., since they remain relatively constant over this period.

Pressure chamber readings vary not only with tree water status but also according to the location of the leaves or spurs in the tree (sun history) and whether they are covered prior to excision. One approach is to cover interior shaded leaves with a small plastic bag for at least 20 minutes prior to excision. Theoretically, this eliminates transpiration as the interior of the bag achieves 100 percent relative humidity and the leaf water potential is in equilibrium with the water status of the adjoining stem, branch, trunk, etc. The entire unit of leaf plus bag is then inserted in the chamber. The resulting measurement is referred to as stem water potential.

An alternative approach is to wrap an interior shaded leaf in a damp cloth just prior to excision, keep the leaf covered during the short time between excision and moving to the chamber location (15 seconds or less), remove the cloth before placing the leaf in the chamber, and then take the reading. We call this measurement shaded leaf water potential.

One of the problems with pressure chamber measurements in pistachio is that at the onset of gas injection in the chamber, nonxylem fluids (presumably from the phloem) appear at the cut end of the petiole. These can interfere with identifying the instant xylem fluids appear. Dabbing away the nonxylem fluids with a cotton swab can eliminate the problem.

An alternative approach to dealing with the nonxylem fluids is to take the pressure chamber reading on a small spur rather than a leaf. Again, an interior, shaded spur is covered with a damp cloth just prior to excision. While walking to the pressure chamber, about ½ inch of bark at the cut end of the spur is removed with the thumbnail. The entire spur is placed in the chamber after the cloth is removed and the reading taken. It is much easier to see the first appearance of xylem fluid using the spur since there is no interference from nonxylem fluids and the cross-sectional area of viewing is larger than the leaf.

Using pressure chamber measurements in pistachio management has certain benefits. Since it is difficult to determine effective winter rainfall and the contribution of soil moisture to ET_c (especially early in the season), using the pressure chamber can help growers determine when to start irrigating. When using a pressure chamber, the units most commonly used to measure pressure are the bar (1 bar = 14.5 psi) and the megapascal (1 Mpa = 10 bars). Research suggests that fully irrigated, mature pistachio trees grown under high evaporative demand conditions should have midday shaded leaf water potential values for Stage I, Stage II, Stage III, and postharvest of 7 to 8 bars, 8 to 10 bars, 10 to 11 bars, and 10 to 12 bars, respectively. While imposing water stress during Stage II, when RDI is used, the target midday shaded leaf water potential is 18 to 20 bars (1.8 to 2.0 MPa).

One is able to convert the different measurements possible with the pressure chamber to each other by using relationships we have established between these stress indicators. Figure 13.9 shows the relationship between midday shaded leaf water potential and midday stem water potential from bagged leaves. A strong linear correlation exists between the two; the absolute readings of the shaded leaf water potential are greater than of the stem water potential. Figure 13.10 shows the relationship between midday shaded spur water potential and midday shaded leaf water potential. We have found no significant differences in water potential measurements among Atlantica, PGI, and UCBI.

Irrigating Young Trees

Estimating the irrigation requirements of immature pistachio trees is difficult for two reasons. First, researchers have yet to develop specific information on young pistachios. We must assume that information from other deciduous trees, such as that for almonds presented in figure 13.3, applies to pistachios. Successful application of experimental results also depends on accounting for differences in irrigation regimes. Since the evaporation component accounts for a much larger percentage of ET_c in young orchards, differences in wetted surface area and frequency of irrigations have relatively large effects on ET_c.

The second problem in determining the irrigation requirements of first- or second-year trees is that application efficiency of the irrigation systems cannot be measured easily. This is because of the small size and uncertain location of the root zone. Since irrigation systems are usually designed to perform optimally in a mature orchard, a significant amount of water can be lost to deep percolation in young orchards just because water is applied outside of the root zone. Even drip or microsprinkler systems that have high distribution uniformities and high application efficiencies in mature orchards have much lower

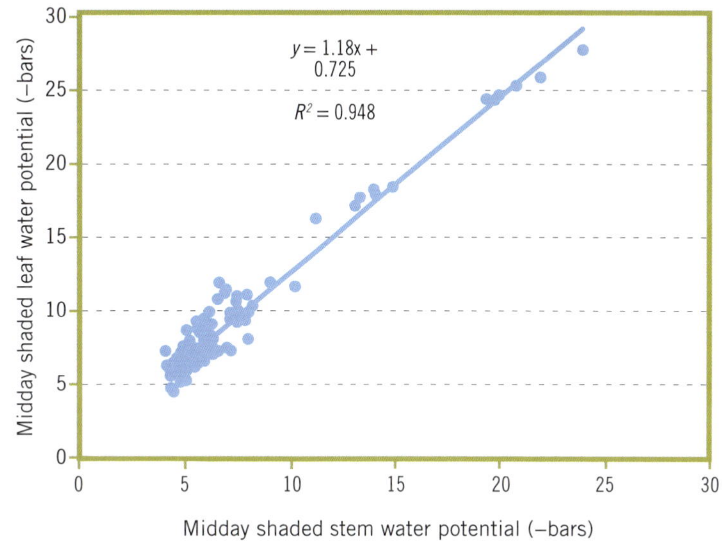

Figure 13.9. Relationship between two tree water status indicators: midday shaded leaf water potential and midday shaded stem water potential. Solid line is linear best fit; R2 is correlation coefficient.

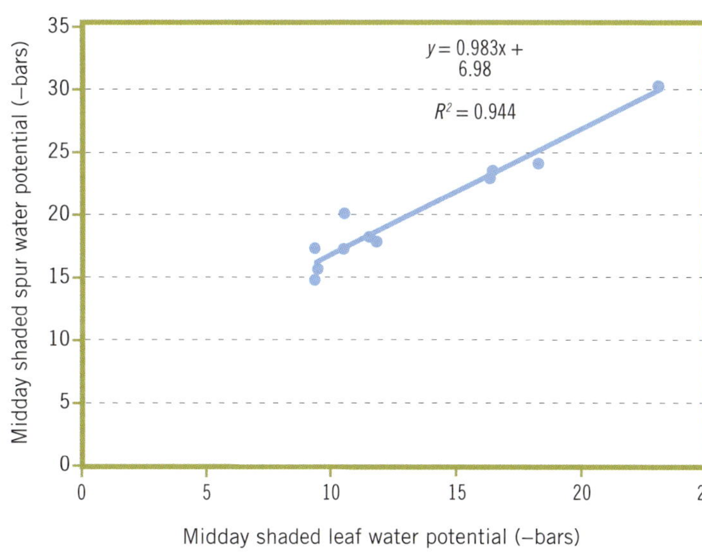

Figure 13.10. Relationship between two water status indicators: midday shaded spur water potential and midday shaded leaf water potential. Solid line is linear best fit; R^2 is correlation coefficient.

application efficiencies in new plantings. (Values of 30–50% are typical in first-year trees). Application efficiencies improve as the root zones develop.

Growers can maximize the application efficiency in young orchards by managing their irrigation systems in a way that limits the amount of water applied outside the root zone. With a drip or microsprinkler system, you can place the emitters close to the trunk, recognizing that they will be moved as the trees mature. An alternative is to operate microsprinklers at reduced pressures to decrease the area wetted, although you should check the system to make sure it maintains a high uniformity of application. With surface systems, small furrows on either side of the tree can improve application efficiency.

Effect of Irrigation on Disease

Water management can influence the predisposition of pistachio trees to infection by *Verticillium dahliae.* Since soil temperature might be a factor influencing infection, early-season irrigation timing may have to be adjusted in orchards with Verticillium wilt. The experience of cotton growers facing this problem can be used as a guide to developing optimal irrigation strategies. For pistachios, refill the soil profile with a winter irrigation, and delay the first post–leaf-out irrigation as long as possible. This practice allows the soil temperature to remain relatively high during the spring, a time when most Verticillium wilt infections occur. It should be noted that withholding water for too long a period can create tree water deficits that may ultimately affect orchard yield more than the disease.

Fungal diseases, primarily Alternaria late blight, can reduce fruit quality due to shell staining. In orchards that have slowly permeable soils and applied water ponds on the soil surface for days after an irrigation, or where sprinkler spray patterns are directed into the tree canopies, fungal disease is more prevalent. Higher humidity levels in such orchards promote disease activity. With sprinklers, the best solution is to use low-angle ones. With soils that have poor infiltration rates, we have shown that buried drip irrigation can reduce orchard humidity, thus reducing incidence of fungal diseases. The key is installing the system such that the soil surface is not (or is minimally) wetted throughout the season.

The presence of perched (shallow) water tables or excessive soil salinity also impacts pistachio ET, and care must be taken to avoid long-term (more than 2 days) saturation of the root zone. Anaerobic conditions encourage Phytophthora root and crown rot and impair water and nutrient uptake.

References

Doster, M. A., and T. J. Michailides. 1995. The development of early split pistachio nuts and their contamination by molds, aflatoxins, and insects. Acta Horticulturae 419:359–364.

Fereres, E., D. A. Martinich, T. M. Aldrich, J. R. Castel, E. Holzapfel, and H. Schulbach. 1982. Drip irrigation saves money in young almond orchards. California Agriculture 35(7, 8): 11–12.

Goldhamer, D. A., and R. Kanber. 2012. Pistachio. In Crop yield response to water. FAO Irrigation and Drainage Paper No. 66. Rome, Italy: Food and Agriculture Organization of the United Nations. 416–429.

Goldhamer, D. A., and R. L. Snyder. 1989. Irrigation scheduling: A guide for efficient on-farm water management. Oakland. University of California Division of Agriculture and Natural Resources Publication 21454.

Goldhamer, D. A., R. K. Kjelgren, L. Williams, and R. Beede. 1985. Water use requirements of pistachio trees and response to water stress. Proceedings of the National Conference on Advances in Evapotranspiration. American Society of Agricultural Engineers: 216–223.

Kanber, R., A. Yazar, S. Onder, and H. Koksal. 1993. Irrigation response of pistachio (*Pistacia vera* L.). Irrigation Science 14:7–14.

Polito, V. S., and K. Pinney. 1999. Endocarp dehiscence in pistachio (*Pistacia vera* L.). International Journal of Plant Sciences 160:827–835.

Spann, T. M., R. H. Beede, and T. M. DeJong. 2009. Contributions of short and long shoots to yield of 'Kerman' pistachio (*Pistacia vera* L.). Scientia Horticulturae 121:495–500.

Walker, R. R., E. Torokfalvy, and M. H. Behboudian. 1988. Photosynthetic rates and solute partitioning in relation to growth of salt-treated pistachio plants. Australian Journal of Plant Physiology 15(6): 787–798.

14 Assessing and Improving Water Penetration

Blake L. Sanden, Terry L. Prichard, and Allan E. Fulton

Supplying adequate water to a mature orchard without stressing the tree from either excessive or insufficient water is the goal of effective irrigation management. Poor water penetration is the major problem for many surface-irrigated orchards in California. It is characterized by the inability of water to readily infiltrate into the active root zone to sustain crop water needs until the next irrigation. The problem is typically associated with flood-irrigated orchards and less frequent irrigation (every 10 to 20 days), but it can be a problem for low-pressure irrigation systems (drip and minisprinklers) as well.

Figure 14.1. Typical vascular discoloration, gumming, and shutdown of root water transport from Phytophthora root and crown rot in *P. vera* due to extremely fine silt soil and adjacent drip irrigation. *Photo:* B. L. Sanden.

Across California, nonstressed mature pistachios use about 40 to 48 inches of water per year. For the southern San Joaquin Valley, this demand can easily exceed 25 inches for June, July, and August together. When poor water penetration occurs, the trees become water stressed due to underirrigation.

Ironically, poor water penetration not only leads to inadequate irrigation; it can also increase the time the soil surface remains saturated. As the lower root zone dries out, the tree depends more on water from roots near the surface, but poor water penetration (and especially high-frequency drip irrigation) may render this zone saturated much of the time, reducing oxygen needed by roots and significantly increasing the risk of root diseases such as Phytophthora root and crown rot. UCBI and PGI rootstocks have proven to be fairly resistant, but *Pistacia vera* on its own rootstock is quite susceptible (fig. 14.1). Common symptoms of poor water penetration are outlined below.

Evidence of Slow Water Penetration

- midseason depletion of deep soil water and inadequate recharge of subsoil water, even after long irrigations
- water that ponds on the soil surface for long periods, disrupting orchard access
- reduced vegetative growth and yield
- reduced split percentage
- higher incidence of root diseases resulting from poor soil aeration

The Water Penetration Process

The first step in determining a solution or remedial practice for poor water penetration is to take a close look at the infiltration process. At the onset of irrigation, water infiltrates at a high rate. Initially the soil is dry and may have cracks through which water can infiltrate rapidly. As the soil wets from the surface into the root zone, the distance from the soil surface (and standing water) to the wetting front of infiltrating water increases. At this point, clay particles swell, closing surface cracks and limiting access to small, drier soil pores beneath. After this point, these initial cracks and pores become less important in sustaining infiltration rates. Infiltration rates decrease significantly, and water moves primarily by the force of gravity through the larger macropores in the soil. Depending on the soil, as much as 50 to 80 percent of the total infiltrated water can occur in the first 3 to 6 hours of a 24-hour irrigation set. It is precisely for this reason that surface irrigation down a quarter-mile run can be done somewhat uniformly even though the tail end may only get 4 to 6 hours of water compared with the head end.

The stability of soil aggregates and these larger pores depends on the interaction of soil minerals and the salinity of the water in these pores (fig. 14.2). As the irrigation continues, the salt composition of the soil water begins to more closely reflect that of the irrigation water, which is generally less saline. Changes in soil water chemistry can also reduce infiltration rates.

Not only does the rate of infiltration slow down after the first few hours of irrigation, but successive irrigations over the season compact the surface and decrease the soil's ability to infiltrate water. Figure 14.3 shows how cumulative infiltration slows way down after the first 6 hours and decreases with each irrigation over the season. In this example, the addition of gypsum to this very low salinity irrigation water improved infiltration on 8/4.

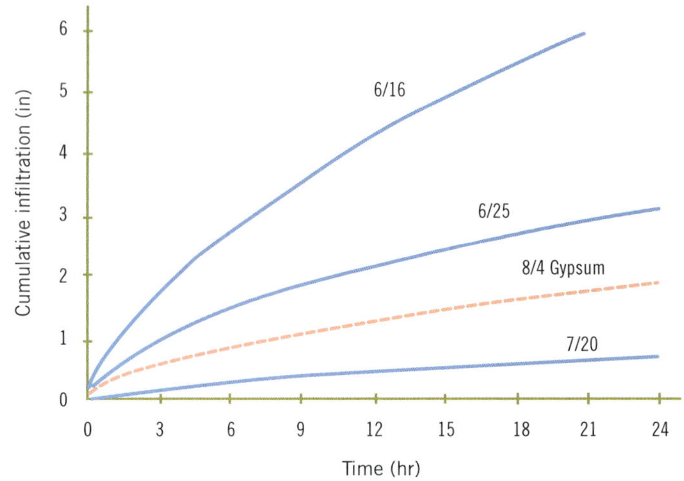

Figure 14.3. Cumulative depth of infiltration for Wasco sandy loam over the season for the same furrows in furrow-irrigated cotton at the Shafter Field Station (1995) with low-salinity canal water (0.02 dS/m).

The following characteristics of the soil and irrigation water have the greatest influence on water penetration.

Soil

- dryness at start of irrigation
- distribution and size of soil particles and pores
- surface access to soil pores
- cracks
- total salinity of soil pore water
- composition of soil pore water salinity
- nonuniformity of root zone soil (layering)

Irrigation water

- total salinity
- composition of salinity
- depth of water applied to the soil surface
- SAR (sodium adsorption ratio)

Water penetration can only be improved by increasing soil pore volume, individual pore size, and access to pores. Total water salinity and the composition of salts influence the way soil particles aggregate, or stick together. Chemical or mechanical modification of aggregation and pore volume will improve water penetration.

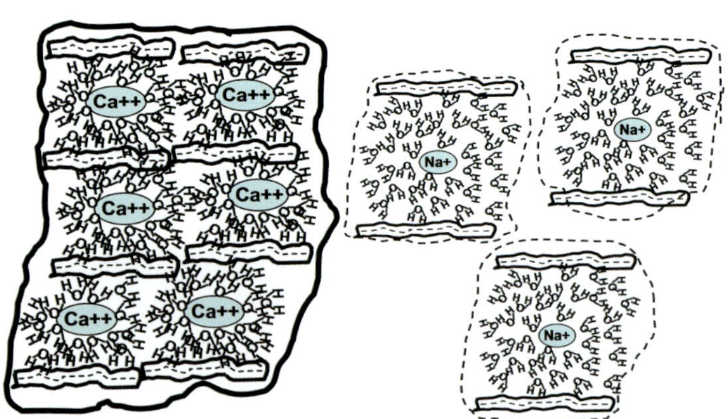

Figure 14.2. Conceptual illustration of soil aggregate stability. Stable aggregates with plentiful calcium on clay exchange sites due to a small hydration shell (left); weak soil aggregates due to a large hydration shell created by low salinity or excessive sodium in the soil pore water (right).

Pore Size, Porosity, and Water-Holding Capacity

Pores are the spaces between soil particles through which water and air move. Porosity is the total volume of all the pores and cracks per total volume of soil. The total porosity is made up of large-, medium-, and small-size pores. Pore size and total porosity determine soil water-holding capacity.

Sandy soils. Soils with high sand (spherical particles) tend to have larger individual pores but lower total porosity. This can result in high infiltration rates (with some exceptions), but less stored water in the root zone. However, these soils also have lower clay content, and, especially if it is a fine sand with a significant amount of silt, they may have very poor development of soil aggregates. The aggregate structure is important for balancing the soil with different pore sizes; the smaller- and medium-size pores are important for water retention and the larger pores for water infiltration and soil aeration. Some of the soils in the southern San Joaquin Valley with the worst water penetration are fine sandy loams and sandy silt loams.

Clay soils. Clay-dominated soil tends to have more small pores and a higher water-holding capacity than sandy soils. Water moves more slowly through the smaller pores, because smaller pores provide more surface area for water to adhere to and thus create more resistance to water flow. But clays are made up of microscopic, electrically charged mineral plates (as illustrated in fig. 14.2) that can group together into much larger aggregates. These aggregates can become larger than sand particles and actually create large macropores. The small pores within an aggregate remain, and larger pores are formed between the aggregates. The pores also shrink as the soil dries, developing cracks that may go as deep as 2 to 4 feet. The cracks are then a major channel for water penetration, usually swelling shut after wetting. Some of these cracking soil types have higher infiltration rates than some other soils. The net effect is more and larger pores, which significantly enhance water penetration and aeration.

Organic matter. Soil organic matter plays a significant role in stabilizing soil aggregates. This is because it increases the number of exchange sites in the soil matrix; encourages microbial activity, which produces waste products that help bind soil particles together; and increases pore size by decreasing soil bulk density.

Soil Crust Formation

Formation of a soil crust decreases infiltration by impeding the access to soil pores beneath the crusting layer. The crust is formed due to the dispersion of soil aggregates and loss of porosity at the soil surface. Weak cementation often follows when the soil dries. The formation of a soil crust, or surface seal, leading to reduced infiltration has been recognized as a problem in California agriculture since the early 1900s.

In arid and semiarid areas, soil crusts are often the result of sodic conditions (excess exchangeable sodium in the soil or irrigation water, or too little total salinity) in fine-textured silty soils. In the early and mid-1900s, when water and land were relatively cheap, these soils were the focus of reclamation and amendment strategies to open up new cropland. These conditions are mainly problems for growers on the west side of the San Joaquin Valley.

More recently, attention has been focused on crusting and penetration problems on many nonsaline and nonsodic soils that are coarse- to medium-textured. After decades of cultivation on many of these orchard soils, a significant decrease of larger pores within the surface profile can result. In some cases, the problem has been made worse by the use of very low salinity irrigation water (via the Friant-Kern Canal, along the east side of the San Joaquin Valley) to replace groundwater pumping (see the section "Structural crusts" below). The increased use of herbicides for no-till orchard floor management can also decrease soil organic matter and soil microbial activity. This also results in decreased soil aggregation and reduced pore size. The resulting surface sealing (crusting) can be categorized as either structural crusts or depositional crusts as defined below.

Structural crusts

A structural soil crust is formed by the destruction of existing soil aggregates and a subsequent reorganization of the resident soil particles into a sealing layer. The destruction of soil aggregates can occur as a result of mechanical energy, such as droplet impact from rain and sprinklers, or a lack of sufficient chemical energy to hold soil particles together, or both. The mechanical breakdown of soil aggregates tends to sort soil particles, leaving a film of finer particles on top that clogs the entry of water into the larger pores beneath. In furrow and flood systems, this process is called slaking and is usually a combination of mechanical and chemical dispersion of soil aggregates. The structural crusting at any given spot is compounded by additional depositional crusting laid down from clays and silt carried in from the upstream part of the furrow (fig. 14.4).

Figure 14.4. Surface sealing of Milham sandy clay loam in a flood-irrigated almond orchard due to low-salinity, sodic well water. Water penetration is only to 7 inches deep after a 32-hour irrigation set (inset). (White pen at bottom indicates depth of penetration.) The surface remains saturated 8 hours after irrigation ceases. *Photos:* B. L. Sanden.

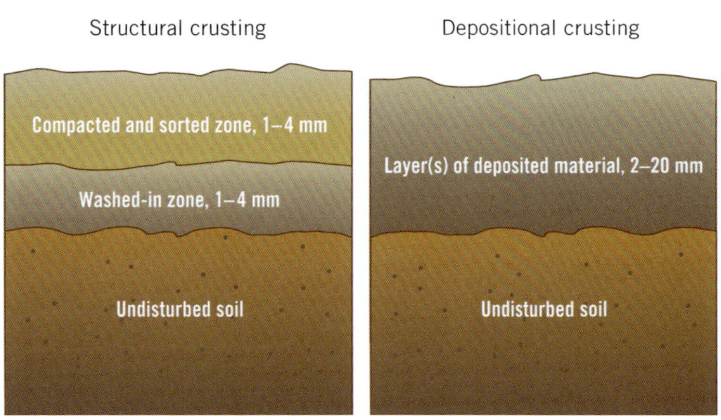

Figure 14.5. Conceptual illustration of structural and depositional crusts.

Figure 14.6. Surface sealing of Wasco sandy loam due to very low salinity in soil and irrigation water. *Photos:* B. L. Sanden.

Depositional crusts

Formed by the sedimentation of fine soil materials over the native soil surface, a depositional crust limits access to the larger resident soil pores. This type of soil crust is most often the result of high-velocity water in the head end of the furrow or watershed eroding fine particles that settle out when the water slows. The size of the particles in suspension is small (the particles are usually clays); their platelike structure forms a very effective barrier to soil pores. Reduced infiltration from this type of crusting can sometimes be made worse when irrigating with effluent waters from dairy lagoons or wastewater treatment plants. Suspended organic particles as well as sediments act as clogging agents and can interfere with the soil's ability to resist dispersion.

Both structural and depositional crusts are thin, characterized by higher-density soils with greater strength and smaller pores than the underlying soil. These crusts are usually less than ¼ inch thick (figs. 14.5 and 14.6), but they often limit infiltration for the entire root zone.

Impact of Salinity and Sodicity

In California orchards and vineyards, structural crusting is by far the greatest impediment to adequate water penetration. As mentioned earlier, the most effective deterrence against poor water penetration and crusting is aggregate stability. The important soil and water factors that we can most easily test and manipulate are

- total salinity, measured as EC (electrical conductivity)
- sodicity, evaluated by the sodium adsorption ratio (SAR)
- pH

(Chapter 15, "Salinity Management in Soil and Water," provides the reader with a detailed explanation of these and other chemical criteria.)

The salinity of the surface soil is determined by measuring its electrical conductivity. The EC is an important factor in determining crusting. However, the water around soil particles (i.e., soil water) is strongly influenced and rapidly modified by the constituents of irrigation water. Reduced EC in the soil water causes clay swelling to increase, reducing the size of soil pores. Irrigation water with an EC of less than 0.3 decisieman per meter (dS/m) can cause problems on most soils. Each soil has a unique amount of soil water salinity (flocculation threshold) at which dispersion of the particles occurs.

Table 14.1. Guidelines to diagnose water infiltration problems based on soil and irrigation water quality

Sodium adsorption ratio (SAR)	Potential for water infiltration problems to develop	
	Likely if EC_e or EC_w is	Unlikely if EC_e or EC_w is
0.0–3.0	< 0.3	> 0.7
3.1–6.0	< 0.4	> 1.0
6.1–12.0	< 0.5	> 2.0

Source: Adapted from Ayers and Westcot 1985.

Dispersion is also dramatically affected by the ratio of sodium to calcium and magnesium, called the sodium adsorption ratio (SAR).

For an EC above 0.3 dS/m, researchers have worked out general guidelines (table 14.1 and fig. 14.7) that can be used to diagnose potential infiltration problems as the EC and the SAR change. In general, aggregate stability increases as the EC increases and the SAR decreases.

These sodium-based guidelines will not necessarily work for all soils. Some California soils outside of the San Joaquin Valley contain serpentine clays (e.g., some areas of Napa Valley and the foothills east of the San Joaquin Valley) that are rich in magnesium and relatively low in calcium. In such an environment, magnesium may behave like sodium, and the result is unstable soil that tends to disperse and become impermeable. Although the diagnostic criteria for such conditions have not been extensively tested, some professional consultants suggest that when the magnesium to calcium ratio exceeds 1:1, serpentine soils may develop infiltration problems. Soils rich in exchangeable potassium may also have infiltration problems. Some reports maintain that when potassium is the predominant cation, it has the same effect on soil stability and porosity as does sodium: the soil becomes less stable, disperses at the surface, and seals over. Soils with a predominance of montmorillonite and illite clays are most easily dispersed by excess magnesium. Hydrous oxides of aluminum, iron, and organic matter components, however, exert a stabilizing force on clay, which acts against the dispersing effect of sodic water or water with very low salinity.

Irrigation Management Options

Reviewing management options for applying and scheduling irrigation water is always the first step in dealing with water penetration problems.

Alternate water supply. For many orchards in the Shafter-Wasco area of Kern County, the simple solution is alternating irrigations between the very pure snow-melt canal water and well water, which has a higher EC and SAR. Although higher EC may be beneficial for water infiltration, the increased salinity and specific constituents in the well water may not be necessary or good for the trees over the long term.

Increase irrigation frequency. For other flood orchards, the best answer may be a laser level of the middles during dormant season and an increase in irrigation frequency (making it every 5 to 7 days instead of 10 to 14 days in July and August). It is imperative to have a very uniform grade to ensure adequate drainage with no ponding in the middle of the orchard and a tailwater return system to exercise this option. Otherwise, waterlogging and disease will cause damage to the trees.

Switch to microirrigation. Obviously this option provides the grower with the greatest flexibility and precision in applying water and chemical amendments. Most California soils will infiltrate an application rate of 2 to 6 inches per day if the water is uniformly applied over the whole orchard floor. The difficulty with microirrigation systems is that this means applying water to a much reduced wetted area directly under the emitter, and runoff can still occur. Furthermore, the root system in established orchards is more extensive, and switching to microirrigation means that a smaller soil volume will be supplying the crop water needs.

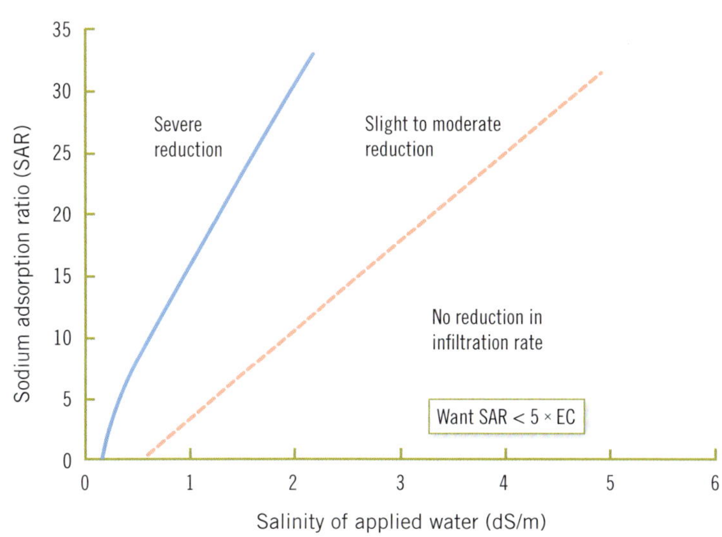

Figure 14.7. Interaction of total salinity as EC and the sodium adsorption ratio of applied water in infiltration. Figure based on Ayers and Westcot 1985.

Example: Drip irrigation system application

Orchard spacing: 19 ft × 20 ft = 380 ft^2
System: single-hose 4, 1 gph dripper/tree
Total application rate: 0.41 in/day
Wetted diameter under drip: 4 ft/dripper
(4 drippers = 50.3 ft^2)
Wetted area of orchard floor: 50.3 ÷ 380 = 13.2%
Infiltration in wetted area: 0.41 ÷ 0.132 = 3.1 in/day

To meet the peak evapotranspiration (ET) of 0.3 to 0.32 inches per day in July, this means irrigating 3 out of 4 days. The bottom line is that it is necessary to infiltrate 9 inches of water through the soil underneath the dripper every 4 days! This is the reason some drip-irrigated orchards still have runoff or low-lying swampy areas in the field. On the other hand, a 12 gph microsprinkler throwing a 12-foot-diameter pattern (ft^2 113) applies 4.09 inches per day to the wetted area but only needs to run 1 day out of 4 to meet peak ET demand.

Solving a penetration problem by modifying irrigation practices and using an auger to check soil moisture is always the starting point, and it will be less costly than various amendments or cover crops. However, many soils still require additional amendments and improved cultural practices to stabilize soil porosity and prevent soil crusting and decreased water penetration. The next sections of this chapter discuss these options.

Whichever options are chosen, the single most important thing you can do where water infiltration is a problem is to make sure that soil moisture in the root zone is completely recharged by the end of February.

Prevention of Soil Crusts

Where soil permeability is low, prevention of soil crusting is often the best course of action and usually the most economical. Prevention includes the application of amendments, use of soil surface covers, management of soil organic matter, and improved management of irrigation. However, once a crust has formed, tillage may be required before other options can be effective.

Tillage

Shallow tillage can disrupt both structural and depositional crusts. In cases of moderate crusting problems, one tillage pass per season can restore infiltration rates. However, in soils with severely reduced infiltration, tilling before each irrigation is common. Shallow tillage to disturb the surface crust is accomplished using shallow disk, harrow, or even rolling cultivator. A spade-type cultivator can be used that actually fractures any disk or plowpan that might be at 8 to 10 inches and then lifts the soil without mixing it.

The middles of some pistachio orchards are cultivated regularly to prevent weeds that attract undesirable insect pests. This tillage operation is often successful in preventing water penetration problems that may be severe in adjacent, noncultivated almond orchards.

Organic Matter Management

All soils contain a small component of organic matter, and in the San Joaquin Valley it ranges from 0.5 to 2 percent. Uncultivated soils contain more organic matter than those cultivated under typical orchard conditions. With this reduction in organic matter often comes reduction of the stability of soil surface aggregates.

The application of soil organic mixtures such as compost and manures can increase porosity, percentage of macropores, and aggregate stability, thus increasing the infiltration rate. However, the organic matter content of soils in arid or semiarid areas does not increase because of the rapid rate at which the organic matter decomposes in California's climate. In a 10-year study conducted at the University of California, Davis, researchers incorporated cover crops into the soil. The percentage of organic matter in the soil did not increase over that time. The infiltration rate, however, did increase.

The implication of this and other research is that organic additions are beneficial by virtue of the products of their decomposition. These products consist of polysaccharides and polyuronides, which act as binders to stabilize aggregates. To be effective, organic matter additions or cover cropping should be continual, because decomposition products are short-lived, especially in California's climate.

Crop residues. Trees provide leaves and prunings that can be left in the orchard for decomposition or incorporation. Interest in chipping or shredding brush is growing as a result of restrictions on burning. Correctly prepared brush can add a significant quantity of organic matter. Be sure to prepare brush properly. Chip prunings into small pieces, and shred brush more than once or disk to incorporate material. Other studies in almonds in Madera (Holtz et al. 2004) showed that tilling in chipped tree prunings nearly doubled the amount of macroaggregates in the soil compared with no chips. In some years, *Verticillium,* as well as some beneficial fungi, increased significantly with increased incorporation of chips, while other years showed no difference. Water infiltration in this study was not measured.

Manure, green waste, and mulches. Animal manures have long been applied to orchards to supply nutrients and improve water infiltration. If you use manure, be aware of the potential effects of salinity and sodicity and have a plan for preventing them. Manure often contains weed seed, so be prepared to handle unwanted vegetation as well. Meek et al. (1982) more than doubled infiltration rates in a Holtville clay in the Imperial Valley with as little as 10 tons per acre of dairy manure. However, the benefit declined by half at the end of the first season and was completely gone at the end of the second year. This was true even for an application of 40 tons per acre.

Green waste is a term used for a mixture of lawn clippings, prunings, and garden materials. The mixture is increasingly available for agricultural use. Currently it is available in raw form and in various states of decomposition. The composted materials offer a high ratio of organic materials per unit volume. Green waste is easily spread, and its viable weed seed content is low. Work in citrus in Ventura County has shown significant reduction in water use by reducing evaporation from the soil with an 8-to-12-inch-thick application of woody green waste mulch (Faber et al. 2001). Increased density of rooting and soil porosity at the soil surface underneath the mulch were also found. A banded application of green waste compost over the hoses of double-line drip in pistachios near Delano improved root zone water content and decreased runoff in the year of application (Blake Sanden, unpublished data).

Cover crops. While pistachio middles in mature orchards are generally kept clean, it is useful to briefly mention the use of cover crops, as they can be a valuable tool in reclaiming marginal ground during the development of a young orchard. Cover crops protect the soil surface from droplet impact under sprinkler irrigation as well as provide significant organic matter biomass for decomposition and microbial stabilization of soil aggregates. In addition, cover crops can slow the velocity of surface water, reducing erosion and subsequent depositional crusting.

Water use. However, cover crops can compete with trees for nutrients and water. If an orchard contains clover as a perennial cover crop or actively growing winter and summer resident vegetation, water use can increase by 10 to 20 percent (Prichard et al. 1989) (table 14.2). The orchard manager must supply additional water, or crop stress will occur. The use of winter annual cover crops and vegetation control strategies during summer months, such as chemical mowing, can reduce the water requirement.

Dry matter production. Cover crops can be planted as annuals or perennials, or they can be resident vegetation. Mature trees with full canopies may shade out cover crops by midseason. Annual cover crops or resident vegetation consisting of winter and summer annuals can produce 2 to 4 tons of aboveground dry matter. The ratio of top portion to underground dry matter (roots) has been estimated at 1.5 to 1.0. Thus, a cover crop that yields 6,000 pounds biomass per acre aboveground yields about 4,000 pounds per acre belowground in the form of roots. Total biomass from the cover crop is then 10,000 pounds per acre (5 T/ac).

Maintaining or improving infiltration. Cover crops can prevent decline of infiltration rates over the season (Prichard et al. 1989). Table 14.3 shows that the cover crops prevented a nearly 50 percent decline in infiltration by the end of the season in this almond orchard study. Increases are attributed to physical factors, such as channels created by roots, surface protection, and increased soil aggregation. Compared with bare ground, soil with a cover crop has greater aggregate stability and more macropores.

Using Chemical Amendments to Improve Infiltration

Chemical amendments to improve water penetration are really doing only one thing—decreasing the resistance to water moving through the profile. This is done in two ways. One is to maintain or improve soil

Table 14.2. Water use by a mature almond orchard with cover crops and bare soil

Treatment	Increase in water use (%)
resident vegetation	120
clover	110
bromegrass	98
bare soil	100
chemical mowing	100

Note: Water use is relative to that of bare soil.

Table 14.3. Accumulated infiltration at 120 minutes through various cover crops and bare soil in a mature almond orchard

Treatment/ground cover	Early season (mm)	Late season (mm)
clover	66.8 a	63.2 a
resident vegetation	52.3 a	54.9 a
bromegrass	52.8 a	65.3 a
chemical mowing	63.0 a	39.1 b
bare soil	53.3 a	32.5 b

Note: Numbers followed by different letters are significantly different at the 0.05 level.

structure through the release of free calcium (salts and acids) or to stabilize aggregates with binders and water-soluble polyacrylamides (PAM) or both. The other is to decrease the surface tension of water with surfactants (PENMAX, etc.) Four types of chemical amendments are used to improve water penetration problems: salts, calcium-supplying materials, acids or acid-forming materials, and soil conditioners, including polymers and surfactants.

Salts. Any fertilizer or amendment that contains salts increases the salinity of the irrigation water when applied to the soil surface or dissolved in irrigation water, and this ultimately influences the

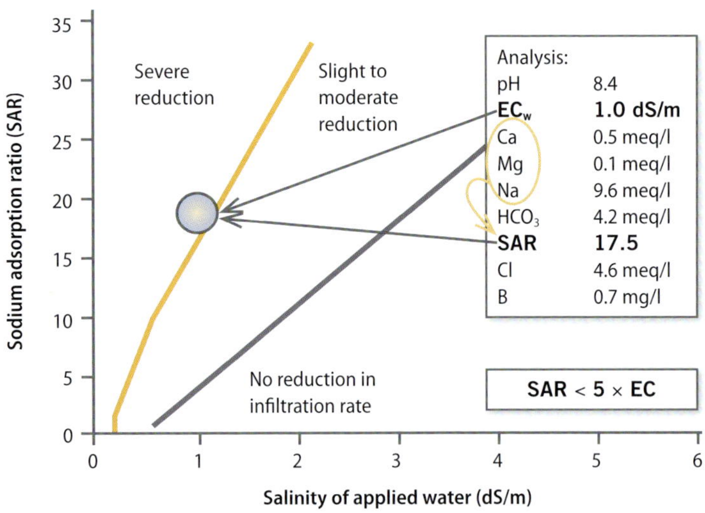

Figure 14.8. Estimating potential infiltration problems and determining amendment options from an irrigation water analysis.

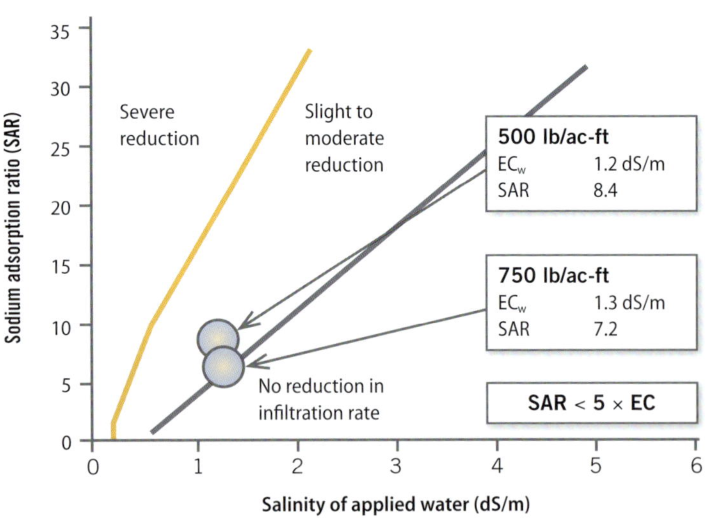

Figure 14.9. Revised infiltration potential after injecting gypsum into irrigation water at rates of 500 and 750 pounds per acre-foot.

soil water. Whether the increased salinity is advantageous depends on the SAR of the irrigation water. As for the effects of salt alone, increasing the salinity above an EC of 4 dS/m has little effect on infiltration.

Calcium materials. Gypsum has been by far the most widely used calcium-based amendment applied to agricultural soils. Adding calcium salts to soil and water increases both the total salinity as well as the soluble calcium, both of which improve infiltration. Calcium salts commonly used on alkali (high-pH) soils include gypsum, calcium chloride ($CaC1_2$), and calcium nitrate ($Ca(NO_3)_2$). These are fairly soluble and can easily be applied through the irrigation water. Lime and dolomite are used only for broadcast applications on acid soil, as they are virtually insoluble in alkali conditions.

Gypsum Injection Rates for Water

Injection of gypsum ($CaSO_4{\cdot}2H_2O$) in the irrigation water is the most common amendment practice in the San Joaquin Valley. Amendment rates from 1.0 to 3.0 milliequivalents per liter (meq/l) of calcium are considered low to moderate; rates that supply 3.0 to 6.0 meq/l of calcium are considered moderate to high. The following example calculations show the reader how to estimate the quantity of gypsum required to improve infiltration. (Tables 15.8 and 15.9 in chapter 15, "Salinity Management in Soil and Water," give detailed information on the chemistry, equivalent rates, and comparative costs for calcium and acid-type amendments. The following example refers to these tables.)

Calculating gypsum rates. A partial water analysis is shown in figure 14.8. This water presents absolutely no salt- or ion-tolerance problems for the pistachio tree. However, the high SAR, especially given the high-pH and -bicarbonate levels, indicates significant infiltration problems, as indicated by the large black circle in the figure. To achieve good infiltration, some of the sodium needs to be offset with calcium. A preferred method of supplying calcium is by injecting gypsum. The following four steps are required to calculate the correct rate:

1. Determine the purity of the gypsum and the actual pounds per acre-foot needed for 1 meq/l calcium. Assume that the gypsum solution purity is 92%; table 15.8 indicates that 234 pounds of gypsum will be required to supply 1 meq/l:

$$234 \text{ lb/ac-ft} \div 0.92 = 254 \text{ lb/ac-ft per 1 meq/l calcium}$$

2. Use the desired application rate to calculate the EC of the water with the additional calcium:

500 lb/ac-ft ÷ 254 ≈ 2 meq/l
New EC = 1.0 + 0.2 = 1.2 dS/m

3. Calculate the new SAR:

$$SAR = Na \div [(Ca + Mg) \div 2]^{0.5}$$
$$= 9.6 \div [(2.5 + 0.1) \div 2]^{0.5} = 8.4$$

4. Locate the intersection of the new SAR and EC on the infiltration chart (fig. 14.9).

You can see that adding another 250 pounds per acre-foot (a 50% increase) gives a very small additional infiltration benefit and is not cost effective.

Example: Practical field application using water in fig. 14.9:

Field size/system: 80 ac, single-line drip
Application rate: 0.45 in/day
Flow rate: 700 gpm, 3.12 ac-ft/day
Required gypsum over 80 ac: 1,556 lb/day
Net gypsum application: 19.4 lb/ac
Total injection days for 25 T silo: 32 days
Total season gypsum: 622 lb/ac
(using table 15. 9 from chapter 15) 622 lb/ac
Cost of solution gypsum: $34.21
Cost of 2 T/ac pit gypsum, applied: $67.28

For most field settings, it is rarely necessary to inject gypsum all the time. Most growers inject every other or every third irrigation (as would be the case in the above example), often ending the season with a total application of 600 to 1,000 pounds per acre of 92 percent gypsum. This may or may not be sufficient for your orchard—but even if you doubled the application frequency in the previous example, the cost of the high-quality bulk gypsum applied at 1,200 pounds per acre would be the same as pit gypsum applied at 2 tons per acre during the dormant season. The benefits of gypsum injection during the season are almost always superior to dormant-season applications. Figures 14.10, 14.11, and 14.12 show different types of setups using a gypsum solutionizer machine.

Water-run gypsum in flood irrigation. Of course, gypsum can also be injected into water that is applied to surface-irrigated borders and furrows, but the expense of a silo and solutionizer machine (plus the high-grade gypsum) is quite high. A few growers have set up silos to auger solution-grade gypsum directly into the standpipe attached to the mainline and field valves. Turbulence at this point is usually sufficient to dissolve the gypsum.

A 4-year trial in almonds in Kern County used the simplest approach for midseason, water-run gypsum application. About 250 pounds of high-grade pit gypsum (75% purity) was applied next to

Figure 14.10. Mobile gypsum solutionizer machine with horizontal mixing bar for injecting gypsum slurry into drip system. *Photo:* B. L. Sanden.

Figure 14.11. Gypsum injection machine attached to large silo for automatically adding bulk solution-grade gypsum. *Photo:* B. L. Sanden.

Figure 14.12. This batch mixer with vertical mixing bar is suitable for dissolving some fertilizers for injection as well as solution-grade gypsum. *Photo:* B. L. Sanden.

Figure 14.13. High-grade pit gypsum (75% purity) side-banded next to orchard alfalfa valves at 250 pounds per check for dissolving into water stream during irrigation. Material and application cost $6 per acre. *Photos:* B. L. Sanden.

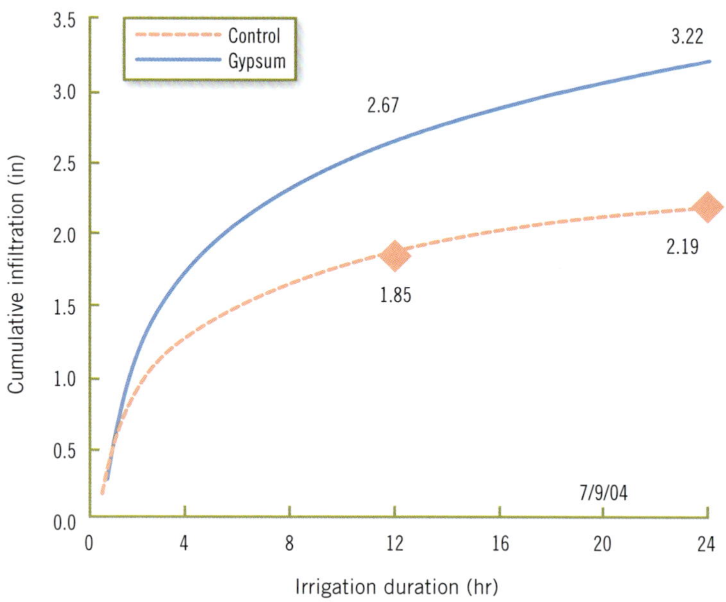

Figure 14.14. Infiltration benefit of water-run, coarse-grade, 75%-purity gypsum in almonds.

the alfalfa valves at the head end of the field using a terragator equipped with a side-throw belt (fig. 14.13) in the beginning of May and again in July. Total calcium applied to the head end was, of course, much higher than at the tail end using this technique, but the benefit to infiltration as determined for the whole check was significant by midseason (fig. 14.14).

Acids and acid-forming materials. Commonly applied acid or acid-forming amendments include sulfuric acid (H_2SO_4) products, soil sulfur, ammonium polysulfide, and calcium polysulfide. The acid from these materials dissolves soil lime, which is native to most California soils, to form a calcium salt (gypsum), which then dissolves in the soil water to provide exchangeable calcium. The acid materials react with soil lime the instant they come in contact with the soil. The materials with elemental sulfur or sulfides must undergo microbial degradation in order to produce acid. This process may take hours or years, depending on the material and particle size (in the case of elemental sulfur). Since these materials form an acid in the soil reaction, they all can reduce soil pH if applied at sufficiently high rates. Fine-ground soil sulfur used to be the most economical way to supply free calcium by acidifying the soil, but high demand by China and India starting in 2007 has sent prices soaring. Some supplies of coarse "popcorn sulfur" are available, but these can take 10 to 20 years to break down. Therefore, the most economical calcium source at this time is ground wallboard gypsum from construction waste. (See table 15.9, chapter 15.)

Water-run acid. The water used for the above gypsum example would be a good candidate for acidifying amendments. Starting in the late 1950s, the Harmon SO_2 Generator (manufactured in Bakersfield) was the first production agriculture "sulfur burner" made available to farmers. These machines feed ground elemental sulfur into a small furnace. The burning produces sulfur dioxide, which combines with water trickled through the machine to make sulfuric acid; the acid is then injected into the irrigation water. In recent years, better pumps and safeguards have been developed to inject concentrated sulfuric acid directly.

Returning to our water in the example above, the pH is quite high (8.4) and the level of bicarbonate (HCO_3) is high at 4.2 meq/l. Adding gypsum to this water and running it through a drip system will significantly increase the chances of plugging the system with lime precipitate. The soil to be irrigated with this water is likely also to be alkaline. If the soil pH is greater than 8, acidification of this water and/

or the soil may be beneficial to crop growth. Neutralizing the bicarbonate will definitely increase free calcium in the soil water matrix and improve infiltration. Table 15.9 in chapter 15 shows that it takes 133 pounds per acre-foot of 100 percent pure sulfuric acid to release 1 meq/l of calcium (assuming that the acid contacts lime in the soil, neutralizing the carbonate molecule and releasing the calcium). This is the same amount of acid required to neutralize 1 meq/l of bicarbonate in this water; the amount would be 532 pounds per acre-foot of 100 percent sulfuric acid. Because the additional acid will rapidly drop the pH, use a pH stick meter or a swimming pool test kit to show how much acid can safely be added to the water. Brass valves, transite pipe, and some membranes and plastics in older systems irrigation systems (pre-1992) are sensitive to pH less than or equal to 4.5. Newer all-plastic systems with fiberglass or epoxy-lined filters are generally safe to a pH of 2.5.

A plastic venturi injector can be used for injecting acid, but noncorroding solid displacement pumps are preferred for accuracy and safety. As a general rule of thumb, however, do not allow the water pH to drop below 5. In the final analysis, a cheaper and less-involved alternative is to band ground sulfur on drip lines or under microsprinklers.

Soil conditioners. Two types of amendments are available in this category: organic polymers and surfactants.

Organic polymers, mainly water-soluble polyacrylamides (PAMs) and polysaccharides, are used to stabilize the soil surface. These extremely long chain molecules bind aggregates together. This action helps resist the disruptive forces of droplet impact and decrease soil erosion and sediment load in furrow irrigation systems. These organic polymers can improve infiltration on soils with illite and kaolinitic clays common in the northwestern United States, but USDA researchers in Fresno have found that infiltration is not improved in soils with mostly montmorillinite clays typical of soils in the San Joaquin Valley. Water-soluble PAM is not to be confused with the crystal-like, cross-linked PAMs that expand when exposed to water. These materials do not influence water penetration; rather, they enhance the water-holding capacity of soils for small-scale applications such as those used at container nurseries.

Organic polymers can have different effects on infiltration. The effect depends on polymer properties—such as molecular weight, structure, and electrical charge—and the salinity of the irrigation water. Interactions between a polymer and a water molecule also affect the flocculation threshold of shrink-swell clays.

Polymers can also be classified as charged (ionic) or noncharged (nonionic). These can behave differently depending on whether they are added to very pure water (like the Friant-Kern, where EC is 0.03 to 0.1 dS/m) or higher-salinity water like the California Aqueduct (0.5 to 0.8 dS/m).

Researchers have noted a correlation between polymer effectiveness and sprinkler irrigation or rain. Polymers have been shown to work best when sprayed on the soil surface at a rate of about 4 pounds per acre, followed with an application of gypsum in soil or water delivered in the form of high-energy droplets.

Other conditioners include synthetic and natural enzymes, microbial soups, and a vast array of products with uncertain scientific validity.

Surfactants, or wetting agents, are amendments that reduce the surface tension of water. They are usually most effective in soils that contain a high percentage of organic matter or are covered with mulch. Such soils include turf soils, forest soils, and burned rangeland. They are usually costly to use for large-scale agricultural applications.

Note of caution: Soils with extremely high silt content (> 50%), mica, or zeolite clays often do not respond predictably when applying amendments. Consult local experts before investing heavily in amending these soils, as you may not see an economic improvement.

Broadcast, Banded, and Surface-Applied Amendments

While adding amendments directly to the water is ideal for managing soils with infiltration problems caused by surface crusting, a variety of other needs may require applying amendments to the soil. Less costly, usually coarser materials can be applied in this manner and serve other purposes besides just improving water penetration. One major use of broadcast amendments is to aid dormant-season leaching for reclamation reduction of root zone salinity. The other purpose is to alter pH to free up micronutrients or supply them directly in the amendment. Chapter 15, "Salinity Management in Soil and Water," briefly discusses these application techniques at the end of the chapter. The nutritional benefit is not addressed.

Deep Tillage in Mature Orchards

In some situations, growers may become desperate and use extreme measures in the dormant season. In most cases, the judicious use of amendments and practices already mentioned will have been put in place years earlier. However, some orchards have been planted to nonuniform layered soils without any deep tillage prior to planting—and in these cases, an examination of backhoe pits reveals significant hardpan and other layers that limit root development. Tillage in orchards with rows up to 18 feet wide is limited to a single pass with a deep ripper. For wider rows, two passes may be possible, but this will be limited by the size of the tractor and the spread of the tree scaffolds.

Caution. Ripping will damage the existing root system, especially in orchards where water penetration has been limited for a long time. However, improved soil characteristics and root pruning will help to encourage new root growth. Shallow chiseling to 15 inches in orchards with swales that collect runoff water with extra silt and clay may help to shatter these structural crusts and aerate surface soils that may be plugged with anaerobic bacterial scum. Roots take time to begin growing, and regrowth varies with the season, soil conditions, and the carbohydrate status of the tree. In any event, do not till all the middle rows at once. Modifying alternate middles each year produces the best results. Ripping is normally most effective in the fall, after harvest, when tree water use is low and soils are dry and easy to shatter and mix. Deep ripping in an existing orchard is not known to be a beneficial practice.

References

Ayers, R. S., and D. W. Westcot. 1985. Water quality for agriculture. FAO Irrigation and Drainage Paper 29, rev. 1.

Faber, B., J. Downer, and J. A. Menge. 2001. Differential effects with mulch in citrus and avocado. UCCE ANR Ventura County website, http://ceventura.ucanr.edu/Com_Ag/Subtropical/Publications/Mulch/Differential_Effects_of_Mulch_-_2001_/.

Holtz, B. A., M. V. McKenry, and T. C. Caesar-TonThat. 2004. Wood chipping almond brush and its effect on the almond rhizosphere, soil aggregation, and soil nutrients. Acta Horticulturae 638:127–137.

Meek, B., L. Graham, and T. Donovan. 1982. Long-term effects of manure on soil nitrogen, phosphorus, potassium, sodium, organic matter, and water infiltration rate. Soil Science Society of America Journal 46:1014–1019.

Oster, J. D., M. J. Singer, A. Fulton, W. Richardson, and T. Prichard. 1992. Water penetration problems in California soils. University of California Kearney Foundation of Soil Science.

Prichard, T. L., W. M. Sills, W. K. Asai, L. C. Hendricks, and C. L. Elmore. 1989. Orchard water use and soil characteristics. California Agriculture 43(4): 23–25.

Sanden, B. L. 2005. Yield and water management impacts of water-run gypsum in border irrigated almonds. Annual Meeting of the American Society of Agronomy, Nov. 6–10, Salt Lake City, UT. Abstract 8788.

15 Salinity Management in Soil and Water

Blake L. Sanden, Allan E. Fulton, and Louise Ferguson

Pistachios are better adapted to moderately saline, calcareous soils than any other commercial deciduous tree crop grown in California. However, proper assessment of potential salinity problems cannot be made simply by looking for visible salt precipitate on the soil surface or around drip emitters (fig. 15.1). Rather, routine soil and water samples are recommended to more thoroughly assess the situation.

This chapter discusses the role of salinity management in establishing and maintaining a successful pistachio orchard. Recommendations are based on the use of soil and water analyses as practical and economical salinity management tools. We will discuss how to

- collect representative soil and water samples
- understand terms used in analytical reports
- check the quality of analytical reports
- diagnose different types of salinity problems
- make remedial management decisions

Soil and Water Analysis

Sampling Approaches

There are two basic philosophies for sampling soils and water to diagnose and manage salinity problems in pistachios. The first involves preplant sampling followed by annual or biannual sampling. This approach is usually required if marginal soil or water quality is often found in the area being developed. The second involves troubleshooting or infrequent analysis only. In areas where soils and water are mostly uniform and of good quality and previous crops show no toxicity symptoms, soil and water sampling is only necessary to troubleshoot problems that may appear after planting or when forced to switch the water supply.

Regardless of which approach you choose, soil and water sampling must be representative for the analytical results to be of value. Results from unrepresentative sampling (e.g., soil pulled from just one or two backhoe pits) may be misleading and costly. Although obtaining representative composite samples involves some effort and expense, the costs are often well worth the effort.

Soil Sampling

Salinity can vary considerably throughout an orchard. You should take at least one composite soil sample for several depths in each zone in the orchard having a similar soil type. USDA soil surveys are good starting points for targeting sampling areas. Even for very uniform fields, a minimum of one composite sample for each 40 acres should be taken. Figure 15.2 illustrates the possible soil variability for

Figure 15.1. Calcium salts precipitated on the soil surface during the winter above a tile drain interceptor and sodic salt ring around a drip emitter (inset), southwestern San Joaquin Valley. *Photos:* B. L. Sanden.

a 160-acre field in Kern County. Fields that have been planted to row crops like cotton and furrow irrigated for many years may have "head and tail zones" that need separate sampling due to differences in leaching. Excavating a backhoe pit to a 6- or 7-foot depth in each area allows a visual examination of the soil profile to assess potential drainage problems and the depth of ripping required. Soil samples for each 1-to-2-foot increment (or layer) can be taken from these pits but should be composited with other subsamples for a given soil type as described below.

Begin by sampling the soil to at least a 4-foot depth in 6 to 12 locations (each spot 20 to 200 ft apart) for a given soil type. An auger with a diameter of 2 to 3 inches works best. Use an open-loop-type Dutch auger for moist clay soils. A standard bucket auger with opposing cutters is required for sandy or dry soils (fig. 15.3). A rubber mallet is very useful for loosening and removing the soil from the auger. Use four buckets, one for each foot of soil, and mix together the samples from the different holes into one bucket for a given depth (e.g., 0 to 1 ft) so that one large sample is formed for that depth. The result will be four composite samples for that soil type. Take about 1.5 pounds of this sample to your preferred agricultural laboratory the same day. If you have to wait a few days to submit your samples, air-dry them by placing each sample in a large paper bag set outside in a dry area exposed to the wind. Compositing minimizes the number of soil samples requiring analysis while achieving the most representative average data for the field.

Figure 15.2. Soil type variability and required sampling areas for a 160-acre orchard development.

Repeat soil sampling in existing orchards at the same time each season every couple of years to evaluate salinity and amendment management strategies with respect to rainfall patterns, irrigation scheduling, and water distribution around the tree. This ensures that salinity distribution and accumulation reflect the long-term trends occurring in the orchard and not just different levels of applied water and crop evapotranspiration (ET, or water use) for that year. Sampling after harvest provides a salinity assessment in an orchard when root zone salinity is usually highest. Irrigation is commonly delayed or deficit during harvest, allowing salts to accumulate in the root zone. Also, fall sampling gives the most advance notice that additional irrigation water may be needed for salinity control during the winter season when the ground is cold, trees are dormant, and extra irrigation for leaching is less likely to cause waterlogging and disease.

The method of irrigation and the ability to apply water uniformly must be considered in order to collect representative samples. In flood-irrigated pistachio orchards, sample 5 to 10 feet to the side of a tree row for best results. In sprinkler-irrigated pistachios, sample across the sprinkler pattern to ensure that two-thirds of the composite includes soils from the center of the wetting pattern, which receive the most applied water, and about one-third from the edges of the wetting pattern, where salts tend to accumulate. The same method applies to microsprinklers and drip irrigation. Sampling fields irrigated with low-pressure systems can be a bit tricky due to the highly variable salt distribution, and consulting local advisors or experts is recommended prior to sampling.

Variable soil textures contribute to nonuniform salinity levels in an orchard. Sandy loam tends to have lower salinity than silt loam and clay loam because infiltration rates are often higher and leaching is greater in sandy loam. Sampling from similar soil types reduces variability in salinity levels. Uniform growth is a good indicator of uniform soil type and water availability.

Why sample soils at multiple depths? If infiltration is a problem or irrigation set duration is too short, salinity in the top 1 to 2 feet will often be higher than at 3 to 4 feet. Surface soils containing salinity and sodium or chloride levels far in excess of the irrigation water usually indicates that soil crusting restricts water infiltration and that irrigation water quality is probably also part of the problem. A soil sample from 0 to 3 inches collected along with a water sample may help diagnose infiltration problems. With adequate irrigation to meet crop

ET, good water penetration, and a small amount of leaching, the root zone salinity (EC_e) of a nonsaline soil at 4 feet deep will be about 2 to 3 times the salinity 0 to 2 feet deep. For microirrigation systems, if the EC_e at 4 feet is four to six times greater than at 1 foot, there is a good chance that the irrigation scheduling has been too efficient, resulting in deficit irrigation and no leaching. In this event, recheck soil salinity levels in the spring after winter rainfall and pre-irrigation recharge. In general, the salinity over all depths can be averaged to determine the potential impact of salinity on pistachio ET and possible specific ion toxicities.

Annual crop with aerial and GPS imagery for field evaluation can provide greater detailed information than zone sampling. This is best done when the field is planted to wheat, alfalfa, or cotton. The resulting map, when coupled with GPS coordinates, more accurately outlines different quality zones in the field than does a USDA soil survey. Composite sampling as described above is then done in these zones and the resulting recommendations for soil amendments applied with GPS-guided variable-rate equipment. For fields with a high degree of variability in salinity, alkalinity, and sodicity, the variable-rate approach might save money compared with conventional application of the same amount of amendment to the whole field. Variable-rate amendments can improve overall orchard development by concentrating the right amount of soil amendments in the right place. However, this overall system is more expensive than conventional sampling and requires the assistance of a professional consulting company.

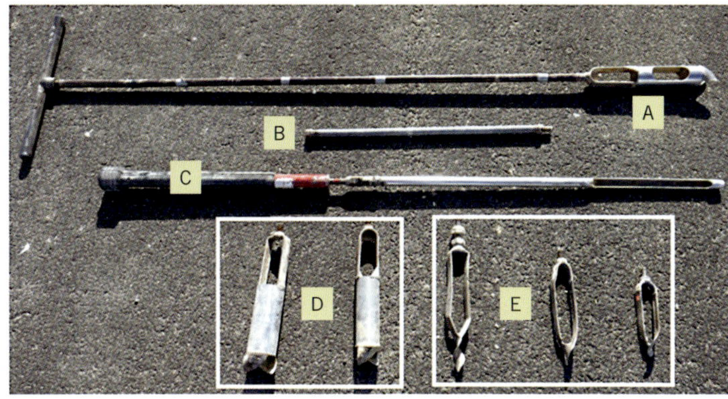

Figure 15.3. Soil augers and probes. The open-sided mud auger (A) is the most versatile auger for moderately sandy to clay loam soils, allowing easier removal of soil through the open sides. Additional extensions (B) allow for sample extraction to a depth of 20 feet. The open-faced Lord-type probe (C) is used for retrieving narrow 12-inch soil cores to a depth of 3 feet using either a simple push "T" handle or a sliding impact hammer as shown. The standard bucket auger (D) is a closed-barrel auger with opposing cutters similar to the mud auger, but it is the only one that works on sands and dry soil. The Dutch auger (E) is best for moist clay soils and cleans out easiest. *Photos:* B. L. Sanden.

Water Sampling

Sampling irrigation water for salinity assessment is much simpler than sampling soils. First, thoroughly rinse a plastic container in the water that is to be sampled. Collect a small sample (4 to 16 oz). Completely fill the container with water; this eliminates air, which would otherwise promote calcium carbonate precipitation.

Before taking a sample from a well, let the pump run for at least 30 minutes (2 to 4 hours in cases where the average EC is greater than 1.5 dS/m). This should be sufficient time to flush the well of static water, stabilize salinity, and establish the pumping water elevation that represents the primary water-bearing strata. If the orchard site is surrounded by other deep wells, sample in July or August when groundwater pumping is the greatest. Collect a representative water sample for laboratory analysis to establish a baseline value. If the pumping depth to groundwater varies considerably over the year and is declining, invest in an inexpensive portable electrical conductivity (EC) meter (usually less than $150) and monitor the total salinity of the well water every 4 to 8 weeks. Submit a new sample for analysis when the total salinity (EC_w) increases about 20 percent. Otherwise, a sample every 3 years is sufficient, unless you suspect the salinity might change over time.

To establish a baseline for surface water, take samples from canals or ditches with flowing water. Submit the first sample for analysis and use this as a reference point. As described for groundwater monitoring, use a portable electrical conductivity meter to determine how often the surface water supply should be analyzed.

If possible, submit a water sample for analysis on the same day that it is collected. If the sample must be stored, refrigerate it to minimize changes in salinity. Storage at room temperature will allow calcium (Ca) and bicarbonate (HCO_3) to precipitate and lower the total salinity level of the water. Usually, irrigation districts test for water quality and can supply you with the lab results.

Reading Laboratory Analytical Reports

An analysis of an irrigation water sample or a saturated soil extract (the most common method of analyzing soil salinity in the alkaline soils of the western United States) measures dissolved salts. Additional salts in the form of ions are attached to electrically charged exchange sites on clay particles in the soil. These ions affect soil structure and infiltration but do not add to possible salt toxicities.

The importance of these ions will be discussed later. Table 15.1 lists the determinations usually provided in an analytical report relating to irrigation water or a soil extract. Analytical results can be lengthy and full of unfamiliar terms; the following section explains what these terms mean.

Terminology and Units

After the composite soil sample is dried and ground, distilled water is added to make a saturated paste. The water is then extracted from the sample by vacuum and used for all the analyses listed in table 15.1.

Saturation percentage

The saturation percentage (SP) equals the weight of water required to saturate the pore space divided by the weight of the dry soil. Saturation percentage is useful for characterizing soil texture. Very sandy soils usually have SP values of less than 20 percent; sandy loam to loam soils have SP values between 20 and 35 percent; and silt loam, clay loam, and clay soils have SP values from 35 to over 50 percent. Also, salinity measured in a saturated soil can be correlated to soil salinity at different soil water contents measured in the field. As a rule, the SP soil water content is about two times higher than the soil water content at field capacity. Therefore, the soil salinity in a saturation extract is about half of the actual concentration in the same soil at field capacity.

pH

The pH of a soil or water supply measures hydrogen ion concentration (activity). Although pH is closely related to bicarbonate concentration and the availability of some macro- and micronutrients, it does not correlate with total salinity. It is, however, important when selecting the most appropriate soil amendments.

Electrical conductivity

The electrical conductivity (EC_e for extracts from soil, EC_w for irrigation water) is a measure of total salinity based on how easily an electric current passes through the extract, but it does not give any indication of the salt composition. However, EC is one of the most important numbers on an analysis because nearly all crop salt tolerance levels are based on EC. The internationally accepted reporting unit for EC is decisiemens per meter (dS/m). This unit is equal to millimhos per centimeter (mmhos/cm), which is still

Table 15.1. Laboratory determinations and units for evaluating irrigation water quality and soil salinity

Test	Symbol	Unit	Water	Soil
saturation percentage	SP	%	no	yes
acidity-alkalinity	pH	none	yes	yes
electrical conductivity	EC_e, EC_w	dS/m	yes	yes
calcium	Ca^{2+}	meq/l	yes	yes
magnesium	Mg^{2+}	meq/l	yes	yes
sodium	Na^+	meq/l	yes	yes
bicarbonate	HCO_3^-	meq/l	yes	yes
carbonate	CO_3^{2-}	meq/l	yes	yes
chloride	Cl^-	meq/l	yes	yes
sulfate	SO_4^{2-}	meq/l	yes	yes
boron	B	mg/l (ppm)	yes	yes
nitrate-nitrogen	NO_3-N	mg/l (ppm)	yes	yes
sodium adsorption ratio	SAR	none	yes	yes
adjusted sodium adsorption ratio	SAR_{adj}	none	yes	no
exchangeable sodium percentage	ESP	%	no	yes
lime requirement	LR	T/ac	no	yes
gypsum requirement	GR	T/ac-6 in	no	yes
lime percentage	$CaCO_3$	%	no	yes

used by some labs. Many labs that do environmental and agricultural testing often report EC_w in micromhos per centimeter (µmhos/cm). Divide these values by 1,000 to obtain dS/m or mmhos/cm.

Total dissolved solids

Another term often found on analytical reports is total dissolved solids (TDS). This is the weight of all soluble salts in milligrams per liter of water (mg/l). About 640 mg/l equals 1 dS/m, but this conversion factor changes as salinity increases and salt composition changes. TDS is not useful in evaluating salinity because crop tolerance thresholds are correlated with EC_e and EC_w rather than with TDS.

Salts

Salts such as sodium chloride (NaCl) and calcium sulfate ($CaSO_4$) consist of positively charged cations and negatively charged anions bonded together by opposing charges. In irrigation water or soil water, many of the bonds are broken, and the water consists of individual cations and anions. To understand the impact of salinity on soil structure and crop tolerance, soil and irrigation water samples must both be analyzed for these soluble cations and anions. Calcium (Ca^{2+}), magnesium (Mg^{2+}), and sodium (Na^+) are the major cations in soil extracts and irrigation water. Although potassium (K^+) is important as a nutrient, it has a low solubility relative to calcium, sodium, and magnesium salts, and soluble potassium is usually a very minor component of salinity. Bicarbonate (HCO_3^-), sulfate (SO_4^{2-}), and chloride (Cl^-) are the main anions in most soil extracts and irrigation water. Along with these anions, boron (B) and nitrate-nitrogen (NO_3-N) are commonly reported in analytical results. Boron does not contribute significantly to total salinity and the osmotic effects (salt stress) on plants, but it is important in diagnosing specific ion toxicity. Knowing the nitrate-nitrogen content of the soil and irrigation water is valuable in making fertilizer decisions, but nitrate does not contribute significantly to the total salinity.

Salinity

The preferred unit for reporting individual cations and anions is milliequivalents per liter (meq/l). This unit is used specifically in salinity evaluation and reporting. Most agriculturists who work with pesticides, fertilizers, and tissue analyses are familiar with parts per million (ppm) and milligrams per liter (mg/l) but are less familiar with meq/l. When all ions are reported in meq/l, this provides the best comparison of the relative ionic strength of the different cations and anions. Clay particles in soil are negatively charged and adsorb the positively charged cations. It is the concentration of the charges, not the weight of the ions, that affects soil structure and eventually the decisions you will make on soil amendments. Reporting cation and anions in meq/l is one of the hallmarks of a quality agricultural lab. These numbers are easily converted back to the weight of the various salts, which is needed when calculating the tonnage of required amendments. Table 15.2 shows how to convert meq/l to mg/l.

SAR

The unadjusted sodium adsorption ratio (SAR), adjusted sodium adsorption ratio (SAR_{adj}), and exchangeable sodium percentage (ESP) are calculated from the individual cation and anion determinations. These indices must be used along with the EC values to evaluate salinity and sodicity accurately. (The calculation of these indices is covered in more detail in chapter 14, "Assessing and Improving Water Penetration," but their importance is described below.)

The unadjusted SAR indicates the ratio of sodium to calcium and magnesium in a soil water extract or irrigation water sample. An increasing SAR value indicates higher levels of sodium in comparison with calcium and magnesium. High sodium concentrations relative to calcium and magnesium reduce soil aggregate stability, decrease water infiltration, and increase the chance of sodium accumulating to toxic levels in leaf tissue. Use the SAR, rather than sodium concentration levels alone, to evaluate sodicity problems. More sodium can be tolerated in a soil extract or water sample when calcium increases proportionally to sodium.

Table 15.2. Conversion factors from meq/l to mg/l

Cation/anion	Symbol	Conversion factor
bicarbonate	HCO_3^-	61
calcium	Ca^{2+}	20
carbonate	CO_3^{2-}	30
chloride	Cl^-	35
magnesium	Mg^{2+}	12
sodium	Na^+	23
sulfate	SO_4^{2-}	48

Note: mg/l = meq/l × conversion factor.

Table 15.3. Example irrigation water quality analysis to demonstrate how to check the quality of a laboratory report

Analysis	Example
pH	8.4
EC_w	1.0 dS/m
Ca	0.5 meq/l
Mg	0.1 meq/l
Na	9.6 meq/l
HCO_3	4.2 meq/l
CO_3	1.0 meq/l
Cl	4.6 meq/l
SO_4	0.1 meq/l
B	0.7 mg/l
NO_3	5.2 mg/l
SAR	17.5
SAR_{adj}	16.6

A high SAR is not the only indication of possible infiltration problems. Irrigation water high in bicarbonate or carbonate can add to possible soil sealing, as these anions can take free calcium out of the water by precipitation of lime.

ESP is closely related to SAR. These ratios differ in that SAR is an index used for water alone, comparing soluble sodium with soluble calcium and magnesium, whereas ESP is for soil only, indicating the amount of sodium ions bound to exchange sites in the soil. Today, most laboratories do not measure ESP directly because this requires extra analyses to measure the total cation exchange capacity and the exchangeable sodium content. Instead, most laboratories report an estimated ESP based on a correlation between SAR and ESP (see chapter 14, "Assessing and Improving Water Penetration").

Gypsum requirement

Most laboratory reports of a soil analysis provide either a gypsum requirement (GR) or a lime requirement (LR). A GR is usually provided for alkaline soils with pH above 7 and SAR above 10 to 15. An LR is only for acid soils with a pH less than 7. The GR reported on the lab analysis is a calculation for the amount of gypsum needed to remove all the sodium from the soil exchange sites. Complete replacement of sodium is unnecessary, so the practical gypsum application needed is usually somewhere between 50 to 75 percent of the lab GR. It is advisable to consult your local UC Farm Advisor for assistance in determining these requirements.

Verification of Lab Results

Lab reports should be checked to make sure that the numbers reported are valid. Sometimes mistakes are made, and simple checks can avoid potential problems. A good report will give the cation and anion concentrations in meq/l. The accuracy of the analysis can be evaluated using the two-step procedure described below.

Check 1: Cation-anion balance

Salts such as NaCl, $NaHCO_3$, and $CaSO_4$ consist of cations and anions bonded together by electrical charges. For each cation, there is an equivalent charge (meq) of an anion bonded to form the salt. This is referred to as the cation-anion balance. Using table 15.3,

$$Na + Ca + Mg \approx HCO_3 + CO_3 + SO_4 + Cl$$

$$10.2 \text{ meq/l} \approx 9.9 \text{ meq/l}$$

When dissolved in a water sample or soil extract, the bonds are broken and the salts exist as individual cations, anions, or neutral ion pairs. The individual cations and ions must be reported in meq/l to perform this check. When reported in meq/l, the sum of the cations equals the sum of the anions. Omit boron and nitrate in conducting this check procedure, because they are reported in mg/l and are usually an insignificant amount of the total salinity. For pH values above 8.5, carbonate is also negligible. Usually the sum of the anions and cations is not exact, but the two ion values should be within 5 to 10 percent of one another.

Check 2: Comparing total salinity with the sum of cations or anions

If the analysis is valid, the salinity level (EC_w) multiplied by a factor of 10 about equals the sum of the cations or anions if the analysis is valid. Using table 15.3,

$$EC_w \times 10 \approx (Na + Ca + Mg)$$

or

$$\approx (HCO_3 + CO_3 + SO_4 + Cl)$$

$$1.0 \times 10 \approx 9.9 \text{ (sum of anions) or } 10.2 \text{ (sum of cations)}$$

In this example, the sum of the individually measured cations or anions is in close agreement to the value calculated from $EC_w \times 10$.

Beware of a report in which the EC multiplied by 10 exactly equals the sum of either the cations or anions, or one in which the cations exactly equal the anions. Such a result may indicate that some of the individual cations and anions were estimated by subtraction rather than determined by direct measurement. Sulfate and sodium are most likely to be estimated, because measuring them requires additional analytical steps and expense.

Diagnosing Potential Salinity Problems

Salinity analyses are used to diagnose three types of salinity conditions in the field: root zone salinity and crop salt tolerance; poor water infiltration rates; and potential accumulation of specific elements to toxic levels.

Pistachio Salt Tolerance

A high soil EC_e or EC_w value indicates high salinity. Excessive salts increase the energy needed by the root cells to extract water. The cells have to concentrate additional solutes to allow water to flow from the soil into the root hairs, thus robbing energy from the tree that would otherwise be available for other growth processes.

Crop evapotranspiration (ET) is directly related to vegetative growth. As water vapor leaves the open stomata on the leaves, carbon dioxide (CO_2) enters for the production of carbohydrates during photosynthesis. When root zone salts are too high, the trees may show reduced shoot growth, smaller nut size, decreased splits, increased sunburn, and kernel shrivel. Necrotic (brown, dead) tissue along leaf tips and margins may indicate excess salt or boron accumulation (fig. 15.4). Decades of salinity research on many crops have established accepted soil salinity thresholds and relative yield decline for many crops (fig. 15.5A).

Pistachios, however, are much more salt tolerant than other nut crops such as almond and walnut. Laboratory studies in Iran and the United States (Ferguson et al. 2002) and a 9-year field study in western Kern County indicate that pistachios can be irrigated with water supplies containing salinity levels up to 8 dS/m without significant yield reduction (Sanden et al. 2004). For this long-term field study, however, it took nearly 5 years to raise the root zone salinity to the 8 to 12 dS/m level with leaching fractions (discussed below) around 30 to 40 percent and 2 to 6 inches of effective winter rainfall. The trees were planted into a soil with an EC_e of about 4 dS/m and established using freshwater from the California Aqueduct (EC of 0.3 to 0.8 dS/m) for 5 years before commencing irrigation with saline water. All trees were budded to Kerman. The UCBI rootstock showed a slight early-yield advantage until the irrigation water salinity reached 12 dS/m (average soil salinity of 13.4 dS/m), at which point the yield of UCBI and Atlantica declined significantly compared with the Integerrima (Pioneer Gold I and II) rootstock. These studies also showed some difference between rootstocks in their ability to exclude

Figure 15.4. Foliar symptoms of excess boron (A) and excess sodium and chloride (B). *Photos:* B. L. Sanden

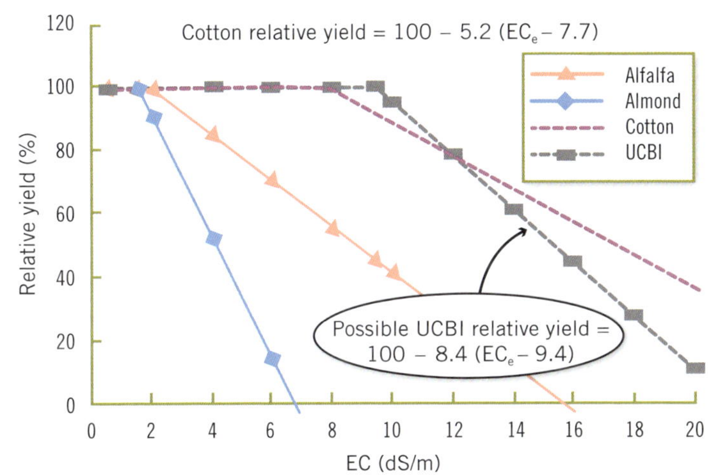

Figure 15.5A. Relative yield (RY) of various crops as a function of soil EC_e. *Source:* Sanden et al. 2004.

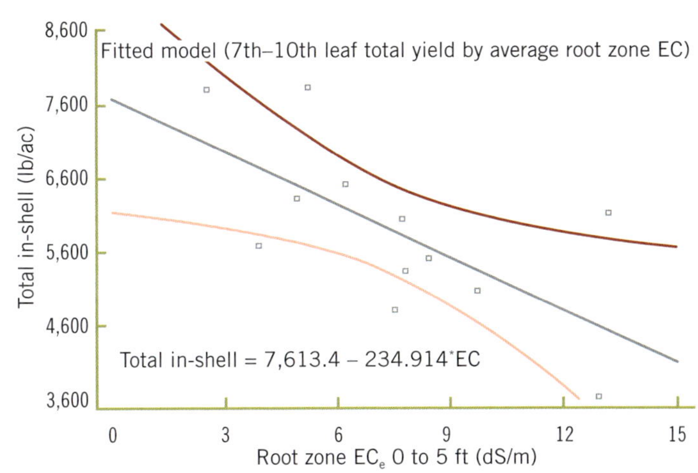

Figure 15.5B. This figure demonstrates how the averaged annual per-acre dry total in-shell yield of 7-to-10-year-old trees (in lb/ac on the vertical axis) decreased as the salinity (in dS/m on the horizontal axis) increased in the root zone. The data is presented as a fitted regression model with yield as a function of increasing salinity, with 95% confidence error bars.

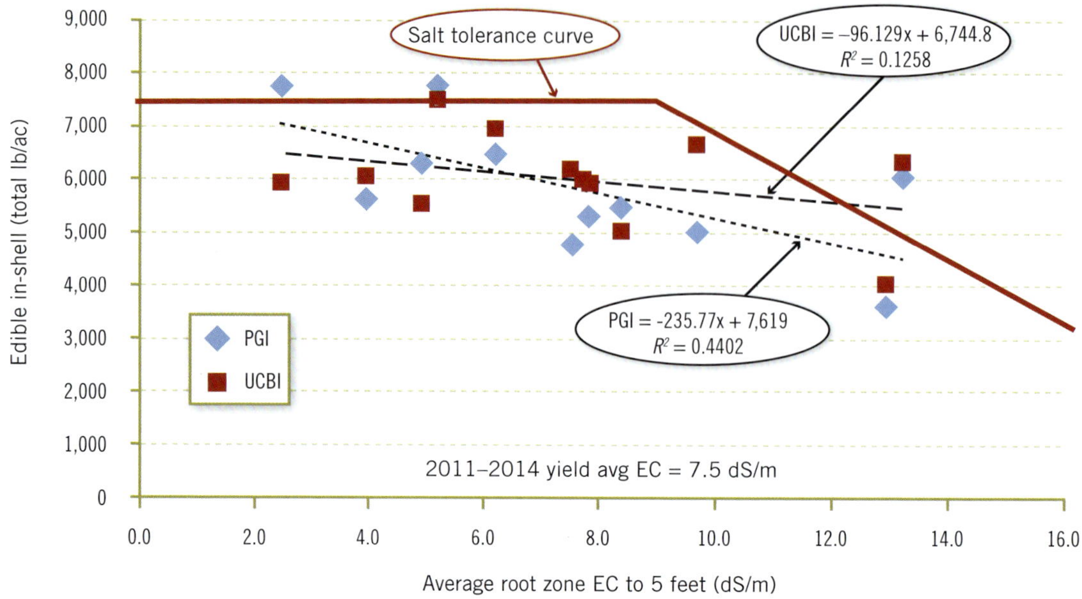

Figure 15.5C. This figure shows the effect of increasing average root zone salinity on dry in-shell edible yield in two long-term salinity trials. The top brown line (the 2004 trial) indicates that pistachios could tolerate a root zone salinity of up to 9 dS/m before yield declined. This data was based on the combined yields of Atlantica, PGI, PGII, and UCBI rootstocks. The black solid and dashed lines (the 2014 trial) indicate that pistachios could tolerate a root zone salinity of 6 to 8 dS/m before yields declined on both PGI and UCBI rootstocks. The yield decline of trees on PGI was greater, and more strongly associated with increasing salinity, than that of trees on UCBI rootstocks. *Source:* Sanden et al. 2004.

sodium and decrease the toxicity symptoms of this ion. Using UCBI as the more sensitive rootstock, the average soil EC_e, and individual tree yield, the study identified a salinity threshold of 9.4 dS/m with a relative yield decline of 8.4 percent per unit increase in EC above this level (see fig. 15.5A). This experiment showed pistachios to be about as salt tolerant as cotton. However, this trial was done in small plots (only four trees of each variety), and it is quite possible that trees in highly saline plots had some roots branching into freshwater plots.

A large-scale (310-acre) study in the Belridge Water District was initiated in 2005 to evaluate the development of newly planted pistachios using saline irrigation water of 3.2 and 5.2 dS/m with 6 and 11 parts per million (ppm) boron. After 3 years and more than 10 times the salt applied to trees watered with saline well water compared with trees watered with the California Aqueduct water, no difference in tree growth was found (Sanden et al. 2008). After 10 years of irrigation with these waters, the saline irrigation water significantly reduced tree canopy and trunk diameter by 7 to 10 percent. After 10 years, the highest root zone salinity was 18.7 dS/m, with an average value of 13.2 dS/m for 7th to 10th leaf. Figure 15.5B reveals that the cumulative in shell yield over those 4 years has a very significant decline above 5 to 6 dS/m—a much lower threshold than the 9.4 dS/m shown in figure 15.5A. Figure 15.5C clearly illustrates this contrast. The good news is that the slope of the yield decline is not the 8.2 percent as shown in figure 15.5A but only 3.1 percent for PGI and 1.4 percent for UCBI. The saline water in this test had a higher ratio of calcium to sodium and sulfate to bicarbonate than that found in most Westside drain waters, and this may provide an extra degree of tolerance to the salt mixture at this site.

Even though these tests indicate that pistachio can tolerate saline irrigation water up to 8 dS/m, experience from more than 600 acres of 25-year-old Atlantica production blocks 2 miles north of this field trial (planted on a 14-by-28-ft spacing) suggests that ultimate tree size may be reduced at long-term soil salinity levels of 5 to 10 dS/m compared with soils with an average EC_e of 2 to 5. However, yields on a per-tree basis are similar to those from the trees in the nonsaline soils.

In the last 10 years, more than 60,000 acres of pistachios have been planted into marginally saline areas. There are always hot spots in these fields, where excessive salinity and alkalinity cause visible tree stress, and the higher salt level may make it more difficult for trees to harden off as they go into the winter. Substantial freeze damage has been seen in some of these hot-spot areas in the last few years, possibly due to this effect. Most orchards planted with drip systems in these areas appear to be doing well—even in some fields where growers had difficulty growing two-bale cotton. Based on these studies and observations and a margin of safety, table 15.4 has been created as a guide to use when reviewing the soil and water EC for mature pistachio trees. These limits assume that irrigation is managed to provide an annual leaching fraction to prevent salt accumulation. (Defining and calculating leaching fractions is addressed later in this chapter in the section "Maintaining Acceptable Root Zone Salinity.")

The field trial used to estimate the above guidelines used saline water where sodium was three to five times the concentration of the calcium. Some Westside tile drainage water and shallow groundwater can run as high as 5 to 10 times the sodium over calcium. These guidelines may overestimate salt tolerance when sodium, chloride, or boron reach these higher ratios when compared with calcium.

Salinity Impacts on Evapotranspiration

Figure 15.6 shows evapotranspiration (ET) data for 2001 from the long-term field trial. Estimates include some drainage for salinity treatments. Full pistachio ET for the season was estimated at 47.8 inches. After starting with a full profile, applied irrigation ranged from 33.5 for the 12 dS/m treatment to 41.5 inches for the 0.7 dS/m control. Field measurements (using the neutron probe) were taken 24 hours after irrigation ceased and the day before the following irrigation. Decreased ET usually means decreased vegetative production, but it does not necessarily mean decreased nut yields.

As a percentage of the 0.7 dS/m ET, the ET for the 4, 8, and 12 dS/m irrigation waters was 92, 84, and 74 percent, respectively. For the 8 and 12 dS/m treatments, this includes several inches of slow leaching throughout the season. Deficit irrigation studies by David Goldhamer et al. (1985, 2012) showed that in-shell yields could still be maintained with water applied at 75 percent full ET. However, the percentage of split nuts can be reduced during sensitive periods.

Toxic Ion Accumulation

Specific ion toxicity, the accumulation of sodium, chloride, or boron to the point of toxicity, can take several years. An excess of these elements in the soil or irrigation water can cause some ions to accumulate in the woody tissue of trees and eventually in the leaves. Burn on leaf margins often means excess chloride or sodium in leaf tissue. The margins of foliage containing excess boron will develop a similar leaf burn that can also have a black edge and expand into interveinal necrosis, with twisting and curling of the leaves. Accumulation of sodium, chloride, or boron ions is likely to reduce production of necessary plant hormones and contribute to nutritional disorders. Figure 15.4 shows the typical leaf symptoms associated with these toxicities.

It is important to diagnose ion accumulation before levels become elevated in the woody and leaf tissues. Once ions accumulate there, the trees have no mechanism to rapidly expel them. Correcting the toxicity in the root zone may require several seasons of proper irrigation management. Analysis of soil, irrigation water, and leaf tissue can diagnose conditions in which toxic ion effects may be a concern. Pistachios have a root physiology that can exclude the uptake of excessive chloride or sodium. Unfortunately, we do not have sufficient data to separate the impacts of salinity in general from the thresholds for specific ion toxicity for sodium, chloride, and boron.

In the previous field experiment in northwestern Kern County, chloride and sodium concentrations were about 40 and 60 meq/l, respectively, for the 8 dS/m irrigation water. The SAR was about 20 and boron concentration was about 1 ppm. These levels in the irrigation water resulted in only slight amounts of marginal leaf burn. In sand tank trials at the USDA salinity lab in Riverside, California, rootstock seedlings showed significant leaf burn after one season of irrigation with water having an EC of 8 dS/m and 10 ppm boron. However, the weight and growth of woody tissue was equal to that of seedlings grown with nonsaline water (Ferguson et al. 2002). In the large Kern County trial begun in 2005, there has been only slight leaf burn by the end of the season of irrigation using well water with an EC of 4.5 dS/m and 11 ppm boron.

Less than half the total boron remains soluble when checking a saturated paste soil analysis. This means that some soils have the ability to tightly adsorb boron and reduce its potential toxicity. This is the case with the Belridge trial on Panoche sandy clay loam.

If soil or water analysis indicates levels of sodium, chloride, or boron higher than the above,

Table 15.4. Guidelines to evaluate orchard soils and water supplies for excess salinity for mature pistachio trees

	Degree of growth or yield reduction			
	Unit	None	Increasing	Severe
average root zone	dS/m	< 6	6–8	> 8–12
irrigation water	dS/m	< 4	4–8	> 8–12

Source: Adapted from Maas 1990.

Note: Guidelines assume leaching fractions of 5–10% for "None;" 10–20% for "Increasing"; and 20–40% for "Severe."

then carefully monitor leaf tissues. It may be only a matter of time before one or more ions accumulate to toxic levels in the trees. Soil and water analyses over time should show declining or steady levels of these ions. If they increase in concentration, then you should rethink your amendment strategy, irrigation scheduling, and leaching. Table 15.5 gives specific ion toxicity limits for sodium, chloride, and boron in July/August leaf tissue.

Nitrate-Nitrogen Toxicity

Toxicity in nitrate (NO_3^-) becomes a concern when too much nitrogen fertilizer is applied. If overapplication is severe, the first effect may be defoliation, but the trees will most likely regrow with tremendous vigor. Foliage may grow so large, in fact, that the leaves curl. Use soil and water analyses to avoid overuse of fertilizer and to achieve efficient nitrogen management practices. The following discussion explains the differences in laboratory reports between total nitrogen and total nitrate, which is the salt form of nitrogen most often taken up by the plant.

Most labs report the amount of nitrogen in the form of nitrate, hence the term is written NO_3-N. This makes it easy to convert the number on the report into the pounds of nitrogen potentially available for crop nutrition. For irrigation water, a nitrate-nitrogen level from 0 to 3 mg/l is considered low; from 3 to 10 mg/l is moderately low to moderately high; and above 10 mg/l is high. For soil samples taken from 0 to 1 feet, a nitrate-nitrogen level from 0 to 10 mg/l is considered low; 10 to 20 mg/l is moderately low to moderately high; and a level exceeding 20 mg/l is high.

To convert the level of nitrate-nitrogen in a water analysis to pounds of nitrogen per acre-foot of water, multiply the concentration reported in mg/l by 2.7. (There are 2.7 million lb of water in 1 acre-foot.) For example, if the analysis reports that a sample of water contains 2.3 mg/l of nitrate-nitrogen, the sample contains 6.2 pounds of nitrogen per ac-ft of water.

Table 15.5. Critical levels of specific ions in pistachio leaf tissue (July/August tissue samples prior to harvest)

Specific ion	Degree of toxicity levels in leaf tissue		
	None	Increasing	Severe
sodium (ppm)	< 100	100–200	> 200
chloride (%)	< 0.2	0.2–0.4	> 0.4
boron (ppm)	< 300	300–700	> 800

Source: Adapted from Ayers and Westcot 1985.

Similarly, you can convert the level of nitrate-nitrogen reported in a soil analysis to pounds of nitrogen per acre-foot of soil. In the lab report, find the level of nitrate-nitrogen (ppm) in a composite sample that represents all samples taken to a depth of 1 foot. Multiply the number by 4 (about 4 million lb of soil in 1 ac-ft). For example, if the analysis reports that the sample contains 9.7 mg/l of nitrate-nitrogen, the amount of nitrogen is 38.8 pounds per acre-foot. Make sure that the lab is reporting nitrate-nitrogen as ppm in dry soil and not the nitrate-nitrogen concentration of the saturation extract. The latter requires a different calculation to estimate the pounds of nitrogen per acre-foot of soil.

Reclaiming Saline Soils and Managing Salinity

The first part of this section deals with reclamation, or correcting an existing salinity problem, while the second part focuses on maintenance of acceptable levels of soil salinity. The former assumes that an orchard or potential orchard site has excess salinity, toxic levels of specific ions, poor infiltration rates, or some combination of these problems. The latter assumes that the orchard has been established on a soil already suitable for pistachio production. In this case, the goal is to manage the soil and irrigation water to avoid salinity buildup and the accumulation of toxic ions and maintain acceptable infiltration rates.

Whether the objective is reclamation or maintenance, proper management of irrigation water is the key to handling salinity. The application of soil and water amendments is also an important component of salinity management, but it is not required in all situations. The following general concepts are critical to effective reclamation or salinity maintenance.

Drainage

Good drainage is the single most effective tool for salinity management. This usually means the absence of hardpans or impermeable clay layers within 12 feet of the soil surface that tend to perch water and prevent the free movement of water and salts below the root zone.

Successful leaching for salinity control is much more difficult on poorly drained soils due to the much slower movement of water and salts below the root zone. Often impermeable clay layers create shallow water tables that can be very salty. The depth of this perched water often fluctuates throughout the season. It is closest to the soil surface in the spring and deepest in the fall. As a result, salinity that may be leached to deeper soil depths in the fall is often transported back into the root zone in the spring when the water table rises.

Pistachios can be grown on soils that are impacted by shallow water, and, if the water is not too saline, even use some of this water for crop ET. But personal experience managing irrigation on thousands of acres of pistachios on the west side of the San Joaquin Valley has shown that maintaining profitable yields on these soils requires more attention to detail than it does on soils with no drainage problems. Many of these soils produced only marginal cotton yields under flood irrigation, but they can be converted to profitable pistachio production using microirrigation systems (fig. 15.7). Pistachios can survive under high levels of salinity and soil saturation, but the combination of these conditions will certainly reduce crop ET and growth. When developing fields with areas prone to these conditions, it is often advisable to increase the amount of dormant irrigation and leaching in low-permeability soils and reduce applied irrigation during the season to avoid oversaturation. Figure 15.8 illustrates this situation, showing reduced tree growth in the row adjacent to a drainage ditch compared with that in the next row over. The perched water depth was about 7 feet, but the root zone under the small trees was saturated up to 2 feet below the surface due to "subbing up" of perched water. If possible, select new fields for development where shallow groundwater stays at least 10 feet below the ground surface.

Some fields are underlain by layers or buried stream channels (stringers) of coarse sand that may intersect adjacent unlined canals. These layers can act as a conduit for freshwater during the season and provide some subirrigation for deep-rooted pistachios. If the field is already well reclaimed and this subirrigation effect does not sub up excess water into the top 3 to 4 feet of the root zone, the trees can be managed so that they achieve optimal growth and yield while also saving the grower as much as a foot of irrigation water. This is the case for several pistachio orchards in the Buttonwillow area, but growers must monitor soil moisture to avoid oversaturation.

Impact of Irrigation Frequency and Method

Small quantities of water (a depth of 1 to 2 in per application) applied frequently over several days are more effective for moving salts below the root zone than an equal depth of water applied in one large application. One-time applications of large quantities of water tend to percolate rapidly through the

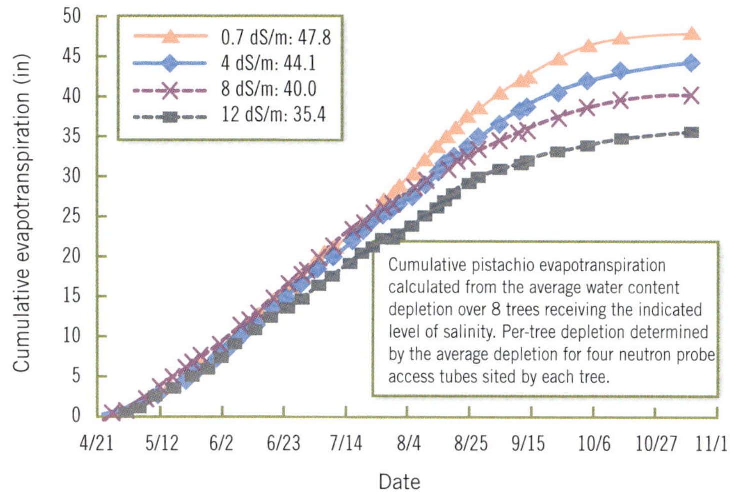

Figure 15.6. Comparative seasonal transpiration for all irrigation salinity levels as determined by soil water content depletion between irrigations.

Figure 15.7. Salinity due to excess saturation, slow drainage, and capillary rise of saline perched water under furrow irrigation in cotton. *Photo:* B. L. Sanden.

Figure 15.8. The effects of poor drainage, reducing tree growth adjacent to an open drainage ditch (saturated soil at the 2-ft depth, left) compared with that in the next row over (good soil moisture and aeration, right) under single-line drip irrigation. The problem is compounded by irrigating to supply ET requirements in the big trees. *Photos:* B. L. Sanden.

larger cracks and pores in the soil. Much less water percolates through the small pores, and the result is a higher salt concentration remaining in the small pores. Sprinkler irrigation, microirrigation, or rainfall allow much more time for salts to diffuse out of the small pores and into the water moving through the big pores and ultimately to leach out of the root zone. Leaching by this method is most effective during the winter, when surface evaporation from the soil and crop transpiration are the lowest. Drip irrigation that wets a minimum of soil on the orchard floor is more effective for leaching during the summer than flood irrigation, but, again, it is important to avoid long periods of saturation.

Reclamation

The depth of leaching required for reclamation is an estimate of the depth of water needed to reduce soil salinity to a level that will not cause yield loss for pistachio. It is commonly expressed as inches of water required per foot depth of soil in the root zone. Reclamation is needed when excessive salts restrict production in existing pistachio orchards or the salinity is higher than acceptable tolerance levels in a soil being considered for development of a new orchard. Table 15.6 shows the depth of leaching required to reclaim various soils. The guidelines in table 15.6 assume that the leaching requirement is applied in several small irrigations or rains, separated by 2 or more days of drainage. Any water lost to evaporation must be estimated and added to the amounts shown in the table. The guidelines also apply to leaching chloride and sodium. Because of its low solubility, leaching excess boron requires as much as five times the water needed to leach other salts, depending on soil type and the residual of precipitated boron in the profile.

Table 15.6. Depth of leaching water required per foot of root zone to be reclaimed given the initial soil salinity and final desired salinity

Desired root zone salinity (dS/m)	Water required (in/ft of root zone) to leach initial salinity			
	6 dS/m	8 dS/m	10 dS/m	12 dS/m
3	1.8	3.9	4.2	5.4
5	0.4	1.1	1.8	2.5
7	0.0	0.3	0.8	1.3

Source: Adapted from Hoffman 1996.

Note: Applicable for all irrigation waters less than 1.0 dS/m.

Example: Reclamation using table 15.6

Assume that a grower has a parcel of land that is being considered for orchard establishment. Laboratory analyses of the irrigation water indicate excellent quality, with an EC of 0.4 dS/m. The average root zone salinity prior to any land preparation is 12 dS/m. The grower would like to have low soil salinity levels for pistachios to a depth of at least 5 feet. Table 15.6 shows that 5.4 inches of water per foot of root zone (in addition to the water needed to bring the soil to field capacity) is required to drop the salinity to 3 dS/m. This equals 27 inches total to reclaim all 5 feet of root zone soil. Due to the high salt tolerance of pistachio, a root zone salinity of 5 dS/m is probably acceptable. Reclamation to this level only requires 2.5 inches per foot, or 12.5 inches for the 5-foot root zone. In practice, some of this leaching can be done after planting (assuming good drainage) in the form of supplying excess water with each irrigation. This can be especially effective with drip irrigation, where only a fraction of the whole root zone (usually 20–40%) is being reclaimed.

If leaching is attempted in one irrigation with a large application of water, the leaching efficiency will decline; it may require more than twice the quantity of water listed above to achieve the same level of reclamation.

If the field overlies shallow water, care must be taken to avoid complete saturation of the root zone. When saturation occurs, it is easier for salts from the lower depths to rise back up into the root zone through capillary action, as evaporation pulls water back toward the soil surface after the irrigation. For this reason winter reclamation is preferred.

Note that all depths of water listed for reclamation or leaching fractions are the depths of water that leach through the soil profile, meaning that the soil is already at field capacity. However, following a crop of alfalfa, cotton, or wheat, the root zone may be depleted by a 5-inch depth to as much as a 12-inch depth of water. This water must be replaced before full leaching can begin.

Maintaining Acceptable Root Zone Salinity

Root zone salinity increases when salts are transported into the orchard with irrigation water. The only way of decreasing salinity is by transporting salts out of the root zone with deep percolation. This is referred to as leaching, and it is an important function of irrigation.

The leaching requirement for maintaining acceptable salinity levels is the fraction of infiltrating water that is not used to refill the root zone or for crop ET but instead percolates below the root zone. It is expressed as a percentage rather than as a specific quantity so discussion of leaching fraction

(LF) can be applied to orchards with various water requirements and water qualities. As the quantity of applied water increases, or as the concentration of the salts in the water increases, more salinity is transported into the orchard. Therefore, more leaching is required to leach salts below the root zone.

Variations in irrigation water quality and soil salinity create the need for different leaching fractions from one orchard to the next. Table 15.7 provides leaching fractions required for irrigation water qualities from 0.2 to 6 dS/m to maintain two different root zone salinity levels.

Example: Leaching fraction using table 15.7

Mature pistachios in the San Joaquin Valley grown in a clean-cultivated orchard planted to a nonsaline soil consume about 42.0 inches of water annually. Assume that the irrigation water has an EC_w of 1 dS/m and that the goal is to maintain an average root zone salinity (EC_e) of 3 dS/m. Table 15.7 shows that an LF of 5 percent is required.

$$\text{Required irrigation} = ET \times (1 + LF) = 42 \times (1 + 0.05) = 44.1 \text{ in}$$

If the EC_w of the irrigation water is 3 dS/m,

$$\text{Leaching fraction} = 42.0 \times (1 + 0.33) = 56 \text{ in}$$

Adequate drainage to handle any excess water above ET is imperative. Monitor soil moisture during the season to avoid saturating the root zone. If significant leaf burn was observed during the season, sample soil salinity by the end of October and use the reclamation information from table 15.6 to estimate needed winter leaching.

These leaching fractions are most appropriate for soils that have already been reclaimed and under continuous cultivation for several years. When this situation exists, it is possible to calculate a desired leaching fraction (as in table 15.7) to achieve a given root zone salinity. The leaching requirement is similar to the leaching fraction, but it can only be established as a requirement when an absolute salinity tolerance threshold has been established for the crop. The leaching requirement is then the leaching fraction that maintains the root zone at or below the threshold level.

Unless your trees begin to decline in the middle of the season and your soil and irrigation water salinity are extremely high (> 6 dS/m), it is usually sufficient to apply water to meet normal crop ET plus some extra for irrigation system nonuniformity. For a new border system with tailwater return, this may be 15 to 25 percent more water than actual ET; for drip irrigation, 10 to 15 percent; and for microsprinkler, 6 to 15 percent. (Note that an actual distribution uniformity, DU, for the orchard should be established by a field evaluation.)

When using this approach, root zone salinity may increase some during the season, but it should return to acceptable levels with postharvest irrigation, winter rainfall, and a light spring irrigation to set your fertilizer. Experience has shown this to be the best program when combined with a continuing sampling program as outlined below.

The Need for Resampling

Every orchard has many unique factors that can impact tree response to water and salinity management. These include nonuniformity of applied irrigation, imperfect scheduling, and often variable ET of the trees, extended harvest cutoff, and variable interaction of soils to different irrigation water quality and fertilizer application. Thus, the effectiveness of the above guidelines can only be verified by resampling the soils on an annual or biannual basis. Repeated sampling will confirm that your salinity management strategy is on track or indicate that adjustments need to be made. Monitoring field moisture levels during the season should always be your first checkpoint.

Root Zone Salinity Modeling Tool

A computer program called WATSUIT (see Oster and Rhoades 1990) predicts long-term EC and concentrations of various salts for a given irrigation water quality and desired leaching fraction. This model has been developed into a user-friendly Windows-based program by Dr. Laosheng Wu at UC

Table 15.7. Leaching fraction (%) above required crop ET to sustain long-term average root zone salinity (ECe, dS/m) for given irrigation water salinity

Irrigation water EC (dS/m)	Desired average root zone EC_e					
	1	2	3	4	5	6
0.2	2	1				
0.5	10	3	2			
1.0	33	10	5	3	2	2
1.5		20	10	7	5	3
2.0		33	17	10	7	5
2.5			24	15	10	8
3.0			33	20	14	10
3.5				26	18	13
4.0				33	23	17
4.5					27	20
5.0					33	24
5.5						28
6.0						33

Source: Adapted from Ayers and Westcot 1985.

Riverside and can be downloaded at the USDA Agricultural Research Service's U.S. Salinity Laboratory website, envisci.ucr.edu/faculty/wu.html.

This model assumes that crop water use follows a 40, 30, 20, and 10 percent pattern from the top to the bottom quarter of the root zone. This is not always true, depending on the soil and the irrigation system or scheduling, but the final average EC and specific ion accumulations are also calculated for the whole root zone.

Additional digital aids

An Excel file, ANALYTICAL CONVERSIONS AND LEACHING CALCULATIONS, is available for download from the Kern County UCCE website, cekern.ucanr.edu/Irrigation_Management/. This spreadsheet can convert lab data from mg/l (ppm) to meq/l, generate a SAR and ESP value to assist in identifying the degree of potential infiltration problems, and calculate the depth of reclamation leaching required by a soil (based on your data).

Finally, one of the best references is *Water Quality for Agriculture* (Ayers and Westcot 1985), published by the United Nations Food and Agricultural Organization and available for free download at the FAO website, fao.org/DOCREP/003/T0234E/T0234E00.htm.

Soil and Water Amendments

Salts cannot be moved without good infiltration and drainage. Some of the field methods for maintaining good water penetration have been discussed previously in chapter 14, "Assessing and Improving Water Penetration." The purpose of this section is to further explain the chemistry and equivalence of different amendment materials that are used for field reclamation and salinity management.

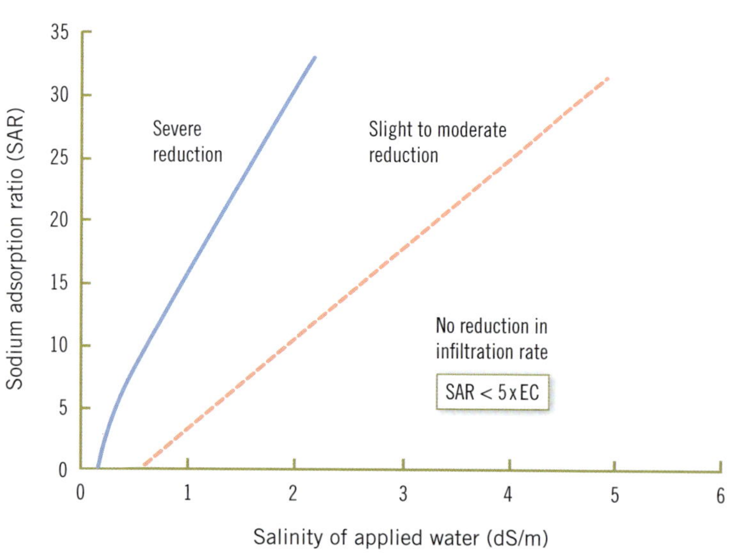

Figure 15.9. Interaction of total salinity as EC with the SAR of applied water for causing potential infiltration problems. *Source:* Adapted from Ayers and Westcot 1985.

Displacing the sodium, and, in some instances, the magnesium and potassium ions attached to exchange sites in the soil, by increasing the calcium ion concentration in the root zone is the major principal of salinity management. Sodium and, to a lesser degree, magnesium and potassium cause swelling and dispersion of clay particles when a soil is irrigated. Calcium, on the other hand, when attached to the electrical exchange sites on clay particles, stabilizes soil aggregates, increases soil porosity, and thereby improves infiltration. Soils that seal up are usually dominated by sodium. Most soil or water amendments contain, or help release, higher amounts of calcium, which helps displace these dispersive ions.

Soils and waters with high SAR values and relatively low EC will be the most responsive to amendments. As EC increases, higher SAR values can be tolerated before causing excessive soil dispersion and sealing, as shown in figure 15.9. In the field trial mentioned in the section "Salinity Impacts on Evapotranspiration," above, the SAR exceeded 20 when the irrigation water EC was 8 dS/m, but infiltration was never a problem.

Irrigation water that is too low in salts (EC < 0.2 dS/m) can often cause problems with dispersion and soil sealing even if the SAR is low. With water above this level, a good rule of thumb is to keep the SAR less than five times the EC. If this condition is met but the root zone salinity is still excessive, leaching may be the most appropriate first step toward correcting the salinity problem before adding amendments.

Types of Amendments

The two general types of amendments are calcium salts and acid-forming amendments. Calcium salts are direct suppliers of calcium, and acid-forming amendments are indirect calcium suppliers through the breakdown of native soil lime that usually exists in high-pH soils.

Common calcium salts include gypsum ($CaSO_4{\cdot}2H_2O$), lime (also called calcite, $CaCO_3$), dolomite ($CaMg(CO_3)_2$), calcium chloride ($CaCl_2$), and calcium nitrate ($Ca(NO_3)_2$). Calcium nitrate and calcium chloride are highly soluble, while gypsum is only moderately soluble over a wide range of pH. When soil and water pH is greater than 7.2, dolomite and lime are only slightly soluble over an extended period of time.

Applying the highly soluble salts through the irrigation water is convenient but typically more expensive. High-grade product and special injection equipment are usually needed for this method. (See chapter 14, "Assessing and Improving Water Penetration," for example calculations and equipment.) Gypsum is reasonably simple to add to irrigation water and is less expensive than calcium chloride and calcium nitrate per unit of calcium. Lime and dolomite are unsuitable as a water-run amendment, as they are nearly insoluble. Gypsum, calcium chloride, or calcium nitrate have negligible effects on soil pH. Lime or dolomite (soil-applied only) is used to increase the pH of acid soils.

Sulfur (S), sulfuric acid (H_2SO_4), urea-sulfuric acid (commonly available as N-PHURIC, $NH_2CONH_2{\cdot}H_2SO_4$), and Nitro-sul ($(NH_4)_2S_2$) are some of the more common acid-forming amendments used in salinity management. Since all contain sulfur or sulfuric acid but no calcium, they supply exchangeable calcium indirectly by dissolving lime that is native to the soil. The sulfur compounds undergo microbiological reactions that oxidize sulfur to sulfuric acid. The acid dissolves soil lime to form a calcium salt (gypsum), which then dissolves in the irrigation water to provide exchangeable calcium. The acid materials do not have to undergo the biological reactions but react immediately with soil lime on application. Acid-forming amendments can also increase the availability of calcium in irrigation water by neutralizing bicarbonate and CO_3 carbonate that otherwise tie up some of the calcium to form lime precipitates. Since these amendments form an acid in the soil reaction, they can reduce soil pH if applied in sufficient quantity.

Amendment Selection

Selecting a soil amendment to reclaim a new orchard site or maintain acceptable infiltration in an existing orchard depends largely on the level of lime in the soil and the relative cost of the materials. As long as lime is abundant in the soil (particularly the surface soil), consider either a calcium salt or an acid-forming amendment.

The choice of amendment depends on the following parameters:

- cost
- how quickly a response is desired
- the depth to which reclamation is desired
- the degree of soil pH correction desired

Table 15.8 lists the various amounts of amendments needed to supply equal amounts of exchangeable sodium to soil and calcium to irrigation water.

Another factor influencing the choice of an amendment is the compound that will be added to the root zone. Some amendments add sulfates (SO_4^{2-}), others chloride (Cl^-) or nitrate (NO_3^-); the nitrate and chloride content will limit how much of these materials can be added. The amount of nitrogen added should not exceed annual crop needs. Concentration of the chloride and its impact on the pistachios should be insignificant when irrigation is sufficient. There have been no reports of sulfates accumulating to toxic levels in pistachios.

Do not use acid-based or acid-forming materials (such as sulfur) in neutral or acidic soils. Only the use of calcium salts is appropriate. Adding lime or dolomite reduces the acidity and makes free calcium and magnesium available. The acid-forming amendments reduce bulk soil pH when applied in sufficient quantity.

Amendment Rates for Irrigation Water

Amendments are most often added to water to improve water infiltration into the surface soil. Amendment rates from 1.0 to 3.0 meq/l calcium are considered low to moderate; rates 3.0 to 6.0 meq/l calcium are considered moderate to high. For example, table 15.8 indicates that an application rate of 468 pounds of pure gypsum, or 266 pounds of pure sulfuric acid, per acre-foot of water supplies the equivalent of 2.0 meq/l calcium. (This assumes that lime is abundant in the surface soil to react with the sulfuric acid.)

As you add calcium, the EC_w increases and the SAR decreases. Remember that the general rule of thumb for minimizing infiltration problems is

$$\text{Desired SAR} < 5 \times EC_w$$

(See chapter 14, "Assessing and Improving Water Penetration," for a specific example on calculating optimal application rates using this guideline.)

Amendment Rates for Soils

Amendment rates for soils are considerably higher than those for water, as the aim is often to achieve a one-time reclamation. The purpose of applying a soil amendment is to reduce the exchangeable sodium throughout the root zone, not just at the soil surface. For soils that have potential for pistachio production, the 100 percent gypsum requirement (GR) may range from 4 to 12 tons per acre (1 to 3 T/ac-ft over a 4 ft root zone). Depending on the choice of amendment, table 15.9 shows that the cost can range from \$60 to \$1300 per acre. If higher amendment rates are

Table 15.8. Amount of amendments required for calcareous soils to replace 1 meq/l of exchangeable sodium in the soil or to increase the calcium content in the irrigation water by 1 meq/l

Chemical name	Trade name: composition	Pounds* per ac-6 in to replace 1 meq/l† exchangeable sodium	Pounds* per ac-ft of water to get 1 meq/l calcium
sulfur	soil sulfur: 100% S	321	43.6
gypsum	gypsum: 23.3% Ca $CaSO_4 \cdot 2H_2O$	1,720	234
calcium polysulfide	lime sulfur: 23.3% S, 7.2% Ca $CaS_5 + H_2O$	1,410	191
calcium hydroxide	Hi-Cal: 13% Ca $Ca(OH)_2 + H_2O$	3,076	418
potassium thiosulfate	KTS: 25% K_2O, 26% S $K_2S_2O_3 + H_2O$	1,890‡ 3,770	256‡ 513
ammonium thiosulfate	Thio-sul: 12% N, 26% S $(NH_4)_2S_2O_3 + H_2O$	807§ 2,470#	110§ 336#
ammonium polysulfide	Nitro-sul: 20% N, 40% S $(NH_4)_2S_5 + H_2O$	510§ 1,000#	69§ 136#
monocarbamide dihydrogen sulfate/ sulfuric acid	N-PHURIC (55%): 10% N, 18% S $H_2NCONH_2 \cdot 2H_2SO_4$	1,090§ 1,780#	148§ 242#
sulfuric acid	100% H_2SO_4	981	133

Source: Adapted from Oster et al. 1992.

Notes:

*Salts bound to the soil are replaced on an equal ionic charge basis and not on an equal weight basis. Laboratory data show that an extra 14–31%, depending on initial and final ESP or SAR, of the amendment is needed to complete the reaction. Elemental sulfur must react with bacteria (thiobacillus) to make sulfuric acid and release Ca from dissolving soil lime.

†The meq of exchangeable sodium to replace = Initial ESP – Desired ESP × Total meq/100 grams soil cation exchange capacity.

‡Assumes 1 meq K beneficially replaces 1 meq Na in addition to the acid generated by the S.

§Combined acidification potential from S and oxidation of N source to NO_3 to release free Ca from soil lime. Requires moist, biologically active soil.

#Acidification potential from oxidation of N source to NO_3 only to release free Ca from soil lime.

needed, the soils may be too costly to reclaim even for a salt-tolerant, higher-value tree like pistachio.

Amendment Application Methods

Adding to irrigation water

Adding amendments directly to the water is ideal for managing soils with infiltration problems caused by surface crusting. Research has shown that, on many soils, as little as ½ inch of irrigation or rainfall can cause a crust that restricts infiltration. Such a crust is only at the soil surface and is often much thinner than 1 inch. Improving the quality of irrigation water with water-run amendments puts the amendments at precisely the point where they are most needed. Such a soil needs relatively small amounts of amendment but frequently applied. Water treatment makes this convenient and accurate. Before applying an acid-forming amendment through the water in this situation, be certain that lime is present in the surface soil or that the water contains high levels of calcium and bicarbonate. Care must also be taken to avoid damage to sensitive parts of the irrigation system when injecting acid directly. Brass valves, transite, and cement pipe, for example, will start to pit when pH is less than 4.5. Some older-style drip emitters (pre-1992) may also have membranes that are sensitive to acid.

Amendments in a liquid formulation are the easiest to apply in this manner and are usually most effective in drip or microsprinkler irrigation systems. Solution-grade gypsum, injected as a slurry, is probably the most common amendment applied by this technique, followed by direct injection of sulfuric acid and N-PHuric fertilizer. A final alternative is a sulfur burner using prilled sulfur pellets to create sulfurous acid (H_2SO_3) on-site for direct injection into the irrigation system (fig. 15.10).

Figure 15.10. Sulfur burner using prilled sulfur pellets to create sulfurous acid to acidify irrigation water in reservoir. *Photo:* B. L. Sanden.

Table 15.9. Approximate bulk purity and moisture content, field tons required, and applied cost/acre for selected calcium-supplying amendments to provide a 100% pure gypsum requirement of 1 to 4 T/ac

Amendment	Purity (%)	Moisture (%)	Approx. cost/T ($)	Field tons and total cost/ac to meet the below 100% gypsum requirements*					
				1 T/ac		2 T/ac		4 T/ac	
				T	$	T	$	T	$
Westside pit gypsum	50	8	15	2.17	73	4.35	137	8.70	274
Lima gypsum (Ventacopa)	75	4	28	1.39	70	2.78	125	5.56	248
bulk solution gypsum† (delivered)	92	3	110	1.12	123	2.24	247	4.48	493
ground wallboard	92	5	35	1.14	68	2.29	122	4.58	236
soil sulfur ("popcorn")‡	99	6	210	0.20	58	0.40	103	0.80	192
soil sulfur (granular)‡	99	2	550	0.19	122	0.38	230	0.77	445
sulfuric acid (applied)‡	98	N/A	450	0.58	269	1.16	537	2.33	1,075
Thio-sul† ‡ (delivered)	N-12, S-26	N/A	475	0.47	223	0.94	446	1.88	892
lime sulfur‡§ (delivered)	Ca-6 S-23	N/A	455	0.67	304	1.34	609	2.68	1,218
N-PHURIC 10/55†‡ (delivered)	N-10, S-18	N/A	530	0.63	336	1.27	672	2.54	1,344
beet lime#	60	10	52	1.08	83	2.15	152	4.31	296

Notes:

Costs determined for Kern County, October 2008. Soils with extremely high silt content (> 50%), mica, or zeolite clays often do not respond predictably when applying amendments. Consult local experts before investing in amending these soils.

*Price assumes freight at $12/T and spreading (where applicable) at $14/ac to spread 3 T/ac.

†Chemigation, no application cost.

‡Free lime must be present in soil.

§Some free Ca, but soil lime is needed for complete reaction.

#Acid soil only.

Broadcasting then irrigating

Broadcasting amendments such as gypsum onto the soil surface and irrigating the amendment into the soil is an alternative to water-run applications. The primary advantage to broadcasting is that the gypsum used is less expensive than that used for water-run treatments. However, for surface applications to be as effective, the application has to be properly timed.

If infiltration is a problem in the summer months, the amendment should be applied at the onset of summer and not during the preceding fall or winter. Applying the amendment too early (winter or spring) often renders it ineffective when it is most needed (July and August) and the soil has again sealed up. Surface applications are most effective when applied monthly during June, July, and August at rates equivalent to 250 to 1,000 pounds of 100 percent gypsum per acre.

Use of finer, evenly graded gypsum works more predictably when used in this manner. If only one application is to be made in the dormant season, it is preferable to use coarse material, as the large chunks will require several irrigations to dissolve. Waste wallboard gypsum is ideal for this application. For orchards using alfalfa-type valves, it is possible to place coarse gypsum in a 100- to 250-pound pile next to the valve and let the water carry it down the field. Growers often find midseason broadcasting of gypsum to be a nuisance (or a cause of crop damage) and prefer to add amendments to the water.

Note that the generation of acid from elemental sulfur is a microbial process that only happens under extended wetting of the soil and sulfur interface and only on the surface of the sulfur particle. Gravel-size sulfur particles can take up to 10 years or more to turn into sulfuric acid, especially if spread only on the surface. The finer the sulfur grind, the quicker will be the acidifying effect. Incorporation as explained below is usually the best technique when using sulfur.

Broadcasting and tillage

Land application of amendments (i.e., the equivalent of 3 to 6 T pure gypsum per ac) is most appropriate when the objective is reclamation of a saline or sodic root zone and not just the surface of the soil. During orchard site preparation, broadcast application of coarser, less-refined amendments is usually the most economic approach. Incorporating the amendment by plowing, shanking, or slip plowing speeds up reclamation by quickly getting the amendment to the deeper soil so the exchange reaction can occur. This practice is often accompanied by the banding technique described below to incorporate acid or sulfur down the tree row prior to planting.

Banded surface application

Banding acid-forming amendments is most often done to correct a micronutrient deficiency in alkaline soils by lowering the soil pH to increase nutrient availability. The major concern is crop nutrition, but also the practice usually benefits soil tilth and infiltration due to the release of free calcium from the soil lime. When preparing ground prior to planting, effective rates of acid applied only to the treated band range from 1 to 4 tons of acid per applied acre, depending on the lime content of the soil. In established orchards, however, applications should not exceed 1,500 pounds per applied acre (a 3 to 4 ft spray band), or crop damage may result. Even smaller amounts of acid or fine sulfur shanked into the soil under drip lines can be effective. This technique is very cost effective for banding gypsum along drip hoses and under microsprinklers.

Furthermore, this technique concentrates these amendments into areas of active rooting. It is not essential to modify the entire orchard floor with the limited wetting patterns of microirrigation. It takes 1 pound of 92 percent bulk sulfuric acid to neutralize 1 pound of lime. This means that 20 tons of acid (or about 7 T of sulfur) is needed to neutralize a 1 percent lime content in an acre of soil 1 foot deep. But it is only necessary to dissolve all the lime in a small area of rooting to increase the availability of micronutrients. Application of acid-forming amendments in irrigation water can also be effective, as long as water is applied with drip or microjet irrigation.

References

Ayers, R. S., and D. W. Westcot. 1985. Water quality for agriculture. Rome: United Nations FAO Irrigation and Drainage Paper No. 29, Rev. 1. FAO website, http://www.fao.org/DOCREP/003/T0234E/T0234E00.htm.

Ferguson, L., P. A. Poss, S. R. Grattan, C. M. Grieve, D. Want, C. Wilson, T. J. Donovan, and C. T. Chao. 2002. Pistachio rootstocks influence scion growth and ion relations under salinity and boron stress. Journal of the American Society for Horticultural Science 127(2): 1994–1999.

Goldhamer, D. A., and R. Kanber. 2012. Pistachio. In Crop yield response to water. FAO Irrigation and Drainage Paper No. 66. Rome, Italy: Food and Agriculture Organization of the United Nations. 416–429.

Goldhamer, D. A., R. K. Kjelgren, L. Williams, and R. Beede. 1985. Water use requirements of pistachio trees and response to water stress. Proceedings of the National Conference on Advances in Evapotranspiration. American Society of Agricultural Engineers: 216–223.

Hanson, B., S. R. Grattan, and A. E. Fulton. 2006. Agricultural salinity and drainage. Oakland: University of California Division of Agriculture and Natural Resources Publication 3375.

Hoffman, G. J. 1996. Leaching fraction and root zone salinity control. In K. K. Tanji, ed., Agricultural salinity assessment and management. ASCE Manuals and Reports on Engineering Practice 71. New York: ASCE. 237–247.

Maas, E. V. 1990. Crop salt tolerance. In K. K. Tanji, ed., Agricultural salinity assessment and management. ASCE Manuals and Reports on Engineering Practice 71. New York: ASCE. 287–290.

Oster, J. D., and J. D. Rhoades. 1990. Steady-state root zone salt balance. In K. K. Tanji, ed., Agricultural salinity assessment and management. ASCE Manuals and Reports on Engineering Practice 71. New York: ASCE. 469–481.

Oster, J. D., M. J. Singer, A. Fulton, W. Richardson, and T. Prichard. 1992. Water penetration problems in California soils. University of California Kearney Foundation of Soil Science.

Sanden, B. L., L. Ferguson, C. Kallsen, D. Corwin, B. Marsh, and B. Hutmacher. 2008. Large-scale utilization of saline groundwater for irrigation of pistachios interplanted with cotton. Western Pistachio Association 2007 production research reports. Fresno, CA: Western Pistachio Association.

Sanden, B. L., L. Ferguson, H. C. Reyes, and S. C. Grattan. 2004. Effect of salinity on evapotranspiration and yield of San Joaquin Valley pistachios. Proceedings of the 4th International Symposium on Irrigation of Horticultural Crops. Acta Horticulturae 664:583–589.

16 Diagnosing and Correcting Nutrient Deficiencies

Robert H. Beede, Patrick H. Brown, Craig E. Kallsen, and Steven A. Weinbaum

Pistachio trees require 17 elements for normal growth and reproduction (table 16.1). Carbon, oxygen, and hydrogen are obtained from air or water; the remaining 14 come from the soil or foliar fertilizers. These essential elements are classified as macronutrients, secondary nutrients, or micronutrients, based on the concentration normally present in plant tissues.

The green plant is, first and foremost, a device for capturing the energy of sunlight and using it, through photosynthesis, to convert carbon dioxide into the food we consume. Plant nutrients are essential for the proper function of many complex biochemical reactions within a plant. The balance between the various elements can affect plant health and productivity. In fact, excessive levels of certain elements in the soil or irrigation water can become toxic to the tree. Optimization of pistachio productivity and quality requires an understanding of the nutrient requirements of the tree, the factors that influence nutrient availability, and the methods used to diagnose and correct deficiencies of these essential nutrients.

Table 16.1. Essential nutrient elements required by pistachio trees for normal growth and reproduction

Macronutrient	Symbol
carbon	C
oxygen	O
hydrogen	H
nitrogen	N
phosphorus	P
potassium	K
Secondary nutrient	
calcium	Ca
magnesium	Mg
sulfur	S
Micronutrient	
zinc	Zn
boron	B
copper	Cu
manganese	Mn
chlorine	Cl
iron	Fe
nickel	Ni
molybdenum	Mo

Supplying Nutrients to the Plant

Nutrients in the soil are absorbed into the roots and then transported to various parts of the tree through the xylem, the water transport system common to all plants. The absorption of water and nutrients by tree roots involves different physiological processes. Water uptake is driven by a physical osmotic gradient created by the loss of water through the leaves into the atmosphere (transpiration) and tension within the soil pores available for water storage. In contrast, nutrient absorption requires the expenditure of energy produced by plant tissues found within the tips of roots. Fine-sized, brown roots also contribute to substantial nutrient uptake because of their length and surface area. Nutrient availability and its rate of uptake are influenced by both soil and plant factors. Soil type, texture, depth, pH, salinity, moisture, and high levels of certain elements affect nutrient availability.

Plant factors such as root distribution, rootstock, fruit load, disease, and nematodes affect the availability of nutrients to the root or their rate of uptake. Similarly, low soil temperature, poor aeration, or the presence of a hardpan can limit the plant's ability to obtain nutrients by limiting root growth and health. Under adverse soil or water conditions, the addition of more nutrients may not improve plant performance. Instead, improving soil conditions that limit nutrient availability to the tree may be the solution. This is often true with modification of irrigation schedules to prevent excessive wetting or drying of the soil. Waterlogged soils cause root suffocation and reduced surface area for nutrient absorption. Conversely, excessively low soil water content decreases the concentration of nutrients (such as potassium) in the soil solution phase, when elements are most available for plant uptake. Amendments intended to change pH closer to neutrality (pH = 7.0) or improve soil structure can also influence nutrient availability to the plant. It is essential that soil quality, irrigation scheduling, water quality, and cropping history be considered before deciding on a fertilization plan. Nutrients are most efficiently used when orchardists subscribe to what is commonly referred to as the four Rs. First, the **right rate** of fertilizer is applied to match the tree demand with nutrient supply. Second, fertilizer is applied at the **right time**, when root activity and crop development optimize uptake. Third, the fertilizer is applied in the **right place** to ensure efficient uptake of the nutrients within the active root zone. Finally, the **right source** of fertilizer is chosen to optimize plant uptake and minimize loss past the root zone or into the environment. This chapter provides guidelines on how to assess orchard nutrient status and correct identified deficiencies using these principles.

Assessing Orchard Nutrient Status

Tree growth and productivity can be improved and maintained by monitoring nutrients in the soil and tree, using soil and plant tissue analysis.

Soil Analysis

Photosynthesis produces carbohydrates, which is the carbon source needed for tree growth and nut production. Soil is monitored to increase the probability that the tree will have adequate access to the nutrients necessary to support maximum photosynthesis. Soil analysis typically provides information on available nutrient levels as well as soil conditions such as pH, cation exchange capacity (CEC, the ability of a soil to retain cations for subsequent release into the soil solution), sodium adsorption ratio (SAR, which impacts water infiltration), and salinity. Soil analysis before orchard establishment enables accurate site assessment for pistachios and identifies soil-related issues requiring correction. For example, high salinity may require correction by soil amendments and leaching prior to planting.

Established orchards also benefit from soil analysis, as it assesses the levels of leachable nutrients. Soil analysis can detect problems involving the macronutrients nitrogen, phosphorus, and potassium. It can also reveal a buildup of elemental salts that are toxic if excessive, such as sodium, chloride, and boron. Soil analysis also assists in assessing imbalances in calcium, magnesium, and sodium, which cause water infiltration issues.

Soil analysis for nutrient assessment is most valuable when conducted in combination with a visual symptom assessment of the tree and tissue analysis. Trees are complex, long-lived perennial plants, whose nutritional status represents an integration of age, previous productivity, tree health, and cultural practices, in addition to soil nutrient availability. Of greatest concern is the nutritional status of the tree, not the soil. Hence, soil analysis may be recommended for greater understanding after a nutrient deficiency or excess is identified through foliar symptoms and tissue testing.

Collecting soil samples representative of the whole orchard is challenging. Pistachio tree roots spread through a large volume of soil, and soil type usually varies within the orchard. Soil characteristics also differ with depth. Surface soil can be quite different from the soil at the 1-to-2-foot depth. Soil samples should be taken where roots are most active, which is commonly the upper 4 feet of the profile. For a thorough analysis, soil samples should be taken from 3 to 10 different locations within the area in question. The samples should be composited and analyzed separately by depth. Replicate samples should then be taken from across the orchard. Nutrient deficiencies can be associated with localized soil differences, such as old river beds, differences in topography, sand deposits, cuts or fills, or old corral or pasture sites. In these cases, it is necessary to sample from within each soil type and relate the findings only to trees growing in that soil type. The differences in soil type can be localized and may only impact a few trees in an orchard (Reisenauer et al. 1983). In addition to auger soil sampling, backhoe pits can be very useful in determining the presence and nature of soil layering, perched water, and fluctuating water tables.

It is also important to consider the effect that irrigation has on root distribution and soil fertility within the root zone. Flood or basin irrigation applies water over a large area relatively uniformly and results in wider distribution of roots and area of nutrient uptake, but it is associated with increased risk of leaf and fruit cluster fungal infections. With microsprinkler systems and drip irrigation, the active area of nutrient uptake is within the wetted pattern.

Interpretive guides

The value of a soil analysis as a guide to fertilization practices is limited by the inability to predict the relationship between soil chemistry and plant nutrient uptake. Soil analysis is best suited for assessment of pH, saturation percentage, CEC, and salinity.

Soil pH. Soil pH is a measure of the hydrogen ions present in the soil nutrient medium readily available for plant uptake. It is a log scale that ranges from 1 to 14, with 1 being highly acidic and 14 highly basic, or alkaline. A pH of 7.0 represents equal amounts of acid and base and is therefore neutral. Soil pH has a significant effect on nutrient availability. In general, determination of soil pH can aid in the diagnosis of nutrient deficiencies. Plants may be deficient in one or several nutrients as a result of the following conditions: high pH (> 7.5), which greatly limits the solubility of many elements (e.g., zinc, copper, manganese, iron), and low soil pH (< 6.0), which can lead to deficiencies of phosphorus, magnesium, or calcium. Aluminum, iron, or manganese toxicity can occur when the pH is less than 4.0. High levels of calcium carbonate (lime) in the soil have been observed to induce deficiencies of magnesium or zinc in developing orchards.

Salinity. Research on the effects of salinity in pistachio indicates that this nut crop has significantly greater salt tolerance than other nut crops (see chapter 15, "Salinity Management in Soil and Water"). Specific upper thresholds for water and soil salinity are discussed in that chapter, but a large-scale field trial conducted in the southern San Joaquin Valley suggests that the present values may not be conservative enough to prevent yield losses over the long term. Research indicates that soil chloride (Cl) and sodium (Na) in excess of 50 milliequivalents per liter (meq/l) and up to 4 parts per million (ppm) boron (B) appear to be tolerated without obvious negative effects. The long-term effect of excess soil boron (> 4 ppm) may be reduced photosynthesis from marginal leaf necrosis, which can affect 50 percent of the total leaf area. Pistachios may tolerate exchangeable sodium percentages (ESP) as high as 15 percent over the long term. However, this level of exchangeable sodium in the surface soil can cause structural deterioration that creates water infiltration problems. Under these circumstances, management strategies (gypsum additions, organic matter amendments, supplemental drainage) should be implemented to enhance soil structure. Hence, water stress can be an indirect but significant effect of high soil sodium levels. Pistachio, or perhaps the choice of rootstock in California, appears to have the ability to block sodium and chloride movement into leaf tissue. Leaf tissue analysis will therefore not provide much knowledge about the sodium and chloride hazard in the soil and water.

Plant Analysis

Leaf analysis may be more useful in diagnosing many mineral deficiencies and some toxicities in tree crops than soil analysis. The mineral composition of a tree is dependent on many factors, such as its age, stage of development, climatic conditions, availability of mineral elements in the soil, root distribution and activity, irrigation, and so on. Leaf samples integrate all these factors and provide an estimate of which elements are being adequately absorbed by the roots. Understanding why a nutrient may be deficient or in excess in leaf tissue requires additional information from soil and water analyses and fertilization practices. Good record keeping is therefore invaluable in understanding and maintaining crop nutrient status.

Sampling procedure

It is essential that sampling techniques be standardized if valid comparisons are to be made among fields and from one season to the next. Choice of sampling method also varies depending on the purpose of the survey. If the aim is to identify a problem in an isolated area, sampling a few poor trees and some good ones should suffice. If a determination of overall nutrient status in a large orchard is required, more extensive sampling of trees from many sites will be required.

The standard sampling procedure is as follows. Fully expanded subterminal leaflets (pistachios typically have three to five leaflets per compound leaf) are randomly collected from **nonfruiting** branches at about 6 feet from the ground. Four to 10 leaflets are typically collected per tree, and 18 to 20 trees are sampled in each orchard block, with each tree 25 yards apart (Siddiqui and Brown 2013). The surface contamination of leaves sprayed with micronutrients cannot be easily removed by washing, so the results from the laboratory should take this into consideration. Additional effort could be made to sample fully expanded leaflets from growth produced after the foliar treatment. Samples should be kept in labeled paper bags, protected from excessive heat, and submitted to the analytical service within 24 hours of collection.

The critical levels established through experimentation and observation (table 16.2) are based on leaf sampling from late July through August. Comparative samples to diagnose problems and identify differences between good and poor trees can be collected at any time of the year.

Table 16.2. Critical and adequate nutrient levels (on a dry-weight basis) for pistachio leaflets sampled from late July to early August

Element	Critical value	Adequate range
nitrogen (N)	1.8%	2.2–2.5%
phosphorus (P)	0.14%	0.14–0.17%
potassium (K)	1.6%	1.8–2.2%
calcium (Ca)	2.0%	2.1–4.0%
magnesium (Mg)	0.45%	0.5–1.2%
sodium (Na)	—	—
chlorine (Cl)	—	0.1–0.3%
manganese (Mn)	30 ppm	30–80 ppm
boron (B)	90 ppm	150–250 ppm
zinc (Zn)	7 ppm	10–15 ppm
copper (Cu)	4 ppm	6–10 ppm

Source: for nitrogen, Weinbaum et al. 1988, 1995; for potassium, Brown et al.1999; for boron, Uriu and Peterson 1984 and Brown, Zhang, et al. 1993; for zinc, Uriu and Pearson 1981, 1983, 1984, 1986; for copper, Uriu et al. 1989; for magnesium and calcium, Brown and Siddiqui 2012.

Early-sampling protocol for in-season nutrient correction

In 2014, an early-sampling protocol called UCD-ESP was developed by University of California researchers (Siddiqui and Brown 2013). This protocol acknowledges that sampling leaves in July is too late to allow for current-season adjustment of fertilization rates, and leaf sampling alone does not provide sufficient information to make fertilizer recommendations. The model uses May leaf analyses (40 days after full bloom) and yield estimations to predict the July percentages of nitrogen and potassium. A May assessment of nitrogen and potassium levels allows for fertilizer rate adjustments prior to the onset of kernel filling in late June, when these macronutrients are in greatest demand. These guidelines are based on extensive data from four high-yielding orchards across California from 2009 to 2012. This model and additional sampling guidelines can be accessed at the UC ANR website, ucanr.edu/sites/scri/Assessment_of_Nutrient_Status_in_Pistachio/PPM/ .

Interpreting Leaf Analysis

Results of plant tissue analysis are reported as the concentration of a nutrient on a dry-weight basis. For macronutrients, concentrations are reported as a percentage of the dry weight, while micronutrients are reported in parts per million (microgram nutrient per gram dry weight). For each element, the laboratory will usually identify the "critical value" (CV) or the "adequate range" to aid in interpretation of the results. Table 16.2 provides the current critical values and adequate range for pistachio leaflets sampled from late July to early August according to the guidelines previously described. Critical value, or critical level, refers to the nutrient concentration at which distinct symptoms of deficiency are present. The expression of visual symptoms for a given element at the reported critical value is not always exact. Tissue nutrient concentrations below this level will result in poor plant growth and potentially reduced yields.

The adequate range refers to the nutrient concentration range at which growth is not limited. Adding more of a nutrient as a fertilizer may not increase leaf tissue concentrations of that nutrient in subsequent sampling. Improved plant performance should also not be expected. Research has shown that the presence of excessive amounts of nitrogen or potassium within the soil does not correlate with plant tissue concentration, because nutrient uptake is driven by plant demand, not concentration gradient. A 6-year nitrogen trial in two locations reported only a 0.2 to 0.3 percent increase in August tissue levels between the 50- and 500-pound nitrogen treatments (Ferguson et al. 1987). Also, tissue concentrations above the suggested adequate range may inhibit plant growth either by nutrient toxicity (such as boron and chloride), by induced deficiencies of other nutrients (high potassium requires more magnesium for a positive yield response), or by soil concentrations so high that plant uptake capacity is exceeded, and the risk of leaching beyond the root zone exists. Nitrogen is the element of greatest concern for leaching because of its potential to reach the groundwater, with subsequent negative effects on human health.

Critical values are crop specific. It is essential that the testing laboratory use the nutrient recommendations specific to pistachio, since its requirements differ significantly from other tree nut crops. An example is boron (B), whose requirement is much higher in pistachio than almond, walnut, or pecan. Although valuable as a tool to assess orchard nutritional status, adequate values are not absolute. They are based mostly on general tree health and not necessarily on yield or crop quality. Some nutrients, such as boron during bloom and potassium and nitrogen during kernel filling, may require supplementation for short periods to optimize production (Brown, Ferguson et al. 1993; Weinbaum et al. 1995) even though tree health appears good.

Ideally, fertilization of mature pistachios would replace only the amount of nutrients consumed by the plant in seasonal growth and crop production. To achieve this objective, the total annual requirement of each nutrient must eventually be determined, as well as the percentage of each nutrient removed from the orchard in the crop. A model has been developed for mature pistachios that estimates the nitrogen and potassium requirement based on the yield history, current conditions, and previous applications (Rosecrance et al. 2000). It can be accessed at the UC Fruit and Nut Research and Information Center website, ucce.ucdavis.edu/rics/fnric2/PistachioNKModel/pistachio_nk_model.html.

Critical values for nitrogen, potassium, calcium, magnesium, boron, zinc, and copper have been established through several research projects. Other nutrient critical values remain estimates from field observation and levels acceptable in other deciduous crops (see table 16.2).

For several nutrients (nitrogen, boron, and chloride), it is equally important to ensure that nutrient concentrations do not exceed the optimum, as impaired plant growth may occur. Exceeding optimal levels is not necessarily toxic. Often there is a range of "luxury consumption," where no harm is done but no benefit is achieved.

Correcting Nutrient Disorders

Correction of common nutrient disorders usually involves applications of fertilizers. Fertilizing the tree should not be confused with feeding the tree in the same sense that we feed animals. As discussed above, applying fertilizers in excess of plant needs will not produce greater yield. The energy needs of plants are met through photosynthesis. Generally, the application of nutrients in excess of meeting the needs of the tree is an unnecessary expense, and it may lead to phytotoxicity as well as surface and groundwater pollution.

Identifying and Correcting Common Nutrient Disorders

Nitrogen (N)

The macronutrient nitrogen is the most widely needed fertilizer element in pistachios. Nitrogen is used by plants to synthesize amino acids and nucleic acids necessary for all plant functions. Nitrogen-deficiency symptoms eventually appear in most orchards if nitrogen is withheld. Annual leaf tissue and soil analyses can ensure that sufficient nitrogen is applied to meet the crop needs without wasting money on unnecessary application, without encouraging vegetative growth at the expense of reproductive growth, and without polluting surface and groundwater supplies.

Shoot growth is reduced in nitrogen-deficient pistachios. Shoots are thinner, shorter, and, in severe cases, they have reddish bark. Nitrogen is mobile (moves readily throughout plant tissues), and new leaf production is at the expense of older leaves if nitrogen is deficient. Young leaves pale (loss of bright green color) as older leaves turn yellow and drop from the tree early (fig. 16.1). Excessive leaf drop results in a tree with sparse foliage. The petioles and midribs of nitrogen-deficient pistachio leaves become red (Gonzales and Uriu 1985), and crop yield will decrease from reduced nut size and number with increasing nitrogen deficiency. Over- or underirrigation can produce similar symptoms.

Figure 16.1. Young nitrogen-deficient leaves are pale and are lower on the shoot. As deficiency increases, older leaves turn yellow and drop from the tree early. *Photo:* C. Kallsen.

Correcting nitrogen deficiency

April is a good time for initial seasonal nitrogen fertilization. Nitrogen uptake from the soil before early leaf-out is essentially zero, because nitrogen uptake occurs simultaneously with water use by the leaves. Consequently, applications made in the winter are subject to leaching beyond the primary root zone at 3 to 4 feet, especially in production areas averaging 15 or more inches of rainfall. The risk of groundwater contamination is also greater.

Nitrogen uptake in pistachio occurs primarily from early May to late August. The efficiency of nitrogen utilization is greatly enhanced with multiple applications at lower rates, using drip or low-volume fertigation, compared with less frequent, larger applications. Light but frequent application minimizes the potential loss of nitrogen past the root zone. Orchards using basin or furrow irrigation should apply granular nitrogen immediately ahead of the irrigation water or inject liquid forms just below the soil surface to minimize losses from volatilization and to improve application uniformity. Nitrogen use efficiency (NUE) is different from application efficiency. NUE pertains to the plant's consumption of nitrogen relative to that available from the soil, water, and fertilizer. In practice, most modern production estimates of NUE in pistachio are calculated as the nitrogen exported from the field—that is, as yield (kernel, shell, and hull)—divided by the nitrogen applied. Application efficiency (Ea) pertains to how uniformly nitrogen can be physically applied throughout the orchard. Fertigation research conducted in microsprinkler-irrigated almond orchards suggests that this method can provide NUEs of at least 70 percent. Water-run nitrogen application in surface irrigation is much less efficient and is subject to excessive leaching below the root zone.

In one University of California study of mature pistachio trees, nitrogen uptake over the course of the season was similar in on- and off-bearing years at 150 and 136 pounds, respectively (Weinbaum et al. 1995). Nitrogen stored in the tree during the off-year appeared to support the greater demand that nut production placed on the tree during the on-year. Additional University of California nitrogen research (Siddiqui and Brown 2013) has determined that 28 pounds of actual nitrogen is required for every 1,000 pounds of pistachio nuts (kernels, shells, and hulls) produced. On average, an additional 25 pounds of nitrogen is needed to support tree growth. This amount will be reduced slightly in poor-growth years and increased in strong-growth years (+/- 10 lb). Hence, an estimated on-year crop of 4,000 pounds would require almost 196 pounds of actual nitrogen for the season (4 times 28 lb per 1,000 lb of nuts, plus 25 lb for growth). As good environmental stewards, the goal should be to achieve 100 percent nitrogen efficiency. This is not currently possible, and hence a modest uptake-efficiency multiplier will be required (multiply estimated N demand by 1.3) .

Shortly after bloom, assess the potential crop load and make an estimate of the total nitrogen requirement for the season, including the amount required for tree growth and application efficiency. Assess the quantity of residual soil nitrogen using recent analyses. Nitrogen availability in the irrigation water (especially groundwater) should be assessed, since it can be a major nitrogen contributor. A concentration of 10 ppm nitrate-nitrogen in well water contributes 27 pounds of actual nitrogen per acre-foot of applied water. The total nitrogen requirement for the year should be estimated by reviewing the most recent plant tissue analysis to confirm that nitrogen is within the adequate range. If this value is deficient or excessive, the nitrogen fertilization rate should increase or decrease appropriately.

Begin nitrogen fertilization in April with approximately one-fourth of the seasonal requirement, depending on the irrigation method. Since off-year trees begin the season lower in stored nitrogen, one might consider applying half the season's nitrogen prior to shell hardening and the remainder in July and August (Weinbaum et al. 1988, 1991, 1993, 1995). Adding high levels of nitrogen to the soil early in the season during an on-year will not result in greater plant uptake unless the tree is deficient. Avoid excessive irrigation so that nitrogen is not leached past the root zone. Kernel filling accounts for 90 percent of the seasonal nitrogen demand, which must be met from storage within the tree and the nitrogen in the soil and water. Nitrogen application should cease by the first week in August. Visual observations of plant growth and leaf color should be included in any nitrogen management program.

Available data indicates that pistachio growth and yield are not improved with July/August tissue levels above 2.5 percent. Results from a 6-year field experiment in which annual nitrogen application rates reached 500 pounds of nitrogen showed no yield increase above the 3,750 pounds of split nuts obtained with only 200 pounds of nitrogen (Ferguson et al. 1987). Table 16.3 provides a hypothetical fertilizer program using common mixes to meet the nitrogen and potassium requirements of a 4,000-pound, dry, in-shell crop with a fertigation application efficiency of 70 percent.

Current regulatory trends suggest that California's Regional Water Quality Control Board regulations will require monitoring of soil, water, and tissue nitrogen levels for the creation of a nutrient management budget for this macronutrient. Nitrogen applications can be reduced if residual soil levels are high at the beginning of the season. Soil concentrations of 25 ppm nitrate-nitrogen would be considered moderately high in deep, well-drained soil with a rooting profile of 4 to 5 feet. Twenty-five ppm of nitrate-nitrogen in the top 4 feet of soil would be equivalent to approximately 400 pounds of nitrogen

Table 16.3. Hypothetical pistachio production nitrogen and potassium fertilization program

Fertilizer	Lb per gal of product	Gal per ton of product	Lb nutrient per gal of product	
			Nitrogen	Potassium
UN32	11.06	181	3.54	0
10-0-10	9.70	206	0.97	0.97
15-0-5	9.70	206	1.46	0.49
Application date	**Fertilizer source**	**Gal (lb) per acre**	**Lb nitrogen per acre**	**Lb potassium per acre**
April	UN32	7.0 (77)	25	0
May	10-0-10	51.5 (500)	50	50
June	10-0-10	51.5 (500)	50	50
July	15-0-5	51.5 (500)	75	25
Total			**200**	**125**

Note: Marketable yield = 4,000 pounds dry, in-shell nuts; uptake efficiency = 70%.

per acre. Five ppm nitrate-nitrogen would generally be low. Soil nitrogen testing provides an indication of residual nitrogen availability. Plant uptake of this residual depends on root health, water management, crop load, and plant demand. In addition to contributing to the nitrogen required for optimal tree performance, monitoring soil nitrogen is important to prevent it from reaching levels that risk its leaching past the root zone and into the groundwater.

Potassium (K)

The macronutrient potassium is an activator of many plant enzymes. Potassium has important functions in plant-water relations, where it regulates ionic balances within cells. Potassium regulates the leaf stomata opening and subsequently the rate of transpiration and gas exchange, which affects photosynthesis. Plants also need potassium for the formation of sugars and starches, the synthesis of proteins, and the division of cells. It increases the oil content of pistachios and contributes to their cold hardiness.

Potassium-deficient leaves begin showing symptoms in early to midsummer. Leaves become pale (resembling those affected by nitrogen deficiency), and the edges of the leaflets later fold upward and curl inward, revealing a grayish cast (fig. 16.2). The symptoms are most pronounced on older leaves of current shoots. Potassium-deficient trees grow more slowly and produce smaller leaves. More severe deficiency symptoms include leaf yellowing, which starts at the tip and progresses down the curled leaf margins. The affected area widens with time. The yellow leaf tissue eventually turns brown and dies, resembling boron toxicity. Trees moderately deficient in potassium (< 1.5%) may look normal but produce yields below their potential.

Figure 16.2. Early symptoms of potassium deficiency. Leaves become paler, and the edges of the leaflets later fold upward and scorch adjacent to the margin. *Photo:* V. Gonzalez-Reyna.

Correcting potassium deficiency

Research has shown that mature, heavy-bearing pistachio trees require large quantities of potassium (Weinbaum et al. 1995). Most of this potassium is in the nuts and is removed from the orchard at harvest (Brown et al. 1999). Initial research indicated that 18 to 26 pounds of potassium is removed per 1,000 pounds of nuts on a dry-weight basis at harvest (Brown et al. 1999; Weinbaum et al. 1995). The most current research has refined the potassium requirement to 24 pounds per 1,000 pounds of nut dry weight. An additional 22 pounds of potassium is required to support mature tree growth and development (Brown and Siddiqui 2012).

Since potassium is used in large quantities by the developing crop, there is considerable potential for potassium deficiency if it is not applied for several years.

Potassium fertilizer rates can be adjusted based on a suggested leaf tissue concentration of 2.0 percent by dry weight. Injecting or soil banding potassium fertilizers in neutral pH loam soils at rates of 100 to 200 pounds per acre of potash (K_2O) have proven effective in correcting potassium deficiency. This application resulted in higher yield from increased nut weight and split nut percentage, and it reduced blank nuts (Brown et al. 1999). Potassium chloride (KCl), potassium sulfate (K_2SO_4), and potassium nitrate (KNO_3) all provided correction. This research was conducted on San Joaquin, Yolo, and Arbuckle soil series, and the amount of fertilizer potassium necessary can vary greatly depending on soil type. Deficiencies are more likely to develop on lighter soils, which have lower cation exchange capacity (and less slowly available, or "fixed," potassium) and possibly low soluble potassium due to years of farming and plant extraction. Additionally, in some soils, potassium may be fixed within the mineral matrix so that the potassium ion is less available for uptake. The extent of potassium fixation is largely determined by soil mineralogy. When vermiculite clay and other mica-based parent materials are present, potassium fixation can be a significant barrier to meeting the potassium requirement. The Columbia soil series in northern California and the San Joaquin series on the east side of the San Joaquin Valley are examples of potassium-fixing soils.

Pistachios have the greatest demand for potassium during the kernel fill stage of nut development (late June through August.) Nut fill accounts for over 90 percent of the seasonal potassium uptake in pistachio (Rosecrance et al. 1996). The potassium requirement from leaf-out to shell hardening (early June) is low and the trees would not likely respond to fertilization unless the available soil potassium was quite low. Fertilization of soils that fix potassium should therefore be made shortly before nut fill, the period of greatest tree demand. Injecting potassium through drip or microsprinkler irrigation water, or using surface application in bands along drip hoses, has been effective in counteracting fixation since the concentration of potassium is sufficient to saturate the soil fixation potential in that area, allowing plant uptake.

Excessive potassium fertilization can interfere with the uptake of other nutrients like calcium and magnesium. Table 16.3 provides a hypothetical fertilization program designed to meet the seasonal nitrogen and potassium requirement, given the demand of 24 pounds of potassium per 1,000 pounds of dry, in-shell nuts and 22 pounds for tree growth per acre.

Magnesium (Mg)

Magnesium, also a macronutrient, is an activator for many enzymes involved in energy transfer and growth processes. It is a component of chlorophyll and thus is essential for photosynthesis. Magnesium deficiency has not been widely reported in pistachio. Deficiencies are more likely to occur in sandy and alkaline soils. On the west side of the San Joaquin Valley, pistachios are grown on high-pH, highly calcareous and boric soils. These soils may also fail to provide sufficient available magnesium for optimal

Fig. 16.3. (A) Magnesium deficiency develops on mature leaves located in the middle of the shoot. Early symptoms appear as paling of the green tissue between the secondary and smaller veins. (B) The margins of the leaf then become brown and brittle (scorch), leaving an inverted "V" of pale green to yellow tissue along the midvein. (C) Scorching of the tissue continues to the midvein of the leaf, resulting in loss of the "V." Affected leaves may then drop. *Photos:* J. Moore.

tree growth due to antagonistic competition for uptake by excessive calcium and other cations present on the soil colloids. In a similar manner, high rates of gypsum may induce magnesium deficiency in some soils.

Deficiency symptoms appear in late spring to midseason on mature leaves. Symptoms can be isolated to the middle portion of the shoot. Early symptoms appear as degreening of the leaf tissue between the lateral and secondary veins (fig. 16.3). The leaf margins become scorched (brown and brittle), leaving an inverted "V" of green or yellow tissue along the main vein. The scorched area can eventually consume almost the entire leaf. Scorched leaves will then drop. Magnesium deficiency might be confused with potassium deficiency, but any doubt can be resolved by leaf tissue analysis. Recent research shows that magnesium uptake can be limited by a high concentration of potassium (Siddiqui and Brown 2013). Where magnesium deficiency is present, leaf deficiency symptoms have been corrected with foliar fertilizers containing magnesium applied at 50 percent leaf expansion in May. Where soils are acidic, correction may also be obtained with applications of dolomitic limestone.

Zinc (Zn)

The micronutrient zinc is the third most common deficiency in deciduous trees. It is needed in a number of enzymes, as well as in membrane integrity and hormone synthesis. Zinc is required to make auxin, the plant hormone responsible for cell elongation and growth. Zinc also is needed to make chloroplasts, the leaf component that contains chlorophyll.

Zinc deficiency is widespread throughout all pistachio-growing areas. It is prevalent in pistachios grown in sandy and older, oxidized soils of the San Joaquin, Sacramento, and southern interior valleys. Old corral sites and orchards with regular applications of animal manure are prone to zinc deficiency due to their high phosphorus levels. Zinc is rendered nearly insoluble and much less available for plant uptake in soils with pH values of 8.5 and greater. Zinc deficiency is most common in younger tissue due to its immobility.

Zinc-deficiency symptoms appear early in the season. Newly planted and early-bearing trees are particularly prone to zinc deficiency. A spring symptom is the delayed opening of vegetative and flower buds by as much as a month. This delay gives the appearance of cold injury to the 1-year-old wood in the upper canopy. When vegetative buds finally open, the terminal leaves are small, often chlorotic (i.e., yellow), and appear in tufts, giving rise to the terms *rosette* and *little leaf*. Severe deficiency also causes dieback of young terminal shoots (fig. 16.4). Marginally deficient leaves are only slightly reduced in size and show chlorotic areas between the lateral veins. These symptoms can be masked by growth later in the season. However, deficient leaves remain so during the entire season. Zinc-deficient leaves have a distinct, wavy leaf margin. Nuts on zinc-deficient shoots are markedly smaller in size and much redder than normal. The number of nuts per cluster is also dramatically reduced, and most of them are blank (i.e., have no kernel) at harvest.

Symptoms of zinc deficiency may be restricted to terminal leaves on individual branches of a tree. A tree may have several zinc-deficient limbs while the rest of the tree appears healthy. This pattern of zinc deficiency further demonstrates the immobile nature of zinc within the tree. Highly available manganese appears to have induced zinc deficiency in sandy soils in the California high desert east of the Sierra and in Arizona (Kallsen, unpublished data).

Correcting zinc deficiency

Foliar sprays applied postbloom, when leaves are 50 percent expanded (end of April to early May), are the most effective at increasing tissue zinc concentrations. Absorption decreases markedly as leaves mature and acquire a fully developed waxy cuticle. Spring applications are made with 2 pounds of 36 percent zinc sulfate per 100 gallons of spray solution per acre adjusted to a pH of 4.5 to improve solubility and absorption. While a single application is usually sufficient for the growing season, foliar sprays of zinc provide only a temporary annual correction (Brown et al. 1996). For trees that have demonstrated low zinc levels in July or August leaf samples, high rates of zinc sulfate have been applied during the late dormant time period in late February and early March, during the swelling of the bud scales (Beede et al. 1991). However, the delayed dormant timing has not proven as effective as the leaf-expansion timing, and orchards prone to low zinc levels should be treated annually in late April. Most soils are not truly low in zinc. Rather, the chemistry of the soil prevents zinc from being sufficiently available to the plant. Correcting zinc deficiency through soil applications can therefore be slow and expensive. Soil texture and pH are the primary factors affecting the efficacy of zinc fertilization. In sandy soils, trenching 2 to 6 pounds of 36 percent zinc sulfate per tree, depending on tree size, has produced good response. Soil treatment becomes less effective in fine-textured soils or soils

Figure 16.4. (A) Early zinc-deficiency symptoms in the spring. Leaf margins become wavy and shoots appear stunted. As zinc deficiency progresses, terminal leaves are small, often chlorotic (i.e., yellow), and distance between internodes on the shoot decreases, giving rise to the terms *little leaf* and *rosette*. Severe deficiency causes dieback of young terminal shoots. In zinc-deficient bearing trees, nuts are small and reddish, and many are blank at harvest. (B) By midseason the zinc-deficient shoots will often be leafless below the terminal rosette and occur in clusters of shoots. (C) By mid- to late season, a tree with moderate to severe zinc deficiency will have multiple bare shoots with terminal rosettes. By midseason, however, zinc-deficient trees also often have copper deficiency (see fig. 16.7). The major difference is that with copper deficiency the shoot tip curls over and dies instead of developing a rosette. *Photos:* (A) and (C) R. H. Beede; (B) L. Ferguson.

with high pH (≥ 8.0). Fertigating 70 pounds per acre of zinc sulfate annually for 2 years in a clay loam soil (pH 7.5) on the west side of the San Joaquin Valley increased soil zinc levels but did not increase zinc concentration in leaf tissue (Kallsen et al. 2000). Treating alkaline soil with acidic or acid-forming chemicals may release substantial amounts of zinc chemically bound in the soil and thus increase tree uptake. Nitrogen fertilizers like ammonium sulfate, ammonium phosphate, urea, and urea-ammonium nitrate have an acidifying effect on the soil, but their free hydrogen content is insufficient to make large changes quickly. Sulfuric acid, urea-sulfuric acid, and phosphoric acid provide a more rapid response, and they can be applied uniformly through most low-volume irrigation systems. Elemental sulfur also acidifies soil, but its reaction time depends on granular size. It is also best applied as a band to localize acidification and improve response.

Boron (B)

The boron requirement in pistachio is the highest known for any tree crop. Boron, a micronutrient, is required for cell wall structure and may also have a role in maintaining cell membrane integrity. Reproductive structures have a higher demand for boron and are particularly sensitive to boron deficiencies (Brown, Ferguson et al. 1993). Boron appears to be extremely important in pistachio flowering, pollen viability, and nut set. Severe boron deficiency in the soil has been recognized in limited areas, including the Sierra Nevada foothills of central California and sometimes in Lake and Mendocino Counties. Moderate deficiencies are common in orchards on the east side of the San Joaquin Valley. Boron-deficiency symptoms appear early in the season as tip dieback of growing points and deformity in young leaves. The terminal buds can also remain dormant. Lateral, normally dormant buds sprout into growth with short internodes. This new growth may then also experience dieback. In some shoots brown-gray areas of bark distributed irregularly over the shoot may precede this dieback. These shoot and growth abnormalities produce a tree that is stunted with a bushy appearance.

Pistachio leaves deficient in boron are green but crinkled, and the midrib and secondary veins of the leaf become elevated over the leaf surface (fig. 16.5). The leaf tips curl upward and sometimes become so misshapen that they appear bractlike in form. Mature leaves are thick and feel leathery to the touch. They can also become brittle and drop prematurely. The petioles and stems may also be thickened, corky, cracked, and shortened. Entire flower clusters on young, boron-deficient pistachios often drop during bloom and litter the orchard floor. Correcting boron deficiency can be slow and require repeated treatment due to its immobility within the tree.

Toxicity resulting from excessive boron is occurring more frequently as more orchards with saline soils and poor-quality water are planted. Symptoms begin showing in mid- to late summer as boron accumulates in the leaf tissue. Levels in excess of 500 ppm are common where boron is high in the water or soil, and concentrations have reached 2,000 ppm on a dry-weight basis. Where boron is in excess, first the tip and then leaflet margins become brown and necrotic (fig. 16.6). In severe cases, the necrotic tissue progresses into the interveinal areas, and the margins of the leaflets curl upward. Every leaf on the tree may be so severely scorched that the tree appears as if it were injured by fire.

Figure 16.5. Pistachio leaves deficient in boron are green but crinkled, and the midrib and secondary veins of the leaf become elevated over the leaf surface. *Photo:* L. Ferguson.

Figure 16.6. Where boron is in excess, first the tip and then leaflet margins become brown and necrotic. Terminal dieback subsequently occurs in late summer. *Photo:* C. Kallsen.

Correcting boron deficiency

Soil or foliar spray application can correct boron deficiency. Similar to other micronutrients, boron is most available in lighter, acidic soil. A number of boron fertilizer materials containing different amounts of boron are available. Soil application of Borax (11% boron) by the first week in September at the rate of 50 to 75 pounds per acre has given correction lasting from 3 to 5 years. Soils lighter in texture and low in buffering capacity require less boron for correction. Applications through drip systems may be reduced by 25 to 30 percent due to higher efficiency from concentrated treatment. When tissue boron is less than 90 ppm in August, 2 ounces of Solubor (20.5% boron) per tree can be safely applied to low-boron soils through the drip system or as a topical application on either side of the tree. Apply at the base of trees up to 2 years old and to the area wetted by irrigation in older trees. Pistachios easily tolerate tissue levels of 250 ppm without foliar injury.

Boron is transported, and accumulates, in leaves in the tree's water transport system (xylem) but cannot be subsequently retranslocated out of leaves (phloem immobile). As a consequence, boron must be available in small amounts at all times of the year. Since flowering in pistachios occurs prior to significant soil uptake of boron, the boron required for flowering and fruit set must be present in the flower buds in the previous season. To reduce the possibility of a boron deficiency at flowering, preventive foliar boron sprays are frequently used (see below). If boron is deficient in July/August leaf samples, late summer soil treatments may be followed by an application of 2.5 to 5 pounds of Solubor in 100 gallons of spray solution per acre in mid-March (bud swell) to supply critically needed boron to developing flowers and pollen for improved fruit set. Boron can also be applied postbloom at 50 percent leaf expansion (mid- to late April). Apply 3 pounds of Solubor in 100 gallons of water per acre on expanding leaves. Depending on what else is in the solution, Solubor may create a strongly buffered solution of 8.2 pH. If mixed with zinc and copper fertilizers, the tank mixture should be acidified to a pH of 4.5 to 5.5 with citric acid to maintain the uptake of zinc and copper by the leaf.

A foliar spray of soluble boron can be applied anytime that boron deficiency is detected. Boron application prior to bloom, between early bud swell and just prior to first bloom, has been demonstrated to increase fruit set even when leaf symptoms of boron deficiency were not observed (Brown, Ferguson et al. 1993). Sprays during bloom are not recommended, as these can interfere with pollination and impose an increased risk of phytotoxicity. Spring applications during shoot extension are more effective in correcting current-season deficiency than treatments in June or July. The best long-term correction can be expected from a combination of soil and foliar treatments applied yearly. Rates of application should be based on trends observed in tissue analysis and soil tests performed prior to boron application.

Copper (Cu)

Like zinc, copper is a component of many enzymes in the plant and plays a role in energy metabolism. Deficiencies of copper are relatively common in pistachio. Like zinc, copper deficiency is more likely in newly planted and early-bearing orchards. Commonly, zinc- and copper-deficiency symptoms occur together. Neither element can be readily remobilized from older leaves.

Copper deficiency can be quickly differentiated from zinc deficiency because its symptoms appear in midsummer, not during early spring. The tips of the immature leaves near the shoot terminal appear not to develop rather than actually burn, which causes the leaflet to be clover shaped. Slight shriveling of the shoots then occurs, and small dark lesions develop near the shoot tip. Scorched leaves defoliate. Terminal dieback subsequently occurs in late summer (fig. 16.7). Some affected shoot terminals curl downward, resembling a shepherd's crook (fig. 16.8). Kernels are often badly shriveled. Canopy development is severely compromised during the deficiency and correction period.

Correcting copper deficiency

Postbloom foliar application of chelated copper (CuEDTA 7.5 to10%) applied at ⅓ to ½ pound in 100 gallons of water per acre when leaves are 50 percent expanded (end of April to early May) is effective in correcting copper deficiency (Uriu et al. 1989). Application of copper sulfate ($CuSO_4$ 25%) at ½ to 1 pound per 100 gallons of spray solution per acre during this same time period has been successful, but it may cause phytotoxicity. In many cases, treatment for copper deficiency is only required during the early years of orchard development. Once the trees are mature, the incidence of copper deficiency diminishes substantially, presumably because of increases in root density.

A limited survey showed no correlation between soil copper concentration and pistachio tree deficiency (Beede 1989). Soil treatments with copper sulfate and CuEDTA over several years provided highly variable responses (Uriu et al. 1989). Some treatments showed no correction whatsoever, some provided correction, and still others showed phytotoxicity. Copper sulfate fertigated through drip irrigation in an alkaline soil at a rate of 40 pounds per acre annually for 2 years increased soil copper concentration but did not increase copper concentration in pistachio leaf tissue (Kallsen et al. 2000). This variability is most likely linked to soil pH and texture. Lighter, slightly acidic soils might be more responsive. The risks associated with soil treatment are also not justified since foliar treatment has proven highly effective.

Occurrence of Other Nutrient Deficiencies

Phosphorus (P) and calcium (Ca) exist in high concentrations in most California orchard soils and have never been reported as deficient. The phosphorus requirement of pistachio is also quite low

Figure 16.7. Copper deficiency. The tips of the immature leaves near the shoot terminal appear not to develop rather than actually burn, which causes the leaflet to be clover shaped. If severe, terminal dieback subsequently occurs in late summer. *Photo:* L. Ferguson.

Figure 16.8. Midseason terminal dieback from copper deficiency can be distinguished from zinc deficiency by the characteristic curling of the copper-deficient shoot into the shape of a shepherd's crook. *Photo:* R. H. Beede.

relative to nitrogen and potassium. Sidiqqui and Brown (2013) determined that about 3 pounds of phosphorus is removed per 1,000 pounds of dry, in-shell harvested nuts. Of all the major nutrients essential for plant growth, calcium, magnesium, and sulfur (as sulfate) occur in the soil solution in greatest abundance. Calcium and magnesium typically occupy 80 to 90 percent of cation exchange sites on clay particles of productive soils. Soils with high sodium concentration may have 50 percent or less calcium and magnesium on the exchange sites. The effect this has on nutrient and water availability can be significant. Iron (Fe) deficiency (also known as lime-induced chlorosis) is common in other tree crops, but it has not been reported in pistachio. Symptoms of overall yellowing of the canopy in the spring are more likely, due to excessively wet soil. Manganese (Mn) deficiency has also not been confirmed in pistachio. Manganese-deficiency symptoms in other tree crops appear on mature basal leaves in mid- to late summer. Affected leaves first begin to pale, as if they were nitrogen deficient. Leaf size is normal, but leaf surface becomes mottled with chlorotic areas similar to effects of zinc deficiency. Tissue analysis would differentiate between the two.

Chloride (Cl) and Sodium (Na) Toxicities

Sodium toxicity is difficult to separate from chloride toxicity, as the two ions often occur together. The hazards that chloride and sodium present to pistachio, as well as ways to ameliorate them, are discussed in chapter 15, "Salinity Management in Soil and Water."

Concentrations of chloride and sodium encountered in California soils, although often sufficient to kill many other crops, do not typically produce pistachio leaf symptoms because of pistachio's ability to limit the concentration of these elements in leaf tissue. However, in other countries where soil chloride and sodium concentrations are extremely high and irrigation is limited, high leaf concentrations of chloride and sodium have been shown to produce leaf symptoms similar to those caused by boron. Salt injury to leaves does not typically occur until early to midsummer, when their uptake through plant water use begins to cause necrosis (browning) of the leaf tips. Continued accumulation within the leaf tissue causes necrosis along the margins—and under highly saline conditions, it progresses toward the midvein of the leaflet. Choride injury alone tends to burn the terminal half of the leaflet, rather than just the margins. The use of *P. integerrima* and its hybrids as rootstocks in California (unlike in most other parts of the world) may account for this difference. In California, leaf tissue analysis provides only limited information on the chloride and sodium hazard of the soil. Analysis of soil and irrigation water is necessary to identify the source and degree of hazard, as well as the amount of leaching required.

Effect of Rootstock and Scion on Tree Nutrition

Pistachio rootstocks differ significantly in their ability to take up nutrients from the soil. Trees planted on *P. atlantica* rootstocks are less likely to show boron, copper, or zinc deficiency than other rootstocks (Brown et al. 1991). *P. integerrima* and UCBI are equally prone to zinc and copper deficiencies, but boron deficiency is more likely in UCBI (Ferguson et al. 2008). Differences have also been found among scions in nutrient uptake. For example, the cultivar Golden Hills has exhibited higher concentrations of boron in leaves and more pronounced leaf margin burn (where boron is excessive in the soil) than the cultivars Kerman and Lost Hills (Kallsen and Parfitt 2008).

References

Beede, R. H. 1989. Effect of orchard soil chemistry on copper deficiency in pistachio trees. In California pistachio industry annual report 1989, 81–82. Fresno: California Pistachio Commission.

Beede, R. H., J. Padilla, and D. Thomas. 1991. Foliar boron and zinc nutrition studies in pistachio. In California pistachio industry annual report 1991, 121–126. Fresno: California Pistachio Commission.

Brown, P. L., and M. I. Siddiqui. 2012. Development of leaf sampling and interpretation methods for pistachio and development of a nutrient budget approach to fertilizer management in pistachio. In California pistachio industry annual report 2012, 329. Fresno: California Pistachio Commission.

Brown, P., L. Ferguson, and G. Picchioni. 1993. Boron nutrition of pistachio: Final report. In California pistachio industry annual report 1993, 57–59. Fresno: California Pistachio Commission.

Brown, P., Q. Zhang, and R. Beede. 1993. Effect of foliar fertilization on zinc nutritional status of pistachio trees. In California pistachio industry annual report 1993, 77–80. Fresno: California Pistachio Commission.

———. 1996. Foliar spray applications at spring flush enhances zinc status of pistachio trees. In California pistachio industry annual report 1996, 101–106. Fresno: California Pistachio Commission.

Brown, P., Q. Zhang, and L. Ferguson. 1991. Nutrient uptake efficiencies of various pistachio rootstocks. In California pistachio industry annual report 1991, 158–162. Fresno: California Pistachio Commission.

Brown, P., Q. Zhang, Z. Huang, B. Holtz, and C. Hornung. 1999. Agronomic and economic responses of mature 'Kerman' pistachio trees to potassium applications in California. In California pistachio industry annual report 1999, 84–85. Fresno: California Pistachio Commission.

Ferguson, L., J. Maranto, B. Blackwell, and P. Dosanjh. 1987. Nitrogen nutrition of pistachios: Fourth year progress report. In California pistachio industry annual report 1987, 63–64. Fresno: California Pistachio Commission.

Ferguson, L, H. Reyes, B. Sanden, S. Grattan, L. Epstein, and B. Krueger. 2008. Pistachio rootstocks. In L. Ferguson, ed., Pistachio production manual: UCCE pistachio production short course. Madison, WI: Omnipress. 74–75.

Gonzales, V., and K. Uriu. 1985. Nutritional deficiency symptoms in pistachio. In California pistachio industry annual report 1985, 69–73. Fresno: California Pistachio Commission.

Kallsen, C. E., and D. E. Parfitt. 2008. Differences in leaf boron concentration among four pistachio genotypes. Poster abstract. HortScience 43:1283.

Kallsen, C. E., B. Holtz, L. Villaruz, and C. Wylie. 2000. Leaf zinc and copper concentrations of mature pistachio trees in response to fertigation. HortTechnology 10:172–176.

Polito, V., C. M. Niven, A. Fabbri, L. Dollo, P. Metheny, L. Ferguson, H. Cruz, W. Bentley, and B. Blackwell. 1993. Investigation of "damage by other means" in developing pistachio. In California pistachio industry annual report 1993, 87–91. Fresno: California Pistachio Commission.

Reisenauer, H. M., J. Quick, R. E. Voss, and A. L. Brown. 1983. Chemical soil tests for soil fertility evaluation. In Soil and plant tissue testing in California. University of California Press Bulletin 1879. 39–41.

Rosecrance, R., S. A. Weinbaum, and P. H. Brown. 1996. Assessment of nitrogen, phosphorus, and potassium uptake capacity and root growth in mature, alternate bearing pistachio *(Pistacia vera)*. Tree Physiology 16: 949–956.

———. 2000. Development of nitrogen and potassium fertilizer recommendation models to improve nitrogen and potassium use efficiency and to alleviate pollution from pistachio orchards. In California pistachio industry annual report 2000, 124. Fresno: California Pistachio Commission.

Siddiqui, M. I., and P. Brown. 2013. Pistachio early-season sampling and in-season nitrogen application maximizes productivity, minimizes loss. In California pistachio industry annual report 2013, 9. Fresno: California Pistachio Commission.

UC Davis Fruit and Nut Research and Information Center website, fruitsandnuts.ucdavis.edu/Weather_Services/Nitrogen_Prediction_Models_for_Almond_and_Pistachio/.

Uriu, K., and J. Pearson. 1981. Diagnosis and correction of nutritional problems. In California pistachio industry annual report 1981, 30. Fresno: California Pistachio Commission.

———. 1983. Diagnosis and correction of nutritional problems, including the crinkle leaf disorder. In California pistachio industry annual report 1983, 79. Fresno: California Pistachio Commission.

———. 1984. Diagnosis and correction of nutritional problems, including the crinkle leaf disorder. In California pistachio industry annual report 1984, 49–50. Fresno: California Pistachio Commission.

———. 1986. Zinc deficiency in pistachio–diagnosis and correction. In California pistachio industry annual report 1986, 71–72. Fresno: California Pistachio Commission.

Uriu, K., R. Teranishi, R. Beede, and J. Pearson. 1989. Copper deficiency in pistachio. In California pistachio industry annual report 1989, 77. Fresno: California Pistachio Commission.

Weinbaum, S. A., and T. T. Muraoka. 1988. Nitrogen usage and fertilizer recovery by mature pistachio trees. In California pistachio industry annual report 1988, 84–86. Fresno: California Pistachio Commission.

Weinbaum, S., P. Brown, and R. Rosecrance. 1993. Assessment of nitrogen uptake capacity during the alternate bearing cycle. In California pistachio industry annual report 1993, 47–48. Fresno: California Pistachio Commission.

———. 1995. Assessment of nitrogen and potassium uptake capacity during the alternate bearing cycle. In California pistachio industry annual report 1995, 56–60. Fresno: California Pistachio Commission.

Weinbaum, S. A., G. A. Picchioni, P. A. Brown, T. T. Muraoka, and L. Ferguson. 1991. Nutrient demand, storage and uptake capacity of alternate bearing pistachio. In California pistachio industry annual report 1991, 148–157. Fresno: California Pistachio Commission.

17 Harvesting, Transporting, Processing, and Grading

James F. Thompson and Adel A. Kader

Substantial crop loss can occur during preparation for harvest, in the harvest operation, and during transport to the processing plant. This chapter discusses how to minimize yield losses and maintain nut quality during postharvest handling, processing, and grading.

Preparing for Harvest and Harvest Timing

Preparation for pistachio harvest begins in early August with special attention to three factors. The first is irrigation. Pistachio shell splitting is particularly sensitive to irrigation deficits. Therefore, irrigation should be carefully monitored, preferably using budget irrigation methods, through the August and early September shell-splitting period. The objective is to maintain adequate soil moisture to maximize shell splitting while ensuring that the orchard rows will be dry enough to support harvest equipment.

The second factor in harvest preparation is monitoring for *Amyelois transitella*, navel orangeworm (NOW). Infestation by NOW results in unsightly damage by the larvae and increases the susceptibility of the nut to infection with *Aspergillus* spp., fungi that produce the carcinogen aflatoxin. Depending on temperatures, calculated as degree-days, the third generation of NOW emerges in early to mid-August and lays eggs in the sutures of the early-split nuts. In these nuts, the shells split before the hulls dehisce. When the shell splits prematurely, the hull also splits and exposes the kernel to infestation. Though early-split nuts are usually no more than 1 to 5 percent of a tree's total nut load, an infestation of even 1 percent of the nuts can result in aflatoxin levels above maximum allowable levels, 20 parts per billion (ppb) in the United States and 4 ppb in the European Union. The first step in monitoring for NOW is to assess the quantity of early-split nuts. If early splits are higher than one or more nuts per cluster, the early-split nuts should be monitored for eggs. If egg laying is present, the trees should be sprayed with an insecticide (see chapter 18, "Navel Orangeworm and Obliquebanded Leafroller," for more information on NOW control). Preharvest intervals vary depending on the insecticide used, and harvest must be delayed until the interval has elapsed. Thus, spraying for NOW control must include a harvest date projection.

The third factor of harvest preparation is determining when to harvest. Maturation of pistachio hulls is signaled by color changes and is closely linked to shell splitting. As maturity proceeds, the hull turns from green to ivory and then to rose, and the shell turns from translucent to opaque. Lack of color changes generally signals a blank or aborted nut within the shell. At full maturation, the in-shell nut ejects easily from the hull when the nut is held between the thumb and forefinger and pressure is applied to the long axis of the nut. As these visible changes occur, increases in fat and sugar content cause the kernel moisture, respiration rate, and total protein content to decrease and the kernel dry weight to increase. The final indication of full maturity is formation of the abscission zone between the nut and its rachis. At maturity, nuts detach readily when the rachis is gently shaken. Optimal harvest time is the 2 to 3 weeks bracketing full maturity.

As with all crops, maturation is uneven throughout the tree. Nuts on the upper southwest quadrant and those on the periphery mature sooner. Light crops also mature earlier than heavy crops. Thus, normal harvesting does not remove all the potential crop. However, most nuts remaining in the tree after shaking are blank (empty) or have nonsplit shells. Ethephon, a foliar compound that hastens and reduces variability in maturity of some crops, is ineffective on pistachios.

The final factor in harvest timing of pistachios is availability of services: harvesters, transporters, and processors. Currently, the majority of California pistachio production consists of a single female cultivar, Kerman. Therefore, allowing for climatic differences in the major production areas, the entire statewide crop ripens at about the same time. However, this situation has been changing since the 2005 release of the new, earlier-maturing California cultivars Golden Hills and Lost Hills, which mature 5 to 7 days earlier than Kerman. The harvest season is usually during the month of September. Because pistachios are mechanically harvested and the equipment is seldom owned by the growers, harvests are contracted and are rapid. Processors receive pistachios from multiple orchards simultaneously and long waits at processor reception are likely. Thus, orchard readiness and availability of the harvest contractor, transportation, and processor must all be coordinated.

Timeliness of harvest is very important in maintaining nut quality. Late harvests increase the potential for damage from a fourth flight of NOW; predation by birds; strong winds that remove nuts; hull deterioration, which stains the nutshells; and early rains that halt harvest, increase fungal infestation (particularly *Aspergillus* spp.), and promote kernel deterioration. Further, any hull deterioration present at harvest is exacerbated during transport delays to the processor.

Harvesting

Harvesting principles are the same for both young and full-bearing pistachio trees. The nuts are removed by knocking or shaking onto a catching frame. Because of the high moisture content (40 to 50% on a fresh-weight basis), fragility of their ripe hulls, and open shells within the hull, pistachios are susceptible to mechanical injury and contamination if they drop to the orchard floor. *Aspergillus* spp. are present in wet orchard soils and have the potential to infect pistachios that contact the soil.

Young Pistachio Trees

Young pistachio trees, 6 years and less, are hand-harvested. Tarps are spread under both sides of the tree to cover at least 5 feet beyond the edge of the canopy on all sides. Nuts are removed by knocking the tree trunk with a padded mallet or striking the branch behind the clusters with a pole. When nuts cease dropping, the tarps are gathered by the edges and luffed vertically. This brings most of the leaves and lightweight debris to the surface for removal prior to dumping into wood or plastic field bins (4-by-4-by-2 ft). Bins hold about 1,000 pounds of freshly harvested nuts and should have at least 5 percent of the vertical surfaces occupied by air vents. Empty bins are distributed in the orchard prior to harvest, filled with harvested nuts, and then transported to the edge of the orchard by mechanical bin carriers. Forklifts are used to load the bins onto trailers for transport to the processor. The bins can also be dumped into hopper-bottomed trailers and the nuts hauled in bulk to the processor.

Mature Pistachio Trees

Unless the crop is too light to justify the cost of machine harvesting, which can occur on light-crop years of this alternate-bearing crop, mature pistachios are mechanically harvested. Pistachio harvesters consist of two separate, self-propelled units about 24 feet in length. One unit contains a shaker head that is clamped onto the tree trunk about 2 feet above the ground. This unit mates with the other machine to form a continuous collection surface under the tree canopy (fig. 17.1). Both units have gently sloping wings extending under the canopies into the row middles. After the tree has been shaken for 6 to 8 seconds (shake patterns and pressures vary with manufacturers), the wings direct the nuts into the bottom of one catch frame. The nuts are then conveyed over an air separator to remove light debris. The major difference in harvester systems is in how the nuts are handled after shaking. In a bin system the nuts are collected in pallet bins that have been distributed in the orchard before harvest. After filling, they are moved by mechanical bin carriers to the edge of the orchard for collection. The bins are either stacked on flatbed trucks or dumped into hopper-bottomed trailers for transport to the processor (fig. 17.2). The bulk system consists of carts carried behind the shaker receiver. The carts often have large, central, continuously revolving screws to distribute the crop in the machine. When filled, the carts are unloaded by a continuous transport system or left at the end of the row to be unloaded into bulk trailers by an elevator system. Larger operations generally use a bulk system, as it is more efficient for large volumes. Most harvesters have the capacity to cover an acre, containing 112 female trees, per hour.

Transport and Postharvest Storage

Shell staining can increase greatly during postharvest transport and storage, particularly if high levels of hull damage were sustained during harvest. Shell staining generally increases with increased temperatures and increased holding times. Even at night the temperature of nuts held in trailers increases up to 1.1°F per hour. Nuts with good-quality, intact hulls may be held for up to 48 hours at ambient conditions without an increase in staining. Nuts with poor-quality hulls show damage after only 8 hours at 104°F, 24 hours at 86°F, and 40 hours at 77°F.

If delays are anticipated prior to transport or at the processor, keeping the bins or hoppers in the shade will provide some protection. Compared with bins, trailers have a greater potential for allowing increases in nut temperature, particularly in the front-bottom area of the trailer. Nuts transported in bins, leaving the orchard at the same time as the bulk trailers, will arrive cooler. Therefore, if travel time between the orchard and the processor is particularly long, or delays at the processor are anticipated, bins with at least 5 percent of the vertical surfaces vented are the preferred mode of transport. Bulk trailers with mesh sides are preferable to solid trailers. If hulling capacity is limited, priority should be given to nuts with the highest internal nut temperatures and poorest hull quality.

If delays over 4 hours occur at the huller, circulating ambient air through bins will prevent temperature increases and nut deterioration for up to 2 days. Hauling trailers at highway speeds induces air ventilation in the unhulled nuts and cools them. If the anticipated delay is longer than 2 days, keeping the bins in cold storage at 32°F with an airflow of 0.1 cubic foot per minute per pound (cfm/lb) and less than 70 percent relative humidity is an alternative. If possible, sort before storage to remove debris and damaged nuts that are more susceptible to decay, as it will minimize storage losses. Under these conditions, fresh unhulled pistachios can be held for up to 8 weeks without degradation. However, few processors currently have the needed sorting and cold storage facilities.

Processing

When pistachios arrive at the processing plant, they move through a series of procedures as outlined in figure 17.3. Previously, most processors used a single-stage drying process, using air at 140° to 160°F for 10 to 14 hours, to achieve a safe storage moisture content of 4 to 6 percent wet-basis moisture content. Drying time was affected by the initial nut moisture and the ambient relative humidity. Now, however, the drying of pistachios is generally a two-stage process. It uses less energy and increases the output of the heated air dryer compared with the single-stage process. The hulled nuts are first dried to 12 to 13 percent moisture content in a column or belt dryer, originally designed for grain drying, or a continuous belt dryer (fig. 17.4). In these dryers, the nuts dry in about 3 hours at temperatures near 180°F. Drying at air temperatures above 180°F causes shells to split open so widely that the kernel drops out. A rotating drum dryer, at the same temperatures, can also be used for the first stage of drying. In the second stage, the nuts are transferred to grain bins equipped with perforated floors, where they are further dried to 4 to 6 percent moisture with unheated, forced air or air heated to less than 120°F (fig. 17.5). Weather in California during harvest is usually warm and dry, and the second stage of drying

Figure 17.1. Typical shake-catch harvester for pistachio trees. *Photo:* A. A. Kader.

Figure 17.2. Pallet bins are moved from the harvester to the edge of the orchard and dumped into hopper-bottomed trailers for transport to the hulling and drying facility. *Photo:* J. F. Thompson.

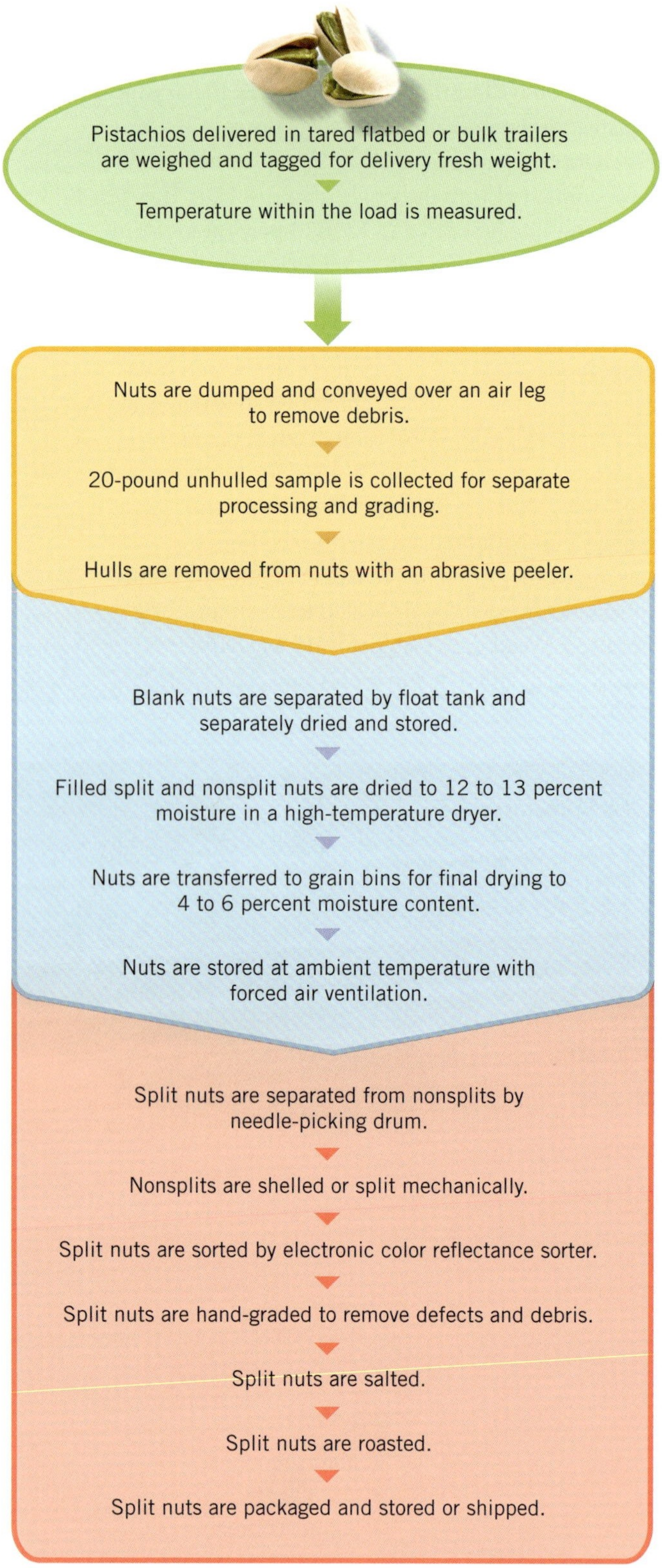

Figure 17.3. Processing procedures.

requires only 24 to 48 hours. The nuts can then be stored in these bins until needed for processing. Smaller operations may use bin dryers for single-stage drying. Eight hours at 140° to 150°F produces the desired 4 to 6 percent moisture content.

In very small operations, sun drying and ambient air drying may also be used. Drying in the sun requires 3 to 4 days, with protective covering to prevent predation by birds and rodents. Drying nuts in bins with ambient air requires 3 days in California. Nut depth in the bin should not exceed 4.5 feet, and air should flow into the bins at a rate of 15 cubic feet per minute per cubic foot of nuts. The major disadvantage to this method is the potential for fungal growth during the early part of drying.

Storage of Dried Pistachios

Once dried to 4 to 6 percent moisture content, but before further processing, nuts can be held at 68°F and 65 to 70 percent relative humidity for up to 1 year. Other factors that slow nut quality loss during storage are storage at low temperatures, exclusion of oxygen through controlled atmosphere storage or packaging under nitrogen, and fumigation or insect-proof packaging to prevent insect damage. Low oxygen and low temperatures slow the development of oxidative rancidity of nuts and reduce insect infestation and activity. However, pistachios are less susceptible to oxidative rancidity than other nuts as a result of their lower ratio of unsaturated to saturated fatty acids.

The Food and Drug Administration (FDA) defines "safe moisture levels" in terms of water activity. This is the equilibrium relative humidity in the headspace air above a sample of nuts. It is expressed as a decimal (unlike relative humidity, which is expressed as a percentage). Water activity levels less than 0.70 at 77°F do not support fungal growth. This equals a whole-nut moisture content of about 7 percent. The relationship between moisture content and equilibrium relative humidity (ERH) is temperature dependent. For any given nut moisture content, water activity rises about 0.03 for every 18°F rise in temperature, in the range of 0.20 and 0.80 water activity.

Lower temperatures slow lipid oxidation (which results in rancidity), prevent mold growth, and greatly reduce insect activity. Temperatures between 32° and 50°F are recommended for pistachios, depending on expected storage duration: the lower the temperature, the longer the storage life. Pistachios are more stable and have a longer storage potential than almonds, pecans, or walnuts.

Low-oxygen atmospheres (< 0.5%) aid in maintenance of flavor quality. This procedure is rarely used commercially, but controlled atmosphere storages are available and are used for a few fruit crops. In the final product package, the atmosphere may be modified by vacuum packaging or by flushing with nitrogen to exclude oxygen.

Insects in stored pistachios can be controlled by fumigation. Alternative methods of insect control include insecticidal-controlled atmospheres or irradiation. Storage of nuts in 0.5 percent oxygen and 10 percent carbon dioxide kills all instars of stored product pests and NOW after 2 to 5 days at 81°F. Very low and very high temperatures (near freezing or between 104° to 120°F) also effectively kill insects. After fumigation, the use of insect-proof packaging is essential to prevent reinfestation.

Figure 17.4. Gas-fueled column dryers are used to dry pistachios after hulling. *Photo:* J. F. Thompson.

Figure 17.5. After drying is finished, nuts are stored in fan-ventilated grain bins. *Photo:* J. F. Thompson.

Quality and Grading

Size and appearance are the most important components of pistachio quality. In the U.S. Grade Standards, quality factors include degree of dryness (7% whole-nut moisture content or lower) and freedom from foreign material and damage caused by mold, insects, spotting, rancidity, and other defects. Defects that result in downgrading can be external (the shell) and internal (the kernel). External shell defects include total or partial nonsplitting, adhering hull material, stained shells, and damage by other means, including deformity and bird damage. Internal kernel defects include damage from insects and fungal pathogens, small immature kernels, rancidity, and decay. Size, designated by the number of nuts per ounce, is also an important quality attribute.

Aflatoxin contamination is associated with particular nut defects: nuts that freely separate from their shell, nuts smaller than 40 count per ounce, severe shell staining, evidence of insect damage, and large nuts that float during density separation prior to drying. Removing these nuts can reduce contamination in a lot. Reprocessing closed-shell nuts by rehydration increases aflatoxin contamination.

Grower payout is calculated from the fresh weight of pistachios delivered to the processing plant, corrected for the weight of foreign material removed prior to hulling. Prior to hulling, a grading sample is drawn from each delivery. This sample is processed individually, the fresh- to dried-weight ratio is calculated, and a third party inspects for the percentage by weight of split nuts, nonsplit nuts, blank nuts, and nuts with adhering hulls, light and dark stain, and other defects. Correction factors for these defects, along with the cost of hulling and drying, are applied to the corrected delivery weight to calculate grower price. The major components in determining return to the grower are weight delivered and the percentages by weight of the filled split, filled nonsplit, and blank nuts.

Chemical Composition and Nutritive Value

The approximate chemical composition of pistachios is given in table 17.1. Actual composition may vary depending on cultivar, maturity at harvest, and moisture content. After appearance, flavor is the most important aspect of pistachio quality. Thus far, the production area within California and the presence of shell staining do not appear to have any effect on the flavor of pistachios. Maturity at harvest, moisture content (4 to 6% is optimal range), and total sugar content do influence flavor. Rancidity development associated with long storage times also reduces flavor quality.

Table 17.1. Pistachio composition and nutritive value

Component (unit)	Amount per 100g edible portion
water (%)	5.6
proteins (%)	19.6
fats (%)	53.2
carbohydrates (%)	19.0
fiber (%)	2.2
ash (%)	3.0
sugars (%)	6.1
calories	594
unsaturated/saturated fatty acids ratio	8.1
Vitamins	
A (I. U.)	230
thiamin (mg)	0.67
niacin (mg)	1.4
C (mg)	0
Minerals (mg)	
Ca	131
P	500
Fe	7.3
K	972
Mg	158

Limited research indicates that cull pistachios, which typically represent 1.5 to 2.5 percent of the marketable nuts, can be fed to sheep and cattle. However, the sharp edges of broken nutshells can harm animal intestines. Rations for cattle can contain up to 20 percent culls and for sheep up to 10 percent. Pistachio hulls have high tannin and phenolic contents that limit their use in animal feed. Some processing plants give the hulls to cogeneration plants for free.

At the top of the following column is a summary checklist for harvesting pistachios.

May

- Establish contracts with harvester, hauler, and processor.
- Select harvest method and choice of hauling in bins or bulk truck, depending on distance to processor.

July

- Make sure irrigation meets the needs of the trees.

August

- Start monitoring for early splits and third flight of NOW.

September

- Begin monitoring nuts for color change and easy dehiscence of hull.

References

Campbell, B. C., R. J. Molyneux, and T. F. Schatzki. 2003. Current research on reducing pre- and post-harvest aflatoxin contamination of U.S. almond, pistachio, and walnut. Journal of Toxicology Toxin Reviews 22(2, 3): 225–266.

Kader, A. A., C. M. Heintz, J. M. Labavitch, and H. L. Rae. 1982. Studies related to the description and evaluation of pistachio nut quality. Journal of the American Society for Horticultural Science 107(5): 812–816.

Thompson, J. F., T. R. Rumsey, and M. Spinoglio. 1997. Maintaining quality of bulk-handled, unhulled pistachio nuts. Applied Engineering in Agriculture 13(1): 65–70.

USDA (United States Department of Agriculture). 2004a. United States Standards for Grades of Pistachio Nuts in the Shell. Agricultural Marketing Service, Fruit and Vegetable Programs, Fresh Products Branch. USDA Agricultural Marketing Service website, http://www.ams.usda.gov/standards.

———. 2004b. United States Standards for Grades of Shelled Pistachio Nuts. Agricultural Marketing Service, Fruit and Vegetable Programs, Fresh Products Branch. USDA Agricultural Marketing Service website, http://www.ams.usda.gov/standards.

Part 5 Insect and Mite Pest Management

18 Navel Orangeworm and Obliquebanded Leafroller

David R. Haviland, Walter J. Bentley, Joel P. Siegel, Brent A. Holtz, Kent M. Daane, and Bradley S. Higbee

Pistachios in California are attacked by two different lepidopterous pests. These include the navel orangeworm (NOW) (*Amyelois transitella* (Walker)) and the obliquebanded leafroller (OBLR) (*Choristoneura rosaceana* (Harris)). Navel orangeworm is currently the most important insect pest of pistachio because it feeds directly on mature pistachio nuts and is associated with aflatoxins. It is a significant concern everywhere that pistachio trees are grown in California. Infestations of OBLR are more sporadic and primarily result in leaf defoliation and damage to nut clusters prior to maturity.

Navel Orangeworm

Navel orangeworm (NOW) was first identified in 1915 from specimens collected in Hermosillo, Mexico, and at that time named *Myelois venipars* Dyar (in the insect family Gelechiidae). It was soon found in the United States, collected in light traps near Tempe, Arizona, in 1920, and later in Orange County, California, in 1942. It quickly spread throughout southern California and, in 1949, NOW was found in northern California in Tehama County. Initially NOW was a pest of walnuts and almonds, but by the mid-1960s, larvae of NOW were detected infesting pistachios in the San Joaquin Valley. Infestations in pistachio can result in reduced crop yield and quality, greater field management and processing costs, and aflatoxin contamination. Navel orangeworm is also commonly found in figs and can be found in apple, apricot, carob, date, grapefruit, lemon, loquat, orange, peach, pear, pecan, plum, and pomegranate. Regardless of the host plant, NOW feeds on the nut or fruit rather than the leaves or shoots.

Description

The female and male moths are similar in appearance, predominantly gray or sometimes black with narrow, wavy, black bands across the front pair of wings, although the male is smaller in size (fig. 18.1). The wing expanse of females can be over 1 inch, with the length of the body up to ½ inch. Moths are characterized by the pointed palpi (part of the insect's mouthparts) at the front of the head that are angled upward by about 30 degrees. Eggs are creamy white in color, reticulated, flat, and approximately the size of a pinhead (about 0.03 in). One study found that NOW lays an average of 84.6 eggs per adult female, with some females laying more than 200 eggs. Egg production varies depending on the developing

Figure 18.1. Navel orangeworm adult on pistachio. *Photo:* J. Siegel.

larva's food quality and the adult's longevity. Another study, in almonds, showed that egg laying (oviposition) peaks on the second and third day after the adult emerges from pupation and continues for 9 days; as with egg production, the oviposition period depends on adult vigor and longevity (which depends on temperature and adult food sources). As the egg matures, it becomes bright orange to red (fig. 18.2), and eventually a small black-headed worm (larva) can be seen inside the egg shell (chorion).

Larvae can hatch in as little as 3 days during the summer but can take as long as 30 days to hatch during late winter and early spring. The newly hatched larvae, or 1st instars, are small, about 0.04 inch in length, with a reddish-orange body and a mahogany-colored head. After the larvae's first molt, to the 2nd instar, the body is off-white (or cream) to pale red, depending on the larval diet. Thereafter, the larval stages look similar for the five or six instar stages, with the exception of size, and at maturity the larvae may reach ½ inch in length (fig. 18.3). The larvae have sclerotized (hardened) dark areas, about the size of a pinhead, that form a half-circle on the upper portion of each side of the middle thoracic section above the middle pair of true legs (fig. 18.4). This characteristic is used to separate NOW larvae from that of other lepidopteran pests in pistachio orchards. A pair of gonads, located dorsally in the middle of the abdomen, identifies male larvae.

Figure 18.2. Navel orangeworm eggs turn reddish orange before hatching. *Photo:* J. K. Clark.

Figure 18.3. Larvae of navel orangeworm and fecal material (frass) inside a damaged pistachio. *Photo:* J. Siegel.

Injury

Navel orangeworm adults do not appear to be attracted to new (immature) pistachio nuts, but when the hull begins to tear or loosen and the shell splits, then oviposition on the nut is more likely. Because the shell splits open prior to the hull tearing, NOW larvae can quickly enter the nutmeat once the hull seal is broken. However, any factor that compromises hull or shell integrity increases the likelihood of successful infestation. Navel orangeworm are attracted to volatiles that are associated with nuts that have been damaged, such as by other worms, birds, or fungi. It is common to find eggs deposited on damaged nuts before the shell splits. On rare occasions, NOW larvae can also enter through the stem.

The first signs of an infestation are small, pinhole-size punctures in the nut kernel. This damage occurs prior to webbing and the deposition of fecal material (frass) and often goes unnoticed. As the larva grows in size, the entire nut is fed upon, and extensive amounts of webbing and frass will appear (see fig. 18.3). Navel orangeworm are not cannibalistic, and as many as six larvae of different sizes can be found infesting a single pistachio nut. When multiple worms are found in a nut, they are probably all NOW.

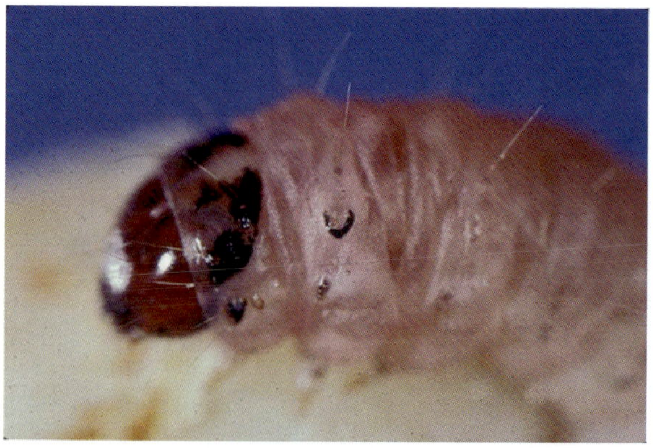

Figure 18.4. Navel orangeworm larvae have a pair of crescent-shaped markings on the second segment behind the head. *Photo:* J. K. Clark.

In addition to direct damage caused by feeding, NOW infestations can result in nut contamination by numerous species of fungi, including *Aspergillus flavus* and *Aspergillus parasiticus*, which produce aflatoxins (see the section "Aflatoxin Contamination" in chapter 26, "Foliar, Fruit, and Branch Diseases"). Pistachios contaminated with aflatoxin, especially aflatoxin B1, face severe trade restrictions, and, for this reason, there is little tolerance for the presence of any aflatoxins in nut crops.

Seasonal Development

Accurate knowledge of NOW populations' seasonal development (or phenology) aids in managing this pest. Navel orangeworm phenology includes the beginning and end of moth flight periods, as well as the egg deposition periods for each generation. The number and occurrence of NOW generations, in relationship to the progression of nut development in each region (and orchard), determine if the NOW flight coincides with periods when the new crop is most susceptible to infestation.

The development rate of NOW larvae, which affects the number of NOW generations, is influenced by temperature and food quality. The lower and upper temperatures that sustain egg and larval development are 55°F and 94°F, respectively. Of course, NOW can survive at colder and warmer temperatures, but the egg, larva, and pupa will not continue to mature. Temperature limits for development are used to determine degree-days (DD), which express the length of time in relation to temperature required for an insect to develop from a newly laid egg to an adult (i.e., the generation time). Degree-days can be calculated using the UC IPM website, ipm.ucanr.edu, combined with either CIMIS weather data from stations near the orchard or temperature loggers placed in orchards.

The rate of development is also influenced by the larva's diet. For example, one study reported that larvae feeding on overwintered almonds (mummies) required about 1,056 DD per generation, whereas larvae feeding on fresh nuts required only 635 to 739 DD (a 30 to 40% faster development period). More recent work compared the development of NOW larvae feeding on pistachio or almond nuts and found considerable variation in the development time within a group of same-aged NOW. For instance, there are typically four annual NOW generations in the southern San Joaquin Valley and three full and a partial fourth generation in the central and northern San Joaquin Valley; however, the number of generations in any orchard may be more or less, depending on the combination of seasonal temperatures and food quality. Another consequence of food quality is that NOW generations overlap within the same region. In August, for example, NOW larvae feeding on fresh pistachio nuts, overwintered almond nuts, fresh almond nuts, or even different nut cultivars, may have different development times. The overlap in generations is more commonly found toward the end of season, which reduces the ability of egg traps or pheromone traps to show discrete generations due to an overlap in the third and fourth flights.

Overwintering and first flight

A description of the seasonal development begins with overwintering NOW larvae in the mummy nuts. Survival of these larvae is highly variable and depends on temperatures, rainfall, and location of the host (i.e., pistachio nut). For example, survival can be twice as great in mummies on the tree as those on the ground, but grounded mummies may be more plentiful. All infested mummy nuts contribute to the resident orchard population in the spring.

Moth emergence from wintered nuts (first flight) begins in February, increases substantially in March, is well under way by April, and usually peaks in late April or early May (fig. 18.5). Moth captures in pheromone traps depend on NOW density as well as temperature, with increased captures during warm weather and minimal captures during cold weather. Fluctuations in moth density and moth activity during the first flight often result in a bimodal appearance to the first flight (see fig. 18.5).

Although some of the earliest-laid eggs develop successfully during warm days in February, most eggs that develop successfully are laid from late March onward. Because the NOW larvae commonly overwinter in different larval stages, adult emergence can be drawn out, and the resulting oviposition period by first-flight moths may not end until mid-June (4 months after it began).

First generation and second flight

Eggs from the first flight, termed first-generation eggs, are laid on the mummy nuts from the previous season. NOW can only survive on the nuts (rather than leaves or shoots), so until the new pistachio crop is susceptible (damaged nuts, pea splits, or early splits), NOW survival in the orchard depends on the presence of the previous season's mummies. Emergence of the first-generation adults (second flight) from mummies occurs from late June throughout July and possibly into August (see fig. 18.5).

Second generation and third flight

Moths from the second NOW flight in pistachios typically lay eggs on mummy nuts that are still in the tree or on the ground from the previous year's harvest. However, in some years the latter half of the second flight can overlap with the presence of small pea-split nuts, damaged nuts, deformed nuts, or early-split nuts that can occur in late July. These damaged and split nuts usually constitute a small percentage of the new crop in July, but, if abundant, they can contribute substantially to NOW population growth because the eggs are laid on a higher-quality food source (new crop). Pistachio growers next to almond orchards should also consider that second-generation eggs in almonds are laid in the new crop at the time that Nonpareil and other early-harvested varieties start to split (see fig 18.5).

Second-generation NOW larvae become adults in late July through August (third flight), when the bulk of the new pistachio crop becomes susceptible. The exact timing of the flight depends on degree-days as well as host quality. For example, second-generation larvae laid on mummy nuts, new pistachios, or new almonds in neighboring orchards will all have different developmental times and emerge at different times as adults of the third flight.

Third generation and fourth flight

In late July through August, third-flight moths lay eggs to begin the third generation (see fig. 18.5). The larvae that hatch from these eggs are highly significant because they infest nuts of the new crop that will be harvested. This makes them a direct pest of the commodity being produced, both in terms of damage to the kernel and the increased likelihood of aflatoxin contamination associated with NOW damages. The primary focus of NOW management programs is to prevent third-generation larvae from infesting the nuts.

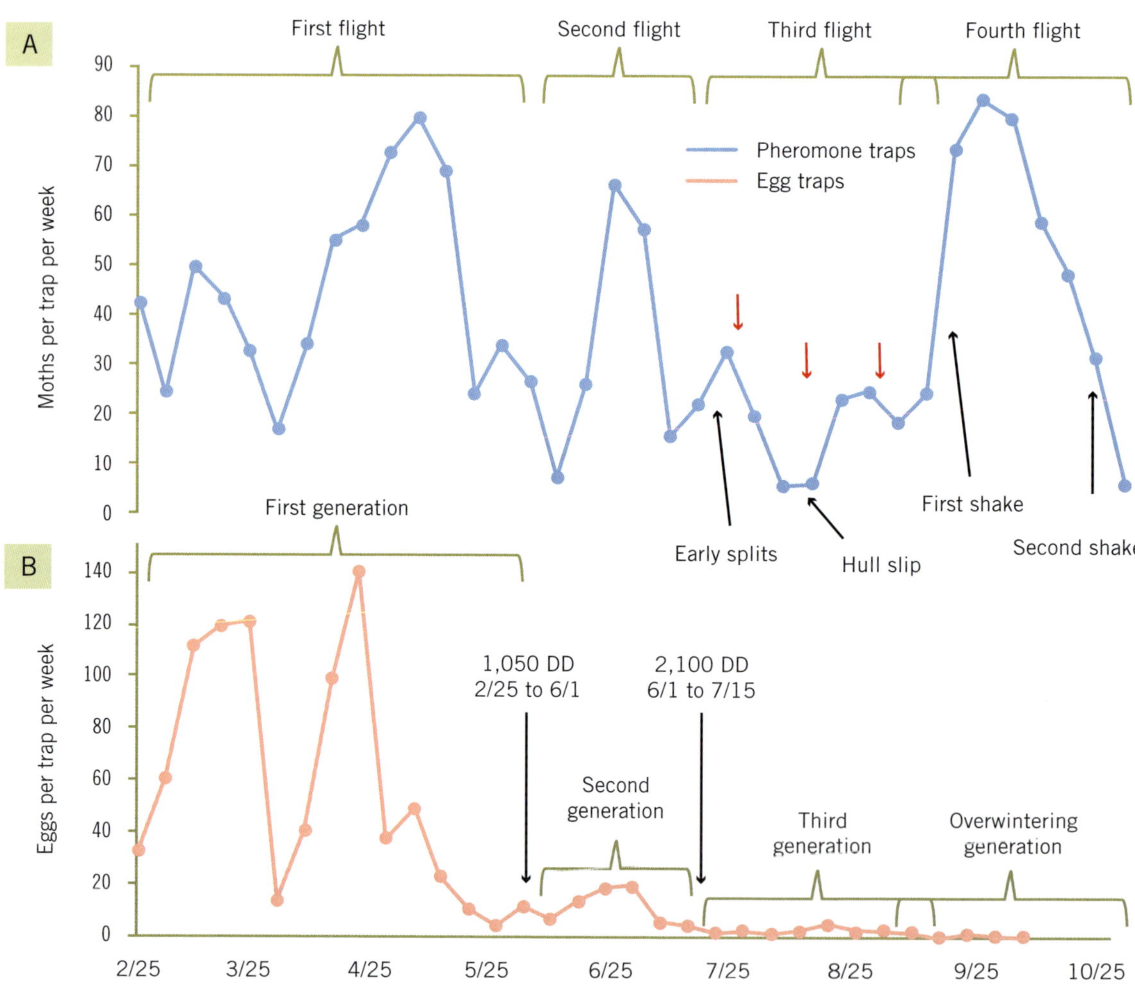

Figure 18.5. Navel orangeworm captures in a 640-acre commercially sprayed pistachio orchard in Kern County during 2014 using (A) pheromone traps and (B) egg traps. Flight periods shown are approximately 2 weeks earlier than normal due to an above-normal accumulation of degree-days (DD) during 2014. Red arrows indicate insecticide applications on July 27, August 12, and August 30. *Source:* B. Higbee.

Many third-generation larvae are removed from the orchard during harvest. Larvae that remain within nuts that are not harvested may overwinter as larvae (particularly in the northern San Joaquin Valley) or may pupate and emerge as adults in September as the fourth flight (particularly in the southern San Joaquin Valley). In the southern San Joaquin Valley, the fourth flight usually starts during the second week of September but can occur earlier or later depending on season-long degree-day accumulations. Fourth-flight moths mate and lay eggs on the new pistachio crop. These eggs hatch into larvae that will be harvested during a late first shake or during a second shake, or they will remain in nuts on the tree or on the ground as overwintering larvae within mummies.

Figure 18.6. Black trap for monitoring egg laying by navel orangeworm. *Photo:* D. Haviland.

Monitoring

Navel orangeworm adult activity can be monitored using egg traps and pheromone traps. Egg traps contain bait that entices adult moths to lay eggs on the trap (fig. 18.6). Pheromone traps contain a pheromone-based lure that attracts male moths into a trap, where the adults get caught on a sticky surface (fig. 18.7). Both traps can be used to help predict key events in NOW activity, such as the start or peaks of future flights. Additionally, pest control advisers can estimate potential NOW pressure by sampling mummy nuts in the tree and on the ground to determine the percentage that are infested during the winter and spring. They can also evaluate early-splitting and main crop nuts for the presence of NOW eggs and larvae.

Figure 18.7. Pheromone trap for monitoring adult male navel orangeworm. *Photo:* D. Haviland.

Egg traps

For several decades, egg traps have been the primary method for monitoring NOW populations. Egg traps are made of a porous black cylinder that contains a mixture of pressed almond meal and almond oil (3 to 10% by weight) to entice adult females to lay eggs on the trap (see fig. 18.6). Egg traps are most effective during the first flight, when there is minimal competition from other possible places for a moth to lay eggs, particularly in orchards using good winter sanitation practices to remove mummies from the tree and ground. Egg traps can also be valuable during the second NOW flight but have little to no value during the third and fourth flight, when the trap is present at the same time as new crop nuts (see fig. 18.5).

Egg traps should be placed into the orchard in mid-February in the lower San Joaquin Valley and mid-March in northern areas. For small orchards, use a minimum of four traps. For most orchards, one trap per 5 to 20 acres is recommended, though a lower trap density of one per 50 acres can be used in regions with large, contiguous blocks of pistachio orchards.

Traps should be checked for eggs once or twice per week in February to April to determine the start of the primary egg-laying period (called the biofix), after which traps can be evaluated weekly until they are removed. The biofix is the start date for accumulating degree-days in the NOW phenology model available at the UC IPM website, ucipm.ucanr.edu/WEATHER/index.html. When using the NOW model, choose a local weather station or enter your own weather data, and set the biofix date as the day to start accumulating degree-days (such as Feb. 25 in fig. 18.5). When running the model, the computer will use actual high and low daily temperatures for all dates between the biofix and current date, and it will use long-term temperature averages for dates in the future. When looking at the results of the model, determine the date where 1,050 degree-days have accumulated from the biofix date. That date is the predicted date that second-flight moths will begin to lay eggs of the second NOW generation. From that point, adding another 1,050 degree-days (total of 2,100 degree-days from the biofix) predicts the date that third-flight moths will begin laying eggs of the third NOW generation. Predicting the start date of third-generation egg laying can help to determine the most effective timing for insecticide applications.

Pheromone traps

Pheromone traps were first commercially available in 2013 as a new tool for monitoring adult male NOW. Pheromone traps contain a lure that is impregnated with a four-component blend of attractants that simulate a female moth emitting sex pheromone to attract male moths. The lure is placed into large delta or wing traps that contain a sticky surface (see fig. 18.7). Males that are attracted to the lure become stuck in the trap, where they can be counted. Make sure not to confuse NOW with the meal moth *(Pyralis farinalis)*, which is also attracted to the trap but is a light brown color with dark brown bands across the wings.

Pheromone traps should be placed into the orchard during mid-February to mid-March, prior to the first flight of NOW. Count the number of moths weekly to determine the start, peak, and end of NOW male flights. These data can be used to estimate the start, peak, and end dates of future NOW flights. For example, if a sharp upswing in captures occurs in late April, this date can be set as a biofix in a degree-day model to predict a sharp upswing in captures that can be predicted in late June or early July, when 1,050 degree-days from the biofix have accumulated. Likewise, calculating 1,050 degree-days from biofix dates at the start, peak, and end of the second flight can help predict the start, peak, and end of the third flight, thus helping determine how these dates correspond to key crop phenologic events, such as the presence of pea splits or the date that hull slip begins for the majority of the new pistachio crop. When using pheromone traps to monitor NOW, distinct generations can be seen in most orchards through the second flight (see fig. 18.5). However, as mentioned previously, discrete generations are not always seen from the beginning of the third flight through the end of the season due to the varying developmental rates of larvae and the overlap of the end of the third and start of the fourth flight.

Early-splitting nuts

During the second half of July, pistachio trees produce small, abnormal nuts that can be referred to as pea splits, damaged nuts, deformed nuts, and early-split nuts. All of these nut types split open a few weeks prior to hull slip in the main crop. Although these nuts only make up a small percentage of the overall crop, they can play an important role in allowing moths at the end of the second flight to lay eggs into high-quality, new-season nuts instead of old, degraded mummies from the previous season. The percentage of early-splitting nuts varies widely from year to year, as does the potential overlap between the second NOW flight and the presence of these nuts. For that reason, during the second half of July and early August, pest control advisers are encouraged to monitor the pistachio crop to determine the prevalence of early-splitting nuts and the prevalence of NOW eggs on these nuts, as a way to determine if an insecticide application timed at early-splitting nuts is justified.

Natural Control

The natural causes of mortality for NOW include prolonged exposure to low temperatures or rain, predation, parasitism, and disease. Orchard cultural practices, such as ground covers, can impact each of these factors. For example, NOW overwintering in mummies on dry berms have higher survival than those in mummies in the wetter row middles, especially if there is a winter ground cover.

Biological agents include parasitoids, predators, and entomopathogenic bacteria and nematodes. Two parasitoids of NOW are commonly found in California pistachio. One of these, *Goniozus legneri* Gordh, is a wasp in the family Bethylidae. The adult is shiny black and about 0.25 inch in length. The adult female first stings the NOW larva to paralyze it, then measures the size of the worm to determine how many parasitoid eggs to deposit (fig. 18.8), and then lays the appropriate number of *Goniozus* eggs. The parasitoid eggs hatch with the larvae attached by their mouthparts to the paralyzed worm, where the larvae feed (fig. 18.9) until reaching the pupal stage. The second parasitoid, *Copidosoma plethoricum* Caltagirone, is in the family Encyrtidae. The adult is shiny black and small, about 0.04 inch. This is a polyembryonic, egg-larval parasitoid, meaning that the adult wasp places her egg inside the NOW egg (fig. 18.10), and when the NOW becomes a late-stage larva, the parasitoid egg begins to divide (polyembryonic, like maternal twins), eventually forming hundreds of siblings that develop and pupate inside the larvae, emerging as adults to chew their way out of the mummified NOW carcass. Other parasitoids that can be found attacking NOW include *Habrobracon hebetor* (Say) and *Mesostenus gracillus* Creson, both in the family Ichneumonidae. Parasitism levels of more than 25 percent are often recorded in orchards with high levels of NOW infestation, but at lower NOW densities these parasitoids are not very effective, and their presence in orchards is often inconsistent.

Generalist insect predators also provide some natural NOW control. The two most common predator groups are green lacewings, such as *Chrysopa* and *Chrysoperla* species (family Chrysopidae), and plant bugs, especially *Phytocoris relativus* Knight (family Rhopalidae). Green lacewing larvae can attack NOW eggs and all larval stages, while *Phytocoris* primarily feeds on NOW eggs. The plant bug also causes epicarp lesion and nut drop early in the season (see chapter 20, "Small Bugs: *Neurocolpus, Phytocoris, Calocoris,* and *Lygus*"), and its value as a predator has not been fully investigated. Other predators exist, but they provide minimal population reduction.

Some vertebrate species also contribute to NOW mortality. Field mice are found in many orchards where ground covers are present. Their impact has been primarily studied in almonds, where they feed on mummies on the ground during the late fall and winter. Birds also feed on infested nuts during the winter. However, the actual contribution of these vertebrates to NOW control is not known.

The most successful entomopathogenic nematode species studied for use in pistachio is *Steinernema carpocapsae* (Weiser). This microscopic pathogen can be artificially introduced into orchards through the irrigation system (chemigation) and, at a concentration of 500 million infective juveniles per treated acre, it can help reduce NOW in mummy nuts on the ground. Success depends on several physical and environmental factors, including sufficient soil moisture prior to application, maximum daytime soil temperature of 90°F, and nighttime temperatures above 36°F. Chemigation cannot be used with drip irrigation because of insufficient water flow and wetted ground surface. The level of NOW mortality provided by resident nematodes is negligible, and that provided by augmented nematodes through chemigation has been variable in field studies and is not generally used commercially. Other NOW pathogen species include several viruses, microsporidia, and the bacterium *Bacillus thuringiensis* Berliner; their impact on the NOW population has not been studied in detail.

Figure 18.8. Adult *Goniozus legneri* measuring navel orangeworm larva before it places eggs on the host. *Photo:* K. Daane.

Figure 18.9. Late larval *Goniozus legneri* just before pupation after consuming navel orangeworm larva. All that remains of the navel orangeworm is the head capsule. *Photo:* K. Daane.

Figure 18.10. Adult *Copidosoma plethoricum* inset ovipositing into the egg of a navel orangeworm. From one parasitoid egg, over 300 adult parasitoids emerge from a single navel orangeworm larva. *Photo:* F. Skinner.

Management

Sanitation

Orchard sanitation is the process of removing mummies from the trees and then destroying them on the ground during the winter. The practice has a simple but critical goal: removal of the NOW habitat from postharvest until the new crop is susceptible. NOW larvae must have either mummy nuts or susceptible new crop nuts to develop and survive in the orchard, so the presence of mummies is the greatest contributor to NOW in pistachio before the new crop is susceptible. If mummies are completely removed, the orchard will be free from NOW the following season unless adults fly in from elsewhere.

In contrast to sanitation efforts in almonds, however, mummy destruction techniques have not been as effective in pistachios. Long-term studies in Kern County have demonstrated a 10 to 70 percent reduction in damage at harvest in orchards that were completely sanitized (mummies removed from the trees and ground) compared with orchards that were only partially sanitized (mummies removed from trees only) or not sanitized. In pistachio orchards, mummies remain in the trees during dry winters. Even by mid-July of the following year, 50 percent of these tree mummies may still be available for NOW larval development, and larvae can still be found infesting tree mummies in August. Overwintering pistachios on the orchard floor also contribute to the NOW population, and grounded nuts are often more numerous than mummies in trees. As fall and winter progress, closed-shell and the lighter blank nuts fall from the tree, and by late February about 50 percent of the grounded mummies are split. Depending on winter climate and orchard conditions, some of these grounded mummies rot and others germinate in late winter. Following a wet winter, there may be as few as 13 percent split nuts remaining on the orchard floor by mid-June.

Sanitation, the cornerstone of NOW management in almonds, was recommended for pistachio as early as 1976, yet the practice was not adopted as quickly in pistachio because the benefit had not been demonstrated as it had been in almonds. Navel orangeworm survival in tree mummies is more than twice as high as it is in ground mummies. For that reason, removing nuts from the tree is imperative. Once the nuts are on the ground, blow them into the drive rows, where they are easier to destroy. Mechanically blowing pistachios from the tree row has a variable success rate, and as much as 50 percent of the nuts can be left behind if early rains stick the nuts to the ground or there are numerous ruts or uneven ground close to the raised berm on which the trees are planted. Therefore, the technique should be repeated to maximize nut removal from the berm. Pistachios can also be hand-raked from the berm, but the cost is prohibitive for most growers. Subsequently, the nuts in the drive row are then buried by disking, which may also be repeated at intervals. Disking mummies results in an approximate 50 percent reduction in moth emergence with each pass. Most of this mortality occurs from the burial of the mummies, although larger larvae are able to crawl to the surface and pupate from as deep as 6 inches. However, many nuts are still left on the surface and are available for reinfestation in the future. Some growers are converting their orchard floors to allow the use of the same mummy shredders that are used in almonds. Results have been similar to that of disking in terms of reduction in moth emergence and number of mummies that remain on the orchard floor. In these converted orchards with a flat orchard floor, it is also possible to use almond sweepers or other sweepers attached to tractors to relocate up to 95 percent of the mummies on the berms to the drive row, where they can be destroyed. Many growers allow pistachios to simply fall from the trees during the late fall and winter before blowing them, in contrast to almond management which involves shaking and poling the nuts off the trees. Since the goal of sanitation is to maximize the destruction of overwintering nuts, it works best when the trees are shaken or pistachios are removed by poling in late October to ensure that most nuts spend the winter on the ground. Pruning also removes some infested nuts and causes additional mortality to NOW, but it is sanitation that causes the greatest reduction in the population.

Early harvest

Early harvest (sometimes referred to as timely harvest) is a NOW management technique whereby harvest is completed early, typically before mid-September, to ensure that nuts are removed from the tree prior to the fourth flight of NOW. Due to the limited availability of harvest equipment and huller capacity, early harvest is not possible on all acreage. For that reason, pistachio farmers should plan their harvest schedules so that the first orchards to be harvested are the ones with a history of NOW rejects, or where poor sanitation has occurred, or where NOW traps and monitoring of nuts indicate that NOW populations are at their highest.

The value of early harvest was documented in an evaluation of over 10,000 processor loads from harvests in 2012 and 2013 throughout the San Joaquin Valley. On average throughout the industry during these 2 years, the percentage of NOW damage doubled every 200 degree-days, which is the equivalent of every 9 to 11 days, depending on the county and temperatures. This means that, on average, orchards that had 0.5 percent damage on September 15 had 1 percent damage by September 25, 2 percent damage by October 5, and 4 percent damage by October 15. It is also important to remember that these increases in damage occurred in commercial orchards where insecticides were used. If insecticide programs had not been used, it would be anticipated that the crop losses would have been much greater. Despite the clear value of an early harvest, however, many growers are using a multiple-shake strategy, and some are harvesting orchards through mid-October because of the current value of the pistachio crop. Nevertheless, in order to ensure the highest quality, the majority of the nuts should be removed at the first shake.

Insecticides

Insecticides play a major role in the management of NOW. Over the past decades, pesticides that contain active ingredients belonging to older chemical classes, such as organophosphates and carbamates, have been replaced by fourth-generation pyrethroid insecticides that disrupt nerve transmission (Insecticide Resistance Action Committee [IRAC] 3A); insect growth regulators that accelerate molting in larvae (IRAC 18); insecticides responsible for the loss of cell function and disruption of nerve impulses (IRAC 6); insecticides that target the motor neurons (IRAC 5); and insecticides that target calcium channels in the skeletal muscle (IRAC 28). The efficacy of these products varies, depending on the amount of coverage achieved, the NOW life stage present, and resistance. For example, some insecticides work primarily against eggs and newly emerged larvae prior to their entry into the nut, while others affect all life stages including adults. All of these insecticides have improved photostability and provide a much greater persistence than that in older classes of insecticides. However, the efficacy of the newer pyrethroids has recently become a concern due to the establishment of resistant populations of NOW throughout the southern San Joaquin Valley. For a list of insecticides that are currently registered for use against NOW in pistachios, consult the online UC IPM Pistachio Pest Management Guidelines, ipm.ucanr.edu/PMG/selectnewpest.pistachios.html.

Navel orangeworm pressure in pistachio orchards varies by location, year, and management practices. Pistachio orchards with low to moderate NOW pressure are typically sprayed one or two times per season for NOW. In regions or orchards with high to severe NOW populations, additional sprays may be necessary. The suggested timing for insecticide application at different levels of NOW pressure is as follows:

Hull-slip timing. Nearly all orchards require an insecticide application approximately 1 month prior to harvest, when the new crop becomes vulnerable to attack. This period begins when the hulls start to slip and, at the same time, the third flight of NOW is present. This timing can be predicted using degree-day models with either egg or pheromone traps. If using egg traps, this timing occurs approximately 2,100 degree-days after the April or May biofix, indicating the start of first-generation egg laying, or 1,050 degree-days from the late June to early July biofix, indicating the start of second-generation egg laying (see fig. 18.5). If using pheromone traps, this timing occurs approximately 4 to 7 days after the start of the third flight as the moths begin to lay third-generation eggs.

Post–hull-slip timing. In orchards with moderate NOW pressure, as well as in orchards with a multiple-shake strategy, a second insecticide application is often needed to protect the new crop. This application, commonly referred to as a post-hull-slip spray, is made approximately 2 to 3 weeks after the initiation of hull slip, based on NOW pressure, harvest date, and harvest strategy. Application of this second spray can help protect new crop nuts at a time when residues from the hull-slip spray have mostly degraded. In orchards with low to moderate NOW pressure, these sprays—in combination with the benefits of worm removal from the trees during the first shake—are usually sufficient to provide adequate control of NOW during the second shake. However, in years with high to severe NOW pressure, particularly if NOW incidence during the first shake was high, and if there is a long delay anticipated between the first and second shakes, then an insecticide application after the first shake may also be considered. The decision to make this additional application should take into account the anticipated yields and quality of a second shake. The costs of an insecticide application after the first shake may be justified in orchards where high yields of split nuts are anticipated during the second shake compared with other orchards where low yields of primarily closed-shell nuts are anticipated.

Early-split timing. The insecticide application strategy stated above focuses on protecting the new crop nut. In orchards that typically experience high NOW pressure, an insecticide application prior to hull slip may also be needed to prevent second-flight moths from laying eggs into pea-split, damaged, or early-split nuts that are vulnerable to NOW earlier in the season but do not have an economic value. When needed, the timing of this spray is typically 2 to 3 weeks prior to the initiation of hull slip. The need for this spray is determined by historical knowledge of NOW pressure, in-season population monitoring, and the prevalence of pea-split, damaged, or early-split nuts. In years when NOW pressure is low and few pea-split or other early-split nuts are present, this spray is not needed. This timing may take on increased importance in the future as the acreage of newer, early-harvested varieties (such as Golden Hills) increases. These early varieties differ in nut phenology from the dominant Kerman variety. Because they split earlier in the season, there is an increased risk that moths from the second flight will be able to oviposit in the new crop.

May timing. Currently, many growers are applying insecticides during the first NOW flight for control of true bugs such as *Lygus* and the small bug complex of *Calocoris* and *Phytocoris* species. The resulting NOW suppression is a secondary benefit of a spray for true bugs. However, these same sprays can have a negative impact on the *Phytocoris relativus* that provide biological control of NOW by eating its eggs.

Field trials on the efficacy of insecticide applications against NOW typically show that one application of an effective insecticide can provide an approximate 50 percent reduction in damage at harvest. Additional sprays can further reduce damage, providing approximately 65 percent reductions with a two-spray program or 75 percent reductions with a three-spray program in some cases. These reductions can be increased or decreased depending on the NOW life stages affected by the active ingredient, application timing (i.e., application of an ovilarvicide when eggs are hatching), the level of synchronization of life stages of NOW, and coverage. Since NOW adults are active at night, applications targeting adults should be most effective when made late at night or at dawn. Although insecticides are one of the best control tools, overuse and poor timing can cause outbreaks of other pests, such as scales and mealybugs. The impact of repeated sprays on important predators of NOW has not been well documented.

Figure 18.11. Mating disruption uses puffers that are hung in trees to release synthetic navel orangeworm sex pheromone. *Photo:* D. Haviland.

Mating disruption

During the early 2000s, significant efforts were made to develop and integrate mating disruption as an option for NOW control in California. Mating disruption utilizes release devices, called puffers, to emit the principal component of the NOW pheromone produced by a female moth. When an orchard is filled with the artificially produced pheromone component, this decreases the ability of the adult male to locate calling females and mate. This decrease in mating success, as well as delays in the time it does take for a male to locate a female, ultimately results in the reduction of NOW offspring.

Commercial mating disruption currently uses two puffers per acre that release pheromones each evening for 200 days. The aerosol can with the principal pheromone component (Z,Z)-11, 13-hexadecadienal is housed in a plastic container (often referred to as a cabinet) (fig. 18.11) that contains the mechanism that controls pheromone release. Puffers are typically placed in the orchard in late March to early April, prior to the initiation of the first NOW flight, and they remain active throughout the season. The cabinets must be securely fastened to a tree limb, and they should be placed in the male trees or removed prior to shaking the trees to ensure that they do not end up in the bins.

Research has shown that mating disruption works best when used on large acreages. Most current usage is on areas ranging from a few hundred to a few thousand contiguous acres, though data also show that effects can be seen in orchards as small as 20 acres. Mating disruption also works best in orchards where good winter sanitation programs have been employed. One of the weaknesses of mating disruption is that it provides little to no protection to orchard edges from gravid females that might migrate into orchards from outside sources; this is particularly true on the upwind side of the orchard.

Due to the presence of synthetically produced pheromone throughout the orchard, monitoring with pheromone traps alone is not sufficient to keep track of navel orangeworm populations. Additional pheromone or egg traps are needed, particularly on the orchard edges, to help determine if moths are migrating into the orchard from an outside source. Growers and pest control advisers (PCAs) should also monitor pea-split and other early-split nuts from late July to harvest to ensure that mating disruption is working. If eggs are found, insecticide treatments are needed. Because mating disruption is a relatively new technology, the current industry practice is to use puffers in addition to insecticide sprays, not in place of them. However, over time it may be possible to reduce the number of insecticide sprays where mating disruption is used, as more experience and research lead to confidence in how to use this relatively new technology.

Obliquebanded Leafroller

Obliquebanded leafroller (*Choristoneura rosaceana* (Harris)) (OBLR) is a native North American moth in the family Tortricidae. The larvae feed on numerous crops in the San Joaquin Valley, including almonds, apples, apricots, cherries, pears, plums, prunes, peaches, and walnuts. Over 70 different plant species have been recorded to support OBLR development, which is not unusual for some leafroller species. Noncrop hosts that are common sources of OBLR in California include Valley oak (*Quercus lobata* Nee), wild plum (*Prunus subcordata* Benth.), choke cherry (*Prunus virginiana* L.), California buckeye (*Aesculus californica* (Spach) Nutt.), rose (*Rosa* spp.), and raspberry (*Rubus leucodermis* Dougl). Prior to 1989, there was little record of damage by this pest in California. In 1989, however, it was found on apples in California's San Joaquin Valley. In pistachio, feeding damage occurs directly on developing nut clusters and leaves. Leaf infestation can result in defoliation prior to nut maturity and harvest. The reasons for the occasional outbreaks of OBLR on pistachio are not known.

Description

Obliquebanded leafroller moths are large, approximately 0.6 inch long, and are the largest lepidopterans found in pistachio. From above, the wings form a bell-shaped outline, with two colors predominating: a background fawn (light brown) and medium mahogany, forming angled strips on the wings (fig. 18.12). They are easily confused with a much smaller species called the garden tortrix (*Clepsis peritana* (Clemens)). The garden tortrix is attracted to the pheromone for the western strain of OBLR but is caught in March and April, much earlier than OBLR. The garden tortrix is a third the size of OBLR, and it does not feed on pistachio.

Obliquebanded leafroller eggs are flat and overlapping, much like fish scales, and they are covered with a cementlike substance that colors the egg mass a light green. They are most often laid on leaves.

Six larval stages take place. The worms are light green with a dark black thorax and head (fig. 18.13). Full-grown larvae reach approximately 1.25 inches in length. They are very active when disturbed, wriggling and dropping on a silken thread to the ground. Obliquebanded leafroller pupae are dark brown, about 0.5 inch in length, and are usually found in leaves that are webbed together on the tree.

Seasonal Development

Obliquebanded leafrollers overwinter within silken hibernacula on the tree as 3rd instar larvae. Emergence from the hibernacula occurs in early March in the Bakersfield area and mid-March from Fresno north. Mature larvae are easily recognized by their light green body coloration with a dark black thorax

Figure 18.12. Obliquebanded leafroller adult. *Photo:* J. K. Clark.

Figure 18.13. Obliquebanded leafroller larva in nut cluster. *Photo:* K. Daane.

and head. The initiation of emergence usually occurs prior to leafing out of pistachio. Once leaves are present, the easiest method of detecting OBLR is by the presence of webbed or tied leaves. When these leaves are disturbed or separated, the worms actively wriggle backward and drop to the ground on a silken web.

Pupation occurs in mid-April. The first moth flight begins in late April and May. Eggs are usually laid in groups of approximately 200, but as many as 900 have been recorded from a single female. Eggs hatch approximately 12 days after being laid, although the rate is temperature dependent.

The lower temperature development threshold of OBLR is 43°F. In California, an upper threshold has been set at 94°F. Using these temperature limits, an average total of approximately 2,000 DD is required for a generation to develop. In northern California, as far south as Madera, there are generally two generations of the pest per year. South of Madera, three generations are common. Figure 18.14 presents the seasonal flight trend of OBLR in a Madera pistachio orchard.

Natural Control

Obliquebanded leafroller abundance is influenced by parasitoids. A number of parasitoid species attack OBLR, but the two most important are *Macrocentrus irridescens* French and *Habrobracon gelechiae* Ashmead (both in the family Braconidae). *Macrocentrus irridescens* is a polyembryonic wasp, with the reddish-brown adults a little larger than a mosquito. The adult wasp stings OBLR larvae, placing an egg inside the worm. The parasitoid can attack any of the larval stages but prefers the 2nd and 3rd instars. Parasitized worms are quite swollen and do not move readily as do unparasitized larvae. The eggs eventually develop into multiple wasp larvae that develop inside the

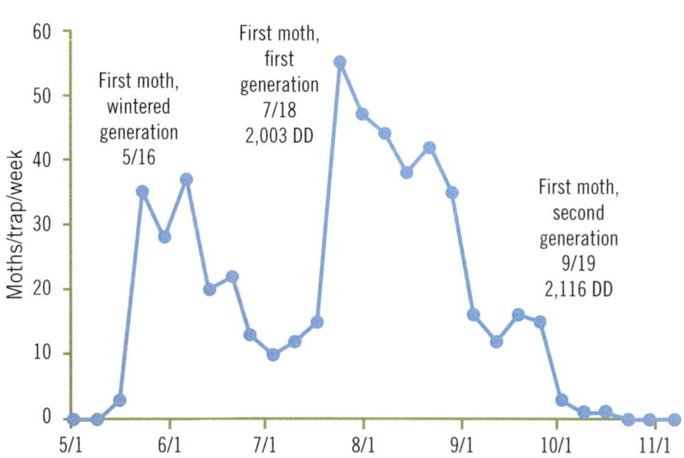

Figure 18.14. Seasonal flight activity of obliquebanded leafroller in pistachio, Madera, CA, 1999. *Source:* W. Bentley.

worm until they are ready to pupate, typically in a webbed mass near the host on the leaf or dropped to the soil. About 27 adults emerge from each parasitized OBLR. *Habrobracon gelechiae* is a generalist parasitoid of caterpillar species. This parasitoid has the advantage of very fast development, from egg to adult in about 7.1 days at 93°F. The adult lays eggs on the worm, where they develop externally, very similar to *G. legneri* on NOW. In orchards where *H. gelechiae* has been found, OBLR numbers have dropped quickly, and parasitism rates as high as 80 percent have been recorded in the second and third OBLR generations.

Several other species of parasitoids, such as egg parasitoids (*Trichogramma* sp.), larval-pupal parasitoids (*Spilochalcis* sp.), and a *Goniozus* species, are also commonly found in pistachios in California. Many additional species found in other areas of the United States have not been found in the San Joaquin Valley.

Monitoring

Moths are monitored with standard pheromone sticky traps that have a rubber pheromone cap baited with the western strain OBLR pheromone. The first trapped moths of the season are used as the biological fix (biofix) to begin calculating degree-days. Traps should be placed in orchards by mid-April and examined weekly. Pheromone septa should be changed every 6 weeks if monitoring throughout the season. Trap bottoms often become covered with moths and wing scales within a week. If this occurs, trap bottoms should be changed weekly. Pheromone traps are useful in establishing the biofix for treating summer populations of OBLR with insecticides.

In the spring, as pistachios begin budbreak and leafing, small worms can be found emerging from the hibernacula where they have overwintered. They are easily observed by examining small leaf clusters for webbing and feeding damage. This method of monitoring can be used to determine when larvae are exposed and susceptible to many nondisruptive insecticides such as spinosad and *Bacillus thuringiensis*. Spraying for OBLR during the spring is quite effective and minimizes damage from summer larval feeding.

Injury

Damage from summer generation larvae differs from spring damage. In the summer, larvae tie two contiguous leaves together with webbing and feed between the leaves. In some cases, they also move to developing clusters when the unfilled nuts are destroyed (see fig. 18.13). Although thresholds for treatment have not been established, the abundance of tied leaves can easily be counted. It is not uncommon to find hundreds of such tied leaves per tree. A level of over 50 tied leaves per tree probably results in some level of cluster damage. Tree size and age greatly influence the need for controlling OBLR during the summer.

Management

A variety of insecticides provide effective control of OBLR. These include *Bacillus thuringiensis* (Bt), spinosad, insect growth regulators, and other reduced-risk insecticides. According to insecticide timing studies, optimal control of OBLR results when applications are made between 800 and 1,000 DD, using 43° and 94°F as the lower and upper temperature thresholds, respectively, after the first moth is caught in the spring. Studies are currently being conducted on OBLR management in March and April, when first larvae are found infesting clusters. Nondisruptive sprays of *Bacillus thuringiensis* appear to work well in reducing the population of overwintering larvae as they emerge from diapause. Further information on insecticides and application timing can be found at the UC IPM Pistachio Pest Management Guidelines website, ucipm.ucanr.edu/PMG/selectnewpest.pistachios.html.

References

Burks, C. S., B. S. Higbee, D. G. Brandl, and B. E. Mackey. 2008. Sampling and pheromone trapping for comparison of abundance of *Amyelois transitella* in almonds and pistachios. Entomologia Experimentalis et Applicata 129:66–76.

Caltagirone, L. E., D. W. Meals, and K. P. Shea. 1968. Almond sticktights contribute to navel orangeworm infestations. California Agriculture 22(3): 2–3.

Doster, M. A., and T. J. Michailides. 1994. *Aspergillus* molds and aflatoxins in pistachio nuts in California. Phytopathology 84:583–590.

———. 1995. The relationship between date of hull splitting and decay of pistachio nuts by *Aspergillus* species. Plant Diseases 79:766–769.

Engle, C. E., and M. M. Barnes. 1983. Developmental threshold temperature and heat unit accumulation required for egg hatch of navel orangeworm (Lepidoptera, Pyralidae). Environmental Entomology 12:1215–1217.

Goodwin, J. A., and H. F. Madsen. 1964. Mating and oviposition behavior of the navel orangeworm, *Paramyelois transitella* (Walker). Hilgardia 35(18): 507–525.

Haviland D. R., W. J. Bentley, R. H. Beede, and K. M. Daane. 2014. UC IPM pest management guidelines for pistachio. Oakland: University of California Division of Agriculture and Natural Resources Publication 3460.

Higbee, B. S., and C. S. Burks. 2008. Effects of mating disruption treatments on navel orangeworm (Lepidoptera: Pyralidae) sexual communication and damage in almonds and pistachios. Journal of Economic Entomology 101:1633–1642.

Krugner, R., K. M. Daane, A. Lawson, and G. Yokota. 2005. Biology of *Macrocentrus iridescens* (Hymenoptera: Braconidae): A parasitoid of the obliquebanded leafroller (Lepidoptera: Tortricidae). Environmental Entomology 34:336–343.

Leal, W. S., A. L. Parra-Pedrazzoli, K. E. Kaissling, T. I. Morgan, F. G. Zalom, D. J. Pesak, E. A. Dundulis, C. S. Burks, and B. S. Higbee. 2005. Unusual pheromone chemistry in the navel orangeworm: Novel sex attractants and a behavioral antagonist. Naturwissenschaften 92:139–146.

Ohlendorf, B. 1985. Integrated pest management for almonds. Oakland: University of California Division of Agriculture and Natural Resources Publication 3308.

Onstad, D. W., H. Reissig, and C. A. Shoemaker. 1985. Phenology and management of the obliquebanded leafroller (Lepidoptera: Tortricidae) in apple orchards. Journal of Economic Entomology 78:1455–1462.

Rice, R. E., L. L. Sadler, M. L. Hoffman, and R. A. Jones. 1976. Egg traps for the navel orangeworm, *Paramyelois transitella* (Walker). Environmental Entomology 5:697–700.

Sanderson, J. P., M. M. Barnes, and W. S. Seaman. 1989. Synthesis and validation of a degree-day model for navel orangeworm (Lepidoptera: Pyralidae) development in California almond orchards. Environmental Entomology 18:612–617.

Sanderson, J. P., M. M. Barnes, R. R. Youngman, and C. E. Engle. 1989. Developmental rates of the navel orangeworm (Lepidoptera: Pyralidae) at various constant temperatures. Journal of Economic Entomology 82:1096–1100.

Seaman, W. S., and M. M. Barnes. 1984. Thermal summation for development of the navel orangeworm in almond (Lepidoptera: Pyralidae). Environmental Entomology 13:81–85.

Siegel, J. P., L. P. S. Kuenen, B. S. Higbee, P. Noble, R. Gill, G. Y. Yokota, R. Krugner, and K. M. Daane. 2008. Postharvest survival of navel orangeworm assessed in pistachios. California Agriculture 62:30–35.

Siegel, J. P., L. A. Lacey, R. Fritts Jr., B. S. Higbee, and P. Noble. 2004. Use of steinernematid nematodes for post harvest control of navel orangeworm (Lepidoptera: Pyralidae, *Amyelois transitella*) in fallen pistachios. Biological Control 30:410–417.

Siegel, J. P., L. A. Lacey, B. S. Higbee, P. Noble, and R. Fritts Jr. 2006. Application rate and abiotic factors on the efficacy of *Steinernema carpocapsae* for control of overwintering navel orangeworm (Lepidoptera: Pyralidae, *Amyelois transitella*) in pistachios. Biological Control 36:324–330.

Wade, W. H. 1961. Biology of the navel orangeworm, *Paramyelois transitella* (Walker), on almonds and walnuts in northern California. Hilgardia 31(6): 129–171.

Zalom, F. G., W. W. Barnett, and G. V. Weakley. 1984. Efficacy of winter sanitation for managing the navel orangeworm, *Paramyelois transitella* (Walker), in California almond orchards. Protection Ecology 7:37–41.

Zalom, F. G., J. H. Connell, and W. J. Bentley. 2000. Validation of phenology models for predicting development of the navel orangeworm *Amyelois transitella* (Walker) in California almond orchards. Proceedings of the 2nd International Symposium on Pistachios and Almonds. Acta Horticulturae 470:525–553.

19 Soft Scales

Robert H. Beede and Kent M. Daane

Soft scales include insects in the family Coccidae (Hemiptera) that use piercing-sucking mouthparts to feed on plant juices. Several coccid species can be found infesting pistachios, including European fruit lecanium, (*Parthenolecanium corni* (Bouché)), frosted scale, (*Parthenolecanium pruinosum* Coquillett), black scale (*Saissetia oleae* (Olivier)), and brown soft scale (*Coccus hesperidium* L.). All four species have a wide host range that includes many important agricultural and ornamental plant species.

Description

European fruit lecanium (fig. 19.1) and frosted scale adult females are light to dark brown in color and hemispherically shaped. Black scale adult females are similarly shaped but black in color (fig. 19.2). The adult insect bodies are about 1/8 to 1/5 inch in diameter, with shallow wrinkles over the surface. Frosted scale adults can be easily distinguished from the other species by a characteristic white wax that thinly covers their bodies at maturity, and frosted scale are slightly larger than European fruit lecanium. Black scale is distinguished from European fruit lecanium by the presence of ridges that form the letter "H" on the back of immature and adult scales. It is considered a primary pest of citrus and olives but can occasionally be found on pistachios. Brown soft scale is a mottled, yellow-brown color when young and becomes darker at maturity (fig. 19.3). Adults are flattened and elongate, resembling a football sliced in half, measuring about 1/8 inch in length. Male scales occur only rarely for European fruit lecanium and frosty scale, and they have not been observed in California.

For each of these soft scale pests, oval eggs (fig. 19.4) are placed under the female shell, or carapace. The small, unsettled 1st instars, or crawlers (named for their ability to disperse), hatch and leave the adult female from thin cracks between the plant surface and adult carapace (fig. 19.5). Immature of these scale pests begin yellow to orange in color, but they become darker as they feed, grow in size, and develop on leaves and twigs of the tree. Immature scales often appear shiny, flat, and light brown to translucent in color.

Seasonal Development

European fruit lecanium, black, and frosted scales possess a relatively simple biology with one generation per year. Brown soft scale has three to five generations per year and is often found on the nut. Females reproduce parthenogenetically (without mating), which is why males are either very rare or never found. The scales typically overwinter on the previous year's fruiting wood. Development slows during the winter but then resumes in the spring with rapid growth to the mature adult female. The adult deposits eggs beneath her body, where they

Figure 19.1. Adult European fruit lecanium soft scale. *Photo:* R. E. Rice.

Figure 19.2. Adult black scale and a black scale nymph in the rubber stage. *Photo:* J. K. Clark.

Figure 19.3. Infestation of brown soft scale on pistachio. *Photo:* D. Haviland.

are protected by the carapace, and the female adult withers and dies after egg laying (oviposition) is complete. More than 2,000 eggs have been found under a single female, although numbers from 500 to 1,000 eggs per female are more common for these soft scale species. Egg hatch typically begins from April to May and can be extended into June and even July, depending on scale species and seasonal temperatures. After hatching, the young crawlers move from the dead female onto expanding leaf tissue, where they settle and begin feeding.

There are typically six stages recognized: egg, 1st instar (which includes the crawler and settled 1st instar), 2nd instar, 3rd instar, preovipositional adult (often called the rubber stage, see fig. 19.2), and adult. The immature stages have two molts (first to second and second to third). As they mature, these soft scales are less likely to move, although it is possible for some species to move until the preovipositional adult stage. For species with one generation per year, depending on the time of emergence, settled crawlers (1st instars) will molt from June to August, to the 2nd instar stage. Development is slow during the summer because of high temperatures. In early September, 2nd instar scales migrate back onto the shoots produced that season.

These different soft scale species will overwinter on different parts of the pistachio. Fruit lecanium scale shows the most obvious movement to overwintering sites on the tree trunks, where it remains until late December or February. Immature soft scales remain motile throughout the winter, depending on temperature and weather. For European fruit lecanium scales, migration back to 1-year-old wood occurs in January when they molt into 3rd instar nymphs. By late February, they have grown rapidly and become rubbery. This represents the preovipositional stage. Soon afterward, they develop the hard carapace, the crustlike outer shell; the mature female begins to deposit her eggs underneath her body and both are protected by the carapace. With the egg hatch, the cycle of one generation per year then begins all over again.

Frosted scale and black scale are not as common on pistachio as European fruit lecanium scale. Their seasonal development is relatively similar to that described for European fruit lecanium scale. In pistachios, these scale species have only one generation per year, although each has been observed to have two generations (or a first and partial second) on ornamentals in coastal regions.

Injury

These soft scale pests are phloem feeders, and as they feed, they exude carbohydrate-rich honeydew, a sugary substance that can accumulate on the leaves and fruit clusters and serve as a substrate for sooty mold fungi. Research suggests that heavy densities of European fruit lecanium scale (10 to 30 3rd instars per inch in early February) can result in a significant reduction in shoot growth, overall tree vigor, and fruit size; the presence of this pest also reduces shell splitting, another stress-sensitive physiological event in pistachios. In one study, removal of European fruit lecanium from vigorous, 12-year-old Kerman trees resulted in 50 and 17 percent longer shoots during the off- and on-years, respectively. Fruit set and kernel filling were not affected, indicating their lesser sensitivity. This data suggests that heavy scale populations—regardless of the species—would cause economic injury by reducing new fruitwood development and nut quality.

Natural Control

European fruit lecanium, frosted scale, black scale, and brown soft scale are parasitized by several species of wasps (hymenopterous parasitoids). Depending on the species and their abundance, parasitic wasps can control soft scale if not disrupted by insecticidal treatments. Parasitized scales have small pinholes on their carapace where the wasps have emerged. Some parasites only attack specific soft scale species and even at specific stages of scale development. Thus, successful biological control often requires that multiple parasitoid species be present. The identification and distribution of scale parasites on pistachios in different regions is not presently known.

In a study conducted from 1994 to 1995, *Metaphycus luteolus* (Timberlake) was the most prevalent parasite of European fruit lecanium, collected during the dormant season on pistachios in western Kings County. The adult parasitoids emerge from 3rd instar and immature adult scales beginning in January. The parasitoid goes through a number of generations, with numbers increasing until the end of April, when crawlers emerge from their mothers. Examination of parasitized scales indicated that no viable eggs were produced compared with 2,500 to 4,500 eggs per unparasitized female. *M. luteolus* was not capable of biologically controlling European fruit lecanium by itself because it does not typically attack the crawler stage and its parasitism of immature adults was insufficient.

Small numbers of the wasps *Metaphycus lounsburyi* (Howard) and *Coccophagus lycimnia* (Walker) have also been identified attacking soft scales on pistachios. Although capable of parasitizing 2nd instars, *M. lounsburyi* typically attack 3rd instar to adult scales. *Coccophagus lycimnia* is important because this parasitoid species can attack 1st instar soft scales. However, the numbers recorded were insufficient for control of European fruit lecanium scale. *Coccophagus lycimnia* has an unusual reproduction (called arrhenotokous), in which mated females act as primary parasitoids that attack and kill the soft scale, while unmated females act as hyperparasites that attack other parasitoid species (e.g., *M. lounsburyi*) inside the soft scale.

These same parasitoid species may attack black scale and frosted scale in the pistachio orchard, but surveys have not yet been conducted. For example, *Metaphycus helvolus* (Compere) is a common parasitoid of black scale, and *Metaphycus californicus* (Howard) is considered the most effective against frosted scale.

Figure 19.4. Eggs of European fruit lecanium scale next to the female scale's covering. *Photo:* J. K. Clark.

Figure 19.5. Crawler stage of frosted scale. *Photo:* J. K. Clark.

Several predators present in pistachios feed on soft scale. In addition to *Phytocoris*, lacewing and lady beetle larvae feed voraciously on the crawler stages. Big-eyed bugs have been observed feeding on eggs. Insecticide applications made against *Phytocoris* in April most likely kill many emerging scale parasites and reduce the predator populations.

Orchards with high tree vigor, dense canopies, and cooler temperatures from low-volume irrigation systems are more predisposed to soft scale outbreaks. Mild winters also favor greater overwintering scale survival. Severe winters with heavy rainfall and wind can cause such dramatic soft scale reductions that treatment is unnecessary. Spring temperatures above 90°F have also been reported to cause significant mortality of newly hatched crawlers.

Monitoring and Management

Monitoring for scale is performed by visual examination of the 1-year-old fruiting wood in early to mid-January. Carefully check randomly selected 12-inch dormant shoots for live and parasitized scales. Light to moderate populations would be from one to five live scale crawlers per inch in early February. Heavy populations range from 10 to 30 per inch. Tree age, vigor, fruitwood, and split nut percentages should be considered when making a treatment decision. Another consideration when *Phytocoris* is present is that removing soft scales, which are a preferred food source for *Phytocoris*, can result in increased levels of *Phytocoris* feeding on the nuts.

Management of soft scales is accomplished by a combination of biological and chemical controls. In most years, natural controls are sufficient to keep black scale and frosted scale below damaging levels. European fruit lecanium scale has been the more important pest. In cases where soft scale levels build over time, insecticide treatments are justified. Treatments from mid-December to mid-February are very effective against soft scale and overwintering *Phytocoris*. Moderate populations have been successfully controlled with oil alone, while heavy populations often require oil plus an insecticide. For optimal control, frosted scale should be treated before its waxy coating develops. European fruit lecanium scale populations should be treated before reaching the rubber stage in late February. Depending on weather conditions and scale pressure, re-treatment may not be necessary for three to five seasons. Oil application also can advance pistachio budbreak and nut maturity by 7 to 10 days. Brown soft scale is usually associated with high levels of parasitism in orchards where broad-spectrum insecticides have been avoided. Additional information on pesticides currently available can be found in the UC IPM Pistachio Pest Management Guidelines, ucipm.ucanr.edu/PMG/selectnewpest.pistachios.html.

References

Bailey, S. F. 1964. A study of European fruit lecanium scale, *Lecanium corni*, on prune. Journal of Economic Entomology 57(6): 934–938.

Beede, R. H., R. E. Rice, K. M. Daane, D. R. Thomas, J. Padilla, S. Soliman, W. Bentley, and G. Weinberger. 1995. Relation of *Phytocoris relativus* to lecanium scale infestations on pistachios. California pistachio industry annual report 1995, 107–118. Fresno: California Pistachio Commission.

20 Small Bugs: *Neurocolpus, Phytocoris, Calocoris,* and *Lygus*

Walter J. Bentley, Peter B. Goodell, and Robert H. Beede

Mirid Plant Bugs

The small plant bugs are all members of the family Miridae. Four species are commonly found in this group: *Neurocolpus longirostrus* (Knight), *Phytocoris relativus* Knight, *Calocoris norvegicus* (Gmelin), and *Lygus hesperus* (Knight). The first two species develop on pistachio, whereas the second two move to pistachio from overwintering host crops in the spring and damage developing clusters. These insects are characterized by piercing-sucking mouthparts that are capable of penetrating and feeding on soft plant tissue. They are unable to penetrate a hardened pistachio shell. Consequently feeding damage is restricted to the period just after bloom through shell hardening, which occurs in early June. Research conducted with *Calocoris* indicates that shell hardening is sufficient to prevent small bug damage when 1,197 DD (> 50°F) has occurred after 75 percent bloom of pistachio. While the nuts are BB-size, feeding results in blackening and dropping of small nuts (fig. 20.1). As the nuts grow, feeding results in epicarp lesion instead of nut blackening and subsequent nut drop. Epicarp lesion is characterized by localized dark sunken areas at the base or side of the nut that often oozes a gummy substance (fig. 20.2). If the nut is sliced in half, dark sunken areas are present where the bug has inserted its mouthpart (fig. 20.3). On close examination with a 14× hand lens, the puncture wound can be seen.

Figure 20.1. Buckeye bug feeding damage on pistachio. *Photo:* W. J. Bentley.

Figure 20.2. Epicarp lesion symptoms shortly after bloom. *Photo:* W. J. Bentley.

Figure 20.3. Internal damage from *Phytocoris* prior to shell hardening. Note dark, sunken internal shell tissue. *Photo:* R. E. Rice.

Neurocolpus longirostrus

The California buckeye bug (*Neurocolpus longirostrus* Knight (Miridae)) is a native insect common throughout the Sierra Nevada and Coast Range foothills of California. In its native habitat, *N. longirostrus* has been found primarily on California buckeye (*Aesculus californica* (Spach) Nutt) and has been collected from Kern County north to Butte and Glenn Counties. Adults and eggs have also been collected from coffeeberry (*Rhamnus* spp.).

Description

Adults of the buckeye bug are approximately ⅜ inch long and are straw colored with mottled light to dark brown markings over the body. The most distinguishing characteristic of *N. longirostrus* adults and nymphs is the enlarged first antennal segment (nearest the head), which is swollen and hairy. Nymphs are gray green with similarly colored wing pads and alternating white and brown bands on the legs and antennae. In pistachio, eggs are laid in the terminal buds or in the base of new leaf petioles. Figure 20.4 shows two stages of *N. longirostrus* development.

Injury

Because this insect has been found only in pistachio orchards that are near its native host, California buckeye, it is not a widespread problem. However, damage caused by it can be quite severe in certain orchards in foothill areas of California. At the same time that small nymphs are developing wing pads, new nuts are developing and are about the size of a BB. If damaged at this time, nuts quickly darken and fall from the tree. If the buckeye bug is the cause, multiple nuts in a cluster will be blackened (see fig. 20.1). This often involves over 70 percent of the cluster. As the nuts continue to develop and reach ½ inch in length, the shell begins to harden. Typical epicarp lesion symptoms can be observed at that time. External symptoms include a brown sunken area where the insect has fed. This often appears as darkened concentric rings. Internally, distinct cavities or pits in the soft shell tissue can be seen (see fig. 20.3). Small nuts exposed to *N. longirostrus* nymphs in late April develop epicarp lesion symptoms 36 to 48 hours after feeding, with severe internal shell pitting most often found at the base of the nut where the stem attaches. Crop loss estimates in impacted orchards have ranged as high as 30 to 35 percent from *N. longirostrus* and in a few instances have been even higher.

Figure 20.4. Adult (A) and 3rd instar nymph (B) buckeye bug. *Photos:* W. J. Bentley.

Seasonal development

Neurocolpus longirostrus overwinters in the egg stage on both buckeye and pistachio. Egg hatch coincides with buckeye flowering and pistachio bloom in March. Nymphs are found in late March through mid-May. There appear to be four or five nymphal stages. Adults begin to appear in pistachio orchards by early May and are present in the orchard only until late May or early June, when they lay eggs that overwinter. These eggs hatch the following spring. *N. longirostrus* has only one generation per year in California.

The nymphs are relatively easy to detect with beating tray samples early in the season at about the time that pistachio bloom is complete and small nuts begin to form. First or second instar nymphs are easily seen on beating trays, and population evaluations can be made at this time. Adults can move from nearby buckeye trees or other hosts into pistachio orchards and lay overwintering eggs in May and June. The adults have been observed to migrate distances over ¼ mile from native buckeye to commercial pistachio plantings. Once established, *N. longirostrus* can maintain itself as an endemic pest in pistachios.

Monitoring and management

Sampling for *N. longirostrus* should begin just prior to bloom, using a beating tray to dislodge the insect from trees (fig. 20.5). As with the sampling of other plant bugs, the tray is held under clusters and the branch supporting the clusters is struck forcefully three times. Various items have been used for this purpose. Trashcan lids, 18 to 24 inches in diameter and painted white, will suffice. Most professionals use a wire-framed circle about 22 inches in diameter. Two cross braces made from ⅛-inch rods divide the circle into quarters. A 6-inch length of

1-inch-diameter tubing is welded perpendicular to the frame at the circle's center, where the support rods cross. This serves as the handle. The wire circle is then covered with white fabric, into which elastic or a drawstring has been sewn around the perimeter for tightening onto the frame. Approximately 10 samples of three beats each are conducted for every 40 acres of pistachios. Pest consultants often increase the sampling intensity on smaller acreage plantings when insect counts increase.

The beating tray is held at eye level beneath a fruiting limb to enable the user to observe whatever lands on the tray. A hand mallet is used to rapidly strike the tree limb sharply three times. The tray is then quickly brought down to a comfortable level for insect counting. Although no threshold for treatment has been established, very low numbers of this insect can cause substantial damage. In conjunction with sampling to assess the insect populations, nut clusters should be examined for early detection of blackened nuts and the development of epicarp lesion. Pheromone traps are available to monitor *N. longirostrus*, though no correlations have been established between populations of adult males in the traps and damage to the pistachio crop. Adults are not found in pistachio orchards after June.

Once *N. longirostrus* invades pistachio orchards, chemical control will eventually be required due to the severe impact of this pest on soft-shelled nuts.

Natural controls

At least four parasites are known to attack *N. longirostrus* eggs on buckeye throughout its native host range. These parasites have not been found in eggs in pistachio orchards, and it is not known whether they would be able to exert any effective biological control on *N. longirostrus* in commercial plantings.

Figure 20.5. Beating tray used to sample for small bugs. *Photo:* J. K. Clark.

Phytocoris

Phytocoris is a genus of true bugs (Hemiptera: Miridae) that is comprised of more than 400 species worldwide, and 160 of these are in western North America. They occur on a wide variety of hosts, including pines, oak, almonds, quince, and prunes. In 1985, two species (*P. californicus* Knight and *P. relativus* Knight) were documented to damage pistachio by causing epicarp lesion to nuts. Without microscopic examination, they are indistinguishable from each other, but *P. relativus* is the more common.

Description

Adults are about ¼ inch in length (fig. 20.6A). When disturbed, they move very quickly to a protected place on the branch. Both adults and nymphs are gray in color with long, slender legs and antennae (fig. 20.6B). The antennae of both are distinctly banded in black and white. Eggs are elongate and tubelike with a white cap that protrudes from the bark surface (fig. 20.7).

Injury

Phytocoris is capable of causing epicarp lesion and loss of immature nuts prior to shell hardening. Damage is usually greatest in April as the insects

Figure 20.6. Adult (A) and 2nd or 3rd instar nymph (B) *Phytocoris. Photos:* W. J. Bentley.

Figure 20.7. *Phytocoris* eggs laid in 1-year-old pistachio wood. *Photo:* J. K. Clark.

grow and egg hatch is completed. Research in southwestern Kings County indicates that abundant *Phytocoris* populations of five per three-beat sample during early nut development in April and May result in lesion levels of 15 to 20 percent. However, there has been no correlation between the percentage lesion and yield of filled nuts in an extended study. This suggests that factors other than *Phytocoris* populations and nut injury prior to shell hardening affect pistachio yield.

Seasonal development

Phytocoris overwinters in the egg stage on 1-year-old wood. Most eggs are laid in the bark tissue (see fig. 20.7), but eggs can also occasionally be found in vegetative or fruit buds. *Phytocoris* has four generations per year, and a fifth generation may occur in hot years. Figure 20.8 shows the seasonal population dynamics of *Phytocoris*. First instar nymphs, hatching from the overwintering eggs, emerge the first week of March and develop into adults by mid-April. The remaining three generations occur during the growing season, with peaks in early June, August, and late September to early October. These peaks are usually less distinct, presumably because of overlapping from previous generations. Egg laying from the April and June generations occurs principally in the soft rachis tissue of the cluster and less frequently in leaflet petioles and current season shoots. In November, adult *Phytocoris* migrate to the trunk bark, where large numbers of 2nd instar soft scales (*Parthenolecanium* spp.) reside. *Phytocoris,* presumably, feed on scale before returning to dormant 1-year-old wood for egg laying.

During March and April, adult soft scales (chapter 19), prior to forming their crustlike covering and laying eggs, are a primary food source for *Phytocoris.* Thus, populations of *Phytocoris* are often proportional to the density of soft scales. After adult scales harden, *Phytocoris* readily feed on scale eggs that are laid by females by probing beneath them with their needlelike mouthparts. *Phytocoris* has not been observed feeding on newly hatched scales in the spring, but they readily feed on 2nd instar stages that have migrated from the leaves to woody shoots in the fall.

Laboratory feeding studies with *Phytocoris* indicate that they feed on southern stink bug eggs and nymphs, as well as navel orangeworm eggs. Navel orangeworm eggs were preferred over frosted and fruit lecanium scale eggs in these studies. *Phytocoris* consumed an average of 25.6 navel orangeworm eggs daily in petri dish feeding studies, where alternative plant food material was provided. When provided only navel orangeworm eggs, daily consumption increased to 40.5 eggs. *Phytocoris* showed no interest in leaffooted bug eggs but were documented to cannibalize nymphs. In addition to their predatory characteristics, research has demonstrated that the species in California pistachios must feed on plant material for optimal survival. In laboratory observations, *Phytocoris* repeatedly probed the ends of pistachio twigs, leaves, and fruit stem tissue with their needlelike mouthpart. Plant tissue probably supplies *Phytocoris* with water and carbohydrates; immature scales, various insect eggs, and smaller insects serve as their protein source. Adult *Phytocoris* that were provided with navel orangeworm eggs and plant material lived 31.5 days compared with only 6.5 days when fed eggs alone. *Phytocoris* appears to cause epicarp lesion in the process of meeting its nutritional requirements. However, it did not appear to cause epicarp lesion as readily as two other true bug species, *Neurocolpus longirostrus* and *Calocoris norvegicus*.

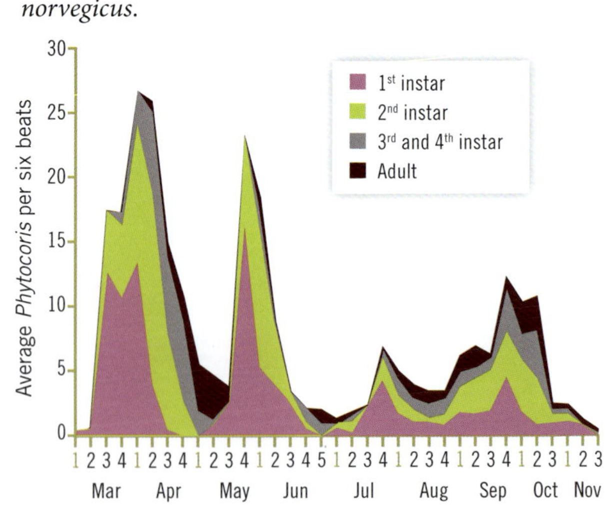

Figure 20.8. *Phytocoris relativus* life cycle from March to November 1992, based on weekly beating of female pistachio trees (Kerman cultivar) receiving no insecticide treatments and averaging counts from six replications. *Source:* R. H. Beede.

Studies on percentage small nut injury versus percentage fruit set suggest that pistachios compensate for the loss of nuts injured prior to shell hardening by setting and filling nuts that would otherwise fall off. This compensation capacity appears greater for trees in their on-bearing year. Artificial removal of up to 80 percent of the nuts on selected clusters in May had no effect on either the total nut number or the number of filled nuts at harvest compared with control clusters, from which nuts had not been removed. Compensation decreases at shell hardening when natural nut drop has ended, thus limiting the number of nuts that can be lost without reducing yield. Nut removal in excess of 20 percent shortly after shell hardening correlated strongly with reduced nut numbers at harvest. However, *Phytocoris* cannot penetrate shell-hardened nuts.

Monitoring

The procedure used for sampling *Phytocoris* is the same as that used for *Neurocolpus longirostrus*. Sampling should begin just prior to bloom, using a beating tray as previously described. *Phytocoris* fall from the cluster and onto the tray, where they can be counted. The percentage of darkened nuts caused by feeding is much less than is found with *N. longirostrus*, usually less than 10 percent of the cluster being damaged. Although a very effective pheromone is available for monitoring adults, research is needed to correlate trap catches with economic injury.

Natural controls

Several insects common to pistachios prey on *Phytocoris*. They include various spiders, assassin bugs (*Zelus* spp.), and green and brown lacewing (*Chrysopa* spp., *Chrysoperla* spp., and *Hemerobius* spp.). Spiders have been observed feeding on *Phytocoris*, where they are often found in spiderwebs. Lacewings are observed in greatest numbers after shell hardening and frequently feed on 2nd instar soft scales and *Phytocoris* nymphs. Spring application of permethrin insecticides adversely impacts these and other beneficial insects for the rest of the season. These insects include big-eyed bugs (*Geocoris* spp.), minute pirate bugs (*Orius* spp.), and hymenopterous wasps, which parasitize soft scales. Rough shield stink bug *(Brochymena)*, considered a beneficial stink bug, is also significantly reduced by permethrin treatment. The economic importance of losing these beneficials early in the season is still not known.

Management

There are no data-supported treatment thresholds for *Phytocoris*. Pest managers should consider trends from week to week and observe nut injury levels. Three *Phytocoris* per 10 beat tray samples would be considered low, and 7 to 10 might prompt treatment.

Insecticide applications, if warranted, should be made before nymphs turn into adults in May to prevent egg laying and subsequent population buildup. This is especially difficult for large-acreage growers since ground applicators typically treat about 40 acres per sprayer per day. Guidelines for how much damage can be tolerated at various nut development stages are not yet available. Growers considering treating *Phytocoris* are therefore advised to experiment with small untreated areas within the orchard and evaluate the impact.

Calocoris norvegicus

Another mirid commonly associated with pistachios (*Calocoris norvegicus* (Gmelin)) is known as the potato mirid in Europe and New Zealand. *Calocoris* is not a native insect, unlike other mirid pests in pistachios, but it is a relatively recent introduction to North America. It first appeared in collections from California during the 1940s and has since spread through most of northern and central California. Field surveys have shown it to be much more prevalent in northern California and the Sacramento Valley than in the southern San Joaquin Valley. However, high populations of *Calocoris* have been found as far south as the San Joaquin River.

Description

Adult *Calocoris* are small, yellow-green bugs, about ¼ to 5⁄16 of an inch long with two prominent black spots on the thorax just behind the head. The outer tips of the wings appear dark when folded over the body. They are often mistaken for *Lygus*. Figure 20.9 compares adult (A) and nymph (B) *Calocoris* with adult (C) and nymph (D) *Lygus*. There are four or five nymphal stages of each species.

Seasonal development and injury

Preferred hosts of *Calocoris* include native or cultivated legumes such as purple vetch *(Vicia dasycarpa)*, clovers (*Trifolium* spp.), and alfalfa *(Medicago sativa)*; broadleaf weeds such as mustard (*Brassica* spp.), fireweed *(Epibobium angustifolium)*, London rocket *(Sisymbrium irio)*, and curly dock *(Rumex crispus)*; and occasionally grass (family *Gramineae*). *Calocoris* does not appear to have a strong affinity or continual association with pistachios, as does *Neurocolpus longirostrus*. Instead, it seems to build to high populations on weeds in and around pistachios. As the native hosts dry down or are cultivated, adults

Figure 20.9. Comparison of adult (A) and nymph (B) *Calocoris* with adult (C) and nymph (D) *Lygus*. Each species has four or five nymphal stages. *Photos:* J. K. Clark.

migrate into pistachio trees and cause early-season epicarp lesion similar to that caused by other mirids prior to shell hardening. A strong direct correlation has been shown between populations of *Calocoris* in weeds, as well as alfalfa, and the epicarp lesion damage in adjacent pistachios. Studies in northern California have shown that uncontrolled populations of *Calocoris* adults could cause yield losses of up to 48 percent. *Calocoris* generally appears to prefer moist, cool environments. It has only one generation per year and overwinters in the egg stage in stems of weed hosts. Nymphs appear in early to mid-March in the Central Valley and in late March in the Sacramento Valley. Adults are present from mid-April through June. Adults are usually found 3 to 9 weeks after the first nymphs are collected in sweep net samples from the orchard ground cover.

Caged *Calocoris* females have been observed laying eggs in the rachis of nut clusters. However, these eggs do not hatch, and thus *Calocoris* does not overwinter in pistachio trees as does *N. longirostrus*. *Calocoris* populations decline and disappear with the maturing and drying of native weed hosts in late spring and early summer. Cover crops and weeds within pistachio orchards can sustain adult populations of *Calocoris* for extended periods. The adults migrate during the day between orchard ground covers and nut clusters in the trees.

Sampling

Sampling for *Calocoris* involves evaluating weed hosts before pistachio bloom and, later, beating tray samples as previously described for *N. longirostrus* and *Phytocoris*. A standard sweep net with a 15-inch-diameter opening and a 24-inch handle is used to dislodge immature *Calocoris* from weeds. The number of sweeps should be sufficient to evaluate the presence of plant bugs; a minimum of 25 sweeps in each of the predominant weed species is suggested. If *Calocoris* is found, determine the stage of development by looking for the presence of small wing pads on the insect. When wing pads are present, the adult stage will soon be reached, and migration to the tree will then occur. No treatment thresholds have been established. Sweep net sampling is used to predict movement to the orchard.

Once adults are found, the orchard should be sampled with a beating tray. Be observant for the progression of darkened nuts caused by *Calocoris* feeding.

Recently, a sex pheromone for *Calocoris* has been identified and synthesized. This pheromone is highly specific for adult male bugs. To date, field recommendations for the use of pheromones in detecting and monitoring adult bugs have not been developed.

Management

Control of *Calocoris* and concurrent reduction of epicarp lesion is best achieved through reduction of weed hosts in and around pistachio orchards prior to the maturation of *Calocoris* nymphs. Weeds should be cultivated or treated with an appropriate insecticide when 4th or 5th instar nymphs are first detected in sweep samples to prevent further development of the insect. Timely and effective weed control should reduce or eliminate the need for most insecticide sprays for *Calocoris* in pistachio orchards. Migrating adult populations usually do not travel more than 300 feet into pistachio orchards from outside sources. Therefore, if chemical controls are under consideration, border treatments on the perimeter of orchards should reduce damage to negligible levels. To date, there are no known predators or parasites that might contribute to biological control of this pest species.

Lygus hesperus

The western tarnished plant bug, or lygus bug (*Lygus hesperus* (Knight)), can be one of the most difficult pest problems in pistachio. The severity of this insect as a pest in orchards depends on the development of the insect on host plants on which it lives and feeds. Unlike *Phytocoris* and *Neurocolpus*, *L. hesperus* does not develop on pistachio. Instead, it moves into the orchard from surrounding crops or weeds and into the tree when understory vegetation becomes unsuitable. Environmental conditions such as rainfall and temperature, which favor the emergence and establishment of many broadleaf weeds, favor *L. hesperus* population development in the general western areas of the San Joaquin Valley.

Description

Lygus hesperus is predominantly brown with a gold, strawlike mottling also present throughout the body. It is about ¼ of an inch in length. Another species, *Lygus elisus* Van Duzee, is a light green color and the same size, but it is less common. Both of these species have a yellowish triangular structure (the scutellum) on the back between the wings. *Lygus* can be confused with *Calocoris*, but *Calocoris* has two distinct spots on the pronotum, just behind the head, and usually has a green (versus yellow for *Lygus*) scutellum (see fig. 20.9). *Lygus* can be distinguished from *Neurocolpus* by its short first antennal segment and uniform leg coloration compared with the long, swollen first antennal segment and alternating white and brown bands on the legs and antennae of *Neurocolpus*.

The immature or nymphal stages of *Lygus* are light green and may resemble aphids. As the nymphs mature, they develop wing pads, and black dots can be seen on the back of the insect. Two dots are located on each of the thoracic segments, and one dot appears on the abdomen. *Neurocolpus* nymphs have black dots and mottling, but they also have black and white stripes on the legs and antennae. *Calocoris* nymphs do not have black dots. Immature stages of *Lygus* are seldom found on pistachios because little egg laying occurs on the crop. It is important to recognize the nymphs as they reach maturity in weed hosts, as this indicates when migration to pistachios may occur. Just prior to migration, small wing pads become visible on the back of the nymphs. While older nymphs can cause substantial damage in many crops, adults are the key in population distribution. Adults can fly away from drying host plants but nymphs cannot.

Injury

Like that of *Neurocolpus*, *Calocoris*, and *Phytocoris*, the damage done by *Lygus* occurs prior to hardening of the shell. These insects only cause epicarp lesion, as they are incapable of piercing a hardened pistachio shell.

Seasonal development

Lygus overwinter as adults and move from ground cover protection to a variety of host plants in spring. During the growing season, five to seven generations per year occur in alfalfa, but only three can develop in cotton due to the shorter growing season. In fall, as day length decreases to less than 9 hours, adults enter into an arrested development termed sexual diapause.

To understand *Lygus* management, it is important to understand its biology in the agroecosystem that pistachios are part of. *Lygus* is a native insect in the San Joaquin Valley that feeds on developing reproductive or growth tissues. *Lygus hesperus* has a reported host range of 110 plant species, with two plant families (Asteraceae and Fabaceae) constituting 39 percent of its recorded hosts. Lygus bugs are pests on cotton (*Gossypium* spp.), seed alfalfa *(Medicago sativa)*, strawberry *(Fragaria × ananassa)*, bean *(Phaseolus vulgaris)*, apple *(Malus domestica)*, peach *(Prunus persica)*, pear *(Pyrus communis)*, celery *(Apium graveolens)*, broccoli *(Brassica oleracea)*, cauliflower *(Brassica oleracea)*, and vegetable seed crops. They generally do not use grasses as reproductive hosts. Although they do not cause economic damage in alfalfa forage fields, *Lygus* reproduce there, and large populations move to neighboring crops when alfalfa fields are cut throughout the year.

Lygus intensity in any given year depends on the following: the abundance of hosts (determined by rainfall patterns); warm temperatures in the early season that allow *Lygus* to complete multiple generations; and early summer hosts. This dynamic is shown in figure 20.10.

In California's Mediterranean climate, most *Lygus* movement from wild areas occurs between April and May, and few noncultivated hosts survive beyond June. Winter rainfall patterns set the stage for the abundance, diversity, and longevity of hosts that are suitable for colonization by *Lygus*. Early-winter rains tend to produce grass in the surrounding hills, while late-winter rains favor broadleaf hosts suitable for *Lygus*. These broadleaf weeds include London rocket *(Sisymbrium irio)*, redstem filaree *(Erodium cicutarium)*, and clovers (*Trifolium* spp.). Abundant rainfall in spring provides deep moisture for Russian thistle *(Salsola iberica)* and tarweed (*Hemizonia* spp.). In years when moisture is adequate for extending these host plants into June or July, additional generations of *Lygus* can build and create severe migrations. Many crops, such as sugarbeet *(Beta vulgaris)*, alfalfa *(Medicago sativa)*, tomato *(Solanum lycopersicum)*, and weeds within orchards, vineyards, and row crops, provide areas of refuge during the winter and spring, regardless of weather conditions.

Summer hosts consist almost entirely of cultivated plants or their associated weed complexes, which in the San Joaquin Valley includes over 200 crops. As these crops mature and are readied for harvest, the habitat becomes less suitable, and *Lygus* migrate into neighboring fields. Alfalfa hay represents a substantial portion of cropland that *Lygus* inhabit, preferring it over many other plants.

Thus, it is the proximity of a pistachio orchard to these sources that determines the intensity of the *Lygus* migration. Since the majority of pistachio production occurs on the western edge of the San Joaquin Valley, native plant growth in the hills of that area is more important as a source of *Lygus* than the cultivated crops bordering the orchard. For example, in years when there is late-winter or early-spring rainfall (February through April), plants such as tarweed or Russian thistle can cover hundreds or even thousands of acres. The insect population density

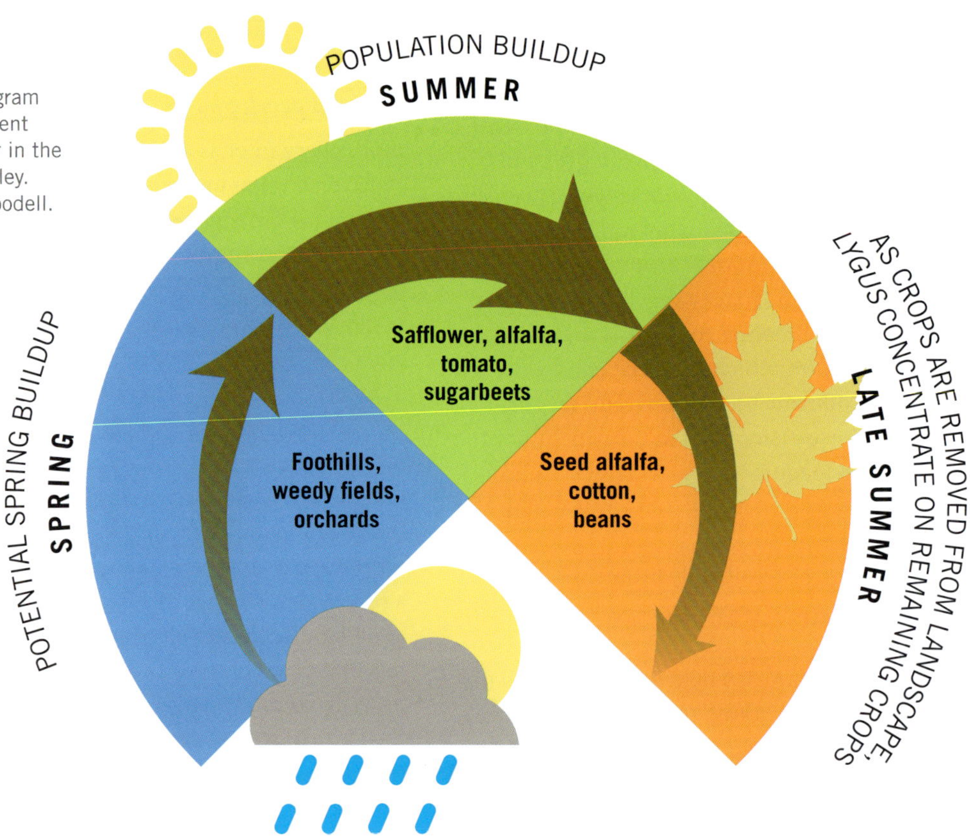

Figure 20.10. Generalized diagram of *Lygus* movement through the year in the San Joaquin Valley. *Source:* P. B. Goodell.

can build to high levels if the plant can draw sufficient moisture to remain a suitable host to *Lygus* for multiple generations.

Once the insects move into the orchard, the understory vegetation can become an important source of *Lygus* that can move into the tree when the vegetation becomes unsuitable. Cover crops that serve as good *Lygus* hosts include vetch *(Vicia sativa)*, bur medic *(Medicago polymorpha)*, crimson *(Trifolium incarnatum)*, rose *(Trifolium hirtum)*, and berseem *(Trifolium alexandrinum)* clovers, and these should be avoided. For seeded cover crops, the use of subterranean clover *(Trifolium subterranum)* has been suggested, as it does not host *Lygus*.

The timing of *Lygus* movement into pistachio orchards determines the severity of the problem. *Lygus* can penetrate the shell tissue only before it hardens, so migration into an orchard between bloom and shell hardening is a critical time. For example, both birdsfoot trefoil *(Lotus corniculatus)* and bur clover (*Medicago* spp.) naturally decline prior to the growth of nut clusters in pistachio. However, other weed hosts that overlap with developing nut clusters include lupine (*Lupinus* spp.), London rocket, various mustards, stinging nettle (*Urtica* spp.), and tarweed, and the development of *Lygus* on these weeds is important. During dry years, these weed species usually become unattractive at the same time that pistachio nut clusters are susceptible to damage from *Lygus*.

Plants that host *Lygus* should be sampled to detect pest presences and estimate the threat of movement. A standard 15-inch-diameter sweep net is useful in sampling *Lygus* from the tops or sides of plants, allowing counting of the insects to monitor population dynamics. As an aid in predicting possible *Lygus* movement, it is valuable to identify the host, note its condition (e.g., vigorously growing, flowering, drying), and determine the stage of *Lygus* collected from the host (e.g., small nymphs, large nymphs, or adults). Grasses are not a host for *Lygus*, and sampling for *Lygus* in grasses will result in a misleading estimate of population abundance.

Monitoring

Monitoring *Lygus* should be done with a sweep net on host weeds at the appropriate time. Numerous sweeps should be made to detect the presence of *Lygus* and their stage of development. As the wing pads become visible on the captured insects, there is potential for migration if the weeds dry. If weeds can be destroyed within pistachio orchards, the insects residing there are easily killed.

Lygus is difficult to detect within trees. The most common method of sampling is the use of a beating tray. This technique is not very efficient, and even two or three insects found is an indication of a damaging population.

Natural controls

Although there are parasitoids and predators that attack *Lygus*, they do little to control the pest in pistachio. One of the common egg parasitoids, *Anaphes iole* Girault, helps in biological control of *Lygus* in its native habitat. The movement of lygus bugs into the crop and their feeding there occur so rapidly that there is little chance for biological control to establish.

Management

If epicarp lesion is increasing within the orchard and adult *Lygus* can be found, it may be necessary to spray the crop (see Pistachio Pest Management Guidelines at the UC IPM website, ipm.ucanr.edu/PMG/). Current thresholds for treating *Lygus* in pistachio have not been established. If the decision to spray is made, timing should be based on the development of adults in the weed hosts nearby. It is important not to disturb weeds within the orchard during the period from bloom through shell hardening. If weeds are disturbed, *Lygus* may be driven into the trees, where feeding results in epicarp lesion.

Due to the migratory nature of this pest, future work should address regional management of *Lygus*. Research has already documented that alfalfa and safflower are preferred hosts, and practices such as strip cropping can help keep *Lygus* in the alfalfa. Future research could potentially develop better ways to use host weeds or plants as trap crops that would keep the insect from migrating to the less preferred pistachios.

References

Beards, G. W., and F. E. Strong. 1966. Photoperiod in relation to diapause in *Lygus hesperus* Knight. Hilgardia 37:345–362.

Beede, R. H., R. E. Rice, K. M. Daane, D. R. Thomas, J. Padilla, S. Soliman, W. Bentley, and G. Weinberger. 1995. Relation of *Phytocoris relativus* to lecanium scale infestations on pistachios. California pistachio industry annual report 1995, 107–118. Fresno: California Pistachio Commission.

Beede, R. H., S. Soliman, J. Padilla, D. Thomas, D. Rose, and A. Champlin. 1996. The effect of nut removal before and after shell hard on the compensation capacity of pistachios. California pistachio industry annual report 1996, 86–91. Fresno: California Pistachio Commission.

Beede, R. H., D. Thomas, R. E. Rice, K. M. Daane, J. Padilla, and W. J. Bentley. 1994. Relation of *Phytocoris relativus* to lecanium scale infestation on pistachios. California pistachio industry annual report 1994, 89–99. Fresno: California Pistachio Commission.

Bentley, W., and S. Mochizuki. 1986. Further investigations in controlling insects causing epicarp lesion and kernel necrosis in pistachio. California pistachio industry annual report 1986, 135–139. Fresno: California Pistachio Commission.

Champlain, R. A., and G. D. Butler, Jr. 1967. Temperature effects on development of the egg and nymphal stages of *Lygus hesperus* (Hemiptera: Miridae). Annals of the Entomological Society of America 60: 519–521.

George, M. R., K. Olson, and J. W. Menke. 1988. Range weather: A comparison of three California range research stations. California Agriculture 42(1): 30–32.

Goodell, P. B. 2009. Fifty years of the integrated control concept: The role of landscape ecology in IPM in San Joaquin Valley cotton. Pest Management Science 65:1293–1297.

Leigh, T. F., S. H. Roach, and T. F. Watson. 1996. Biology and ecology of important insect and mite pests of cotton. In E. C. King, J. R. Phillips, and R. J. Coleman, eds., Cotton insects and mites: Characterization and management, No. 3. The Cotton Foundation Reference Book Series.

Mueller, S. C., C. G. Summers, and P. B. Goodell. 2003. A field key to the most common *Lygus* species found in the agronomic crops of the Central Valley of California. Oakland: University of California Division of Agriculture and Natural Resources Publication 8104.

Purcell, M., and S. C. Welter. 1991. Effect of *Calocoris norvegicus* (Hemiptera: Miridae) on pistachio yields. Journal of Economic Entomology 84:114–119.

Rice, R. E., R. A. Jones, and R. Reese. 1986. Biology and control of insects and mites on pistachios. California pistachio industry annual report 1986, 131–134. Fresno: California Pistachio Commission.

———. 1987. Biology and control of insects and mites on pistachios. California pistachio industry annual report 1987, 58–62. Fresno: California Pistachio Commission.

Rice, R. E., T. J. Michailides, R. A. Jones, and J. M. Ogawa. 1985. Biology and control of insects causing epicarp lesion and kernel necrosis in pistachio. California pistachio industry annual report 1985, 64–68. Fresno: California Pistachio Commission.

Scott, D. R. 1977. An annotated list of host plants for *L. hesperus* Knight. Bulletin of the Entomological Society of America 23:19–22.

Stern, V. M. 1969. Interplanting alfalfa in cotton to control *Lygus* and other insect pests. Proceedings of the Tall Timbers Conference on Ecological Animal Control, No. 1.

Thomas, F., C. Ingels, and L. Hendricks. 1998. Cover cropping in nut orchards. Nut Grower Magazine (August): 6–7, 17.

21 Stink Bugs and Leaffooted Bugs

Kent M. Daane, Glen Y. Yokota, Walter J. Bentley, Gary Weinberger, Jocelyn G. Millar, and Robert H. Beede

Though once considered virtually pest-free, California pistachios are now attacked by a variety of insect pests, including the true bugs (Rice et al. 1985; Daane et al. 2005). For convenience, these pests are commonly grouped as small and large bugs. The first group includes several species of Miridae and Rhopalidae, most importantly *Calocoris norvegicus* (Gmelin), *Phytocoris relativus* Knight, and western tarnished plant bug (*Lygus hesperus* Knight). The small bugs may be abundant early in the season, but they usually cease to cause damage after the pistachio shell begins to harden (Bobstock et al. 1987; Michailides et al. 1987; Beede et al. 1996).

This chapter focuses on the second bug group, the large bugs, which potentially can be more damaging, in part because they cause damage over a greater period of the season. The large bugs include species of Pentatomidae, most notably *Thyanta pallidovirens* (Stål) (a redshouldered stink bug); *Chlorochroa uhleri* (Stål) (referred to hereafter as the Uhler stink bug); the Say stink bug (*Chlorochroa sayi* Stål); and the green stink bug (*Chinavia hilare* (Say)). The large bug group also includes some species of Coreidae, commonly called leaffooted bugs, including *Leptoglossus clypealis* Heidemann, *L. zonatus* (Dallas), and *L. occidentalis* Heidemann. These large bugs can cause the same damage as their smaller relatives early in the season. However, their larger and stronger mouthparts enable them to puncture through the hardening shell and into the nutmeat later into the season. For this reason, large bugs generally pose a greater threat to the pistachio crop than small bugs.

Most of the large bugs found in pistachios are native to North America and, in fact, were identified long ago: *T. pallidovirens* (a redshouldered stink bug) in 1859, the green stink bug in 1832, the Uhler and Say stink bugs in 1872, and the leaffooted bugs in 1910. Since these species existed in North America long before pistachios, they have other host plants to feed upon, providing a refuge from any controls implemented inside the pistachio orchard. Stink bug host plants include crops such as cherry, peach, tomato, bean, wheat, soybean, cotton, and alfalfa; ornamentals such as juniper, redbud, dogwood, elderberry, and honey locust; and weeds such as peppergrass, wild lettuce, mustard, cheeseweed, and Russian thistle. In South Carolina, for example, researchers showed that stink bugs move into cotton fields from nearby, untreated vegetation in midsummer (Greene et al. 2001).

As mentioned, most of the large bugs in pistachios are native to North America, but the brown marmorated stink bug (*Halyomorpha halys* [Stål]) is native to Asia; this pest is now present in California. The brown marmorated stink bug was first found in the eastern United States in 1998 and has since been reported to cause damage to a wide range of fruit and vegetable crops. Damage to these crops includes puncture wounds and "cat facing" (a deformation), which can render the crop unmarketable; currently, it is not known what level of risk or types of damage the brown marmorated stink bug might cause to California pistachios.

Description

Stink Bugs

Stink bug adults. Adults of the four stink bug pest species commonly found in the pistachio canopy are, at first glance, difficult to separate from each other—they appear to be large green bugs, about ½ inch long, with a distinctly shield-shaped body when viewed from above and forewings with a clear membrane (the last section of the wing). Closer examination reveals differences among the adults that can be used to separate the species (fig. 21.1).

Of the adult stink bug pests in pistachio, *T. pallidovirens* (a redshouldered stink bug) is smallest, about ¼ to ½ inch from front to back (fig. 21.2). This species is closely related to the redshouldered stink bug *Thyanta custator accerra* McAtee. The adults of these redshouldered stink bugs are pale green and have a distinctive red (and faint yellow) stripe running across their widest shoulder section, called the pronotum (see bug parts in fig. 21.1). However, adults can have different color morphs, or phases. For example, this stink bug can be a much brighter green just after molting to the adult stage, and a pale brown form (without the red stripe) is also found, especially in winter. Nevertheless, redshouldered stink bugs are easily separated from the other green-colored stink bugs by their smaller size and the reddish stripe (when present). *Thyanta pallidovirens* is the species commonly found in orchards on the east side of the Central Valley.

The two most similar species are the Say *(Chlorochroa sayi)* and Uhler *(C. uhleri)* stink bugs. The adults are relatively large, about ½ to ⅗ inch. The body is a bright green and edged with a yellow-orange line on the lateral margin (the edge of the bug when looking from above) of the head, pronotum,

Figure 21.1. A flat green stink bug showing the major insect parts. *Photo:* K. M. Daane.

Figure 21.2. Redshouldered stink bug adult. *Photo:* K. M. Daane.

and wings. The body also has a rough texture with tiny white specks and three ivory-colored spots along the base of the scutellum (the triangular segment on the middle of the back), and the antennae are typically dark. Unlike the other stink bugs, these *Chlorochroa* species appear to have a distinctly rounded (rather than flat) back, when seen from the side. The Uhler and Say species can be distinguished by the membranous portion of the wing, which is transparent, pale, or white in the case of the Uhler stink bug (the more common of the two species in pistachio orchards), whereas the wing's membranous portion in the Say stink bug has a purple hue.

The green stink bug is also large, about ½ to ¾ inch long. It can be distinguished from the *Chlorochroa* species (fig. 21.3) by its shape, which is flat rather than oval. Slight but consistent differences in the color pattern are also evident. The green stink bug is always completely deep green, and it has five faint yellow dots across the top of the scutellum (between the scutellum and pronotum). The third and fourth antennal segments are blackish while the other antennae segments are green, the edge of the pronotum can have a faint yellow border, and the abdominal margin has conspicuous dark dots. The green bug is very easily confused with the southern green stink bug (*Nezara viridula* (Linnaeus)); however, the latter is rarely found in pistachio orchards (though it is an important pest in tomatoes and other row crops).

Figure 21.3. Uhler's stink bug *(Chlorochroa uhleri)* adult. *Photo:* K. M. Daane.

Stink bug nymphs. Stink bugs have five nymphal stages. The 1st instars are small (about $\frac{1}{20}$ in), brown or black in color, and remain clustered in aggregations around the egg mass. The relatively uniform color and pattern make it difficult to separate the species of the smaller stink bug stages, but with each consecutively larger instar, the stink bugs develop different color morphs and patterns—and once their striking patterns are recognized, they can be easily separated.

The redshouldered stink bug found in California pistachios has an immature that is initially brown with white stripes on its abdomen and pronotum (fig. 21.4). Some authors have reported that the 1st instar of this stink bug is not seen but instead remains inside the egg. As the nymph develops to the 3rd instar, the white, pale red, and brown pattern continues to develop. By the 4th instar, this redshouldered stink bug's head, pronotum, scutellum, and developing wing pads are mottled brown and white, while the abdomen is white striped with brown and green markings (fig. 21.5). The green and brown colors predominate by the 5th instar.

The Say and Uhler stink bugs have a relatively simple pattern. The smaller instars are blue-black in color, with a faint lateral orange margin. This pattern generally continues until the 3rd instar, after which each successive stage becomes greener in color but still retains the lateral yellow to orange margin.

The green stink bug is anything but green in the early stages. While the color pattern can vary, 2nd instars typically have an orange head, pronotum, and scutellum, with a black and white striped abdomen (fig. 21.6). The 3rd instar's head, pronotum, and scutellum turn black, often with an orange lateral margin; the abdomen may have a striped pattern of white, green, orange, or yellow with a black center region. The 4th and 5th instars become increasing green, with dark spots on the lateral margins of the abdomen becoming apparent (which is also seen in the adult) (fig. 21.7).

Stink bug eggs. Stink bug eggs are barrel shaped or drum shaped with circular lids and typically laid in tightly packed clusters of 10 to 40 eggs. Eggs are difficult to tell apart in the field, but color variation does provide some indication: the redshouldered stink bug eggs are tan to brown, the green stink bug eggs are orange to yellow, and the Say and Uhler stink bug eggs are white to tan.

Leaffooted Bugs

Leaffooted bug adults. The two more common leaffooted bugs, *L. clypealis* and *L. zonatus* (there are no common names for these leaffooted bugs), along with the western conifer-seed bug (*L. occidentalis*), are easily separated from the stink bugs. They are longer, narrower, and darker, with a distinctive white stripe that transects the body and leaflike flattened extensions on their hind legs (fig. 21.8). *Leptoglossus clypealis* is more commonly found in pistachios than is the western conifer-seed bug. *Leptoglossus clypealis* has a sharp point on the front of its head (the clypeus is a part of the face of an arthropod) and a distinctive white stripe that zigzags across the back, whereas *L.*

Figure 21.4. Redshouldered stink bug small nymph. *Photo:* K. M. Daane.

Figure 21.5. Redshouldered stink bug large nymph. *Photo:* K. M. Daane.

Figure 21.6. Green stink bug small nymph. *Photo:* K. M. Daane.

Figure 21.7. Green stink bug large nymph. *Photo:* K. M. Daane.

zonatus does not have a sharp point on the front and has two distinctive yellow to orange spots above the white zigzag stripe. The western conifer-seed bug ("occidental" means "western" in Latin) also lacks this sharp point, does not have the yellow to orange spots, and the white line transecting its body can be quite faint.

Leaffooted bug nymphs. Like the stink bugs, the leaffooted bugs have five nymphal stages. Outwardly, the nymphs have the same general shape as the adults—without the wings. Small immatures have an orange and brown head and thorax and a green abdomen when seen emerging from eggs (fig. 21.9); as the bugs mature, they may have a reddish-brown appearance, but they become darker as they develop and their wing pads (developing wings) enlarge. The last larval stage and the adult look generally the same except for the absence (nymph) or presence (adult) of fully developed wings with the white stripe (fig. 21.10).

Figure 21.8. Leaffooted bug adults. The two most common species, *Leptoglossus clypealis* (A) and *L. zonatus* (B), aggregating on pomegranate. *Photos:* (A) K. M. Daane; (B) G. Y. Yokota.

At first glance, immature leaffooted bugs and immature assassin bugs, which are beneficial predatory insects, can be difficult to separate. These two groups can be distinguished by the shape of the mouthparts: assassin bug nymphs have a short, stout beak that often is thrust forward in a stabbing motion (to pierce insect prey) and which, when retracted, does not extend back much beyond the front legs. This contrasts with the long, narrow beak of the leaffooted bugs, which extends all the way under the body, reaching to the hind coxa, or the beginning of the last pair of legs.

Leaffooted bug eggs. Adult females lay dark brown, barrel-shaped eggs that are placed end to end, typically forming a line of 5 to 30 eggs. The eggs are often found on leaves, twigs, and nuts. Nymphs emerge from a round hole atop the egg.

Injury

Direct Damage

Stink bugs and leaffooted bugs feed on various plant parts. However, seed and immature fruit are preferred—and in pistachio, damage occurs only to the nuts. These large bugs have needlelike, piercing-sucking mouthparts called stylets. They use their mouthparts like sharp straws to puncture pistachio leaves, stems, and fruit and suck out the plant juice. Most stink bugs inject a watery saliva containing digestive enzymes that help liquefy the plant content. Excess saliva solidifies around the stylet to form a stylet sheath, which can remain in the plant tissue

Figure 21.9. Leaffooted bug nymphs emerging from egg mass. *Photo:* R. E. Rice.

Figure 21.10. Leaffooted bug medium nymph. *Photo:* K. M. Daane.

after feeding. Rice et al. (1985) demonstrated that hemipteran feeding on developing nuts causes epicarp lesion (darkened area of the hull), the result of probing or feeding by any of the bugs in this group and the oxidative effect of their saliva. The epicarp lesions often involve mesocarp and endocarp tissues, and they can result in stained shells later in the season. When the bugs' mouthparts probe deeper into the nut and stab the nutmeat, a pocket of necrotic material forms directly on the kernel, and this damage is called kernel necrosis.

Feeding is often also associated with molds that may infest and stain or blacken the nut, a condition referred to as stigmatomycosis (Michailides and Morgan 1990). This occurs if the bugs' mouthparts are contaminated with fungi or if fungi on the nut surface gains entrance into the fruit when punctured. In nuts affected by stigmatomycosis, the entire kernel may become dark, wet, and foul-smelling.

Two major factors determine the level of large bug damage. First, at what time of the season does the insect feed? Second, how large is the insect (or better stated, how large are the insect's needlelike mouthparts)? While this seems simple enough, these two factors can result in many different scenarios and changes in the optimal spray timing. For example, differences among seasons, regions, and pistachio cultivars lead to variation in the pistachio tree's development, as well as the bug species and stages present (affecting both the seasonal period and the insect size). This means that control decisions based on a simple calendar date to time spray applications may be impractical.

Time of the season and shell hardening. A description of the seasonal changes in bug damage requires an understanding of the nut's development and response to the insect's punctures. Early in the season, when the nuts are small (½ in diameter), feeding punctures through the fruit pericarp (the outer wall of the nut comprised of the outer epicarp, the fleshy middle mesocarp, and the inner endocarp) quickly result in darkened areas of the hull—that is, epicarp lesions (Bostock et al. 1987). These early-season (April to mid-May) damaged nuts often drop from the cluster (Purcell and Welter 1991) and typically do not result in yield loss because the natural nut drop compensates for the loss of insect-damaged nuts, similar to that described for *Phytocoris* (Beede et al. 1996).

As fruit load begins to establish (typically in May), insect-damaged nuts often remain in the cluster for a short time. This is when the blackened nuts showing epicarp lesions become very evident and can best be used to monitor bug activity (fig. 21.11).

In midseason (mid-May through June), fruit load is set but shell lignification (hardening) is not complete. The damaged nut remains in the cluster, and the resulting epicarp lesion can stain the outer shell, lowering market value (Rice et al. 1985; Uyemoto et al. 1986). During this time, the kernel is also developing, and feeding punctures can result in necrotic spots on the nutmeat (kernel necrosis), aborted nuts, or fungal contamination (such as stigmatomycosis) (Daane et al. 2005; Michailides et al. 1988).

By July, as the shell continues to harden, the smaller (first- to third-stage) nymphs of most large bugs cannot penetrate to the nutmeat. Also, the necrotic spots formed at the puncture site no longer form quickly, if at all, leaving no trace of feeding damage. This is because epicarp lesions no longer rapidly form in response to insect wounds after peroxidase activity in the pericarp declines during fruit maturation and shell lignification (shell hardening) (Bostock et al. 1987). Therefore, most lesions found after mid-July are the result of earlier feeding wounds. For that reason, new epicarp lesions cannot be used effectively to monitor insect activity after shell hardening.

These seasonal changes in feeding damage have been illustrated in studies using a leaffooted bug (Daane et al. 2005). To manipulate the bug and nut development, pistachio clusters were isolated with organdy cages (before budbreak), into which a single insect was placed at different periods throughout the growing season and allowed to feed during a 7-day period. The resultant numbers of epicarp lesions

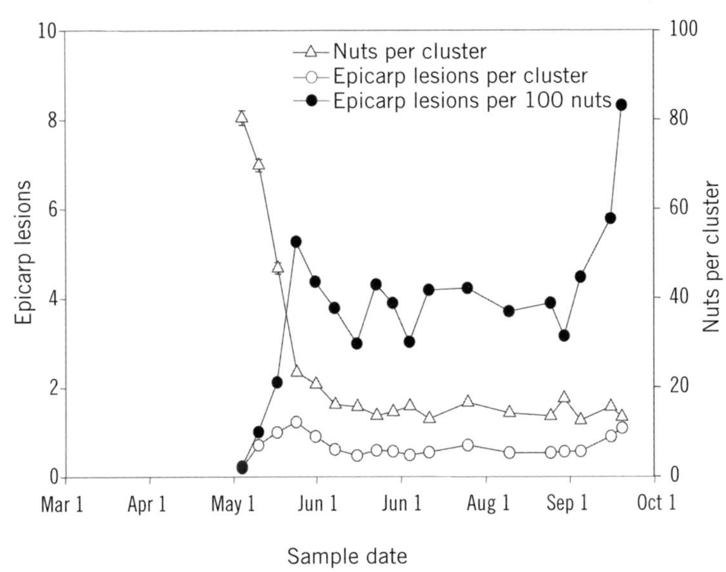

Figure 21.11. The average number of nuts per cluster (right Y-axis) and epicarp lesions (left Y-axis) from pistachio orchards surveyed in Fresno and Madera Counties, showing the natural drop of nuts from April to early May and the number of epicarp lesions increasing in late May and then declining as the number of visible epicarp lesions leveled off to about 4 per 100 nuts (< 1 per cluster) in mid-June. *Source:* Daane et al., unpublished.

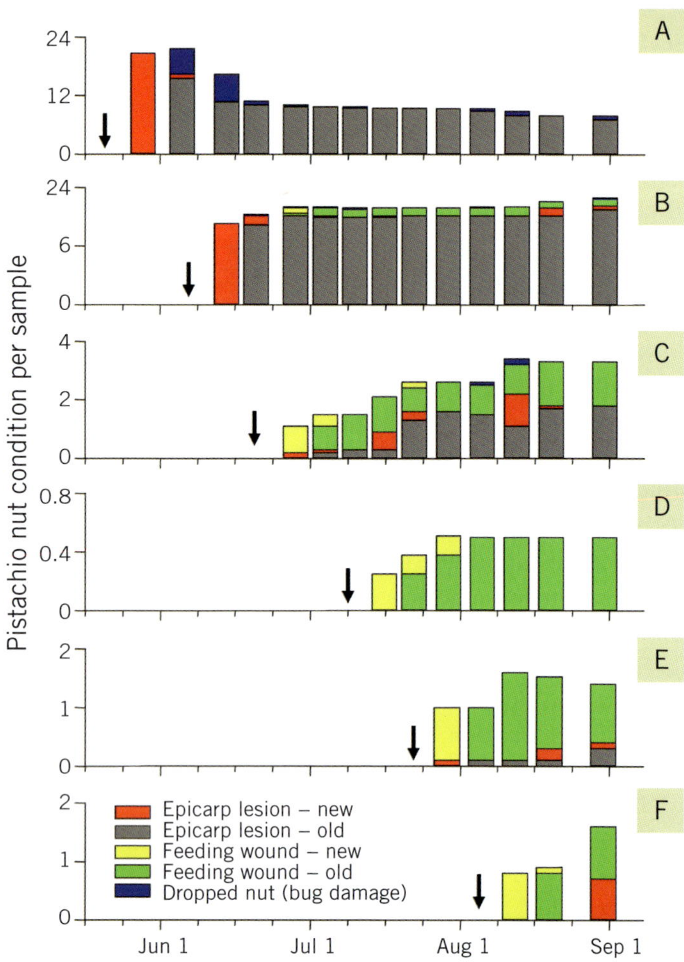

Figure 21.12. Leaffooted bug damage to nuts. The arrows point to the date when the cages were inoculated with leaffooted bug adults. Each stacked bar represents averages for the number of nuts that showed the following damage signs: old epicarp lesion, new epicarp lesion, old feeding wound, new feeding wound, and dropped nut with bug damage.

(after 2 weeks) and levels of kernel necrosis (at harvest) were compared. Results show that during a May exposure period, feeding wounds quickly turn into new and then old epicarp lesions; and although there is later some nut drop, many of the damaged nuts stay with the cluster until harvest (fig. 21.12A and 21.12B). In contrast, when feeding occurred in late June (fig. 21.12C), the puncture wounds did not readily result in epicarp lesions, and by late July and August, leaffooted bug feeding was rarely evident as an "old" epicarp lesion (fig. 21.12D, 21.12E, and 21.12F). Results from similar cage studies with a redshouldered stink bug, the green stink bug, and a leaffooted bug (fig. 21.13) indicate that the seasonal period when epicarp lesions are no longer apparent occurs in mid-June.

The practical application of this information is significant. It is important to note that before mid-May most of the damaged nuts are dropped from the cluster, but the plant naturally compensates for these lost nuts in most years (assuming an average crop load and bug density). Mid-May to mid-June is the important period—the damaged nuts stay in the cluster, affecting the crop quality. Also, the damaged nuts will show epicarp lesions, but these signs of feeding damage may not occur for many weeks. After mid-June, shell hardening prevents most bug damage, and the feeding that does occur often does not result in an epicarp lesion.

Bug species and size. The second important factor is bug size. Basically, the larger the bug, the more damage it can cause because it has longer and stronger mouthparts to penetrate the nutshell, and it feeds more.

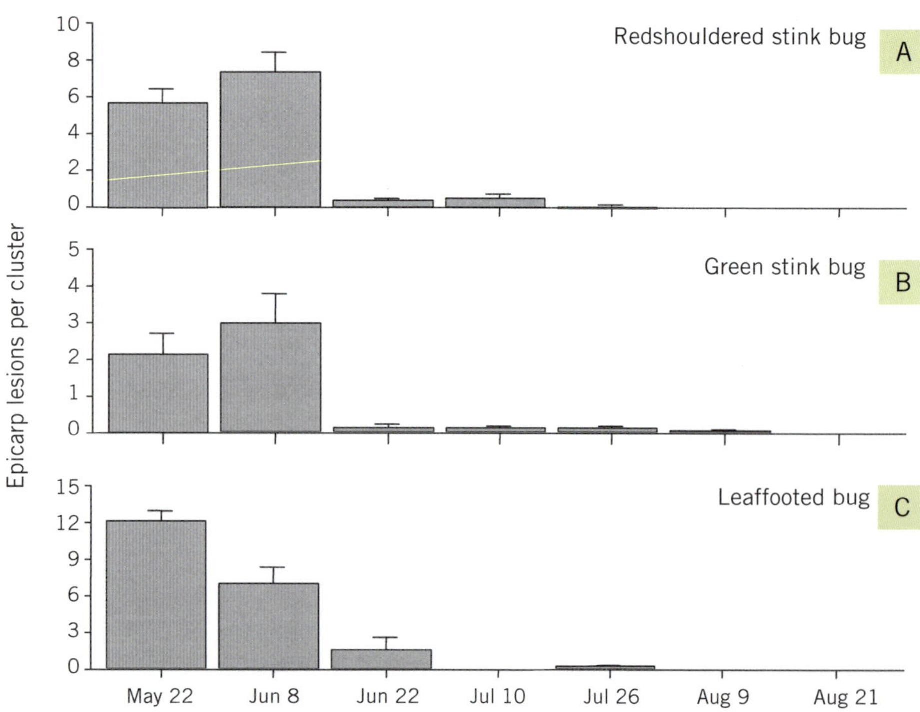

Figure 21.13. Damage to nuts after large bugs were caged with clusters for a 7-day period with an adult of (A) a redshouldered stink bug, (B) the green stink bug, and (C) leaffooted bug.

To illustrate the relationship between bug size and nut damage, different bug species and developmental stages were caged with pistachio clusters for a 7-day feeding period (Daane et al. 2005). Here, we present only results from the critical midseason period (early to mid-June) with a redshouldered and the green stink bug. The pattern among treatments shows that the number of dropped nuts (fig. 21.14A) and epicarp lesions (fig. 21.14B) increased when adults and large nymphs were used, in comparison with smaller nymphs.

The most critical measure of insect crop damage is kernel necrosis. In this study, levels of kernel necrosis were low in the early trial (May, data not shown) and not different among insect species or insect sizes tested—again, a result of damaged nuts dropping from the cluster. In the mid- (June) and late- (July) season periods, there was more damage in treatments that used the larger insects. Thus, insect size does matter in terms of nut damage.

These results for the early- and midseason trials are similar to those of earlier studies (Michailides et al. 1987; Rice et al. 1985), although our results suggest that the hardened shell provides considerable protection from even the largest bugs tested. The damage that was caused after shell hardening is best explained by a Michailides et al. (1988) study, which showed that late-season feeding by the leaffooted bug was successful when the insect's mouthparts penetrated the shell near the "Achilles heel" of the pistachio—that region near the peduncle and the shell division.

Indirect Damage

As mentioned earlier, hemipteran feeding wounds are also associated with fungal infections and can result in a condition called stigmatomycosis (Michailides and Morgan 1990). One fungus is particularly devastating: *Botryosphaeria dothidea* (hereafter referred to as Bd). This is a panicle blight of pistachios (Michailides et al. 1987, 1999), which can result in epidemiclike loss in pistachio.

Just how important are insects in the spread of Bd among and within orchards? Researchers surveyed the ground cover and tree canopy in pistachio orchards in Glenn, Colusa, Contra Costa, Madera, Fresno, and Kern Counties—and all of the orchards sampled had some level of Bd infection. A subsample from the 30,000 hemipterans collected showed that the Bd inoculum was cultured from about 0.2 percent (or 2 out of 1,000). Although the percentage may seem small, some hemipterans do carry Bd spores.

For this reason, a series of experiments were conducted to determine the bugs' role in the movement of Bd (Steffan et al. 1999). Researchers found between-species differences in the ability of stink bugs and leaffooted bugs to carry Bd. For example, after exposure to nut clusters contaminated with Bd, adults of a redshouldered stink bug and the green stink bug carried Bd spores for 3 days, while adult leaffooted bugs carried spores for up to 10 days. These differences are most likely related to the ability of the Bd spores to cling to the outside of the insect (dissection of the bugs' mouthparts showed that the relatively large Bd spores are carried externally, not internally). Researchers also showed that when the Bd spores are already on the cluster, feeding from large bugs increased Bd infection levels, although infection rates resulting from insect feeding were still lower than those resulting from mechanical injury with a pinprick (fig. 21.15).

The accumulated evidence suggests that while only a small percentage of the bugs found in the orchard carry Bd, these bugs can still be an important factor in disease transmission when present in

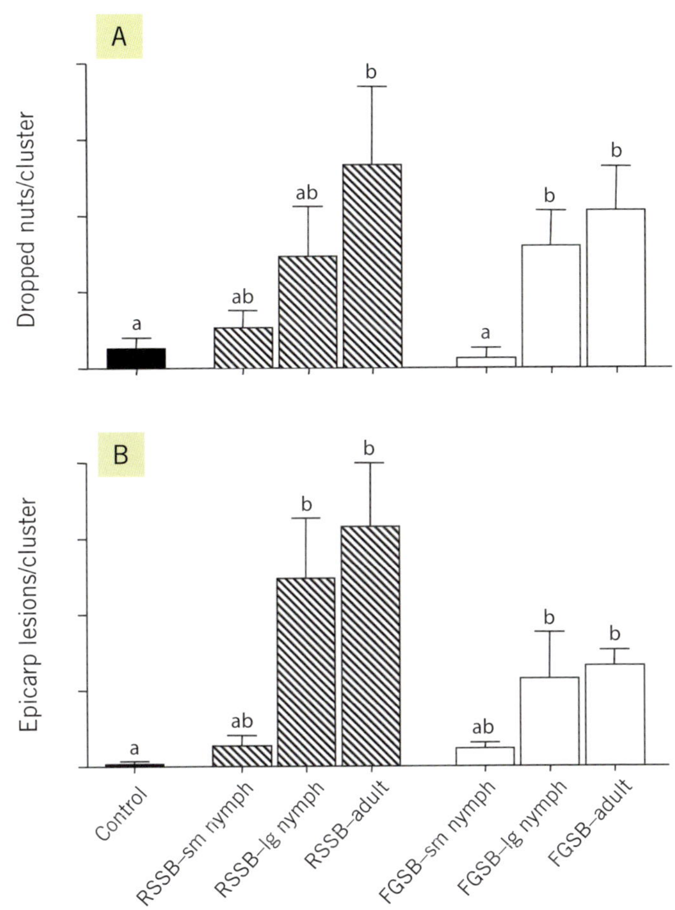

Figure 21.14. Average number of dropped nuts (A) and epicarp lesions (B) after a 7-day exposure period to different bug species (RSSB = a redshouldered stink bug, FGSB = the green stink bug) or development stages (small nymph, large nymph, and adult) in a midseason trial during the critical period near shell hardening (June 10–17). In each graph, different letters above each bar indicate a significant difference between treatments.

large numbers. Typically, if the disease is already present in the orchard, large bugs play a role in facilitating disease entry into the pistachio tree by puncturing the epicarp with their mouthparts, which causes fluid to exude from the nut and onto the epicarp surface. Thus, the bugs create an entrance wound for Bd spores that are already present on the nut surface, and the fluid from the feeding also helps spore entrance and germination.

Seasonal Development

Stink bugs and leaffooted bugs found in the pistachio orchard during the season will overwinter primarily as adults. Adults overwinter on the ground under leaves, in orchards, legume crops, blackberries, or on certain weeds such as Russian thistle, mustards, and little mallow (cheeseweed). Leaffooted bugs often overwinter in aggregations located in protected areas—for example, in woodpiles and barns, near pump houses, under the bark of eucalyptus trees, in cypress, olive, citrus, or juniper trees, under the fronds of palm trees, and in plant debris.

While most of these large bugs overwinter primarily outside of the pistachio orchard, the green stink bug commonly overwinters in the orchard and can be easily spotted on warm days when it moves into sunlit areas on the pistachio trunk and scaffolding or irrigation materials.

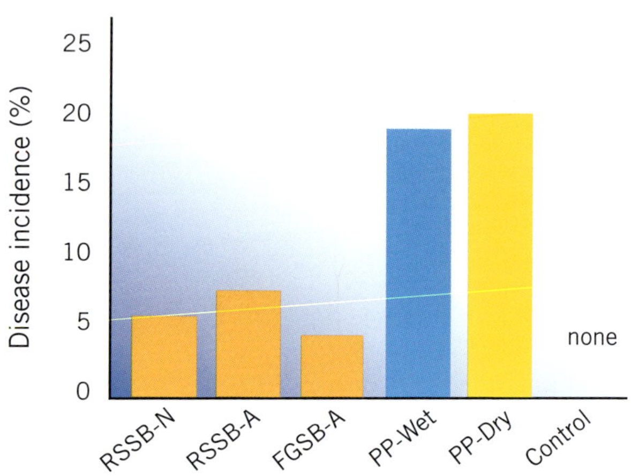

Figure 21.15. Diseased clusters at harvest in cages sprayed with a liquid Bd spore suspension and provided with either of the following: a redshouldered stink bug nymph (RSSB-N) or adult (RSSB-A); the green stink bug adult (FGSB-A); a pinprick immediately after the liquid spray (PP-Wet); a pinprick 24 hours after the liquid spray (PP-Dry); and a no-spray control (Control).

As the weather warms from March through May, the adults disperse to find food sources. These insects are primarily seed feeders, and, once in the orchard, they feed directly on the developing nuts or on ground vegetation seed. The adult stink bugs are adequate fliers, while the leaffooted bugs are strong fliers and can disperse from overwintering sites not directly adjacent to the orchard and move quickly within the orchard.

Stink Bugs

The overwintered stink bug must mate before it can deposit viable eggs. The adults find each other with the help of sex pheromones, which have been identified for *T. pallidoverins* (a redshouldered stink bug) (Millar 1997), Say stink bug, and *C. uhleri* (Uhler stink bug) (Ho and Millar 2001a, 2001b). The overwintered adults feed during this late-winter to early-spring period. Stink bugs can feed on different plant parts, but seed are preferred; and better food increases egg production. For example, when fed peas or lentils with mature pods, a redshouldered stink bug produced more eggs per adult female and lived longer than those stink bugs fed only pea or lentil seedlings or flowering plants (Schotzko and O'Keefe 1990a, 1990b).

After mating there is a 5-to-7-day preoviposition (egg-laying) period. Stink bug egg masses are difficult to find in the field. The redshouldered stink bug found in California's pistachio orchards, for example, typically lays its eggs on sheltered, rough-textured sites (Wang and Millar 1997), such as the pistachio bark rather than the leaves.

After egg hatch, the 1st instar nymphs remain clustered together. In fact, there is a chemical cue that promotes the aggregation behavior of these immature stink bugs. When researchers placed 1st instars of different stink bug species together in a petri dish, they readily formed heterospecific (different species) aggregations similar to their natural aggregations in the field (Fucarino et al. 2004). While in the aggregations, these small stink bugs probably do not feed; the aggregation may help the newly hatched nymphs acquire symbionts, reduce desiccation, increase developmental rate (especially at lower temperatures), and reduce predation.

The stink bugs are generally assumed to have two generations per year (fig. 21.16). A typical life cycle for stink bugs shows an overwintering adult population that deposits eggs from March to May. The resulting nymphs hatch and become adults in

May and June and begin the second generation. A great deal of overlap between the first and second generation takes place during the summer months, and some researchers have suggested that a partial single generation may also occur in some regions.

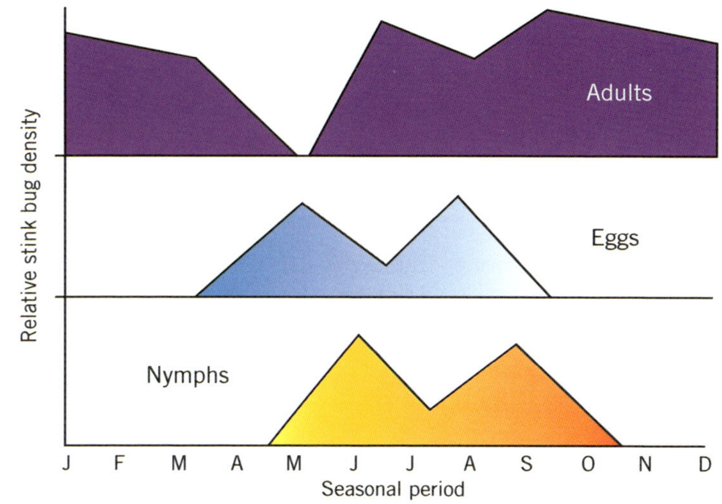

Figure 21.16. Typical life cycle of stink bugs in pistachio orchards.

Leaffooted Bugs

Similar to stink bugs, leaffooted bugs overwinter as adults, but typically in aggregations. In California's Central Valley, mating probably occurs in these aggregations in late February to early March, before the adults disperse to feeding sites.

During their active season, which typically runs from March through November, leaffooted bugs have three full generations. This was illustrated in a yearlong study with leaffooted bugs held at ambient temperatures (Fresno County) and supplied a constant diet of food (green beans, peanuts, Italian cypress) and water. Twenty-five female adults were used to begin the trial (fig. 21.17A). The overwintering adults began depositing eggs on March 15, although the timing varies depending on the annual temperatures, with maximum egg production between April 1 and May 15. Each egg mass contained about 15 individual eggs. Nymphs hatched within 20 days of egg production. Development from 1st to 5th instar stages required about 45 days during the spring period.

A portion of the adults produced from the first generation was used to begin the next series (second generation) of cages (fig. 21.17B). These second-generation adults lived, on average, from mid-June to mid-August, producing eggs during this period, with peak production in July. Nymphs hatched from late June through August, almost all becoming adults before mid-August, and these were used for the next series (third generation) of cages (fig. 21.17C). The second-generation adults began to produce egg masses (third-generation offspring) during the first week of August, and, though they remained alive in the fall, they ceased egg production at the beginning of October. The third-generation nymphs began to hatch in mid-August, and egg hatch continued into December. The resulting nymphs became adults from early September through October, forming the bulk of the overwintering generation, although some adults from the second generation also overwintered. Many of the third-generation nymphs that hatched in October never developed to adults. The third-generation adults to be produced were placed in cages and considered to be the fourth development set; however, few of these adults produced viable eggs.

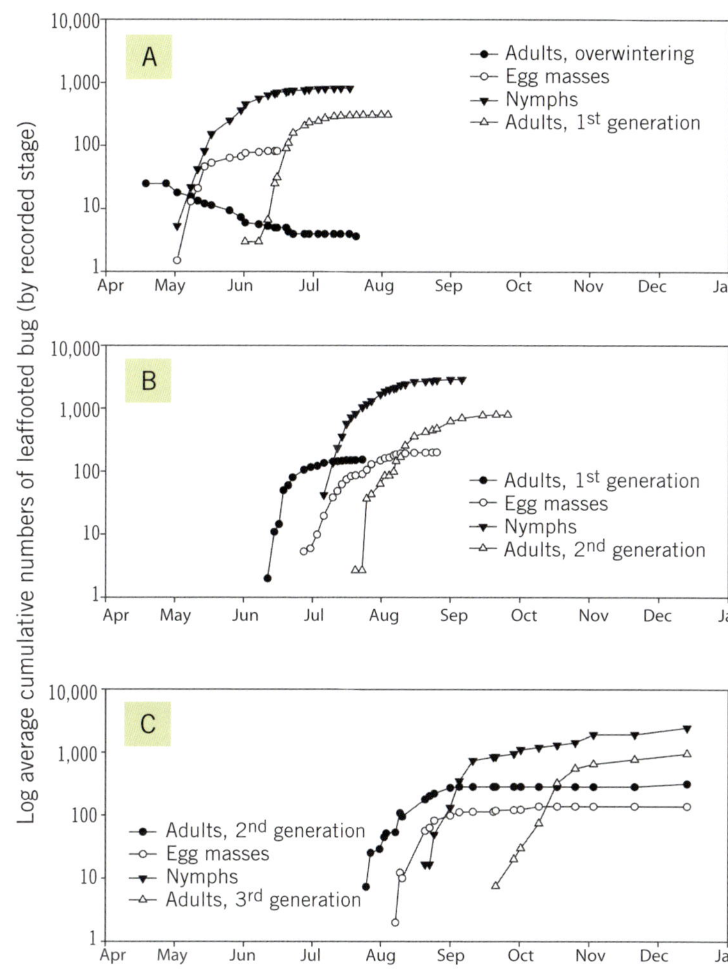

Figure 21.17. The average cumulative leaffooted bug density by development stage for each of three 2007 generations. The bugs were held in cages and provided food (green beans, peanuts, and Italian cypress) and water. Note that the average number of overwintering females (A) is provided, showing the gradual decrease in their density.

Monitoring

As mentioned, these large bugs can overwinter both inside and outside the orchard and can immigrate or emigrate in large numbers as adults (nymphs do not fly, but adults are often strong fliers), causing severe damage in a short time. This overwintering behavior complicates monitoring decisions, and for many orchard managers the goal of monitoring for stink bugs and leaffooted bugs is not well defined.

Field Sampling Techniques

Some growers look for the first sign of specific pests so that annual treatments can be timed, while others attempt to estimate bug densities in order to determine if treatments are needed. The latter strategy relies on both detection and a prediction of damage by the population of bugs present. However, economic injury levels have not been established, due in part to the complexity of the different bug species and the seasonal changes in nut susceptibility. Therefore, most managers base control decisions on their experience with relative bug densities over the years, gained through their repeated annual monitoring and familiarization with bug pests in each region—setting personal standards for what species and population levels produce a reliable early warning for low- or high-density bug years.

Adult large bugs are the stage to look for during the April through May period, when much of the damage occurs and control decisions are often made. Sampling is typically accomplished using the circular beating tray, which is held under one to several clusters of nuts while the branch above is hit sharply with a mallet. Provided that samples are taken frequently and from a large number of trees, the beating tray gives a good indication of changes in the numbers of bugs in the pistachio canopy.

To sample using a beating tray, take 50 beats per 40-acre block, with sampled trees concentrated on the edge rows closest to possible overwintering sites. Each beat consists of two strikes of the branch directly above the beating tray. Give one beat per tree only. It is best to take samples at least every 2 weeks starting in April and continuing into mid-June; sampling should still occur after that period, but damage to the kernel drops considerably as the pistachio shell hardens. Note that damaged pistachios often stay in the cluster after mid-May, so the visible damage may be from a population long gone or dead from treatment application.

There are drawbacks to the beating tray. First, it takes a good deal of time to accurately sample a large field. Second, samples taken by this method underestimate the numbers of adult bugs present, especially on warm days when adults fly away upon being disturbed (counts of immature bugs, which do not fly, are more accurate, but the mature bugs appear in the orchard first). Silva et al. (1995) evaluated different sampling programs and showed that the beating tray does not accurately represent pest densities. Their work suggests that if a primary goal is to determine when bug pests first arrive in the orchard, a sweep net in ground vegetation may be more efficient (fig. 21.18). Still, this method also has drawbacks: it provides little information about bug presence in the canopy, it requires ground vegetation to be present for sampling, and it is better suited for some species of small bugs (e.g., *Lygus*) than for large bugs.

The easiest way to determine the presence of large bugs is to look for their damage. The benefit of looking for damage rather than the bugs themselves is that searches can be made over a larger area, providing a better indication of a low-density population because one bug may be hidden in the tree but can leave a lot of damaged nuts as a signal.

The earliest damage is to the buds, which develop poorly after being fed upon, but this damage does not provide a good signal because it is not easily seen and therefore is typically not used to indicate bug presence. By April, damage to pistachio nuts is more evident; the small BB-size pistachio nuts turn brown (epicarp lesion) and quickly drop to the ground. Initially, the very small nuts damaged from bug feeding from mid-April to early-May can drop from the cluster within 24 hours, without any visible signs of insect damage. This is because epicarp lesions require more than 24 hours to produce

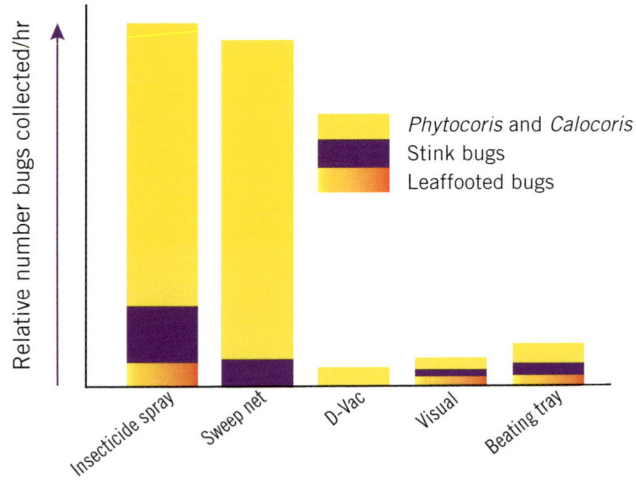

Figure 21.18. The relative quantity of bugs detected (per hour sampling) using the following methods: whole-tree spray and collection; sweep net of ground vegetation; D-Vac suction machine; visual examination of canopy; and a beating tray.

a visibly darkened area (Bostock et al. 1987). The result is that insect feeding on nuts in their earliest development period is easily overlooked. For this reason, it is best to move through the orchard looking for groupings of damaged nuts, which are often easier to see. A good sampling program for a 40-acre block would involve sampling two clusters on each of 200 trees every 1 to 2 weeks during the critical period of mid-April through May. Note that the sample is biased because the trees are searched for damage rather than randomly selecting nuts. While there are no economic injury levels and the tree can compensate for some nut drop, the experience of the pest control adviser (PCA) and orchard manager can determine years with large nut drop or damage.

As mentioned previously, by mid-June epicarp lesions may take weeks to form, if they form at all, creating a delay between the feeding event (and bug presence) and the feeding signal (epicarp lesion). Therefore, the use of damaged nuts as a monitoring tool is best suited for the critical period of April through May.

Because leaffooted adults are so mobile, some additional steps can be taken to determine their early-spring densities. First, monitor overwintering aggregations whenever possible. During the winter, the leaffooted bugs leave the pistachio orchard, but if an overwintering aggregation of leaffooted bugs is found nearby, it can be an excellent sampling tool. In late fall, as the temperatures cool, adult leaffooted bugs begin to leave the feeding area to form winter aggregations. When there is a known leaffooted bug population in September or October, search the nearby sheltered areas, such as juniper, the fronds of palm trees, pomegranates, citrus, olives, and other sheltered sites. Aggregations are often on the largest tree and usually on the sunny side of the tree. Once an aggregation forms in November, there is little movement. During this overwintering period, feeding is probably not necessary, except for water. We do not know yet what cues are used to form aggregations, but they probably include site characteristics, pheromones, and vibrations or sounds that call individuals together. If you find an aggregation, you can monitor it throughout the winter to estimate mortality and the timing of the spring population's movement from winter aggregations into the pistachio orchard.

Second, use appropriate sampling techniques and at the appropriate time. Aggregations in the Central Valley start to break up in mid-March and the adults move out to find food. Almonds are the first concern because the small almond fruit form in early- to mid-March and are food for the hungry overwintered adults. The standard beating tray does not work as well for leaffooted bugs because on warm days (> 65°F) the adults, which are strong fliers, fly away from the sampler. Instead, use a long pole (about 6 ft) to beat or shake the branches while looking (or listening) for the adults flying away. The adults are often in the upper half of the tree, and shaking that region can be productive because the dispersing adults are silhouetted against the sky. Sample close to the edge of the block near the most likely overwintering habitats of the adults, such as eucalyptus or cypress windbreaks. For our studies, we sample 100 trees per block, spending about 30 seconds per tree.

Sex Pheromones

Pheromones recently have been identified for most of the important plant-feeding stink bugs (McBrien and Millar 1999). In contrast to many other insect species, stink bug pheromones are produced by the male bugs, not the females, and depending on the species, the pheromones attract females only, both males and females, or, in the case of the Consperse stink bug, males, females, and immatures. Field trials with the stink bug pheromones have shown that they are not as powerful an attractant as the pheromones for other insects, such as moths, and that supplemental mechanisms, such as the use of vibrational signaling, assist in attraction. Therefore, mate-finding is actually a two-step process. In the first, longer-range step, females are attracted to the vicinity of pheromone-producing males. In the second, shorter-range step, once the male and female are on the same plant, they begin producing vibrational signals, which are transmitted through the plant stem. The bugs use these vibrational signals as homing beacons toward the signal of a potential mate; at each branch point, the bug stops and "listens" with its feet and legs to determine which direction to take. The stink bug pests commonly found in pistachios are likely to only be attracted to pheromone lures when the bugs reach sexual maturity, which in some species can be as long as 10 to 14 days after the final molt, and bugs are unlikely to respond to pheromone lures during the fall when they have gone into reproductive diapause in preparation for overwintering. For these reasons, pheromones are not currently a viable tool for monitoring stink bugs.

For the leaffooted bugs, no long-range attractant pheromones have been identified yet, although there have been hints that these bugs may use attractant pheromones in two contexts. First, there is evidence that adult males produce a group of sex-specific volatiles that may play a role in attracting mates, although none of these chemicals have been completely identified yet. Second, there is evidence from field bioassays that the formation of the overwintering aggregations of adults may be mediated in part by pheromonal signals. However, because these aggregations form only in the fall (long after pistachios are harvested), it does not appear likely that the aggregation pheromone could be used for crop protection purposes.

Natural Control

The natural enemy complex of the stink bugs and leaffooted bugs includes both parasitoids and predators. Egg parasitoids are the most important natural enemies and include *Trissolcus basalis* (Wollaston), *T. utahensis* (Ashmead), and *T. euschisti* (Ashmead) for the eggs of different stink bug species and *Gryon pennsylvanicum* (Ashmead) (fig. 21.19) for the eggs of leaffooted bugs. However, for both the stink bugs and leaffooted bugs, most damage is caused early in the season when egg parasitism is low. There are no evident cultural practices (e.g., cover cropping) that can be used to manipulate these egg parasitoid numbers, and low early-season egg parasitism may be simply the result of migrating adult bugs moving into the orchard ahead of the parasitoids. A fly, *Trichopoda pennipes* (Fabricins), has been reported as a parasitoid of adult stink bugs. Lacewings (*Chrysopa* and *Chrysoperla* spp.), assassin bugs (*Zelus* spp. and others), damsel bugs (*Nabis* spp.), ants (*Dorimyrmex* and *Solenopsis* spp.), earwigs (*Forficula auricularis* Linnaeus), and spiders are the most abundant predators in pistachio orchards. However, these predators are not believed to provide extensive control of any large bugs in pistachio. They are most likely to attack the eggs and small nymphs.

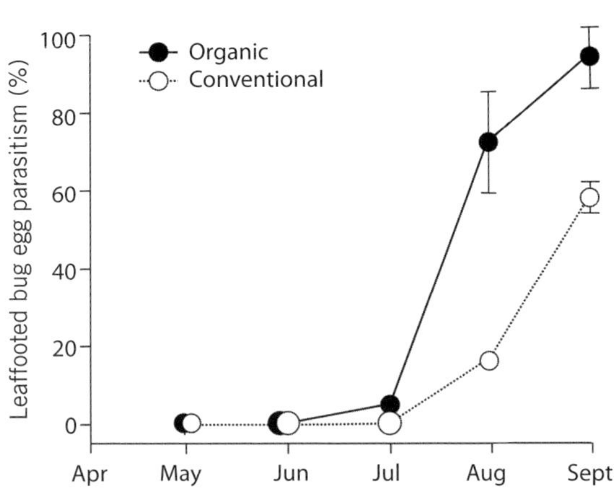

Figure 21.19. Levels of leaffooted bug egg parasitism by the *Gryon* species egg parasitoids in organic and conventionally managed pistachio orchards (2000).

Management

Rational control strategies for these bugs must take into consideration three important characteristics of the species involved. The first is that they are *direct* pests, causing damage to the crop as soon as they appear in the canopy. The second is that they can be *migratory* pests, whose arrival in the canopy is merely the last step in a series of events that may have begun as far away as many miles or as close as the ground vegetation below the canopy. The third is that they are mostly *native* pests, adapted to host plants other than pistachios; they are able to reside outside the pistachio orchard but can move into it at any time during the season.

The fact that these bugs are direct pests has led to common use of insecticidal sprays to combat them. These sprays have been effective in killing the bugs. However, in the absence of additional information on the bugs' density (which changes each season), their feeding habits, and the changing susceptibility of the crop throughout the season, it is difficult to predict how much damage the bugs will cause before a spray can be applied.

The fact that these bugs are migratory has led to efforts to avoid triggering the process of migration to the canopy. How to achieve this goal remains a matter of controversy. One view is that alternate host plants outside the orchard should be identified and destroyed as soon as possible so that the bugs can be eliminated. A contrasting view is that the alternate hosts should be maintained in good condition for as long as possible to keep the bugs in them and to delay migration into the orchard. A third view is that growth of some of the alternate hosts (i.e., a cover crop) should be encouraged on the orchard floor to attract the bugs and prevent their migration from the floor to the canopy. In all cases, it seems prudent to give careful thought to the stage of development of the crop and the level of infestation of alternate host plants when considering any management practice, such as tilling or mowing, that may affect alternate hosts.

Recommendations

1. All stink bugs and leaffooted bugs found in pistachio orchards should be identified correctly so that records of patterns of attack and successful strategies of control can be kept separately for each species.
2. Monitoring of populations of large bugs, predators, and parasites within orchards should be conducted throughout each season, from budbreak to harvest.
3. Surveys of weedy vegetation in and surrounding orchards should be conducted throughout each season to determine the abundances and developmental stages of the different plants and insects present.
4. Careful attention should be paid to the possible impact of any management activities, such as tilling, mowing, or burning, on the bugs within and surrounding the orchard.
5. Decisions to spray should be based on a comparison of numbers of insects found, damage observed, developmental stage of the crop, and previous yields under similar conditions.
6. Records of pest bugs should be evaluated along with those of numbers of natural enemies, development of the crop, abundance of alternate host plants, and yield and damage found at harvest so that recommendations for the future can be improved.

References

Beede, R. H., R. E. Rice, K. M. Daane, D. R. Thomas, J. Padilla, S. Soliman, W. J. Bentley, and G. Weinberger. 1996. Relation of *Phytocoris relativus* to lecanium scale infestations on pistachios. California pistachio industry annual report 1995–1996, 107–118. Fresno: California Pistachio Commission.

Bentley, W. J., R. H. Beede, K. M. Daane, and D. R. Haviland. 2007. Insects and mites. In B. Ohlendorf, ed., UC IPM pest management guidelines: Pistachio. Oakland: University of California Division of Agriculture and Natural Resources Publication 3461.

Bolkan, H. A., J. M. Ogawa, R. E. Rice, R. M. Bostock, and J. C. Crane. 1984. Leaffooted bug implicated in pistachio epicarp lesion. California Agriculture 38:16–17.

Bostock, R. M., C. S. Thomas, J. M. Ogawa, R. E. Rice, and J. K. Uyemoto. 1987. Relationship of wound-induced peroxidase activity to epicarp lesion development in maturing pistachio fruit. Phytopathology 77:275–282.

Daane, K. M., G. Y. Yokota, R. Krugner, P. G. da Silva, S. A. Steffan, R. H. Beede, W. J. Bentley, and G. B. Weinberger. 2005. Large bug damage to San Joaquin Valley pistachios. California Agriculture 59:95–102.

Fucarino, A., J. G. Millar, J. S. McElfresh, and S. Colazza. 2004. Chemical and physical signals mediating conspecific and heterospecific aggregation behavior of first instar stink bugs. Journal of Chemical Ecology 30:1257–1269.

Greene, J. K., S. G. Turnipseed, M. J. Sullivan, and O. L. May. 2001. Treatment thresholds for stink bugs (Hemiptera: Pentatomidae) in cotton. Journal of Economic Entomology 94:403–409.

Ho, H.-Y., and J. G. Millar. 2001a. Identification and synthesis of male-produced sex pheromone components of the stink bugs *Chlorochroa ligata* and *Chlorochroa uhleri*. Journal of Chemical Ecology 27:2067–2095.

———. 2001b. Identification and synthesis of a male-produced sex pheromone from the stink bug *Chlorochroa sayi*. Journal of Chemical Ecology 27:1177–1201.

McBrien, H. M., and J. G. Millar. 1999. Phytophagous bugs. In J. Hardie and A. K. Minks, eds., Pheromones of non-lepidopteran insects associated with agricultural plants. Wallingford, UK: CABI Publishing.

Michailides, T. J., and D. P. Morgan. 1990. Etiology and transmission of stigmatomycosis disease of pistachio in California. California pistachio industry annual report 1989–1990, 88–95. Fresno: California Pistachio Commission.

Michailides, T. J., J. M. Ogawa, and R. E. Rice. 1988. Sites of epicarp lesions and kernel necrosis in relationship to symptoms and phenology of pistachio fruit. Journal of Economic Entomology 81:1152–1154.

Michailides, T. J., R. E. Rice, and J. M. Ogawa. 1987. Succession and significance of several hemipterans attacking a pistachio orchard. Journal of Economic Entomology 80:398–406.

Michailides, T. J., D. P. Morgan, D. Felts, and E. Necker. 1999. Infection of pistachio fruit by *Botryosphaeria dothidea* and the effect of insect puncturing. California pistachio industry annual report 1998–1999. Fresno: California Pistachio Commission.

Millar, J. G. 1997. Methyl (2E,4Z,6Z)-deca-2,4,6-trienoate, a thermally unstable, sex-specific compound from the stink bug *Thyanta pallidovirens*. Tetrahedron Letters 38:7971–7972.

Purcell, M., and S. C. Welter. 1991. Effect of *Calocoris norvegicus* (Hemiptera: Miridae) on pistachio yields. Journal of Economic Entomology 84:114–119.

Rice, R. E., T. J. Michailides, R. A. Jones, and J. M. Ogawa. 1986. Biology and control of insects causing epicarp lesion and kernel necrosis in pistachios. California pistachio industry annual report 1985–1986, 64–68. Fresno: California Pistachio Commission.

Rice, R. E., J. K. Uyemoto, J. M. Ogawa, and W. M. Pemberton. 1985. New findings on pistachio problems. California Agriculture 39(1): 15–18.

Schotzko, D.J., and L.E. O'Keeffe. 1990a. Effect of pea and lentil development on reproduction and longevity of *Thyanta pallidovirens* (Stål) (Hemiptera: Heteroptera: Pentatomidae). Journal of Economic Entomology 83:1333–1337.

———. 1990b. Ovipositional rhythms of *Thyanta pallidovirens* (Hemiptera: Pentatomidae). Environmental Entomology 19:630–634.

Silva da, P. A., K. M. Daane, and N. M. Korfanta. 1996. Biological control of hemipteran pests of pistachio: Appearance, activity and movement of insects in the orchard. California pistachio industry annual report 1995–1996, 61–66. Fresno: California Pistachio Commission.

Silva da, P. A., K. M. Daane, R. H. Beede, W. J. Bentley, C. Anderson, and C. Moutou. 1995. Biological control of hemipteran pests of pistachio: Inventory and sampling. California pistachio industry annual report 1994–1995, 56–61. Fresno: California Pistachio Commission.

Steffan, S. A., K. M. Daane, K. Weir, G. Y. Yokota, T. J. Michailides, and B. A. Holtz. 1999. The role of hemipterans in the spread of the fungus *Botryosphaeria dothidea*. California pistachio industry annual report 1997–1998, 15. Fresno: California Pistachio Commission.

Uyemoto, J. K., J. M. Ogawa, R. E. Rice, H. R. Teranishi, R. M. Bostock, and W. M. Pemberton. 1986. Role of several true bugs (Hemiptera) on incidence and seasonal development of pistachio fruit epicarp lesion disorder. Journal of Economic Entomology 79:395–399.

Wang, Q., and J. G. Millar. 1997. Reproductive behavior of *Thyanta pallidovirens* (Heteroptera: Pentatomidae). Annals of the Entomological Society of America 90:380–388.

22 Mealybugs

David R. Haviland, Robert H. Beede, and Kent M. Daane

Mealybugs are small, soft-bodied insects (family Pseudococcidae) that are covered in white wax and have slender, strawlike mouthparts to feed on plant phloem. On pistachio trees, *Ferrisia gilli* Gullan is the most common species of mealybug. It was first recognized in Tulare County in the late 1990s and by 2008 had spread to over 5,000 acres of pistachios statewide. It has also been found infesting almond, stone fruit, and persimmon trees, grapes, and several types of ornamental trees and shrubs, such as fruitless mulberry *(Morus alba)*. If left uncontrolled, this mealybug can reduce both yield and quality of pistachios.

A second species, grape mealybug (*Pseudococcus maritimus* (Ehrhorn)), also can be found in pistachio orchards. However, its presence in pistachios is rare and limited to low densities that do not impact the crop.

Description

Adult females of *F. gilli* are between 2 and 5 mm in length and are covered in a white mealy wax (fig. 22.1). They have two dark areas or stripes on the dorsum that run the length of the insect, and they possess two white wax tails. When females remain in one location for a period of time, they produce long, glassy, white, waxy, threadlike filaments that extend over the top of the body. These threads can reach a centimeter in length and distinguish mealybugs in the genus *Ferrisia* from most other mealybugs. Large numbers of filaments from aggregations of adult females, particularly in the winter, can give the trunk and main scaffolds a white bearded appearance (fig. 22.2).

Adult males of *F. gilli* measure about one-half the length of the females, have wings, and look like tiny wasps with two long tails. They are rarely noticed in the field.

Immature mealybug stages, called nymphs, look like adult females but are smaller. They have two white wax tails and feed in aggregations. They often have a shinier appearance than the adult females due to their reduced amount of white mealy wax and number of filaments.

The grape mealybug, which is sometimes found in pistachios (fig. 22.3), does not cause any harm and can easily be distinguished from *F. gilli*. The female of the grape mealybug has four slender white tails (two long ones and two short ones), whereas the female of *F. gilli* has two broad white tails. Grape mealybug also has short, white, lateral projections that extend from the sides of the body, whereas *F. gilli* has none. Immature grape mealybugs hatch from egg sacs and are usually distributed in a scattered pattern, whereas *F. gilli* has live birth with immature stages remaining aggregated around the mother. Lastly, when poked, adult females of grape mealybug extrude a bright red liquid through structures called ostioles toward both the rear and front of the top of the body. *Ferrisia gilli* does not extrude such a liquid.

Figure 22.1. Adult females of *Ferrisia gilli* have a striped appearance, two white tails, and are often covered in long, glassy, white filaments. *Photo:* D. R. Haviland.

Injury

Mealybugs use their strawlike proboscis to feed from the phloem tissue of their host. They extract plant juices, digest portions of the fluids, and eliminate the remainder of the sap as honeydew. Honeydew is primarily composed of sugars, which can completely cover the surface of tree limbs, leaves, and clusters. Honeydew supports the growth of black sooty mold that renders the leaves photosynthetically inactive.

The greatest impacts of *F. gilli* on pistachio are from its affinity to feed on hulls within clusters. Mealybug feeding intercepts carbohydrates and other nutrients intended for kernel development, thus reducing kernel size and overall yields. Smaller kernels also affect fruit quality due to increases in the percentage of closed-shell nuts that do not split before harvest. Feeding on the hull also increases the amount of hull breakdown and shell staining, and infested hulls are typically covered in thick layers of honeydew and sooty mold (fig. 22.4). Harvest is also affected when severe hull damage causes nuts to dry up and shrivel on the tree. These nuts often remain on the tree during harvest.

Seasonal Development

The seasonal biology of *F. gilli* has been documented in Tulare County, where the mealybugs have three discrete generations per year. The overwintering generation begins during or just after harvest in September or October. At this time adult females migrate from clusters into cracks, crevices, and knotholes on the undersides of the main scaffolds and on the main trunk. Adult females produce large numbers of 1st instar nymphs, called crawlers, which seek out protected areas to hide for the winter.

In the spring, when temperatures begin to increase, the overwintering nymphs that survived the winter (a small percentage) emerge from their hiding locations and migrate to the tree canopy. This migration is well synchronized to the period of bud swell, usually in March. During March and April, the mealybugs feed on woody areas at the interface between old and new growth. By May, the mealybugs develop into adults and begin feeding on the rachis of the cluster (fig. 22.5).

The first in-season generation begins during late May through early June, when adult females on the rachis give live birth to crawlers that move to the pistachio hulls. During the next 6 weeks, mealybugs feed on the pistachio hull and complete another generation. The second in-season generation begins around mid-July, when adult females produce large numbers of crawlers, which can be seen aggregating around their mother. It is this generation that has the highest number of mealybugs and is recognized to cause the most damage to the crop (fig. 22.6).

Mealybugs of this second generation reach the adult stage at about the same time that nuts are harvested. Adult females then migrate to the trunks and produce crawlers that become the next overwintering generation.

Figure 22.2. Aggregations of mealybugs on the trunk and scaffolds during the winter. *Photo:* D. R. Haviland.

Natural Control

The most common predators of *F. gilli* in pistachios are green lacewings. Lacewing larvae (fig. 22.7) feed by piercing the outer shell of their prey and sucking out the body fluids, giving the mealybug a dried-out, shriveled appearance. Adult stages of some lacewing species are also predaceous. Lacewing development takes approximately 1 month in the summer, and the highest populations are observed from mid- to late summer through early winter. Lacewings are likely the most abundant predator in pistachio due to their relative lack of susceptibility to insecticides used for the control of true bugs (Heteroptera).

A small predatory coccinellid beetle, *Nephus (Scymnobius) sordidus* (Horn), can also be found feeding on mealybugs. Beetle larvae (fig. 22.8) are covered in white tufts of hair that allow them to

Figure 22.3. Grape mealybug is not a pest of pistachios and can be distinguished from *F. gilli* by the presence of four long white tails and shorter white projections on the sides of the body. *Photo:* D. R. Haviland.

mimic the appearance of a mealybug. Adult beetles are approximately 2 mm long, oval shaped, and have a brown, mottled color. They also prey on mealybugs. Larvae and adult predatory beetles are not common in the tree canopy, but they can often be found on the trunk in the fall during mealybug aggregations and in cracks in the bark or soil or under dirt clods at the base of the trunk.

In other crops where pyrethrins are not used, at least two species of parasitoid wasps have been found to reduce mealybug populations. The most important of these is an *Acerophagus* sp. Two other parasitoids, a *Chrysoplatycerus* sp. and *Anagyrus pseudococci,* also have been shown to attack *F. gilli.* Parasitoids overwinter inside adult mealybugs on the trunks and main scaffolds of the trees. Parasitized mealybugs can be distinguished from those that are not parasitized by their round, slightly yellowed, bloated appearance. When the warm weather increases in the spring, adult wasps emerge by chewing a round hole in one end of the mealybug mummy (fig. 22.9). Once emerged, they mate and lay eggs that hatch and parasitize the overwintered mealybug nymphs that have begun to feed in the tree canopy during the spring. Several generations of parasitoids occur per season, and at least one of the parasitoid species can produce up to six offspring in a single mealybug.

Figure 22.4. Pistachios from a mealybug-infested tree (right) compared with those from a tree that was not infested (left). *Photo:* D. R. Haviland.

Figure 22.5. During the early summer, mealybugs feed on the rachis. *Photo:* D. R. Haviland.

Monitoring

In orchards not known to have *F. gilli*, or where the pest is still spreading within an orchard block, the easiest time to survey is after harvest through the end of the year. During this time, look for white tufts of aggregated mealybugs on the trunks and undersides of major scaffolds. Pruning crews can help facilitate surveys for mealybugs if instructed on how to recognize and report this pest.

In-season monitoring in orchards known to be infested should be conducted at budbreak and in mid- to late May. At budbreak, search for mealybugs at the base of new buds on trees previously known to be infested. This helps determine how well the mealybugs overwintered. Return to these locations in mid-May and begin monitoring weekly for crawler emergence. This facilitates optimal timing for application of insecticide treatments where needed.

Figure 22.6. Damage from mealybug feeding in a cluster prior to harvest. *Photo:* D. R. Haviland.

Management

Preventing Infestation

Equipment plays a major role in the spread of *F. gilli.* Mealybugs are notorious hitchhikers, and cultural practices, particularly at harvest, can facilitate this

Figure 22.7. Lacewing larva feeding on mealybug nymphs. *Photo:* D. R. Haviland.

movement. During harvest, mealybug populations are at their highest, and trees are sticky with honeydew. Equipment passes through every row of an orchard as infested trees are shaken, causing infested nuts and foliage to contact shakers, catch frames, and then bins. Harvest equipment is quickly moved from location to location throughout the state. Once at a new site, mealybugs can easily crawl off equipment and initiate new infestations.

Equipment, particularly during harvest, should be washed prior to leaving an infested orchard. Water from a garden hose with a pressure nozzle should be sufficient since honeydew dissolves easily in water and mealybugs do not hold on to metal surfaces very well. High-pressure washers are even more effective. Growers who contract to bring harvest equipment to their orchards should insist that the equipment be cleaned prior to arrival (or clean the equipment themselves prior to letting it enter the orchard). Be sure, however, to wash equipment as far away as possible from the orchard so that any mealybugs that are washed from the equipment are not able to initiate a new infestation.

Birds, such as crows that move into orchards after harvest, can also spread mealybugs. Crows are known to pick up and move infested nuts that were left unharvested. Crawler stages of mealybugs can also inadvertently crawl onto the feet and legs of crows and be moved to another orchard or ornamental tree host, such as mulberry trees, which are commonly used as crow roosts and serve as an excellent alternate host of *F. gilli*.

Figure 22.8. Predatory beetle larva (left) compared with an adult female mealybug (right). *Photo:* D. R. Haviland.

Figure 22.9. Parasitized mealybugs (mummies) showing parasitoid exit holes. *Photo:* D. R. Haviland.

Managing Infestation

Current pest management programs in pistachio require the use of chemical insecticides. This is likely to continue until broad-spectrum insecticide use for true bugs can be reduced so that biological control programs can become established. Treatments are often needed so that damage at harvest can be minimized. One research project suggested that densities of one adult female mealybug per 6 to 10 clusters in May is sufficient to justify making a treatment.

Time the insecticide applications for late May through early June when crawlers are present. If this timing is missed, consider a treatment at the beginning of the second generation in mid-July. Actual dates may be later in climates cooler than Tulare County, where studies have been conducted. Early-season mealybug infestations by the overwintering generation do not cause any economic injury, and there is no need to treat these pests prior to late May or early June, when crawlers of the first in-season generation are born. Postharvest treatments are likewise not recommended, even when mealybug densities are very high, since adult females on the trunk will all die anyway and more than 99 percent of the crawlers do not survive the winter. Additionally, adult females on the trunk serve as overwintering hosts for parasitoids.

References

Haviland, D. R., R. H. Beede, and K. M. Daane. 2012. Seasonal phenology of *Ferrisia gilli* (Hemiptera: Pseudococcidae) in commercial pistachios. Journal of Economic Entomology 105(5): 1681–1687.

———. 2015. Crop loss relationships and economic injury levels for *Ferrisia gilli* (Hemiptera: Pseudococcidae) infesting pistachio in California. Journal of Economic Entomology 108(6): 2683–2690.

Haviland, D., R. Beede, K. Godfrey, and K. Daane. 2006. *Ferrisia gilli*, a new mealybug pest of pistachios and other deciduous crops. Oakland: University of California Division of Agriculture and Natural Resources Publication 8207.

23 Mites

Robert H. Beede and David R. Haviland

Citrus flat mite (*Brevipalpus lewisi* McGregor) (Tenuipalpidae) is commonly found feeding on the rachis and hull tissue of pistachio nuts (fig. 23.1). Left untreated, it can cause nuts and entire clusters to turn brown and dry up on the tree. Twospotted spider mite (*Tetranychus urticae* Koch) and Pacific spider mite (*Tetranychus pacificus* McGregor) (Tetranychidae) also occasionally feed on pistachios. However, their presence and damage to foliage in pistachio are minor compared with their impact on other California nut crops such as almond and walnut.

Citrus Flat Mite

Citrus flat mite is commonly associated with the production of citrus and was first reported scabbing the rinds of lemons near Porterville (Tulare County) in 1942. Similar injury has been observed on pomegranates, resulting in fruit cracking and high cull percentages. It was identified as an injurious pest of pistachios in northwestern Kern County in 1979.

Description

Citrus flat mites are very small and cannot be seen with the naked eye. When seen under magnification, such as with a 10× hand lens, mites are somewhat flat and oblong in shape and are wider at the head end. They move very slowly, and body color ranges from reddish brown to bright red or pink. Citrus flat mite eggs are very small, oval, and bright red in color.

Seasonal Development

Citrus flat mite adults are thought to overwinter in colonies found under loose bark or bud scales. As temperatures approach about 70°F in the spring, they emerge and begin feeding on fruit and fruit stem areas of their host plants. Populations do not normally reach injurious levels until July or August, and high summer temperatures do not appear to inhibit their feeding. Citrus flat mite has one generation per year.

Injury

Damage from citrus flat mite to pistachio is similar to that in citrus and pomegranate. Initially, the feeding of low populations causes minor scabbing of the surface tissue of the rachis (stem) and nut tissue from their rasping-sucking mouthparts. Early damage levels often go unnoticed. As the population rises, the brown discoloration of these tissues becomes readily visible (fig. 23.2). Left untreated, citrus flat mite continues feeding and causes dried hull tissue and completely dehydrated clusters. This results in reduced harvestability, higher staining, and greater percentages of closed-shell nuts that do not split. Citrus flat mite does not damage leaf tissue, which distinguishes it from webspinning mites.

Natural Control

Studies of citrus flat mites have revealed the presence of the predaceous phytoseiid mite (*Metaseiulus*

Figure 23.1. Adult citrus flat mites. *Photo:* J. K. Clark.

Figure 23.2. Citrus flat mite causes scabbing to the pistachio rachis and hulls. *Photo:* R. E. Rice.

flumensis (=*mcgregori*) (Chant)), which is common in several crops throughout California. Its numbers begin increasing in August and reach their highest level at the same time as those of citrus flat mite, but experience has been that their numbers are insufficient to reduce citrus flat mite populations below economically injurious levels. Mites in the family Tydeidae, which feed on fungi and pollen, are also observed in low numbers in pistachios. It is thought that they serve as an alternative food source for the phytoseiid mites.

Monitoring and Management

Look for infestations of citrus flat mite by examining clusters for scabbing on the rachis and hulls from late spring through summer. If damage is present, confirm the presence of flat mites with a hand lens. Mites can most easily be found at the margins of damaged tissue. Treatments should be made if it appears that damage is going to reach economically significant levels.

Citrus flat mite is typically controlled with a single application of sulfur in the late spring through early summer. Dusting and wettable sulfur are both highly effective, though wettable sulfur is more common due to the ability to tank-mix it into an April or May pesticide application for small bugs like *Phytocoris* and *Lygus* (see chapter 20, "Small Bugs"). Experiments with application rates up to 50 pounds of dusting or 15 to 20 pounds of wettable sulfur per acre have failed to result in any phytotoxic effects to pistachio, even in hot weather.

Citrus flat mite can also be managed with horticultural mineral oil. Tests with 415°F midpoint boiling range horticultural mineral oil showed that it performed similarly to sulfur.

Webspinning Spider Mite

Webspinning spider mite is the common name given to mites in the family Tetranychidae. In pistachios, two species of webspinning spider mite can be found. These are the twospotted spider mite (*T. urticae* Koch) and the Pacific spider mite (*T. pacificus* McGregor) (fig. 23.3). While neither of these mites is common in pistachios, both have the potential to cause defoliation, which in turn causes problems related to sun exposure of the fruit.

Description

Spider mites are larger than citrus flat mite, and they appear as tiny moving dots to the naked eye. Adult females are about 1⁄50 of an inch long and lay spherical, amber-colored eggs. Immature and adult stages vary in color from slightly amber to greenish in the spring and summer and from reddish to orange in the fall and winter. Spider mites get their name from the characteristic spiderlike webbing they produce. They also have characteristic feeding spots on the dorsum, which develop soon after the immature stages hatch and begin to feed. Twospotted spider mites develop a single dark spot on either side of the dorsum. Pacific mites have two large, diffuse spots on the forward area of their dorsum and two smaller ones in the rear. However, the spots are highly variable and are not a reliable method for distinguishing the species. The only way to tell the species apart is by slide-mounting the males on their sides and examining the shape of the male reproductive organ, called the aedeagus, under a microscope.

Seasonal Development

Pacific and twospotted spider mites overwinter as mature females under the rough bark at the base of the host tree, on ground cover plants, and in orchard floor debris. Male mites do not overwinter and are less common than female mites.

Figure 23.3. Twospotted spider mite adults and egg. *Photo:* J. K. Clark.

When the weather warms in the spring, overwintering females become active and begin feeding by rasping surface tissue with their mouthparts to release the tissue's cellular contents. They eventually move up into the canopy if natural predators are insufficient. Twospotted spider mites initiate egg laying on the underside of leaves until populations reach higher levels. Pacific spider mite females lay eggs initially on both sides of the leaf. Translucent when first laid, mite eggs become opaque before hatching into immature mites. They molt three times before reaching the adult stage. They have an extremely wide host range.

Pacific and twospotted spider mite populations usually reach their most severe levels in summer, especially where predators have been killed by broad-spectrum insecticides. These mites reproduce rapidly in hot, dusty conditions and on water-stressed trees. If temperatures and food supply are favorable, a new generation can occur every 7 days and 8 to 15 generations can be completed in a season. Mites disperse from areas of heavy infestation by wind.

Injury

Spider mite injury to pistachios is rare but can occur. Damage initially appears as a loss of green color on infested leaflets. Continued feeding causes the leaf to develop small necrotic spots with small amounts of webbing. All life stages of the mites can be found residing and feeding within this webbing. Low mite levels of three to four per leaflet are sufficient to cause defoliation in pistachio. Damage levels can be enhanced by plant stress from high salinity or lack of water.

Natural Control

Sixspotted thrips (*Scolothrips sexmaculatus* (Pergande)) is the most common and most effective mite predator of spider mites in pistachios and can be effective at very low mite populations. Since webspinning spider mite populations develop very slowly on pistachios, thrips often provide effective biological control, and chemical treatment is not justified. Predatory mites such as the western predator mite *(Galendromus (=Typhlodromus, =Metaseiulus) occidentalis)* can also be found on pistachios. However, western predator mites are not known to become abundant, even in orchards with low insecticide use.

Monitoring and Management

Monitor the pistachio tree canopy for damage from spider mites during the late spring and summer. Look for bleached-out areas on leaves that are associated with spider mites and their webbing. Give particular attention to the edges of orchards, especially where dusty conditions exist, and orchards that are stressed from high salinity or insufficient water. If spider mites and their damage are present, look for predators of mites, such as sixspotted thrips, which usually keep spider mites under control.

Pesticide treatments for webspinning spider mites are rarely needed in pistachios. However, when exceptions do occur, several different miticides, as well as horticultural mineral oil, are effective. For treatment options, consult the UC IPM Pistachio Pest Management Guidelines at the UC IPM website, ipm.ucanr.edu/PMG/selectnewpest.pistachios.html.

References

Beede, R. H., R. Ritchie, and H. Tutschulte. 2001. Comparison of Orchex 692 horticultural mineral oil to wettable sulfur for control of citrus flat mite. California pistachio industry annual report 2001, 82–83. Fresno: California Pistachio Commission.

Beede, R. H., D. Reese, F. Gonsalves, R. E. Rice, J. Padilla, and H. Tutschulte. 2000. Miticide efficacy trials in pistachio. California pistachio industry annual report 2000, 135–136. Fresno: California Pistachio Commission.

Elmer, H. S., and L. R. Jeppson. 1957. Biology and control of the citrus flat mite. Journal of Economic Entomology 50:566–570.

Jeppson, L. R., H. H. Kiefer, and E. W. Baker. 1975. Mites injurious to economic plants: Tenuipalpidae. Berkeley: University of California Press.

Schuster, R. O., and A. E. Pritchard. 1963. Phytoseiid mites of California. Hilgardia 34:223–225.

Weinberger, G. B., R. H. Beede, and R. Cortez. 1984. Citrus flat mite control on pistachios. California pistachio industry annual report 1984, 69–71. Fresno: California Pistachio Commission.

Zalom, F. G., W. J. Bentley, L. G. Varela, L. J. Bettiga, and D. R. Haviland. 2013. Spider mites. In L. J. Bettiga, ed., Grape pest management. 3rd ed. Oakland: University of California Division of Agriculture and Natural Resources Publication 3343. 405–421.

24 Minor Pests

Robert H. Beede and David R. Haviland

Pistachio has several pests that are not widespread in their occurrence but nevertheless have the potential to cause significant damage. Most of these pests are migratory in nature and exist on neighboring crops or weed hosts adjacent to or within the orchard. These pests are particularly damaging when they migrate into new orchards 1 to 3 years old, with recently budded orchards being the most susceptible. Any damage that occurs during this period can result in death of the bud or can affect the development of structural branches. Once trees are about 4 years old and these structural branches have been developed, even moderate to high populations of these minor pests are generally not a concern. The following are brief descriptions of these pests and the damage they cause to pistachios. Additional information for some of them can be found in the UC IPM Pistachio Pest Management Guidelines at the UC IPM website, ipm.ucanr.edu/PMG/selectnewpest.pistachios.html.

Ants

The most common ant found in pistachios is the southern fire ant (*Solenopsis xyloni* McCook) (fig. 24.1). These ants, which prefer to feed on protein-based food sources, occasionally damage split pistachios near harvest by chewing out the tips of the nuts. This can be mistaken for bird damage. The likelihood of ants migrating into mature trees is enhanced by pests such as mealybugs or soft scale, which produce honeydew, a sugar-based by-product of their feeding that ants find attractive. However, damage to pistachios at harvest is generally negligible due to harvest operations that shake and catch nuts before they touch the ground, where ants prefer to feed.

Ants can also damage newly budded pistachios by directly feeding on the emerging bud 3 to 4 weeks after insertion or by girdling the base of the bud. Additional problems can arise when ants plug the orifices of microsprinklers in the spring. Baits are the preferred method for managing ants. Baits are broadcast on the field, collected by foraging ant workers, and then fed to the queen. This causes the queen to die and the colony to collapse. Since all of this takes time to work, survey the orchard floor for ant mounds and colonies 2 to 3 days after irrigation in May in the southern San Joaquin Valley and through June in the northern San Joaquin Valley (fig. 24.2). If treatments are necessary, apply baits 45 to 60 days prior to when damage is anticipated.

Aphids

Aphids cause damage by feeding on newly budded pistachios or on young Kerman shoots as they emerge from newly developed trunks during the second season. The species may vary depending on orchard location and surrounding crops, but the

Figure 24.1. The southern fire ant has variable-size workers with red heads and black abdomens. *Photo:* J. K. Clark.

cotton aphid (*Aphis gossypii* Glover) is reported as being the most common (fig. 24.3). Cotton aphid overwinters in the egg stage on alternate hosts such as pomegranate. In the spring, eggs hatch into aphids that develop wings, migrate into crops like pistachios, and then reproduce parthenogenetically (meaning that females bear live young without the need to mate). Large numbers have been reported on the terminals of spring growth in mature trees, but rarely have they been treated because the economic damage is negligible. Aphids on newly budded pistachios or very young trees are of greater concern, since they aggregate on the emerging terminal and can potentially stunt its growth. The treatment decision depends on aphid density and the degree of predation and parasitism evident. A number of predators exist, such as green and brown lacewings (*Chrysoperla* and *Chrysopa* spp.), lady bird beetles (*Hippodamia convergens* and *Coccinella novemnotata franciscana*), minute pirate bugs (*Orius tristicolor*), big-eyed bugs (*Geocoris* spp.), and damsel bugs (*Nabis* spp.). Predation is generally more effective in the spring than summer. The parasitic wasp *Lysiphlebus testaceipes* is an important natural enemy of the cotton aphid. These wasps lay their eggs inside the body of the aphid, and following hatch and development, the adults emerge, leaving a tan-colored, swollen mummy (see fig. 24.3). In cases where biological control is insufficient, soaps, oils, and a variety of selective insecticides can effectively control aphids.

Figure 24.2. Volcano-shaped mound of the southern fire ant. *Photo:* J. K. Clark.

Figure 24.3. Cotton aphid and tan, parasitized aphid mummies on an unopened pomegranate flower. *Photo:* L. Strand.

Beet Armyworm

Beet armyworm (*Spodoptera exigua* Hübner) is a major pest on cotton, alfalfa, vegetables, sugarbeets, beans, and many other crops. Its limited occurrence in pistachio suggests pistachio is not a preferred host, and infestations result from migrations from a neighboring crop or weeds within the orchard. The adult moths are mottled gray and brown, with a wingspan of about 1¼ inches, and they lay eggs in clusters with white, hairlike scales covering them (fig. 24.4). Scout for these egg masses to detect a possible problem, especially when preferred hosts are nearby. The larvae are dull green with wavy lines down their back and a broad, pale green stripe down each side. The most distinguishing feature is a black spot on either side of the body, just behind the head and above the second true leg (fig. 24.5). Beet armyworm larvae are most commonly found in August, shortly after pistachio rootstocks have been budded. They feed on young leaves, emerging Kerman buds, and at the base of young, developing Kerman shoots. Infested trees can suffer enough injury to require rebudding. Beet armyworm can also infest split pistachio nuts, causing damage very similar to that of navel orangeworm. To avoid damage, watch for beet armyworms in surrounding crops and on weeds within the orchard, such as morningglory and pigweed. The parasitic wasp *Hyposoter exiguae* attacks young worms and is an important biocontrol agent. Each wasp is capable of controlling up to 100 worms. Many generalist predators also feed on eggs. Ubiquitous polyhedrosis viruses can also attack armyworms and have been known to decimate populations. In cases where insecticide treatments are needed, most insecticides that are effective against navel orangeworm can also control beet armyworm.

Cankerworms

Cankerworms can occasionally be found on pistachios in the spring. They are sometimes called inchworms, since their characteristic way of moving is by arching the front portion of their body and then placing it a small distance forward. The spring cankerworm (*Paleacrita vernata* (Peck)) (fig. 24.6) has been identified on pistachios in the southern San Joaquin Valley, but the similar-appearing fall cankerworm (*Alsophila pometaria* (Harris)) may also be found (fig. 24.7). Larvae of both species have two dorsal knobs a short distance behind their head. Spring cankerworm larvae have two pairs of abdominal prologs, whereas fall cankerworms have three pairs of abdominal prolegs. Worms can be found standing upright on their prolegs near the tip of their abdomen, motionless to mimic a leafless twig. When disturbed, worms drop and hang from a silk thread. The simultaneous presence of very small (approximately ½ in) and large (1½ in) worms in the spring suggests an extended overwintering emergence. The smaller worms appear to skeletonize leaves, whereas the large ones eat portions of the entire leaf. Cankerworms do not appear to feed on developing buds or on the nuts, but they have been observed feeding on rachis tissue. Fall cankerworm eggs are laid in November and December; spring cankerworm eggs are laid in January and February. Eggs of both species hatch in early spring. After maturing, larvae pupate in the soil. Cankerworms go through one generation per year. Cankerworms are easily controlled with any insecticide registered for navel orangeworm or obliquebanded leafroller.

Figure 24.4. Beet armyworm egg mass. *Photo:* J. K. Clark.

Figure 24.5. Beet armyworm larva, showing a black dot behind the head and pale green lateral stripe. *Photo:* J. K. Clark.

Figure 24.6. Spring cankerworm on a pistachio leaf. Spring cankerworms have two anal prolegs. *Photo:* D. Haviland.

Figure 24.7. Larva of fall cankerworm, with three abdominal prolegs. *Photo:* J. K. Clark.

Darkling Ground Beetle

Darkling ground beetles of the genus *Blapstinus* (Coleoptera: Tenebrionidae) damage young pistachios by crawling up the rootstock or rough support stake and eating the Kerman bud 3 to 4 weeks after budding. They can also damage emerging Kerman shoots during the spring following their first dormancy. Adults are black to bluish black, with reddish legs (fig. 24.8). They are about ⅕ inch long. Adults reside 2 to 3 inches below the soil surface. Beetle larvae also live in the soil. There is one generation per year. Ground beetles are active at night, so scouting during the day will not reveal their presence. Pistachio budders report ground beetles are strongly attracted to the Kerman female, such that budsticks discarded onto infested orchard floors are often covered with adults in 20 to 30 minutes. Ground beetles can also migrate in large numbers from recently

Figure 24.8. Adult darkling ground beetle. *Photo:* J. K. Clark.

Figure 24.9. European earwig on a new walnut shoot. *Photo:* J. K. Clark.

Figure 24.10. Adult false chinch bug. *Photo:* J. K. Clark.

disked alfalfa fields. Where high populations are suspected and newly inserted buds are low, Scotch tape can be wrapped around the base of the rootstock to prevent adults from eating out the bud. Some beetles may still reach the bud by crawling up the stake and crossing over on the tie tape. Carbamate baits applied to the soil are also effective.

European Earwigs

Earwigs (*Forficula auricularia* L.) are relatively small (½ in), slender, shiny-brown insects, which are easily identified by a pair of pinchers on the tip of their abdomen (fig. 24.9). They search for food during the night, and conceal themselves during the day in orchard debris, decaying weeds, or other undisturbed sites in or near the orchard. Pearly white, oval eggs are laid in masses from December through the spring in the soil. They overwinter as adults. Earwigs are only a problem on newly budded trees and second-year trees, where they climb up and eat the terminal out of the emerging shoot. The risk of earwig damage is greatly reduced by ensuring that the orchard floor is free of weeds and debris. Monitor for the presence of earwigs by placing boards or rolled-up newspaper on the orchard floor; they will use either to hide in during the day. If available, carbamate baits placed at the base of each tree are very effective if applied soon after detecting the need for treatment.

False Chinch Bug

False chinch bugs (*Nysius raphanus* Howard) breed in huge numbers in weeds such as London rocket, shepherd's-purse, purslane, and mayweed. They also favor grains and migrate in vast numbers out of rangelands, pastures, wheat fields, or weedy ditches when their hosts dry down. Adults are about ⅛ inch long and are light or dark gray (fig. 24.10). They might be easily mistaken for grassbugs. The eggs are laid at random in crevices in the soil. Nymphs are believed to be the primary overwintering stage, although all stages may be found. Six generations per year are estimated. Weeds within the orchard serve as the breeding ground in the spring. Once these cycle out or are eliminated, countless numbers of bugs migrate onto young pistachios, especially in May and June. When forced from a drying host, most of the false chinch bugs are wingless, but even those with wings join in the ground migration to a new host. False chinch bug feeding on the bark and leaves of young pistachios causes collapse of the tree in about 10 days due to a suspected toxin produced by the nymphs. Heavily infested trees usually die. For this reason, false chinch bug should be considered a serious pest on young trees. Control of false chinch bug is difficult because of their tremendous numbers and constant movement. When needed, broad-spectrum insecticides can provide temporary control. Repeated applications are periodically needed in cases where relatively cool weather in May and June causes a slow drydown of spring hosts and therefore a prolonged migration into the orchard. The best control for false chinch bug is elimination of all weeds within the orchard during the first 3 years.

Grape Bud Beetle

Grape bud beetle (*Glyptoscelis squamulata* Crotch) is primarily a grape pest in California, but periodically it feeds on buds of young pistachio trees. Grape bud beetles have been collected on poplar, willow, tamarisk, mesquite, cheeseweed, beans, corn, alfalfa, rose, peach, and both wild and domestic grapes. Populations in grape have varied with changes in pesticide usage in that commodity since the 1920s, when it was considered a serious pest. Today, grapes occasionally require treatment in the San Joaquin Valley to avoid feeding damage to the primary and most fruitful bud. Adults are about ¼ to ⅜ inch long and 3/16 to ¼ inch wide. Young adults are usually covered with gray scales, which are often rubbed off over time as they crawl through soil to emerge, or in and out of grapevine bark, where they hide during the day. The female grape bud beetle has a long, thin ovipositor and lays batches of 12 to 44 eggs within the thin layers of grape bark (fig. 24.11). There is one generation per year. Adults can be found on grapevines by mid-January. Adult numbers increase toward mid-March, when grape budbreak typically begins. Young pistachios adjacent to grapes can experience loss of the lateral buds used to create the primary scaffolds in 2nd-leaf trees. Loss of these buds in the 36- to 42-inch zone of the new pistachio trunk requires that a lower bud be retrained up the stake and headed-in season at 42 to 44 inches. In-season heading of current growth for development of the primary scaffold branches is less desirable than dormant heading due to the lower number of lateral branches that grow. Often, only two lateral buds push from heading current growth, resulting in a less balanced future canopy prone to splitting at the crotch. For this reason, grape bud beetles may require treatment in young pistachios. Assess the potential for such damage in young orchards by removing bark from the trunks of grapes neighboring the pistachios. Begin monitoring in mid-January and continue until mid-April, when young pistachios should have completed bud push. Alert the vineyardist to the presence of grape bud beetle, since they will return the following year and eventually cause economic loss, especially in table grapes. Refer to the UC IPM Grape Pest Management Guidelines at the UC IPM website, ipm.ucanr.edu/PMG/selectnewpest.grapes.html, to help identify treatment options that might currently be available also for pistachios.

Grasshoppers

Four species of grasshoppers are commonly found in California agriculture, of which the devastating grasshopper (*Melanoplus devastator* Scudder) is the most damaging (fig. 24.12). It breeds in the drying grasses of the Sierra Nevada foothills as well as the Coastal Range near the west side of the San Joaquin Valley. It has increased in importance with the continued expansion of pistachio acreage into the Sierra foothills and valleys above the Westside floor. The valley grasshopper (*Oedaleonotus enigma* Scudder), found in the Bakersfield area, is similar in size and damage to the devastating grasshopper. Both are medium size (about 1 in) and light to dark brown in color. The devastating grasshopper has dark markings on the thorax and a row of dark spots on its front wings. The femur (large portion) of the hind leg is blue at its attachment to the body, at which point it changes to an amber color at its tip. The tibia (lower portion of the leg) is red to grayish. In contrast, the similar-size valley grasshopper has chevron markings on its femur, which are broken, and the tibia is bluish. Both have only one generation per

Figure 24.11. Grape bud beetle adult and eggs. *Photo:* M. E. Badgley.

Figure 24.12. Adult devastating grasshopper. *Photo:* J. K. Clark.

Figure 24.13. Adult vagrant grasshopper. *Photo:* D. Rosen.

Figure 24.14. Adult forktailed katydid. *Photo:* J. K. Clark.

year. Devastating and valley grasshoppers lay eggs in the soil after the first fall rains. Adults die before winter. Overwintered eggs hatch in May through July, and nymphs develop on rangeland vegetation until it dries. Mass migration to a preferred host then occurs. Both species return to the rangeland areas in the fall. Cool spring temperatures and early wet spring weather favor rangeland vegetation; these conditions also delay egg hatch and increase diseases, which reduce grasshopper populations. Dry, warm spring weather, on the other hand, accelerates egg hatch and makes it more uniform. Winter weather has little effect on egg survival. Fungi and bacteria kill many young grasshoppers during cool, wet springs.

The two other common grasshopper species are the green valley grasshopper (*Schistocerca alutacea shoshone* (Thomas)) and the vagrant grasshopper (*Schistocerca nitens nitens* (Thunberg)) (fig. 24.13). The green valley grasshopper is usually found from Bakersfield northward, and the vagrant grasshopper is found from Fresno southward. The green valley grasshopper is large (2½ in long), with a distinctive leaf-green color, red hind legs, and a yellow stripe along the middle of the head and thorax. The vagrant grasshopper is as large as the green valley, but it has a brown body and legs as well as a tan stripe along the midline of the head and thorax. After hibernation in the adult stage, both initiate defoliating feeding in the spring. The adult female lays eggs in May by boring a hole into the soil with her abdomen. The eggs are then inserted within a frothy material that hardens into a spongelike structure called a pod. After hatching in a few weeks, young, wingless grasshoppers burrow to the soil surface and begin feeding on foliage. They reach the winged stage in late summer. Both the green valley and vagrant grasshopper have one generation per year.

Pistachio orchards adjacent to rangeland are most commonly infested. Grasshoppers are indiscriminate in their feeding, stripping trees of leaves and shoots. If possible, monitor adjacent rangeland with a sweep net in April and May for nymphs. Disking of rangeland nearest the orchard edge destroys grasshopper habitat and eggs. Baiting rangelands or orchard edges can reduce migrating populations and minimize their entry into the orchard. Once in the orchard, broad-spectrum insecticides, including carbamate sprays or baits, are usually effective.

Katydids

Katydids occasionally damage newly budded pistachios by feeding on the emerging bud or growing terminal. This results in either loss of the Kerman bud or stunting of the shoot that is being trained up the stake for the main trunk. Since budding typically does not occur until early August, loss of Kerman growth from katydid or other pest feeding could prevent the shoot from achieving the minimum height of 46 inches required for dormant heading. Two katydid species are found in California. The most common is the angularwinged katydid (*Microcentrum retinerve* (Burmeister)). The adult and nymph of this species have a distinctive humpbacked appearance. The forktailed bush katydid (*Scudderia furcata* Brunner von Wattenwyl) has no humpback and is smaller (fig. 24.14). The nymphs of both species have very long antennae, which have alternating black and white color bands (fig. 24.15). Katydids lay their disk-shaped eggs in the fall, often in two overlapping rows. They hatch in April and May. Nymphs feed on foliage and make irregular holes in the middle of the leaves. This may also occur on certain weed species within the orchard. There are no reports of katydid feeding on any stage of pistachio nut development. Katydids reach the adult stage in midsummer, in time to feed on trees in the early budding and training stages. Refer to the UC IPM Pest Management Guidelines for citrus or stone fruit at the UC IPM website, ipm.ucanr.edu/PMG/selectnewpest.citrus.html, to help identify treatment options that might currently be available for pistachios.

Figure 24.15. Forktailed katydid nymph. *Photo:* J. K. Clark.

Figure 24.16. Adult sixspotted thrips is a predator of spider mites. *Photo:* J. K. Clark.

Thrips

Three thrips species are reported as common to pistachios in California. The sixspotted thrips (*Scolothrips sexmaculatus* (Pergande)) is a beneficial species and is very effective in controlling occasional infestations of Pacific spider mite (fig. 24.16). The adult has three distinct dark spots on each wing, which distinguishes it from all other thrips species. It is tiny and slender. Both nymphs and adults feed on mites and mite eggs. The nymphs are clear white in appearance, in contrast to other species, which have some color.

Western flower thrips (*Frankliniella occidentalis* (Pergande)) (fig. 24.17) is found on many plants. It has five or six generations per year. This thrips overwinters in the adult and nymphal stages on weeds or other hosts on the orchard floor, in wild areas, or on crop hosts such as alfalfa and wheat that remain green during the winter. In the San Joaquin Valley, a population peak is usually reached in May, when adult flower thrips from foothill rangelands migrate into the valley. Immature western flower thrips are yellow or brown and opaque. Reproduction may be either sexual or asexual. The minute eggs are laid in soft tissues such as leaves, stems, and hulls. Once laid, eggs hatch in about 5 days. Damage to pistachios is a surface scarring of plant tissue caused by the insect's rasping mouthparts, which injure the surface cells so that the thrips can ingest the cell contents. Their preferred feeding sites in most plants are tender growing shoot tips and flowers. There have been no field reports of heavy thrips populations in blooming pistachio inflorescences. Scarring of newly emerged leaves, the hulls of expanding nuts in May, and distortion of growing tips of young trees have all been reported. Except under extreme situations, these types of damage are generally superficial and do not result in economic losses. Thrips can also be found inside developing nuts that split open during the spring from sudden increases in humidity or plant water status. Research shows that thrips do not cause epicarp lesion.

Citrus thrips (*Scirtothrips citri* (Moulton)) (fig. 24.18) has also been observed on pistachios grown in the eastern citrus belt of the San Joaquin Valley. Treatment for thrips in mature trees is very rare, and it would only be justified if populations are causing extensive early fruit scarring or severe stunting of shoots on young trees. Treatment of newly budded trees being trained up the stake might be necessary if shoot growth has been reduced from terminal feeding or loss of chlorophyll from the lateral leaves. Scarring of the hull tissue in May and June could also increase *Alternaria* infections in orchards with high levels of inoculum. Guidelines for managing thrips in pistachios have not been developed. If treatments are needed, refer to the UC IPM Pest Management Guidelines for citrus, stone fruit, or blueberries at the UC IPM website, ipm.ucanr.edu/PMG/, to help identify treatment options that might currently be available for pistachios.

Western Tussock Moth

Western tussock moth (*Orgyia vetusta* Boisduval) has a colorful caterpillar stage that occasionally attacks pistachio leaves and fruit clusters (fig. 24.19). Feeding damage by larvae has rarely reached economically significant levels in recent times. Tussock moths overwinter in egg masses covered with tan-colored hairs deposited by the female, which give them a fuzzy appearance. The first larvae appear in March. The black or dark green caterpillars are covered with long hairs and have bright red and yellow spots along the body. They have two long tufts of black hair protruding from the head and one from the tail. Adults are present from May through July. The adult males are gray, and the wingless females are grayish white (fig. 24.20). Tussock moths should not be confused with gypsy moths, which are bluish in color. Monitoring for tussock moth is best done during pruning in the winter. If present, alert pruners to the appearance of the tan, fuzzy egg mass, and have them tag infested trees with flagging tape. This allows easy assessment of its density and distribution. If treatment is deemed necessary, insecticides registered for navel orangeworm or obliquebanded leafroller, including *Bacillus thuringiensis,* are very effective against young larvae.

Figure 24.17. Adult western flower thrips. *Photo:* J. K. Clark.

Figure 24.18. Nymphs of citrus thrips. *Photo:* J. K. Clark.

Figure 24.19. Western tussock moth larva. *Photo:* J. K. Clark.

Figure 24.20. Adult female western tussock moth. *Photo:* J. K. Clark.

Part 6 Disease Management

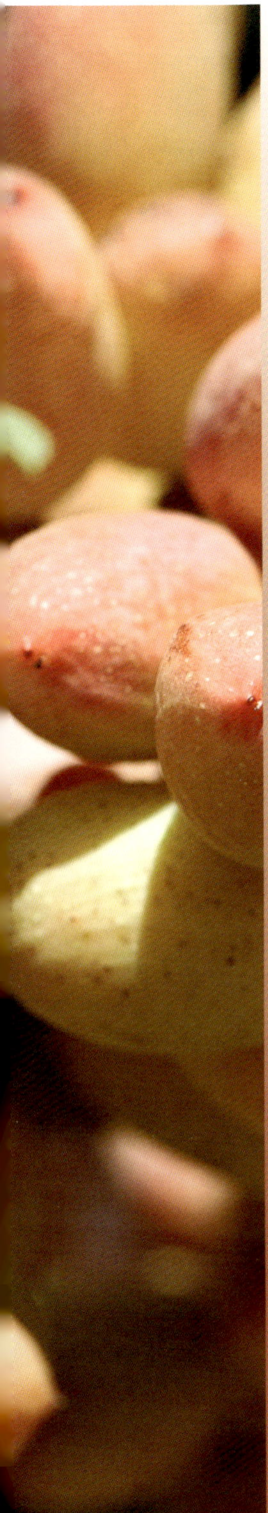

25 Soilborne Diseases

Brent A. Holtz and Beth L. Teviotdale

Verticillium wilt, Armillaria root rot, and Phytophthora root and crown rot are three soilborne diseases known to afflict pistachio trees in California. All three are fungal diseases that occur in most pistachio-growing regions of the state. Of these diseases, Verticillium wilt has killed the most trees and caused growers the most severe economic losses; early pistachio orchards were planted on rootstocks highly susceptible to the pathogen. A fourth soil-borne fungal disease of pistachios, crown gall, affects seedlings grown in nurseries but not field-grown trees (table 25.1). Crown gall is a serious soilborne bacterial disease of many other perennial tree crops; however, pistachios are relatively resistant, and this disease has never been reported as affecting them significantly. The relative importance of soilborne diseases varies with changes in choice of rootstocks. As new rootstocks are developed and introduced into pistachio culture, the array of root diseases may differ.

Verticillium Wilt

Historically the most destructive disease of pistachio in California, Verticillium wilt is caused by the fungus *Verticillium dahliae* Kleb. Trees of any age are subject to infection from the pathogen, especially if on non-resistant rootstocks such as *Pistacia atlantica* and *P. terebinthus* (fig. 25.1). *Verticillium* also infects cotton, tomato, melon, stone fruit, and olive in addition to pistachio in the southern San Joaquin Valley. Non-cultivated plants, both indigenous and weed species, also serve as hosts, and these enable the pathogen to

Figure 25.1. Pistachio trees with severe Verticillium wilt that were removed from the orchard. *Photo:* B. A. Holtz.

Table 25.1. Symptoms and control of soilborne diseases

Disease	Symptoms	When symptoms appear	Control
Verticillium wilt	sudden collapse of branches	spring, early summer	plant resistant rootstocks
	poor vigor that may be accompanied by slow thinning of canopy	growing season over several years	plant resistant rootstocks
Phytophthora root rot	slow thinning of canopy	growing season over several years	
	reduced growth, early senescence	growing season over several years	irrigation management
crown rot	sudden collapse of entire tree	spring or early summer	irrigation management
trunk canker	branch death, profuse gumming	growing season	keep sprinkler water off trunk
Armillaria root rot	reduced growth, early senescence, rhizomorphs on infected roots	growing season	preplant fumigation; plant resistant rootstocks
seedling blight	leaves turn brown, wither, and cling to shoots; seedlings are stunted, blighted, and die	seedling stage in the nursery greenhouse	sterile potting soil and clean nursery benches

become established in noncultivated land. Verticillium wilt of pistachio is more common in the southern half of the San Joaquin Valley than in other areas of the state, and the disease has been most destructive where pistachio trees were grown in fields previously planted to other susceptible crops. The disease is present in the Cuyama Valley, located between the southern San Joaquin Valley and Ventura coast, and occasionally is found in the Sacramento Valley.

Symptoms

The first symptom of Verticillium wilt is interveinal patches of yellowing or scorching of the leaves on affected branches. Collapse of the branch or tree ensues with the first warm weather because the pathogen causes blockage of vital water-conducting tissues, and the tree demands more water than the root system can provide. The classic symptom of Verticillium wilt is rapid desiccation and death of one or more scaffolds or the entire tree, usually in late spring or early summer.

In another manifestation of Verticillium wilt, called thin leaf decline, symptoms develop slowly over several years before the tree becomes economically unproductive or succumbs to the disease. Thin leaf decline is characterized by slow loss of vigor and reduction in growth and yield. A gradual thinning of the canopy occurs until most of the remaining leaves are clustered in tufts at the ends of branches, and shoots may also be seen (fig. 25.2). Why *Verticillium* causes some trees to collapse suddenly and others

Figure 25.2. Pistachio tree showing thin leaf decline due to infection with *Verticillium dahliae*. *Photo:* T. J. Michailides.

to decline slowly has not been explained. An overall lack of vigor may be the most common symptom of Verticillium wilt.

Verticillium dahliae invades and colonizes the plant's vascular system. The water-conducting tissue, the xylem, is composed of small vertical tubes that together form a cylinder just inside the cambium. Only the young xylem, that which is formed in the current year, is able to transport water. Xylem elements infected by *Verticillium* become clogged by substances elaborated by the fungus and the tree. These substances are darkly pigmented, and a cross section of an infected root, trunk, or branch reveals the pattern of *Verticillium* infection in the new xylem: scattered black speckles or a somewhat discontinuous black ring, depending on how many vessels are infected (fig. 25.3). The uniform discoloration of the innermost wood is a consequence of the natural aging process. The discolored xylem elements are nonfunctional and cannot transport water to the upper portions of the tree. The branches that die are the ones connected to the infected root and vascular system and no longer receiving adequate water. The history of *Verticillium* infection in older trees is recorded in the necrosis of each year's ring of xylem (fig. 25.4). Verticillium wilt does not cause rot of bark or wood and does not invade through the aboveground parts of the plant.

Disease Cycle

Verticillium has only limited saprophytic activity in soil, so it does not readily grow and reproduce there. Instead, it either lives on the roots of alternate hosts or it survives in soil in the form of resting structures called microsclerotia. Each microsclerotium is a tiny cluster of approximately 30 dark-colored cells that are extremely hardy. A percentage of the microsclerotia can remain viable in soil for many years. In a soil that was previously planted with an alternate host of *Verticillium*, such as cotton or a solanaceous crop, several hundred microsclerotia can be present in a cubic inch of soil. Most microsclerotia are found in the upper 6 inches of soil, where there is the most biomass from a previous host.

The microsclerotium is also the infective unit of the fungus. It remains dormant in soil until a plant root grows nearby, releasing substances (nutrients) that stimulate it to germinate. These substances can be produced by the roots of both host and nonhost plants, but only young or wounded roots produce sufficient quantities of nutrients to cause the microsclerotium to germinate. Many infections remain in the outer, cortical tissues of roots and never invade the xylem. However, especially if infection occurs just behind the root tip, where the young cells are undifferentiated, the fungus may grow into the new xylem. Inside the xylem element, the fungus produces mycelium and conidia (spores). The conidia are carried upward with water. As conidia become lodged in xylem cells, they germinate; more mycelium is formed, which produces more conidia that again move upward with the water flow. In this manner, the fungus spreads from root to shoot, plugging the xylem vessel along the way. Neighboring xylem vessels apparently are not easily infected. Thus, many independent infections may be needed to involve a large portion of the water-conducting system.

Infections located away from the root tip usually follow another course of entry. The fungus invades a few cortical cells outside the vascular tissues and there forms a new microsclerotium, which later is returned to the soil. This refreshing of the

Figure 25.3. Pistachio branch showing discoloring of the xylem due to Verticillium wilt. *Photo:* B. A. Holtz.

Figure 25.4. Cross-sectional view of a pistachio tree showing the necrosis of each year's ring of xylem due to Verticillium wilt. *Photo:* T. J. Michailides.

microsclerotial population occurs with nonhost as well as host plants and helps the fungus to persist, especially in the absence of susceptible species. A few such cortical colonies later establish xylem infections, which then proceed as described above.

Inoculum Production

Production of microsclerotia in infected plant parts has been studied in annual crops such as cotton and tomatoes but not in tree crops. In annuals, many thousands of microsclerotia are formed in leaves, stems, and roots as the plant withers and dies. When the dead plant is incorporated into the soil, the tissues decay, freeing the microsclerotia. It is not known whether the incorporation of diseased pistachio leaves and branches into the soil contributes to the inoculum in the soil. Microsclerotia are formed in infected pistachio roots, which may contribute to the inoculum level in the soil. However, in a pistachio field trial in California in *Verticillium*-infested soil and with a relatively weed-free planting, the number of microsclerotia per gram of soil dropped from 40 in the first year to 4 by the fifth year and stayed at 2 from the eighth to tenth years. Thus, although a pistachio planting may maintain a relatively low population of microsclerotia, pistachios in a weed-free field do not support a high inoculum concentration.

Conditions That Affect Disease

Verticillium wilt is favored by cool temperatures. Extended spring weather and mild summers often are accompanied by severe losses to this disease. The fungus apparently is eliminated from aboveground portions of trees during hot summer weather. Survival in root tissue and reinvasion of the upper stock is possible but has not been demonstrated. Instead, repeated attacks of wilt apparently represent new infections each year.

The disease is more prevalent in wet soils—presumably partly because they are cooler than dry soils—and irrigation influences the incidence of Verticillium wilt in some annual crops. In an experiment with pistachio in a drip-irrigated orchard (Epstein et al. 2004), the disease increased more down than across rows. Roots were more concentrated down rows and between trees than in the aisle between rows. This suggests that the denser root system down rows improved the chances for infection. Also, inoculum buildup would increase where more infected roots are present. An irrigation system that would discourage root growth in this area might alleviate damage from the disease. Drip irrigation systems buried at different depths fostered changes in root density. More roots developed deeper in the soil, but no decrease in the microsclerotial population or incidence of *Verticillium* was measured.

The population of *V. dahliae* is genetically diverse. Many differences exist in the pathogenicity and virulence of isolates of the fungus, and several strains and races have been identified. To date, isolates cultured from pistachios are also pathogenic to cotton and can be further separated according to their effects on tomato. The significance of this grouping is unclear, but no strong pattern of "wilt" and "decline" isolates has emerged. Nonetheless, variability within the fungus population is an important factor in breeding and screening plant material for resistance.

The nature of resistance to Verticillium wilt by pistachio is not well understood. The fungus is able to invade roots of both susceptible and resistant plants, but once inside the xylem, movement of the fungus up the vascular tissue is slower in resistant plants than in susceptible ones.

Control

The best defense against Verticillium wilt is the use of the resistant or tolerant rootstocks, Pioneer Gold *(P. integerrima)*, PGI, or UCBI (a *P. atlantica* × *P. integerrima* hybrid). UCBI may exhibit slightly less tolerance to Verticillium wilt when compared with PGI; however, UCBI shows more frost and salt tolerance. *P. atlantica* and *P. terebinthus* rootstocks are very susceptible and should be avoided where *Verticillium* is present. Another pistachio hybrid, PGII *(P. integerrima × P. atlantica)*, which has not been commercially available since the 1990s, had mixed susceptibility to *Verticillium*.

Criteria for site selection for an orchard should include a crop history and an assessment of the inoculum level in the soil. The microsclerotial population of the soil can be determined, and this service is available through private laboratories. Preplant or planting site fumigation can reduce soil inoculum levels but will not eradicate the fungus or prevent its re-establishment. Plastic mulches (soil solarization) on fallow soil for several weeks in summer can lower inoculum levels. Application of plastic mulches to planted pistachio orchards is limited in effectiveness and does not work well in shade. Chemical treatments to soil or by tree injection have not protected trees from Verticillium wilt.

Armillaria Root Rot

Armillaria root rot, also called oak root fungus, is caused by the soilborne fungus *Armillaria mellea* (Vahl.) Quel. *Armillaria mellea* is a pathogen of many cultivated and wild plant species, including some species of oak. Orchards planted where oak trees once grew often are severely damaged by the disease, hence the name oak root fungus. Plants growing in the soils of floodplains and along rivers and streams generally are more at risk than are those in other soils outside of such areas. Thus, the disease is most frequently encountered in orchards planted in the Sacramento Valley and along the eastern side of the San Joaquin Valley. Armillaria root rot limits cultivation of several tree crops in California but only occasionally affects pistachios.

Symptoms

The first symptoms usually begin on one side of the tree and include reduced shoot growth, early yellowing, and defoliation. Within a few years, the entire tree becomes thus affected and eventually dies. The disease progresses to adjacent trees, leaving areas of dead trees in its wake. Internal tissues of bark and outer wood of the upper roots and crown are discolored, and distinctive leathery, white, fan-shaped sheets of fungal mycelium, called plaques, develop between the bark and wood (fig. 25.5). With the onset of winter rains, clumps of honey-colored mushrooms may develop at the base of infected trees (fig. 25.6). Rhizomorphs, which are flat, rootlike structures composed of white mycelium covered by a dark, protective coating, grow along the surface of infected roots. Rhizomorphs are not always present or may be difficult to find and can be confused with small roots. The presence of the characteristic mycelial plaques or rhizomorphs on the diseased root systems is sufficient evidence to confirm a diagnosis of Armillaria root rot.

Disease Cycle

Armillaria mellea does not live freely in the soil. It is always associated with infected or dead woody tissues, with which it can survive for decades if not subjected to desiccation. Deep-rooted plants, such as oak, when infected by *Armillaria*, leave a legacy of many potential sources of inoculum distributed through the soil.

Infection of healthy plants occurs when roots grow near dead or diseased roots that harbor the fungus. Close proximity between infected and healthy roots is necessary for infection because rhizomorphs are infective only for short distances away from their food source. Rhizomorphs present on such material penetrate the nearby root and establish an infection. Both the formation of rhizomorphs and infection are favored by moist soil. The fungus also moves from infected to healthy individuals through root grafts. The pattern of disease spread in the orchard—enlarging circles of affected trees—reflects these methods of transmission.

Armillaria mellea is spread when infected roots are moved, which is caused by floods or human activity. Soil or water that does not contain infected plant parts does not carry the fungus. The spores produced by the mushrooms have not been implicated in natural spread of the disease. Often, Armillaria root rot appears to be naturally restricted

Figure 25.5. *Armillaria* mycelium plaque. *Photo:* J. K. Clark.

Figure 25.6. Armillaria mushrooms growing at the base of an infected tree. *Photo:* J. K. Clark.

to certain areas of an orchard and in such cases does not threaten an entire planting.

Control

Control is extremely difficult and usually not successful. Removal of infected roots followed by fumigation as preplant treatment can slow progress of the disease, but eradication is not likely. Nothing is available for use in established plantings. Other soil microorganisms, chiefly *Trichoderma* spp., are antagonistic to *A. mellea* and provide some measure of biological control. This biocontrol occurs only when the fungus has been weakened by sublethal doses of fumigant or by heat or drying. When it is healthy, *Armillaria* is not harmed by *Trichoderma*.

Resistant rootstocks would offer the best protection against oak root fungus disease. Field trials indicate that *P. terebinthus* and UCBI (a *P. atlantica* × *P. integerrima* hybrid) are tolerant, whereas *P. atlantica* and *P. integerrima* are susceptible. However, pathogenicity and virulence in the natural population of *A. mellea* range from weak to severe. Thus, disease response may vary with different combinations of host and pathogen. Breeding programs may incorporate some resistance in future stock.

Phytophthora Root and Crown Rot and Trunk Canker

Root and crown rot can devastate stone fruit, almond, and walnut orchards, but it is not common on pistachio trees in California. Trunk cankers occur occasionally in sprinkler-irrigated orchards. These diseases, caused by soilborne fungi, belong to the genus *Phytophthora*. Most of our knowledge of Phytophthora disease on pistachio is borrowed from research on other crops.

Symptoms

All parts of the root system may be attacked by *Phytophthora*. The term *root rot* refers to infection of small roots, including the fine feeder roots, which results in reduced uptake of water and nutrients. This leads to a weakening and decline of the tree. Aboveground symptoms of root rot usually develop slowly, and diseased trees show reduced growth, thinned canopy, and early defoliation for several years. By contrast, crown rot girdles the tree by destroying the crown and large roots that anchor the tree in the soil. Trees with crown rot typically die within 1 to 2 years after infection; they leaf out in spring, then collapse with the first hot weather. The bark and outer wood of infected roots and crowns are discolored, but the mycelium of *Phytophthora* in the infected tissues cannot be seen by the naked eye.

Phytophthora infections on the trunk or scaffolds of trees form cankers in the bark, and the tree gums profusely. If the trunk or scaffold is girdled by the infection, the portion of the tree above the site also dies.

Disease Cycle

Phytophthora species live independently in the soil and survive long dry periods as oospores, the sexual fruiting bodies of the fungus, or chlamydospores, which are resistant asexual spores. The infective units of *Phytophthora* are called zoospores. These are one-celled structures that are capable of swimming through free moisture (i.e., liquid water, present in or on the soil during or after rainfall or irrigation). In moist soil at moderate temperatures, the pathogen's mycelium can produce reproductive structures called sporangia. As many as 100 zoospores develop inside each sporangium, and when the soil is flooded, the zoospores are released. They are attracted to root exudates and swim to the root surface. There they attach themselves and invade the tissues to establish an infection. Free moisture is typically required for both the production and dissemination of the zoospores. Consequently, root and crown rots are associated with frequent or prolonged periods of high soil moisture. Alternating cycles of wet and dry soil are known to exacerbate Phytophthora diseases in some crops. Stressed plants often are more susceptible to infection.

Many *Phytophthora* spp. are present in canal and river water and can be introduced to orchards irrigated from these sources. To date, *Phytophthora* has not been found in well water.

Phytophthora parasitica has been cultured from rotted roots and crowns and usually is found in trunk and canker infections of pistachio trees. *Phytophthora cactorum* and *P. cryptogea* also have been isolated from trunk and scaffold cankers. In other crops, certain species of *Phytophthora* are associated with root rot, others with crown rot, and some may cause both. Species differ in their host range and temperature, moisture, and nutritional requirements. Much is known about the interaction of different *Phytophthora* species with various rootstocks and the environment for other crops but not for pistachio.

Control

Because prolonged periods of high soil moisture promote disease, water management is the basis for control of Phytophthora root and crown rot. Cultural practices such as planting on berms, shortening irrigation time, and improving water penetration help to lessen root and crown rot. Where root or crown rot is of concern, soil should not be wet longer than 24 hours at a time. The relative susceptibility of pistachio rootstocks to *Phytophthora* species is unknown. No chemicals are registered for control of this disease on pistachios. Sprinklers should be set to avoid water hitting the tree or causing puddling around the crown. This is especially important where irrigation water is taken from canals or rivers that may be contaminated with *Phytophthora*.

Seedling Blight

Seedling blight was first reported in California in 1995. In one nursery, over 10,000 seedlings (consisting of *Pistacia atlantica*, *P. integerrima*, and *P. atlantica* × *P. integerrima*) were lost. The disease has not been observed in orchard trees.

Symptoms

Leaves turn brown, wither, and cling to the shoots. Brown cortical lesions develop on roots, which become discolored. Seedlings are stunted, blighted, and die (fig. 25.7). The rotting tissues decompose and dry, forming a sunken area filled with dried plant parts mixed with fungus mycelium and sclerotia.

Causal Organism

Rhizoctonia solani Kühn is a collective species of several strains that are distinguished from each other by anastomosis between isolates of the same anastomosis group. Anastamosis (fusing of hyphae) does not occur between isolates in different anastomosis groups. Isolates causing seedling blight belong to anastomosis group AG-4. *Rhizoctonia solani* exists primarily as a sterile mycelium that is hyaline when young, but turns yellowish or light brown with age. Hyphal branches of the fungus tend to grow at right angles to the main hypha and have a cross wall near the slightly constricted junction. Under certain conditions, the fungus produces hyphal tufts that form brown to black sclerotia or short, barrel-shaped monilioid cells that function as chlamydospores. The teleomorph *Thanatephorus cucumeris* (A. B. Frank) Donk has not been produced by pistachio isolates.

Figure 25.7. Pistachio seedlings showing seedling blight. *Photo:* B. A. Holtz.

Disease Cycle and Epidemiology

Rhizoctonia solani overwinters in soil as mycelium, sclerotia, or chlamydospores. The fungus is present in most soils and, once established, remains indefinitely. Disease is usually more severe when soils are moderately wet and when plant growth is slow due to adverse environmental conditions. *Rhizoctonia solani* spreads with rain, irrigation or floodwater, equipment, and anything carrying contaminated soil or infected propagative materials. In greenhouses, where pistachio rootstocks are germinated and initially planted, infested benches can be a source of inoculum. Optimal temperature for infection is 59° to 65°F, but the disease is more common in warm areas (75° to 95°F). Seed transmission has not been demonstrated and is unlikely.

Control

Wet, poorly drained areas should be avoided, and seed should be planted in clean soil on raised beds under conditions that encourage rapid growth. Greenhouse benches should be cleaned, especially if a susceptible crop was grown previously. Drenching soil or planting mix with pentachloronitrobenzene helps reduce damping-off in seedbeds and greenhouses.

References

Browne, G. T., and M. A. Doster. 2002. Phytophthora diseases. In B. L. Teviotdale, T. J. Michailides, and J. W. Pscheidt, eds., Compendium of nut crop diseases in temperate zones. St. Paul, MN: American Phytopathological Society (APS) Press. 3–6.

Epstein, L., R. Beede, S. Kaur, and L. Ferguson. 2004. Rootstock effects on pistachio trees grown in *Verticillium dahliae*-infested soil. Phytopathology 94:388–395.

Holtz, B. A. 2002. Pistachio seedling blight. In B. L. Teviotdale, T. J. Michailides, and J. W. Pscheidt, eds., Compendium of nut crop diseases in temperate zones. St. Paul, MN: American Phytopathological Society (APS) Press. 70–71.

Holtz, B. A., T. J. Michailides, L. Ferguson, J. D. Hancock, and A. R. Weinhold. 1996. First report of *Rhizoctonia solani* (AG-4) on pistachio rootstock seedlings in California. Plant Disease 80(11): 1303.

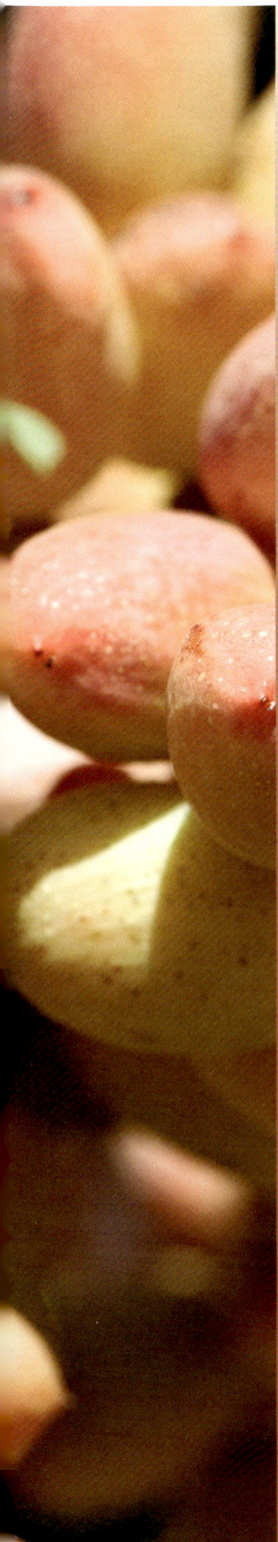

26 Foliar, Fruit, and Branch Diseases

Themis J. Michailides, David P. Morgan, and Mark A. Doster

For many years, Verticillium wilt, caused by *Verticillium dahliae* Kleb., was considered the only fungal disease of pistachio in California causing significant losses of mature pistachio trees. This disease continues to be a problem in older pistachio plantings where *Pistacia atlantica* was used as the rootstock. The development of resistant rootstocks (such as interspecific hybrids PGI, PGII, and UCBI) has reduced the importance of Verticillium wilt of pistachio in California. In the spring of 1983, after a period of heavy and prolonged rains and cool weather, a disease characterized by blighted blossoms and shoots was first observed and described as Botrytis blossom and shoot blight, caused by *Botrytis cinerea*. In the summer of 1984, another disease, characterized by blighted panicles, shoots, and buds, was first observed in an orchard (one of the oldest commercial pistachio orchards in California) near Durham in Butte County and later in several other orchards in northern California and the Chico Tree Improvement Center (converted now to a U.S. Forest Service Station). In subsequent years, the same disease was observed sporadically in orchards in the southern part of the Sacramento Valley and later in central and southern California. Repeated isolations from blighted panicles, shoots, and buds revealed the asexual stage of the ascomycete *Botryosphaeria dothidea* (anamorph *Fusicoccum aesculi*). In 1985, a fourth fungal disease, Alternaria late blight, caused by *Alternaria alternata* and other *Alternaria* spp., was first reported. By the late 1990s, Alternaria blight was considered third in importance, after Botryosphaeria panicle and shoot blight and Verticillium wilt, in causing significant losses. Some of the above-mentioned diseases have also been reported in countries other than the United States. Summaries of the most important diseases with diagnostic symptoms and signs of each disease and of sporadic diseases recorded on pistachio in California are given in tables 26.1 and 26.2. An index of pistachio diseases and associated fungi reported on pistachio in the United States is described in table 26.3. Verticillium wilt and root diseases of pistachio are discussed in chapter 25, "Soilborne Diseases."

Infectious Disorders

Alternaria Late Blight

The first record of Alternaria blight of pistachio is from Egypt in 1974. The disease has since been reported in California, Italy, and Australia. When serious, it causes staining of the shell and severe early defoliation, which reduces yield and weakens the tree.

Symptoms

Symptoms on leaves appear in midsummer and begin as dark brown to black, angular or circular lesions, 3 to 7 mm in diameter, that develop anywhere on the leaf blade. Lesions enlarge or coalesce to form tan blotches 2.5 to 3 cm in diameter. Later in the season, dense sporulation, especially in the lesion center, can turn the lesion black. Multiple infections on leaf blades can cause leaf blight and severe early defoliation. Infections on immature fruit are small reddish spots with a black center about 1 mm in diameter and are associated with lenticels. On mature fruit, hull lesions are black, 1 to 5 mm in diameter, sometimes with concentric zones, and often surrounded by a reddish-purple margin. Lesions along the broken edges of the ruptured hulls of fruit with early splits or with hull cracking intensify shell staining. Occasionally the fungus produces spores on the inner and outer shell surfaces and the inner surface of the hull (figs. 26.1 to 26.6).

Table 26.1. The most important diseases of pistachios *(Pistacia vera)* in California

Disease	Fungal pathogen	Diagnosis: Main symptoms and signs
panicle and shoot blight	*Botryosphaeria dothidea* (Anamorph: *Fusicoccum* aesculi) and other species of Botryosphaeriaceae	panicle, leaf, shoot, and bud blight; black angular or circular lesions on green fruit and immature leaves; large black lesions on fruit, large brown lesions on mature leaves with light beige margins. Pycnidia on blighted shoots, bud scales, base of petiole, fruit, and on leaf lesions
Verticillium wilt	*Verticillium dahliae*	yellowing foliage and shoot; foliage wilting of certain portions of trees; defoliation and sudden death of trees; after sectioning, shoots showing brown to black discolored xylem tissues (very diagnostic for trees infected by *V. dahliae*)
Alternaria blight	*Alternaria alternata, A. tenuissima*, and *A. arborescens*	small black spots on leaves or necrotic large areas with sporulation in the center; leaf blight and early defoliation; black lesions surrounded by red-colored halo on epicarp of fruit; fruit blight; shell staining
blossom and shoot blight	*Botryotinia fuckeliana* (Anamorph; *Botrytis cinerea*)	blossom and shoot blight; cankers initiated from male inflorescences common on 02-16 and 02-18 cultivars; blight and flagging of tender shoots (shepherd's hook) of both male and female trees; sporulation on the base of infected shoots

Causal organism

Alternaria leaf blight is caused by several closely related species of *Alternaria. Alternaria alternata* (Fr.:Fr.) Keissl is the most common species causing the disease, although *A. arborescens* E. Simmons and *A. tennuissima* (Kunze:Fr.) have also been isolated from pistachio and proved to be pathogenic. Sporulating cultures are greenish black or grayish in color. The optimal temperature for growth and sporulation is 81° to 86°F.

Disease cycle and epidemiology

High relative humidity and dew in the orchard favor the disease. The disease starts on leaves of the exposed shoots in the upper canopy and eventually moves into the lower parts of the tree. Blight is generally worse in late summer because longer periods of dew and higher relative humidity above 95 percent are recorded in August and September than in June and July. At the same time, aging leaves and fruit become more susceptible. Leaves stressed by crop load appear to be susceptible to *Alternaria* because infections occur earlier and are more severe on leaves of fruit-bearing shoots than on those of nonbearing shoots. Furthermore, fruit-bearing shoots defoliate sooner than those nonbearing. Subsequently, Alternaria blight is generally more severe in on-crop years than in off-crop years. Severe Alternaria blight is most likely in orchards with higher humidity, such as those located close to rivers, in lower areas, those with cover crops, or those irrigated by sprinklers or flooding versus drip irrigation systems. High-density orchards under any irrigation system can also have higher humidity and subsequently higher levels of disease.

The pathogens are common in nature and survive and develop on crop debris and senescing leaves and weeds. The necrotic areas of leaves with Peters scorch (described in the section "Peters Scorch" later in this chapter) (fig. 26.7), fruit damaged by sunburn or birds, and early-split fruit are often colonized by *Alternaria* spp., probably including the pathogens causing Alternaria blight. Spores are present on developing fruit and leaves from early April until harvest. In early August, there is a rapid increase in inoculum on leaves and fruit. The number of conidia (spores) on leaves and fruit remains high through September and October. The period of sudden rise in spore counts is the critical period for disease development and management.

Control

Alternaria blight of pistachio, like other Alternaria blight diseases, is difficult to manage and requires a combination of fungicide treatments and proper cultural and irrigation practices. All commercial cultivars grown in California are susceptible. QoI (strobilurins – FRAC Group 11) fungicides (e.g., azoxystrobin, pyraclostrobin, and trifloxystrobin, which are registered on pistachio in California) applied two or three times from June to early August can reduce the disease if *Alternaria* strains resistant to QoI fungicides are not present. Resistance to azoxystrobin and cross-resistance to the other registered QoIs was reported in several orchards with Alternaria blight in 2004. Premixtures of carboximides (FRAC Group 7) with the above QoIs are very effective against Alternaria late blight, although resistance to QoIs and carboximides has been recorded in recent years. In addition, chlorothalonil, tebuconazole, and other triazole (dicarboximides

Table 26.2. Sporadic diseases recorded on pistachios in California

Disease	Fungal pathogen	Main symptoms and signs	Year recorded
Phomopsis blight	*Phomopsis* spp.	shoot and leaf wilting (flagging); fruit blight	1985
Sclerotinia blight	*Sclerotinia sclerotiorum*	shoot and leaf wilting and blight (flagging)	1985
root and crown rots	*Phytophthora* spp.	black lesions on the crown of scion but not extending into the rootstocks	1989
	Armillaria mellea	decay of roots; white mycelial plaques and distinct, black rhizomorphs	1990
hull and kernel molds	*Aspergillus niger*	soft, water-soaked hull covered with black sporulation on the surface touching the shell; bright yellow upon drying	1995
	Penicillium expansum and other *Penicillium* spp.	soft water-soaked hull covered with bluish, greenish powdery sporulation	1995
anthracnose hull rot	*Colletotrichum gloeosporioides* and *C. acutatum*	black and pinkish well-developed lesions on the hull; entire surface of hull covered with bright pink and powdery sporulation	1997, 2016
shoot blight	*Botryosphaeria rhodina*	typical *Botryosphaeria* shoot blight symptoms; affected shoots wilt and become dark brown to almost black	2000
powdery mildew	*Oidium* sp.	netted white lesions on fruit that eventually scar and misshape fruit; lesions on young shoots, petioles, and leaf blades initially white but then leaving dark netted scars.	2001

Note: These diseases were recorded throughout California, 1985–2008.

– FRAC Group 9) fungicides can provide some control of the disease if applied several times, but these fungicides are not as effective as the strobilurins or mixtures of carboximides and strobilurins. For pistachios in California, the best timing for spraying is the end of June or early July. For information on past and current fungicides registered for pistachio, their efficacy, and timing against Alternaria blight, visit the "Fungicide and Bactericides Efficacy and Timing for Various Crops Tables" at the UC IPM website, ipm.ucanr.edu. Curtailing irrigation during the critical period in early August lessens disease symptoms, and irrigation systems that allow the ground surface to dry quickly are recommended. Orchards with no cover crop or those irrigated by drip, microsprinkler, or subsurface systems have less severe Alternaria blight than orchards with cover crops or flood or sprinkler irrigation. Delayed harvest should be avoided to minimize fruit infection and the consequent shell staining. A Disease Severity Value (DSV) model, based on wetness, duration, and temperature, is being developed to help growers time the fungicide application and probably reduce the number of applications.

Anthracnose Hull Rot

Although species of *Colletotrichum* have frequently been isolated from washings of pistachio fruit and leaves in California, anthracnose symptoms on fruit (fig. 26.8) caused by *C. gloeosporioides* (Penz.) Penz.

Figure 26.1. Alternaria late blight symptoms on fruit in the field. *Photo:* T. J. Michailides.

Figure 26.2. Early stages of Alternaria late blight, showing small lesions. *Photo:* T. J. Michailides.

Table 26.3. Index of fungi recorded on pistachios (*Pistacia vera* L.) in the United States

Disease	Causal agent	State reported
Armillaria root rot	*Armillaria mellea*	California
blossom and shoot blight	*Botrytis cinerea*	California
leaf and fruit spot	*Alternaria alternata, A. tenuissima, A. arborescens*	California
leaf necrosis; root rot	*Phytophthora nicotianae* var. *parasitica*	California
leaf spot	*Phyllosticta lentisci*	Texas
leaf spot	*Hyalotia pistacina*	Maryland
panicle and shoot blight	*Botryosphaeria dothidea*	Arizona, California
Phomopsis blight	*Phomopsis* spp.	California
Phytophthora root rot	*Phytophthora* spp., *Pythium* spp.	California
root rot	*Phymatotrichopsis omnivorous*	California, Texas
Sclerotinia shoot blight	*Sclerotinia sclerotiorum*	California
Septoria leaf blight	*Septoria pistaciarum, Septoria pistaciae*	Arizona, California
sapwood rot	*Schizophyllum commune*	California
thread blight	*Pellicularia koleroga*	Texas
thread blight	*Rhizoctonia solani*	California, Florida
various hull and kernel decays	*Aspergillus* spp., *Aspergillus flavus, A. parasiticus, A. niger, Cladosporium* spp., *Fusarium oxysporum, Fusarium* spp., *Penicillium* spp., *Eurotium* spp., *Phomopsis, Pestalotiopsis* sp.	California
Verticillium wilt	*Verticillium dahliae*	California
wood rots	*Pleurotus ostreatus*	California

Sources: Farr et al. 1989; the USDA Fungal Database website, http://nt.ars-grin.gov/fungaldatabases/.

Figure 26.3. Late stage of Alternaria late blight with large leaf lesions and sporulation. *Photo:* T. J. Michailides.

Figure 26.4. Large marginal lesions of Alternaria late blight on leaves. *Photo:* T. J. Michailides.

& Sacc. in Penz. (sexual stage *Glomerella cingulata* (Stoneman) Spauld. & H. Schrenk) have only been found once in a commercial pistachio orchard in Tulare County following a wet spring in 1998. Koch's postulates have been completed on pistachios in California, and this is the first report of *C. gloeosporioides* (fig. 26.9) causing hull rot in California pistachio. Another species of *Colletotrichum, C. acutatum* J. H. Simmonds, has been found causing a leaf spot and hull rot of pistachio in Australia. This species causes anthracnose of almonds and strawberries in California.

Symptoms

Black or pinkish, well-developed lesions were observed in the hulls. In some fruit, the entire surface of the hull was covered with pink and powdery sporulation (see fig. 26.8). Lesions on leaves were not observed. Not much is known about the life cycle of this disease on pistachio, and its sporadic occurrence does not necessitate any specific treatment.

Control

No specific control measures are recommended. However, fungicides used for the control of Alternaria blight or Botryosphaeria panicle and shoot blight should also reduce Anthracnose hull rot.

Aspergillus Fruit Blight

This disease is sporadic, but in very hot, dry years it can cause some damage, with occasionally as much as 2 percent loss. The disease develops over the summer on maturing fruit and is associated with fruit injuries that provide sufficient conditions to allow fungal growth.

Symptoms

The early symptom of Aspergillus fruit blight is a water-soaked discoloration of the hull, which becomes soft and light beige. At harvest, infected fruit dry, and their hulls become papery, densely wrinkled, and bright yellow, which also stains the shell bright yellow. All tissues under the epidermis of the epicarp are converted to black powdery spore masses of the pathogen. After washing off the spores, bright brown or yellow stains are seen on the shell surface. This staining differs from the lighter yellow staining caused by delayed harvest and from the uniform brown staining of hull infections caused by *A. alternata* and other species of *Alternaria* (fig. 26.10).

Causal organism

The pathogen is *Aspergillus niger* van Tiegh or other *Aspergillus* spp. in section *Nigri. Aspergillus niger* grows quickly on all substrates. On acidified potato dextrose agar, colonies are initially white with sometimes a yellow felt (i.e., fungal mycelium), then develop a compact layer of dark brown to black conidiophores; colonies are cream to yellow on the reverse side. Large cream to buff sclerotia—globose or oval—are occasionally produced. Optimal temperature for growth is 77° to 95°F.

Disease cycle and epidemiology

The fungus attacks early-split fruit through the ruptured hull or normal fruit with insect, bird, or mechanical damage. Early-split nuts are the nuts that have the hull split along the shell suture where the shell has split (fig. 26.11). The incidence of *A. niger* in early-split fruit is higher (up to 30%) than that in injured normal fruit, and it is about three times higher in early-split fruit that is older than in those formed during the 2 weeks preceding harvest. Sources of inoculum include sporulation of *A. niger* inside early-split fruit, on male inflorescences wetted by rains or irrigation water, or pistachio residues or seed of weeds that have dropped to the ground. Under favorable conditions, the fungus can also grow on leaves and shoots on the orchard floor.

Figure 26.5. Alternaria late blight in the field. *Photo:* T. J. Michailides.

Figure 26.6. Alternaria late blight lesions on fruit. *Photo:* T. J. Michailides.

Figure 26.7. Lesions with necrotic areas of Peters scorch colonized by *Alternaria* spp. *Photo:* T. J. Michailides.

Figure 26.8. Fruit covered with acervuli containing conidia of *Colletotrichum gloeosporioides*. *Photo:* T. J. Michailides.

Figure 26.9. Culture of *Colletotrichum gloeosporioides* from pistachio. *Photo:* T. J. Michailides.

Figure 26.10. Shells of pistachio stained from infections by *Alternaria* spp. *Photo:* T. J. Michailides.

Figure 26.11. Pistachio cluster bearing early-split fruit. *Photo:* T. J. Michailides.

Control

No specific control measures are recommended. However, fungicides used for control of Alternaria blight and Botryosphaeria blight may also reduce Aspergillus fruit blight.

Aflatoxin Contamination

Aflatoxins are naturally occurring toxins produced by specific molds growing on certain commodities. Although the highest risk for aflatoxin contamination is in corn, peanuts, and cottonseed, other crops (including tree nuts) are also sometimes contaminated. Aflatoxin contamination of pistachio nuts grown in California is typically low, with incidences of about 1 in 10,000 nuts. Nevertheless, the tolerances for aflatoxin contamination established by governments are usually very low, typically 20 parts per billion (ppb) or lower.

Symptoms and causal pathogens

Aflatoxins are odorless and colorless chemicals, so they cannot be observed directly in pistachio nuts. Nonetheless, aflatoxin contamination develops during the growth of aflatoxin-producing fungi, and signs of these fungi can be observed. In California, aflatoxins can be produced by three closely related species of fungi: *Aspergillus flavus* Link ex Fries., *A. parasiticus* Spear, and *A. nomius* Kurtzman & Hesseltine. Typically, both *A. flavus* and *A. parasiticus* are present in commercial pistachio orchards, with *A. flavus* being more common than *A. parasiticus*, while *A. nomius* is only very rarely observed. All three species produce abundant yellowish-green spores, resulting in greenish-appearing colonies on the surface of kernels and on other substrates (fig. 26.12). In addition to using the presence of these fungi to identify aflatoxin-contaminated nuts, the appearance of a nut can be used to indicate the likelihood of aflatoxin contamination. For example, nuts with dark brown discoloration on the shell exterior are more likely to be contaminated with aflatoxin, even though the aflatoxin-producing fungi do not cause shell discoloration. Thus, associated symptoms such as extensive dark discoloration of pistachio shells can be indicative of aflatoxin-contaminated nuts; however, some contaminated nuts have little discoloration, including nuts that have a characteristic discoloration restricted to along the shell suture (see fig. 26.58). Also, small nuts are more likely to be contaminated than large nuts.

Disease cycle and epidemiology

The aflatoxin-producing fungi can grow on a wide variety of substrates but cannot grow well on living plant tissue. Therefore, besides decaying pistachio kernels, these fungi are usually found in pistachio orchards growing on plant debris on the orchard floor (fig. 26.13). The debris can be from weeds, cover crops, or the pistachio trees themselves. Examples of debris colonized by aflatoxin-producing fungi are dead leaves of weeds and pistachio, male inflorescences, and fallen immature pistachio nuts. The aflatoxin-producing fungi grow very well at 75° to 105°F but not at less than 55°F. Therefore, they are most active during the summer and have little or no growth during the winter. These fungi typically produce abundant spores (conidia), which are spread by wind and probably by certain cultural practices such as disking or mowing.

As the pistachio nuts mature in the orchard, the kernels become susceptible to decay by aflatoxin-producing fungi. The intact hull, however, serves as an effective barrier limiting the decay of the kernel. Nonetheless, the hull of the pistachio nut can become ruptured, which exposes the kernel to invasion by fungi and insects. An important example of this is early-split nuts. Early-split nuts are the main source of aflatoxin contamination in pistachio nuts. Early-split nuts typically account for about 2 to 3 percent of the nuts on the tree, although some orchards may have over 10 percent of this type of nuts. Early-split nuts form throughout summer, beginning in mid-July and continuing to harvest.

Navel orangeworm (*Amyelois transitella* L.) (NOW) may have an important role in the aflatoxin contamination of pistachio nuts. Most of the aflatoxin found in pistachio nuts in California is in nuts damaged by navel orangeworm larvae. The reason for the association of navel orangeworm with aflatoxin contamination is not understood at this time. Possibly the navel orangeworm larva removes parts of the testa (seed coat), thereby exposing the kernel and enabling the aflatoxin-producing fungi to grow better. In addition, recent research has shown that both larvae and adult NOW are heavily contaminated with *Aspergillus* fungi, including the aflatoxigenic fungi of *A. flavus* and *A. parasiticus* (figs. 26.14 and 26.15). Preliminary results showed that NOW moths emerged from mummies dropped onto the ground had more *A. flavus* than those emerged from mummies that overwintered in the tree.

Aflatoxin contamination is considered to occur preharvest and not postharvest. The toxin, however, continues to be produced during transport from the field to the processor. The heat of the drying process probably kills the aflatoxin-producing fungi in the nuts but has little effect on the aflatoxin already present. Even if the heat does not kill these fungi, they are not able to grow and produce aflatoxin at the low moisture levels of the dried nuts. And as long as the dried nuts are stored at low moisture (e.g., < 8%), these fungi are not able to grow and produce more aflatoxin.

Control

The three main ways to reduce aflatoxin contamination are to control navel orangeworm, minimize early-split nut formation, and avoid harvesting late. Ways to control navel orangeworm are presented elsewhere in this manual (see chapter 18, "Navel Orangeworm and Obliquebanded Leafroller"). The best way to minimize the number of early-split nuts seems to be to ensure that the trees receive sufficient water in late spring. Studies have shown that drought stress in May substantially increases the number of early-split nuts at harvest. Trees on the rootstock *P. atlantica* tend to have more, whereas trees on UCBI tend to have the fewest early-split nuts. Delaying harvest allows more time for aflatoxin-producing fungi to produce more aflatoxin in already-colonized nuts, besides infecting additional nuts (fig. 26.16A). Also, delaying harvest can result in substantial increases in navel orangeworm damage, an important factor in aflatoxin contamination (fig. 26.16B). Thus, late harvests can result in relatively high levels of aflatoxin contamination. Use of atoxigenic strains of *A. flavus* to manage aflatoxin contamination in pistachio have been shown to reduce the incidence of aflatoxin contamination. A registration is available for use of the atoxigenic AF36 strain of *A. flavus* in commercial orchards in California. The AF36 strain is the most common (approximately 5.5%) naturally occurring strain of *A. flavus* in California pistachio orchards, while the incidence of other strains ranges from 0.2 to 1.5 percent.

Figure 26.12. Pistachio fruit infected by *Aspergillus flavus*. *Photo:* T. J. Michailides.

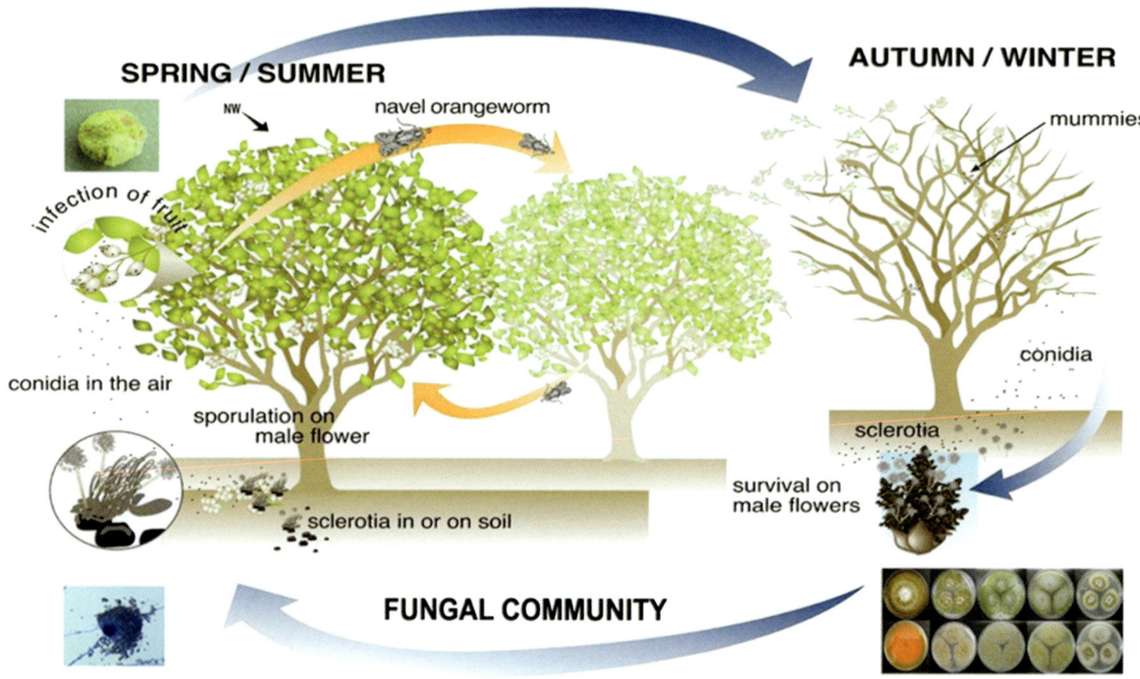

Figure 26.13. Life cycle of *Aspergillus flavus* in a pistachio orchard. *Illustration:* G. Conville.

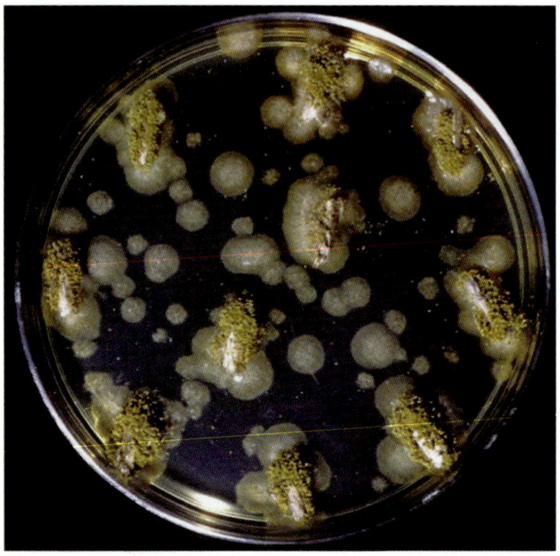

Figure 26.14. Navel orangeworm moths on a plate with *Aspergillus flavus. Photo:* T. J. Michailides.

Figure 26.15. Navel orangeworm moth with *Aspergillus flavus* sporulating on the carcass. *Photo:* T. J. Michailides.

Botryosphaeria Panicle and Shoot Blight

Botryosphaeria blights of woody plants have been known since the early 1900s. Species in the genus *Botryosphaeria* can attack more than 50 plant species, representing 34 genera and 20 families that include almond, avocado, walnut, giant sequoia, coastal redwood, incense cedar, and willow. Botryosphaeria panicle and shoot blight of pistachio was first described in California in 1984 and developed to an epidemic in 1998 that threatened the California pistachio industry. The disease, however, can be effectively managed due to new effective fungicides, selective pruning, and irrigation management. The disease also occurs in Greece (reported there as Camarosporium blight), Italy, and South Africa (although in South Africa two species of *Botryosphaeria* have been reported as causing stem cankers in *Pistacia* spp.). Camarosporium blight also causes panicle and shoot blight, and cankers are formed with Botryosphaeria panicle and shoot blight, too.

Symptoms

Symptoms appear in mid- to late spring as black, circular spots 1 to 2 mm in diameter on shoots, rachises, and leaves. Shoots originating from contaminated or partially infected buds develop black lesions at the base. Lenticel infections in 1-year-old shoots are circular, black lesions 5 to 7 mm in diameter, and they usually do not enlarge in size. In mid-May, leaves on infected shoots wither within 3 to 5 days, and brown blighted shoots and leaves soon are distinct among the healthy, dark green foliage (fig. 26.17). Lesions on petioles kill individual leaflets, and entire leaves die (fig. 26.18) when the petiole is infected and drop beginning as early as July. Early leaflet infections are somewhat elongated spots that aggregate along the midrib. Blade infections appear next and are visible as small, circular black spots. Blade and midrib lesions enlarge as the summer progresses, and they become irregular brown spots up to 15 mm in diameter that are surrounded by a diffuse, slightly chlorotic halo (fig. 26.19). The lesions and halos often coalesce, creating blotches of various shapes that ultimately dry to a tan color. The leaves die and fall from the tree. Most defoliation happens in late summer and can be very severe (fig. 26.20).

Rachis infections occur at the base or at branching points, and the rachis collapses partially or entirely, depending on the location of the lesion and time of infection (fig. 26.21). Initial fruit infections appear in midsummer and are pin-size, circular black spots, usually associated with lenticels. There may be a hundred or more such lesions, most of which develop no further, on each fruit. On one to several fruit in a panicle, a lesion enlarges, the fruit turns black, and the infection moves into the peduncle, rachis, and eventually the shoot on which the fruit cluster is borne, killing part of or the entire cluster (figs. 26.22 to 26.24). Most of the blighted panicle is light tan, the consequence of girdling; only infected fruit are black, and these turn a silver gray in fall, when the fungus produces pycnidia under the fruit epidermis (fig. 26.25). The fruitless dead black rachises, which remain on the tree for 3 or 4 years, are a good diagnostic symptom.

Infections on shoots (or infections that move into shoots from rachises or petioles) create cankers (fig. 26.26). Sunken cankers 1 to 10 cm in length develop around infected leaf and bud scars and do not enlarge in subsequent years. The pathogen sporadically causes cankers that are up to 30 cm in length and are covered with dark exudates on trunks or extending into the main scaffolds. The infection leads to cankers that usually do not kill trees. Gumming may or may not accompany *B. dothidea* infections and is not a diagnostic characteristic. Some cankers may be associated with pruning wounds, but their size is limited.

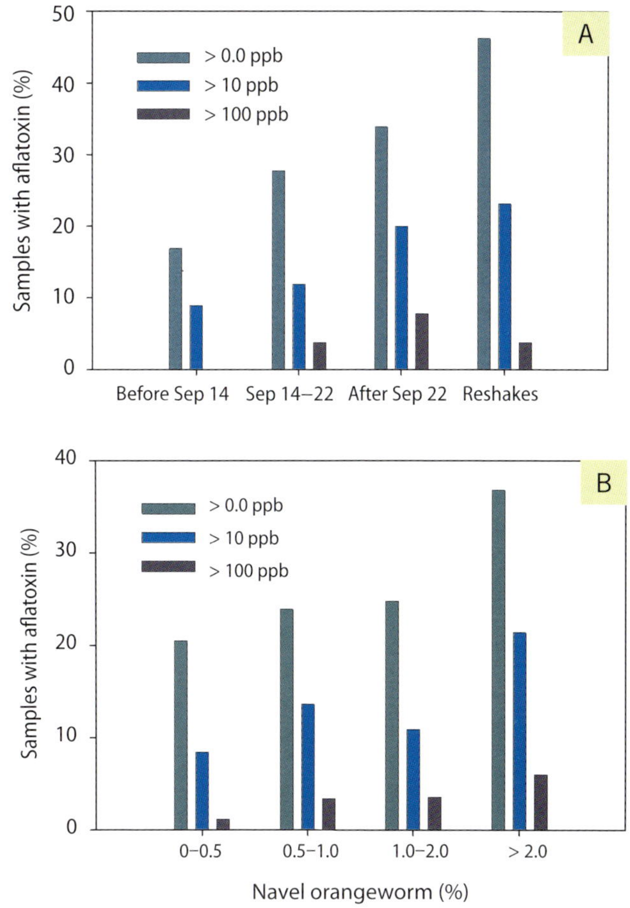

Figure 26.16. Relationship between harvest date and incidence and amount of aflatoxin contamination in pistachios (A) and relationship between navel orangeworm infestation and aflatoxin levels in pistachios (B).

Causal organisms

Fusicoccum aesculi (asexual stage of the ascomycete *Botryosphaeria dothidea* (Moug.:Fr.) Ces & DeNot.) produces black, asymmetrical pycnidia that are solitary or arranged in groups of five to eight, each with an apical ostiole through which the conidia extrude in a gelatinous matrix. Another pathogen that causes shoot blight is *Lasiodiplodia theobromae*, the pycnidial stage of *Botryosphaeria rhodina*, but it is not as common as the anamorph of *B. dothidea*. In cross section, the contents of the pycnidia are shiny white and those of *L. theobromae* shiny black. Conidia of *Botryosphaeria* are hyaline, nonseptate, fusiform, and measure 15–29 × 5–8 μm (fig. 26.27). However, conidia of *L. theobromae* are ovoid, dark brown, two celled, and have a restriction along the equatorial septum. Isolates of *B. dothidea* from pistachio grow well on regular and acidified potato dextrose agar at 68° to 96°F, with an optimum of 81° to 86°F. Initially, colonies are white, later changing to light gray,

Figure 26.17. Shoot blight caused by *Botryosphaeria dothidea. Photo:* T. J. Michailides.

Figure 26.18. Leaf blight caused by *Botryosphaeria dothidea. Photo:* T. J. Michailides.

Figure 26.19. Leaf lesions caused by *B. dothidea. Photo:* T. J. Michailides.

Figure 26.20. Severe panicle and shoot blight and defoliation of trees. *Photo:* T. J. Michailides.

then almost black (fig. 26.28). Many isolates do not produce pycnidia in culture, and some produce reddish pigments. Two morphologically distinct strains (a gray one and a black one) have been recorded since 1994 from infected pistachio fruit, but the importance of these two strains is not known. Characterization of Botryosphaeriaceae isolates in 2014 by means of molecular analysis showed that eight distinct species exist, among which is *B. dothidea*. This study showed that the most widespread species is *Neofusicoccum mediterraneum*, which produces conidia very similar to those produced by *B. dothidea*. Only the pycnidial stage(s) of these pathogens have been found on pistachio; however, pseudothecia, the sexual (ascosporic) stage of *B. dothidea*, was found on other hosts such as almond, walnut, avocado, olive, pyracantha, and California blackberry *(Rubus ursinus)*. Many conidial isolates from various hosts and ascosporic isolates from blackberry routinely infected pistachio in inoculation studies.

Disease cycle and epidemiology

Conidia released from pycnidia present on the previous year's blighted buds, shoots, petioles, rachises, and fruit cause the primary infections in spring and summer (figs. 26.29 to 26.31). New pycnidia in current-season infections develop in late summer and fall and contribute inoculum for late-season infections. In addition, pycnidia in old *B. dothidea* cankers produce conidia for at least 6 years. The pathogen also colonizes deadwood, as a saprophyte, including old cankers caused by *Botrytis cinerea* and shoots and branches killed by Verticillium wilt or *Phytophthora* spp. Buds may be infected as soon as they develop (fig. 26.32).

Conidia are spread mainly by rain and sprinkler irrigation and probably by insects and birds (fig. 26.33). Fall and winter rains spread conidia to leaf and bud scars and buds that may become infected and develop cankers. Some buds may be killed, some partially damaged, while others may remain healthy but are contaminated with conidia of the pathogen. Rain anytime during the growing season moves inoculum to growing tissues (primary and secondary infections). Primary infections occur from overwintered conidial inoculum, while secondary infections are caused by conidia produced during the growing season. Depending on conditions in the fall (rains and warm temperatures), a third cycle of secondary infections can occur during the growing season before winter. Because the pathogen grows best at relatively high temperatures, the disease becomes severe in late spring to summer. Some spring infections (particularly on leaves and immature fruit) apparently remain latent until summer. The factors that trigger their development are unknown, although nutritional and drought stresses have been reported to predispose trees to infection by this group of fungi (Botryosphaeriaceae).

Airborne ascospores have been trapped in low numbers in pistachio orchards, even though pseudothecia (structures producing ascospore inoculum) have not been observed. Pseudothecia were found on trunks and branches of almond and walnut, avocado leaves, dead olive shoots, cankers on pyracantha, and blighted blackberry canes. Pycnidial isolates from these and several other hosts, as well as ascosporic isolates from blackberry-induced blight on pistachio after inoculation, suggests that other hosts can also serve as sources of inoculum for Botryosphaeria panicle and shoot blight of pistachio. Ascospores could function in long-distance dissemination because they become airborne, but pycnidia are the major sources of inoculum for the destructive epidemics in orchards.

Control

Botryosphaeria panicle and shoot blight requires extensive efforts for its control, especially if allowed to increase over several years. The best approach employs fungicides, pruning, and irrigation management. Two or three applications of QoI fungicides (azoxystrobin, pyraclostrobin, or trifloxystrobin) or mixtures of carboximides and QoI fungicides during summer will control the disease, although in orchards with high pressure, sprays should start at bloom and continue during summer using rotational strategies with fungicides that each have a different mode of action. Sprays before or 2 to 3 days after rains are the most effective in managing the disease. Available fungicides for managing Botryosphaeria panicle and shoot blight can be found in "Fungicides, Bactericides, and Biologicals for Deciduous Tree Fruit, Nut, Strawberry, and Vine Crops" at the UC IPM website, ucipm.ucanr.edu. Pruning of infected plant parts 2 inches below the blighted margins in late summer and fall reduces disease the following year. In orchards irrigated by sprinklers, lowering the sprinkler angle so that water does not reach the tree canopy or shortening the duration of irrigation from 48 to 24 hours is recommended. Irrigating only during daytime for 12 hours each on 2 consecutive days is also effective. Two monitoring methods have been developed: BUDMON is based on the percentage of buds harboring spores during the dormant season, whereas ONFIT (Overnight Freezing Incubation Technique) detects latent infections of the pathogen in immature fruit in June and July. These

Figure 26.21. Infection of rachis at the base by *B. dothidea. Photo:* T. J. Michailides.

Figure 26.22. Botryosphaeria panicle blight disease, preventing the filling of the nuts. *Photo:* T. J. Michailides.

Figure 26.23. Panicle blight later in season, resulting in major yield losses. *Photo:* T. J. Michailides.

Figure 26.24. Characteristic silver-colored nuts bearing pycnidia of *B. dothidea* in blighted clusters. *Photo:* T. J. Michailides.

Figure 26.25. Pistachio fruit-bearing pycnidia of *B. dothidea* (silver color). *Photo:* T. J. Michailides.

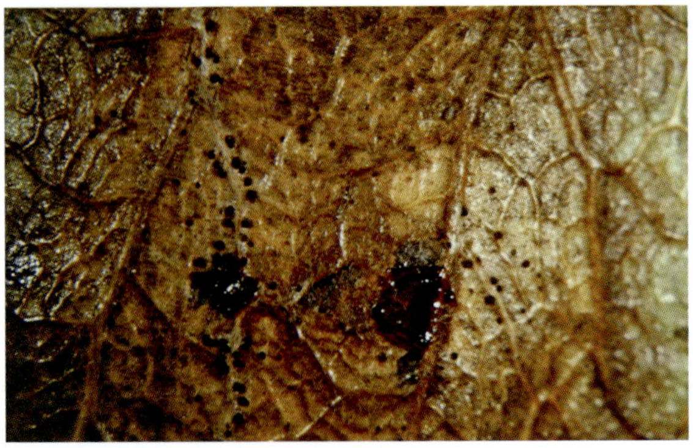

Figure 26.29. Pycnidia of *B. dothidea* in the center of a leaf lesion. *Photo:* T. J. Michailides.

Figure 26.26. A canker initiated by invasion of *B. dothidea* from an infected petiole. *Photo:* T. J. Michailides.

Figure 26.30. Pycnidia of *B. dothidea* on an infected shoot of pistachio. *Photo:* T. J. Michailides.

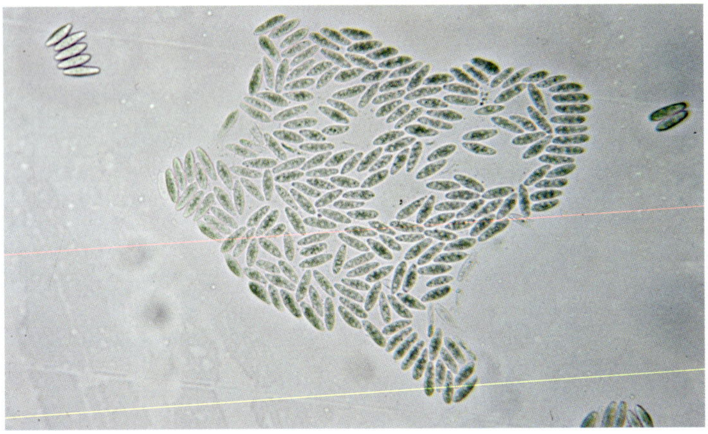

Figure 26.27. Hyaline, one-celled, spindle-shaped pycnidiospores of *B. dothidea* from pistachio. *Photo:* T. J. Michailides.

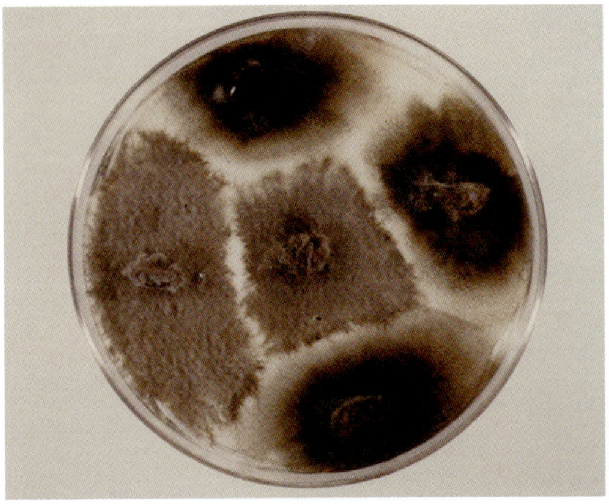

Figure 26.28. Petri plate with two colonies of *B. dothidea* (light gray color) in contrast to three black colonies of *Alternaria* spp. *Photo:* T. J. Michailides.

methods can predict disease risk and aid in disease management decisions. Because the pathogen infests buds, grafting material should be gathered only from trees in healthy orchards.

Botrytis Blossom and Shoot Blight

Botrytis blossom and shoot blight of pistachio was first reported in California in 1984. The disease is sporadic but can cause significant damage in cool, wet springs. Even though direct loss to Botrytis blight usually is minor, the colonization of old Botrytis cankers by *Botryosphaeria dothidea*, a more serious pathogen of pistachio, elevates the importance of Botrytis blossom and shoot blight. Furthermore, rains in May and June can cause partial cluster blight that may be more detrimental in reducing yields in the field.

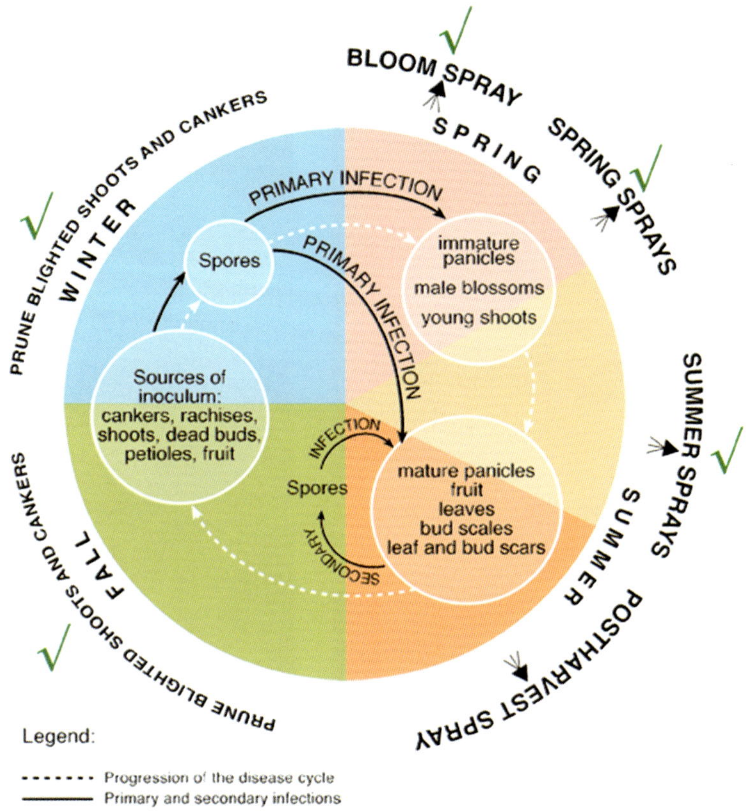

Figure 26.31. Cycle of Botryosphaeria panicle and shoot blight on pistachio. *Illustration:* G. Conville.

Symptoms

The disease occurs in spring and begins with the wilting and death of tender shoots on male and female trees (fig. 26.34). The dead shoots, with leaves attached, are bent in a characteristic shepherd's hook shape. The bases of blighted shoots become brown and swollen, and mycelium is sometimes present in the shoot pith. Infections of inflorescences on male pollinator trees of the cultivar Peters cause blossom blight but do not extend into older wood. Expansion into older wood of the male selections 02-16 and 02-18 often results in cankers up to 25 cm in length and death of 2-year-old wood (fig. 26.35). In very wet years, blossom infections on female trees occasionally move into wood produced the previous year and girdle twigs, thereby blighting young panicles. Diseased blossoms and basal portions of shoots are generally covered by the buff-colored sporulation of the pathogen (fig. 26.36). Infections on leaflets at the midrib and upper part of the blade cause characteristic V-shaped lesions (fig. 26.37), whereas those on leaf blades away from the midrib are large, distinct, circular lesions with wide chlorotic margins (fig. 26.38). Leaves injured by hail can also be infected. The disease does not cause defoliation, but infections on the rachis can kill parts of developed fruit. Seedlings of rootstock nursery plants can also be infected, killing the tender tips (figs. 26.39 to 26.43).

Causal organism

Botrytis cinerea Pers.:Fr., the conidial stage of the ascomycete *Botryotinia fuckeliana* (deBary) Whetzel, produces smooth, lemon-shaped, light brown conidia that measure 11.7 by 9.3 μm. Two strains are commonly isolated from pistachio shoots. One produces abundant conidia and a few large sclerotia, if any; the other one produces sparse conidiophores and conidia as well as abundant small black sclerotia on acidified potato dextrose agar media. Occasionally, sclerotia form on shoots remaining in the tree or on pruned shoots left on the orchard floor that were infected the previous year. Sclerotia incubated over wet sand at temperatures below 68°F produce conidiophores and conidia, and presumably they perform similarly in nature. Apothecia (sexual fruiting structures) have not been observed on sclerotia in culture or in pistachio orchards.

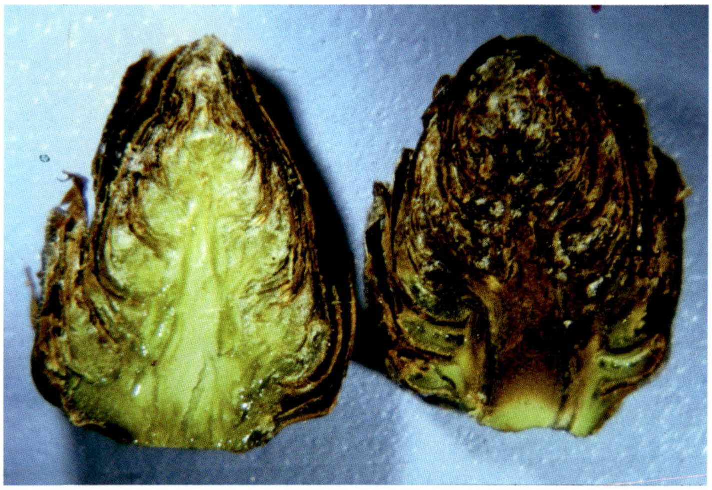

Figure 26.32. Sectioned bud infected by *B. dothidea* (right) versus healthy bud (left). *Photo:* T. J. Michailides.

Figure 26.33. Infection of rachis by *B. dothidea,* initiated from bird-damaged nuts. *Photo:* T. J. Michailides.

Figure 26.34. Blossom and shoot blight caused by *Botrytis cinerea. Photo:* T. J. Michailides.

Figure 26.35. Canker initiated from infected blossom of a male tree by *B. cinerea. Photo:* T. J. Michailides.

Figure 26.36. Sporulation of *B. cinerea* and infection of the base of a succulent pistachio shoot. *Photo:* T. J. Michailides.

Figure 26.37. V-type lesions caused by *B. cinerea* (usually on male trees). *Photo:* T. J. Michailides.

Figure 26.38. Circular lesion caused by *B. cinerea* (usually on female trees). *Photo:* T. J. Michailides.

Disease cycle and epidemiology

Inoculum sources include 1-year-old infected shoots, dead bud scales, infected female and male inflorescences from the previous and current season, conidia produced on leaf lesions and blighted fruit clusters, sclerotia, and dead and senescing plant parts on the orchard floor. Conidia are airborne and also may be blown in from outside sources. Infections start in early spring, when trees bloom and shoots emerge, and they continue for about 3 to 4 weeks or until weather turns warm and dry. In California, the disease is prevalent during cool, wet springs. Losses usually are insignificant, but in very wet years the killing of fruit clusters may reduce yields (see figs. 26.39 to 26.41). Frequently, however, excessive killing of current-season shoots can reduce the next season's fruiting wood, and blight of male flowers creates pollination problems.

Control

One application with thiophanate-methyl during bloom lessens Botrytis blossom and shoot blight. New fungicides such as fenhexamid or a premixture formulation of pyraclostrobin and boscalid also provide good control ("Fungicide Efficacy and Timing for Various Crops" at the UC Kearney Agricultural Research and Extension Center website, kare.ucanr.edu/). Pruning and removing blighted shoots from the orchard reduces inoculum and perhaps also disease. Pruning blighted and cankered shoots to prevent colonization by *B. dothidea* is important. Most cultivars are susceptible, and the male selections 02-16 and 02-18 are more susceptible than the commonly used Peters pollinator. Also, the cultivar Kerman can suffer significant damage in excessively wet years.

Kernel Decay

The severity of kernel decay varies from year to year and averages 1 to 2 percent of the produced nuts. Increased concern for contamination by mycotoxins, especially aflatoxins, has raised the importance of kernel decay (for more details, see the section "Aflatoxin Contamination" above). Decay by *Aspergillus flavus* and *A. parasiticus* (the fungi producing aflatoxins) is rare and occurs with decreasing frequency in pistachio, almond, walnut, and pecan.

Symptoms and causal agents

Common saprophytic fungi (including species of *Alternaria, Aspergillus (Eurotium), Cladosporium, Fusarium, Penicillium, Trichothecium, Ulocladium, Epicoccum*, and *Rhizopus*) may cause decay of pistachio kernels. Kernel decay is associated with dark

Figure 26.39. A few fruit blighted in a cluster due to *B. cinerea* infection. *Photo:* T. J. Michailides.

Figure 26.40. Large portions of a cluster blighted due to *B. cinerea* infection. *Photo:* T. J. Michailides.

Figure 26.41. Blighted fruit in a cluster due to *B. cinerea* infection. *Photo:* T. J. Michailides.

Figure 26.42. Tip of a nursery seedling blighted by *B. cinerea. Photo:* T. J. Michailides.

Figure 26.43. Close-up of tip of a nursery seedling blighted by *B. cinerea* (note shepherd's hook). *Photo:* T. J. Michailides.

Figure 26.44. Characteristic bright yellow staining caused by *Aspergillus niger* infection. *Photo:* T. J. Michailides.

brown discoloration or staining of the shell exterior, but most shell discoloration is probably caused by rupture of the hull and not fungal decay. If decay is extensive, the kernel is discolored beneath the seed coat. The shells of fruit infected by *A. niger* are stained a characteristic bright yellow (fig. 26.44).

Disease cycle and epidemiology

Little is known of the infection biology of fungi associated with kernel molds. Most are common saprophytes that live on dead organic matter in the orchard, and some develop on senescing weeds. Spores are probably spread by wind and blowing dust. Based on observations, kernel mold severity varies by orchard and season and is often highest if significant rains occur in spring or just before and during harvest. Nearly all kernel decay and aflatoxin contamination occur before harvest and are associated with early-split fruit or fruit damaged by navel orangeworm *(Amyelois transitella)* larvae. Improper storage conditions that allow moisture to build up can result in significant contamination by several *Aspergillus* species, including *A. flavus*.

Control

Treating problematic orchards with the biopesticide AF36 (atoxigenic strain of *A. flavus*) to displace the toxigenic *Aspergillus* strains section *Flavi* can reduce aflatoxin contamination. Irrigation to avoid drought stress in May can minimize early splitting. Also, harvest as soon as the nuts are mature to avoid navel orangeworm damage and reduce levels of kernel molds. Pioneer Gold I (PGI) and UCBI rootstocks result in lower incidence of early splits than *P. atlantica* and PGII and thus can contribute to reducing kernel infections by *Aspergillus* fungi.

Phomopsis Blight

Phomopsis blight occurs sporadically on pistachio trees in northern California and Greece and causes little or no economic damage. Symptoms are similar to those caused by Botrytis blossom and shoot blight and Sclerotinia blight. Shoots turn brown in late spring and are distinctly visible among healthy, dark green foliage (fig. 26.45). *Phomopsis* spp. (at least one identified as *Diaporthe chamaeropis*) were also isolated from diseased fruit and cankers caused by *Botrytis cinerea* or *Botryosphaeria dothidea*. Leaf and petiole wilting occurred within 1 week and shoot wilting within 3 weeks after inoculation of current-season shoots in April. Various color isolates of *Phomopsis* spp. were recovered from pistachio and wild blackberry *(Rubus ursinus)*, elderberry

(*Sambucus caerulea*), and willow (*Salix* sp.) growing close to pistachio orchards. Pycnidia with α (elliptical) and β (filiform) pycnidiospores can be found on blighted pistachio. Pycnidiospores are dispersed by rain or sprinkler water, thus contaminating buds. The fungus may also be spread by ascospores of the *Diaporthe* sexual stage (*D. chamaeropis*), which has not been found on pistachio but was found on other hosts. Occasionally, mycelia of *Phomopsis* spp. can be found inside the shell cavity before shell splitting.

Powdery Mildew

Powdery mildew is a minor disease and occurs wherever pistachio is grown.

Symptoms

Powdery mildew attacks hulls, rachises, petioles, the undersides of leaves, and young stems (figs. 26.46 and 26.47). In California pistachios, the pathogens are active only in spring and produce colonies usually less than 2 cm in diameter. A brown scar or russeted look develops on infected tissues as the fungal mycelium dies (fig. 26.48).

Causal organism

Powdery mildew fungi reported on pistachio include *Uncinula verniciferae* P. Henn on *Pistacia chinensis* and an *Oidium* sp. on *P. vera*. Identification of the powdery mildew species on pistachio in California has been difficult because chasmothecia have not been observed. *Phyllactinia guttata* (Wallr.:Fr.) has been reported on *P. atlantica* in Greece, however, and it has erect conidiophores 50 to 350 μm long, which produce singular, clavate, or somewhat rhomboid conidia. One or more germ tubes usually occur on the side of the conidium with an unlobed appresorium. The chasmothecia are scattered, bearing 3 to 15 equatorial appendages. Appendages have a distinct bulbous swelling at the base and are one to one and a half times as long as the chasmothecial diameter. They may contain 6 to 30 broadly clavate to slender asci with two elliptical ascospores. Few studies have investigated the host range or possible biological specialization of this fungus.

Disease cycle and epidemiology

The epidemiology of powdery mildew on pistachio has not been examined. As with other powdery mildew fungi, the conidia are readily wind-disseminated. Chasmothecia develop among the external hyphae and form appendages late in their development. For *P. guttata*, the uneven wall thickness and bulbous nature of the appendages allow for the entire chasmothecium to be dislodged from the host.

Figure 26.45. Shoot blight caused by a *Phomopsis* sp. *Photo:* T. J. Michailides.

Figure 26.46. Powdery mildew infections on petiole. *Photo:* T. J. Michailides.

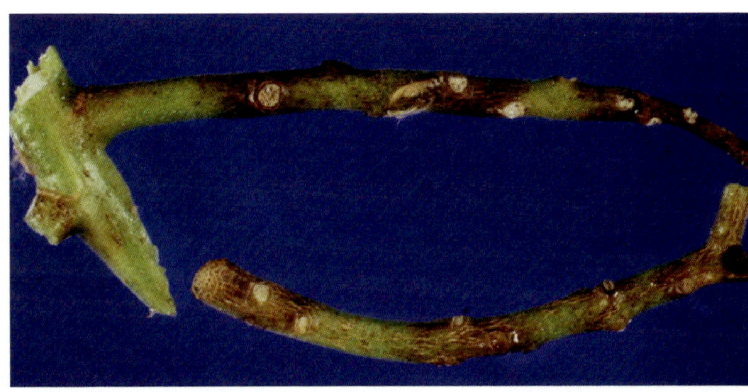

Figure 26.47. Powdery mildew infections on rachis. *Photo:* T. J. Michailides.

Figure 26.48. Brown, netted scars on mature fruit created by powdery mildew infection. *Photo:* T. J. Michailides.

When dry, the appendages fold and lift the entire fruiting structure from the plant surface, allowing it to be dislodged and carried away by wind. These fungi may overwinter as chasmothecia or as mycelium within buds infected the previous season.

Control

The disease occurs sporadically on the pistachio cultivar Kerman but is more widespread on Trabonella and Red Aleppo cultivars. The disease is not serious enough to warrant control measures.

Sclerotinia Blight

Sclerotinia blight is a minor disease of pistachio and occurs primarily in early spring in northern California.

Symptoms

The symptoms are wilting of tender shoots and flagging, shriveling, and drying of the leaves, similar to the symptoms caused by *Botrytis cinerea* (fig. 26.49). The base of the shoot turns a dark gray, and unless sporulation of *B. cinerea* is present on infected tissue, correct diagnosis requires isolation.

Causal organism

Sclerotinia sclerotiorum (Lib.) de Bary produces white mycelium and large, black sclerotia on the colony surface, mainly around the edge of potato dextrose agar plates. Sclerotia are not found on infected pistachio tissues. Apothecia (sexual fruiting structures), which arise from the sclerotia, are brown, relatively flat, up to 10 mm in diameter, and produce ascospores. The fungus does not produce asexual conidia.

Figure 26.49. Shoot blight caused by *Sclerotinia sclerotiorum*. *Photo:* T. J. Michailides.

Disease cycle, epidemiology, and control

Sclerotinia sclerotiorum survives as sclerotia in the soil and can be disseminated by contaminated soil, organic amendments, and water. The development of apothecia is favored by moderately cool (57°F) temperatures and wet conditions. Once established, the fungus grows between 41° and 86°F, with an optimum between 59° and 68°F. Ascospores, which are ejected into the air from apothecia on the orchard floor, are the only inoculum and can travel considerable distances. Inoculum probably originates in nearby vine and prune orchards, where sclerotia and apothecia have been frequently observed. Sclerotia have not been found in pistachio orchards. The infection court is not known, but it is believed that ascospores infect buds as they emerge. Wilting occurs 2 days after infection, while shoot death may take up to 7 to 10 days. The disease is infrequent and sporadic, and control measures are not recommended.

Septoria Leaf Spots

Septoria leaf spots are caused by three species of *Septoria*. *Septoria pistacina* Allescher and *S. pistaciarum* Caracciolo are common pathogens of pistachio in Mediterranean countries and the Middle East. *S. pistaciarum* has been reported on *P. vera* but not *P. atlantica* and *P. terebinthus* in Texas and Arizona in the United States. *S. pistaciae* is extremely rare and has been reported in France, Italy, and California. Severe epidemics of *Septoria* spp. can defoliate and reduce vigor of the trees.

Symptoms

Septoria pistaciarum first causes angular, brown necrotic lesions 1 to 2 mm in diameter that are limited by the smaller veins of the leaf on both leaf surfaces (fig. 26.50). Up to 20 pycnidia form in the center of the lesions, which often merge and become large necrotic areas (fig. 26.51). Similar spots are produced on fruit.

Septoria pistaciae, recorded in one orchard in California, produces numerous subcuticular brown spots 0.5 to 1.5 mm in diameter on both sides of the leaf. Numerous small black pycnidia are aggregated and immersed in the spots. In California, *S. pistaciae* causes distinct grayish to light brown fruit lesions 1 to 4 mm in diameter that are surrounded by a bright, distinctly reddish halo and mostly located near the peduncle (fig. 26.52). Leaf infections have not been observed.

Causal organisms

Mycosphaerella pistaciarum A. Chitzanidis (pycnidial stage *Septoria pistaciarum* Caracciolo) produces black ascostroma with 50 to 60 cylindrical, clavate, short-stipitate asci in each ascostroma. The ascospores, eight in each ascus, are oblong and two celled. Pycnidia have a typical ostiole and form on both sides of the leaf. Conidia are hyaline, filiform, and irregularly curved, with two to five (occasionally up to nine) septa. Spermogonia (male gametangia) produce hyaline, rod-shaped spermatia. Pycnidia of *S. pistaciarum* bear curved to falcate conidia with three to nine septa.

The teleomorph (sexual stage) of *Septoria pistaciae* Desmaz. has not been identified. Pycnidia have a typical ostiole. Pycnidiospores are filiform, straight or curved, and most have one septum, though some have two or three.

Mycosphaerella pistacina A. Chitzanidis produces black ascostroma 90–112 × 80–106 µm in size, each bearing about 20 clavate, short-stipitate asci. Each ascus has eight ascospores measuring 26.2–40 × 3.2–4.8 µm; ascospores are hyaline, two-celled, and have a constriction at the septum. The anamorph *Septoria pistacina* Allescher produces pycnidia (160–200 × 96–160 µm) that develop on upper and lower leaf surfaces. Conidia are 31.8–47.0 × 3.6–4.8 µm in size, hyaline, filiform, and curved, with one septum at the middle. Spermogonia measure 80–118 × 77–106 µm, and spermatia are hyaline, rod shaped, and 3.9–5.6 × 1.0–1.4 µm. This species has not been encountered in the United States. It is common in Turkey and Greece and probably in other Mediterranean countries (fig. 26.53).

Figure 26.50. Leaf infected by *Septoria pistaciarum* from Arizona pistachios. *Photo:* T. J. Michailides.

Figure 26.51. Oozing pycnidia of *Septoria pistaciarum* on a pistachio leaf lesion. *Photo:* T. J. Michailides.

Figure 26.52. Lesions on fruit from California caused by *Septoria pistaciae. Photo:* T. J. Michailides.

Figure 26.53. Lesions on fruit and leaves from Turkey caused by *Septoria pistacina. Photo:* T. J. Michailides.

Epidemiology and disease cycle

Septoria pistaciarum and *S. pistacina* overwinter in fallen leaves that were infected while on the tree in previous seasons. Ascostroma primordia appear on fallen leaves soon after they are on the ground, and young asci develop from them until February to early March. Most ascospores are mature and ready for discharge from late April through May, becoming released during or after rain. The optimal temperature for the development of ascostroma and asci of *S. pistaciarum* is 50°F. Infections due to ascospores of *S. pistaciarum* have an incubation period of about 10 days. Pycnidia with viable conidia of *S. pistaciarum* have been found on fallen leaves throughout winter and up to July, but their role in the disease is unknown. Secondary infections are caused by *S. pistacina* conidia, which spread by rain or sprinkler water, and these infections may continue until late fall. Spermogonia production begins on leaves on the tree in mid-September and continues on fallen leaves until December. The role is unknown. The epidemiology of *S. pistaciae* is presumed to be similar to that of the other species.

Control

Septoria leaf spots are controlled with preventive fungicide sprays. Ethylene-bis dithiocarbamates (zineb, mancozeb) are recommended but are not registered in the United States on pistachio. Chlorothalonil and copper fungicides are also effective but must be applied after fruit have reached 1 cm in size to avoid phytotoxic damage. Methyl-benzimidazole carbamate fungicides are also effective. Applications should begin when the first leaves unfold and be repeated monthly, if necessary, until early June.

Stigmatomycosis

The term stigmatomycosis is the general name for a disease that develops inside externally sound fruit as a result of punctures made by plant-feeding hemipteran insects; and it occurs on a number of crops including cotton, soybean, pecan, pomegranate, citrus, and pistachio in the United States. It has been reported on pistachio in Greece, Iran, and Russia, and it is frequently a problem in California pistachio orchards severely infested by hemipteran insects. In a 1989 survey in California, fruit with stigmatomycosis were found in 90 percent of samples collected from late June to mid-September and from all growing areas.

Symptoms

Nuts infected by stigmatomycosis have a wet, smelly, rancid, slimy kernel. Kernels can be small, dark green, and partially developed, with a brown funiculus; well developed, dark green, and rancid (figs. 26.54 and 26.55); or full size but abnormal, white or light yellow, and jellylike, with a lobed appearance. In contrast, kernel necrosis (see the section "Epicarp Lesion and Kernel Necrosis"), which is caused by large hemipterans, is characterized by dry, punky, brown areas in the kernel. Sometimes kernel necrosis and stigmatomycosis symptoms are present in the same fruit.

Causal organisms

Eremothecium coryli (Peglion) Kurtzman (syn. *Nematospora coryli* Peglion), an ascomycete yeast, and *Aureobasidium pullulans* (de Bary) G. Arnaud cause stigmatomycosis. *E. coryli* grows on potato dextrose agar as yeastlike oval or spherical budding cells, either isolated or in short chains, and has few hyphae, which are septate at maturity. In addition to buds, the yeast produces many asci that are cylindrical to naviculate, with two to eight ascospores arranged lengthwise. Ascospores are apiculate to fusiform, with a distinct septum at or near the center, and the upper cell slightly broader at the septum; after liberation they are held together in a mass by long appendages. *E. coryli* colonies are creamy in color and perfectly circular. The yeast grows at 50° to 99°F, with an optimal range of 85° to 95°F. More asci form at 59° to 68°F than at 77° to 95°F.

Aureobasidium pullulans on potato dextrose agar produces smooth, pale pink, yeastlike colonies that are covered with a slimy mass of spores. Older colonies change to black due to chlamydospore production. Primary conidia are hyaline, smooth, ellipsoidal, one celled, and variable in shape and size; secondary conidia resulting from budding of primary conidia are smaller. Conidiophores are undifferentiated, intercalary or terminal, or they arise as short lateral branches. Septate hyphae are hyaline, smooth, and thin walled. The fungus grows at 50° to 95°F, with an optimum of 86°F.

Disease cycle and epidemiology

The causal agents of stigmatomycosis are associated with hemipteran insects of the families Pentatomidae (stink bugs) and Coreidae (leaffooted bugs). Hemipterans are common pests in pistachio orchards, and up to 10 species have been found in California pistachio orchards. Three common stink bug pests, *Thyanta pallidovirens*, *Chlorochroa uhleri*, and *C. ligata*, and leaffooted bugs *Leptoglossus clypealis* and *L. zonatus* experimentally transmitted *E. coryli* and caused typical symptoms of stigmatomycosis in pistachio. Symptoms first appear in late June after pistachio shells have hardened, but the disease becomes frequent in July through September, a period that coincides with kernel development. Smaller hemipterans, such as *Lygus* and *Calocoris* spp., may carry but not transmit the pathogens because they are unable to puncture the firm fruit pericarp after the second part of May. Stigmatomycosis and kernel necrosis are worse in orchards irrigated by sprinklers than in those irrigated by drip, microsprinklers, or flood. This suggests that requirements for infection are either greater humidity, higher activity of hemipterans in sprinkler-irrigated orchards, or higher abundance of pathogen propagules.

Control

Fungicides do not control stigmatomycosis, but insecticides reduce the hemipteran insect vector and subsequently the incidence of stigmatomycosis.

Diseases Reported in Other Countries

The following diseases have been reported on pistachios in countries other than the United States.

- Anthracnose (leaf spot and hull rot) caused by *Colletotrichum acutatum* J. H. Simmonds, has been reported in Australia.
- Eutypa dieback (or gummosis), caused by *Eutypa lata* (Pers.:Fr.) Tul. & C. Tul., asexual stage: *Libertella blepharis* A. L. Smith, has been reported in Greece.
- Gum canker, caused by *Cytospora terebinthi* Bres., has been reported in Italy and Iran.
- Rust, which attacks *Pistacia vera*, *P. terebinthus*, *P. palaestina*, and *P. lentiscus*, is caused by *Pileolaria terebinthi* Castagne (synonym *Uromyces terebinthi* (DC.) G. Winter) and occurs commonly in the Mediterranean countries.
- Stem canker caused by *Botryosphaeria obtusa* (Schwein.) Shoemaker (pycnidial stage *Sphaeropsis sapinea* (Fr.:Fr.) Dyko & Sutton) has been reported in South Africa pistachios.
- Dieback and canker occurs in the Kerman province of Iran. The suspected pathogen is *Paecilomyces variotii* Bainier, but pathogenicity tests have been inconsistent.
- A bacterial trunk and branch canker caused by a *Xanthomonas* sp. has been reported in the Sorora cv. of pistachios in Australia.

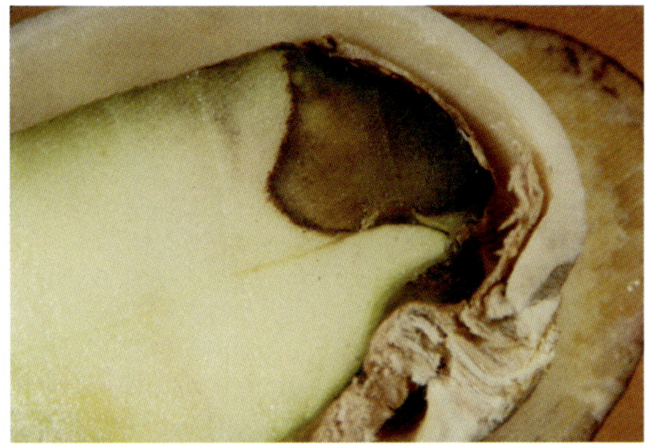

Figure 26.54. Partial lesion on pistachio kernel caused by the stigmatomycosis pathogens. *Photo:* T. J. Michailides.

Figure 26.55. Stigmatomycosis of mature pistachio fruit. *Photo:* T. J. Michailides.

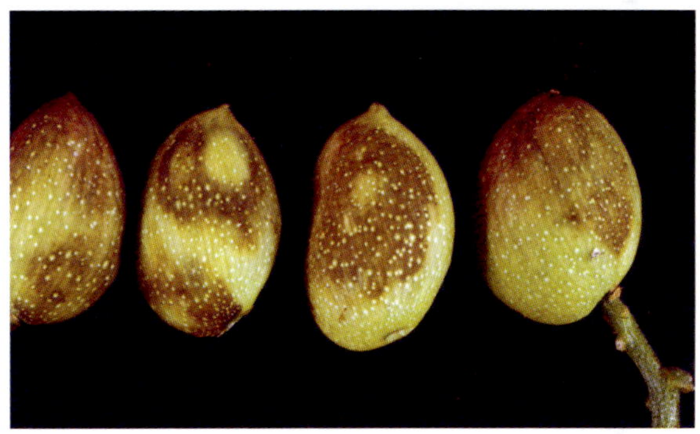

Figure 26.56. Epicarp lesion caused by hemipteran insects. *Photo:* T. J. Michailides.

Noninfectious Disorders

Epicarp Lesion and Kernel Necrosis

Epicarp lesion was initially thought to be a genetic or nutritional disorder until it was associated with feeding by hemipteran insects. It occurs wherever pistachio trees are grown and hemipterans are present. Losses can be significant.

Symptoms

In young fruit, the first symptoms of epicarp lesion appear in late April and during May as a light brown diffuse area with dark brown zones on the outer surface of the green fruit, with or without a black spot where the lesion originated. The affected area varies in size but may eventually cover the entire surface. Generally only a portion of the fruit surface is affected (fig. 26.56). In many cases, resin is found in the center of the lesion on the insect puncture. The lesions darken and dry as they age, remain soft in texture, and become sunken as the underlying tissues collapse. The discoloration generally extends into the shell, which is often thin and poorly developed in that area. Many affected fruit fall. On more mature fruit, epicarp lesion is an extended dark brown to black area on the hull, but the fruit does not fall. Lesions are on the side of the fruit, occasionally just on the tip, and can extend to the base of the pedicel.

Although external symptoms are alike, internal symptoms vary with the species of hemipteran insect that caused them. In immature fruit, mirids cause internal pits 1 to 3 mm in diameter with blackened rims, while larger hemipterans cause a white netting within 24 to 48 hours after insect feeding. As the fruit mature, and after late May, the smaller hemipterans can no longer puncture the lignified shell or cause epicarp lesion, but larger hemipterans can feed on maturing fruit through harvest. As the shell hardens, the white net symptom is sometimes replaced

Figure 26.57. Kernel necrosis caused by large hemipteran insects on mature fruit later in the season. *Photo:* T. J. Michailides.

Figure 26.58. Stain along the shell suture of early-split nuts. *Photo:* M. Doster.

Figure 26.59. Immature nut splitting in a young fruit cluster. *Photo:* T. J. Michailides.

by a small black internal spot (fig. 26.57), where bugs feed with or without a protruding microscopic stylet sheath, or spike. As the shell hardens, damage from attacks by leaffooted bugs and stink bugs shifts to the kernel, causing kernel necrosis. Such kernels are discolored, spongy, and punky. In commercial orchards, where there are successions of various hemipteran insects of the families Miridae, Pentatomidae, and Coreidae, all symptoms are found.

Causal agents

Among the insects known to cause epicarp lesion are *Leptoglossus clypealis* Heidermann, *L. occidentalis* Heidermann, *L. zonatus* (Dallas), *Thyanta pallidovirens* Stål, *Phytocoris relativus* Knight, *P. conspurcatus* Knight, *Calocoris norvegicus* Gmelin, *Neurocolpus longirostris* Knight, *Lygus hesperus* Knight, *Psallus vaccinicola* Knight, *Boisea rubrolineata* Barber, *Liorhyssus hyalinus* L., *Nezara viridula* (L.), *Chlorochroa uhleri* (Stål), *C. ligata* (Say), *Chinavea hilare* (Say), and the boxelder bug (*Leptocoris trivittatus* Say). The squash bug *Anasa tristis*, species of *Brochymena*, and harlequin *Murgantia histrionica* are found on pistachio but do not cause epicarp lesion or kernel necrosis. However, large hemipterans need to be controlled to reduce epicarp lesion and kernel necrosis. Control measures include insecticide application and biological control in the orchard, as well as cultural practices. Cover crops planted along an orchard border can function as trap crops for insects that cause epicarp lesion and can be sprayed to kill the insects.

Hull Rupture of Pistachio

A normal pistachio fruit splits its shell while on the tree prior to harvest, but the hull remains intact and acts as an important physical barrier, protecting the kernel from insect and fungal invasion. The hulls of a small percentage of fruit rupture before harvest. These are usually poor-quality nuts because their kernels are frequently moldy and insect damaged and their shells discolored. Early splitting and hull cracking are distinct types of hull rupture.

Early Splitting

The hull of an early-split pistachio fruit splits along the shell suture (see fig. 26.11). Because both the hull and shell are split in the same place, the kernel is exposed and frequently moldy and insect infested. In most orchards, approximately 2 to 3 percent of the fruit will be early splits, although the percentage can range from less than 1 to more than 10 percent. Early splitting begins in late July and continues into September. Moldy kernels are more likely to occur in fruit whose hulls rupture in early summer. Early-split fruit weigh less than normal fruit and have smaller shells that are stained yellow brown or dark brown, especially along the shell suture (fig. 26.58). Early splits with rough, shriveled hulls often contain substantially more aflatoxin, moldy kernels, and insect infestation than early splits with smooth hulls, which split later. The cause of early splitting is unclear, but irrigation is an important factor. Early splitting is increased substantially by insufficient irrigation during shell formation in late spring (May) and slightly by excessive irrigation in late August. The kind of cultivar can also have an effect on early splitting. Expansion or growth of the kernel in a shell cavity that is normally smaller than the size of the developing kernel is another possible cause of early splitting.

Cracking of the Hull

Cracking of the hull occurs anywhere on the hull except along the shell split, and it exposes the shell. Typically, fewer than 5 percent of fruit have cracked hulls at harvest. Because hull cracking begins later in the season than early splitting, fruit with cracked hulls are less likely to have moldy kernels, and their shells are not as severely discolored as are those of early-split fruit. Expansion or growth of the hull while part is still attached to the shell is a possible cause of hull cracking. The relationship of irrigation to hull cracking is not known.

Immature Nut Splitting

Immature nut splitting occurs sporadically in spring or early summer before the kernel has developed (fig. 26.59). Although the cause is unknown, sudden changes in temperature and moisture may be involved.

Shell Staining

Various degrees of shell stain result from decay or natural deterioration, injury, and oxidation of hulls, thus reducing the quality of pistachio fruit. Decaying hulls stain the shell surface light to dark brown either partially or entirely, depending on the extent of hull decay. A severe brown shell stain usually develops in fruit infected by *Alternaria alternata* and other species of *Alternaria* (see fig. 26.10), particularly in orchards with sprinkler or flood irrigation. *Aspergillus niger* infections discolor shells a bright yellow. *Epicoccum purpuresens* Ehrenb. Ex Schlecht. (syn. *E. nigrum* Link) stains shells rusty to reddish. Natural deterioration of the hull produces a diffuse light yellow color and affects more fruit in late-harvested orchards than in those harvested early. Even in fruit harvested early, the initial signs of staining due to hull deterioration are present as blemishes 2 to 4 mm long near the stem side of the fruit. Failure to remove hulls from fruit within 24 hours of harvest leads to a high incidence of oxidative yellow shell staining.

Harvesting as soon as fruit are mature, as well as immediate hulling, reduces shell staining. Converting irrigation systems to subsurface irrigation reduces relative humidity and shortens the period of dew formation on leaves, resulting in less Alternaria blight and thus less shell staining.

Peters Scorch

This disorder is very common in the male cultivar Peters, less so in other male cultivars, and only occasional in female trees. Because female trees are seldom affected, this disorder's potential for economic damage is undetermined. It has been reported from Arizona and California.

The first symptoms appear in late June and during July, when air temperatures have increased. Initially, one or a few small, black, water-soaked, angular lesions up to 5 mm in diameter appear on each leaflet blade, usually with exuded black gum in the center. These lesions become light green and diffuse, then turn brown and necrotic (figs. 26.60 to 26.62). If rain, dew, or high humidity prevail in summer, these lesions may be colonized by *Alternaria alternata*, other *Alternaria* species, and *Cladosporium* spp., and their sporulation turns the lesions black or dark green (see fig. 26.7). Peters trees showing scorch symptoms defoliate earlier than nonscorched trees. Colonization of scorch lesions by *A. alternata* can serve as an early source of *Alternaria* inoculum and also increases defoliation. Occasionally, lesions caused by *B. dothidea* and Peters scorch may occur on the same leaf. The cause of Peters scorch is unknown, and control measures are not available.

Figure 26.60. Initial symptoms of Peters scorch (cause unknown). *Photo:* T. J. Michailides.

Figure 26.61. Peters scorch later in season. *Photo:* T. J. Michailides.

Figure 26.62. A leaf with characteristic Peters scorch lesions. *Photo:* T. J. Michailides.

Figure 26.63. Pin-size lesion (cause unknown). *Photo:* T. J. Michailides.

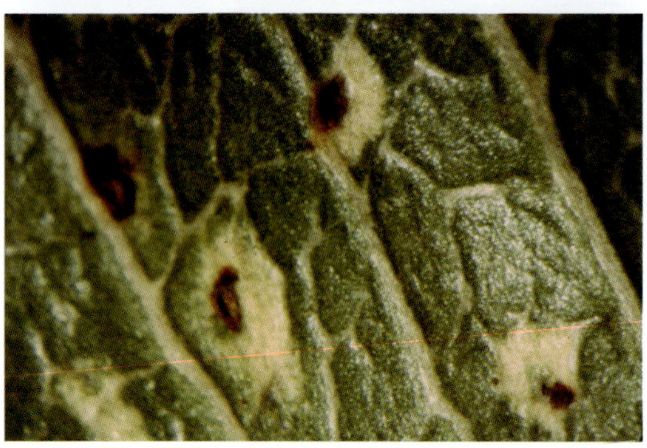

Figure 26.64. Close-up of pin-size lesion (cause unknown). *Photo:* T. J. Michailides.

Pin-Size Lesion

Pin-size lesion occurs infrequently on pistachio trees in California and causes no economic damage. In spring and summer, small chlorotic spots 2 to 3 mm in diameter surrounded by a yellow halo appear on the foliage of both male and female trees (fig. 26.63). At the center of these lesions is a pin-size puncture, which usually penetrates the leaf (fig. 26.64). Insects are suspected of causing these punctures, but the cause is unknown.

Acknowledgments

We thank A. Chitzanidis for information on Septoria leaf spots. We also thank laboratory technician R. Puckett, along with postdocs H. Avenot, Z. Ma, N. Ahimera, G. Driever, and Y. Luo, for contributions to research on various diseases of pistachio reported in this chapter. Finally, the long-term financial support of the California Pistachio Research Board is truly appreciated.

References

Ahimera, N., G. F. Driever, and T. J. Michailides. 2003. Relationships among propagules numbers of *Botryosphaeria dothidea*, latent infections, and severity of panicle and shoot blight in pistachio orchards. Plant Disease 87:846–853.

Ahimera, N., S. Gisler, D. P. Morgan, and T. J. Michailides. 2004. Effects of single-drop impactions and natural and simulated rains on the dispersal of *Botryosphaeria dothidea* conidia. Phytopathology 94:1189–1197.

Aminaee, M. M., and D. Ershad. 1989. Die-back of young shoots of pistachios in Kerman. Proceedings of the 9th Plant Protection Congress of Iran: 79.

Ash, G. J., and V. M. Lanoiselet. 2001a. First report of *Alternaria alternata* causing late blight of pistachio *(Pistacia vera)* in Australia. Australasian Plant Pathology 50:803.

———. 2001b. First report of *Colletotrichum acutatum* causing a leaf spot and hull rot of pistachio. Australasian Plant Pathology 30:365–366.

Assawah, M. W. 1969. *Pileolaria terebinthi* Cast. on pistachio in U.A.R. (Egypt). Phytopathologia Mediterranea 8:157–161.

Avenot, H. F., and T. J. Michailides. 2007. Resistance to boscalid fungicide in *Alternaria alternata* isolates from pistachio in California. Plant Disease 91:1345–1350.

Avenot, H. F., D. P. Morgan, and T. J. Michailides. 2007. Resistance to pyraclostrobin, boscalid, and multiple resistance to Pristine (pyraclostrobin + boscalid) fungicide in *Alternaria alternat*a causing Alternaria late blight of pistachio in California. Plant Pathology 57:135–140.

Avenot, H. F., A. Sellan, G. Karaoglanidis, and T. J. Michailides. 2008. Characterization of mutations in the iron-sulfur subunit of succinate dehydrogenase correlating with boscalid resistance in *Alternaria alternata* from California pistachio. Phytopathology 98:736–742.

Avenot, H. F., H. van den Biggelaar, D. P. Morgan, J. J. Moral, M. Joosten, and T. J. Michailides. 2014. Sensitivities of baseline isolates and boscalid-resistant mutants of *Alternaria alternata* from pistachio to fluopyram, penthiopyrad, and fluxopyroxad. Plant Disease 98:197–205.

Bolkan, H. A., J. M. Ogawa, and H. R. Teranishi. 1984. Shoot blight of pistachio caused by *Botrytis cinerea*. Plant Disease 68:163–165.

Bolkan, H. A., J. M. Ogawa, R. E. Rice, R. M. Bostock, and J. C. Crane. 1984a. Leaffooted bug (Hemiptera: Coreidae) and epicarp lesion of pistachio fruits. Journal of Economic Entomology 77:1163–1165.

———. 1984b. Leaffooted bug implicated in pistachio epicarp lesion. California Agriculture 38(3, 4): 16–17.

Carter, M. V. 1991. The status of *Eutypa lata* as a pathogen. Phytopathological Paper No. 32. London: International Mycological Institute.

Chen, S.-F., D. P. Morgan, and T. J. Michailides. 2014. Botryosphaeriaceae and Diaporthaceae associated with panicle and shoot blight of pistachio in California, USA. Fungal Diversity 67:157–179.

Chen, W.-Q., D. P. Morgan, D. Felts, and T. J. Michailides. 2003. Antagonism of *Paenibacillus lentimorbus* to *Botryosphaeria dothidea* and biological control of panicle and shoot blight of pistachio. Plant Disease 87:359–365.

Chen, W.-Q., N. Ntahimpera, D. P. Morgan, and T. J. Michailides. 2003. Mycoflora of *Pistacia vera* in the Central Valley, California. Mycotaxon 63:147–158.

Chitzanidis, A. 1956. Species of *Septoria* on the leaves of *Pistacia vera* L. and their perfect states. Annales de l'Institute Phytopathologique Benaki 10:29–44.

Corazza, L., G. Granata, D. Avanzato, and G. Chilosi. 1991. Main phytopathological aspects of pistachio nut in Italy. Rivista di Frutticoltura 4:37–40.

De Sousa da Camara, E. 1948. Mycetes aliquot lusitaniae. Agronomia Lusitana 294.

Doster, M. A., and T. J. Michailides. 1994a. *Aspergillus* molds and aflatoxins in pistachio nuts in California. Phytopathology 84:583–590.

———. 1994b. Development of *Aspergillus* molds in litter from pistachio trees. Plant Disease 78:393–397.

———. 1994c. Relationship between shell discoloration of pistachio nuts and incidence of fungal decay and insect infestation. Plant Disease 83:259–264.

———. 1995a. The development of early split pistachio nuts and their contamination by molds, aflatoxins, and insects. Acta Horticulturae 415:359–364.

———. 1995b. The relationship between date of hull splitting and decay of pistachio nuts by *Aspergillus* species. Plant Disease 79:766–769.

Doster, M. A., P. J. Cotty, and T. J. Michailides. 2014. Evaluation of the atoxigenic *Aspergillus flavus* strain AF36 in pistachio orchards. Plant Disease 98:948–956.

Doster, M. A., T. J. Michailides, D. A. Goldhamer, and D. P. Morgan. 2001. Insufficient spring irrigation increases abnormal splitting of pistachio nuts. California Agriculture 55(3): 28–31.

Ershad, D., and M. Barkhordary. 1974. Host range and vectors of *Nematospora coryli* Peglion in Kerman of Iran. Iranian Journal of Plant Pathology 10(3, 4): 34–39.

Ershad, D., and M. Barkhordary. 1976. Investigations on stigmatomycosis (massu disease) of pistachio. Iranian Journal of Plant Pathology 12(1, 2): 17–23.

Farr, D. F., G. F. Bills, G. P. Chamuris, and A. Y. Rossman. 1989. Plants and products in the United States. St. Paul, MN: APS Press.

Goldhamer, D.A., T. J. Michailides, and D. P. Morgan. 2002. Buried drip irrigation reduces fungal disease in pistachio orchards. California Agriculture 55(4): 133–138.

Granata, G., and A. M. Pennisi. 1984. "In vitro" activity of some fungicides against *Cytospora terebinthi*, agent of a pistachio decline. Notiziario sulle malattie delle piante 105(32): 130–135.

Granata, G., A. Sidoti, and L. Corazza. 1993. New acquisitions on control of *Septoria pistaciae* Desm. and *Cytospora terebinthi* Bres. Proceedings of the 9th G.R.E.M.P.A. Meeting—Pistachio: 64–65.

Grigoriou, A. C. 1992. First detection of spermagonia, aecia and basidia of *Pileolaria terebinthi* (D. C.) Cast. [Syn. *Uromyces terebinthi* (D. C.) Winter] on pistachio trees. Proceedings of the 6th National Phytopathological Congress, Hellenic Phytopatholgical Society: 72.

Holtz, B. A., T. J. Michailides, L. Ferguson, J. G. Hancock, and A. R. Weinhold. 1996. First report of *Rhizoctonia solani* (AG-4) on pistachio rootstock seedlings in California. Plant Disease 80:1303.

Korf, R. P., and K. L. Dumont. 1972. *Whetzelinia*, a new generic name for *Sclerotinia sclerotiorum*. Mycologia 64:248–251.

Kouyeas, H. 1979. "Stigmatomycosis" of the nuts of the pistachio tree (*Pistacia vera* L.). Annales de l'Institute Phytopathologique Benaki 12:147–148.

Luo, Y., Z. Ma, H. C. Reyes, D. P. Morgan, and T. J. Michailides. 2007. Using real-time PCR to survey frequency of azoxystrobin-resistant allele G143A in *Alternaria* populations from almond and pistachio orchards in California. Pesticide Biochemistry and Physiology 88:328–336.

Ma, Z., and T. J. Michailides. 2002a. A PCR-based technique for identification of *Fusicoccum* sp. from pistachio and various other hosts in California. Plant Disease 86:515–520.

———. 2002b. Characterization of *Botryosphaeria dothidea* isolates collected from pistachio and other hosts in California. Phytopathology 92:519–526.

———. 2004a. An allele-specific PCR assay for detecting azoxystrobin-resistant *Alternaria* isolates from pistachio in California. Journal of Phytopathology 152:118–121.

———. 2004b. Characterization of iprodione-resistant *Alternaria* isolates from pistachio in California. Pesticide Biochemistry and Physiology 80:75–84.

———. 2004c. A real-time PCR assay for the detection of azoxystrobin-resistant *Alternaria* populations from pistachio orchards in California. Crop Protection 23:1259–1263.

———. 2005a. Advances in understanding molecular mechanisms of fungicide resistance and molecular detection of resistant genotypes in phytopathogenic fungi. Crop Protection 24:853–863.

———. 2005b. Genetic structure of *Botrytis cinerea* from different host plants in California. Plant Disease 89:1083–1089.

Ma, Z., D. Felts, and T. J. Michailides. 2003. Resistance to azoxystrobin in *Alternaria* isolates from pistachio in California. Pesticide Biochemistry and Physiology 77:66–74.

Ma, Z., Y. Luo, and T. J. Michailides. 2003. Nested PCR assays for detection of *Monilinia fructicola* in stone fruit orchards and *Botryosphaeria dothidea* from pistachio in California. Journal of Phytopathology 151:312–322.

———. 2004. Spatiotemporal changes in the population structure of *Botryosphaeria dothidea* from California pistachio orchards. Phytopathology 94:326–332.

Ma, Z., E. W. A. Boehm, Y. Luo, and T. J. Michailides. 2001. Population structure of *Botryosphaeria dothidea* from pistachio and other hosts in California. Phytopathology 91:665–672.

Ma, Z., D. P. Morgan, D. Felts, and T. J. Michailides. 2002. Sensitivity of *Botryosphaeria dothidea* from California pistachio to tebuconazole. Crop Protection 21:829–835.

Michailides, T. J. 1989. The "Achilles heel" of pistachio fruit. California Agriculture 43(5): 10–11.

———. 1991a. First report of Septoria spot in California pistachios. California pistachio industry annual report 1990–1991, 93. Fresno: California Pistachio Commission.

———. 1991b. Pathogenicity, distribution, sources of inoculum, and infection courts of *Botryosphaeria dothidea* on pistachio. Phytopathology 81:566–573.

———. 1991c. Susceptibility of pistachio male cultivars to Botrytis blossom and shoot blight caused by *Botrytis cinerea*. Plant Disease 75:410–415.

Michailides, T. J., and D. P. Morgan. 1990. Etiology and transmission of stigmatomycosis disease of pistachio in California. (Abstract) Phytopathology 80:973.

———. 1992. Effects of temperature and wetness duration on infection of pistachio by *Botryosphaeria dothidea* and management of disease by reducing duration of irrigation. Phytopathology 82:1399–1406.

———. 1993. Spore release by *Botryosphaeria dothidea* in pistachio orchards and disease control by altering the trajectory angle of sprinklers. Phytopathology 83:145–152.

———. 1994. Occurrence of Sclerotinia and Botrytis shoot blights on pistachio in California. Plant Disease 68:163.

———. 1995. Spread of *Botryosphaeria dothidea* in pistachio orchards irrigated with drip or microjets. California pistachio industry annual report 1994–1995, 102–106. Fresno: California Pistachio Commission.

———. 1996. Principles for the control of Alternaria late blight of pistachio caused by *Alternaria alternata* in California. Proceedings of the 9th G.R.E.M.P.A. Meeting—Pistachio.

———. 2004. Panicle and shoot blight of pistachio: A major threat to the California pistachio industry. American Phytopathological Society (APS) website, http://www.apsnet.org/online/feature/pistachio/.

———. 2016. Association of Botryosphaeria panicle and shoot blight of pistachio with injuries of fruit caused by Hemiptera insects and birds. Plant Disease 100:1405–1413.

Michailides, T. J., and J. M. Ogawa. 1987. Botrytis blight of pistachio: New findings. California pistachio industry annual report 1986–1987, 108–112. Fresno: California Pistachio Commission.

Michailides, T. J., D. P. Morgan, and D. Felts. 1997. Spread of *Botryosphaeria dothidea* in central California pistachio orchards. Acta Horticulturae 470:582–591.

———. 2002. First report of *Botryosphaeria rhodina* causing shoot blight of pistachio in California. Plant Disease 86:1273.

Michailides, T. J., D. P. Morgan, and D. A. Goldhamer. 1995. Using subsurface drip irrigation to reduce Alternaria late blight of pistachio caused by *Alternaria alternata* (second year report). California pistachio industry annual report 1995–1996, 129–136. Fresno: California Pistachio Commission.

Michailides, T. J., D. P. Morgan, and W. Krueger. 1995. Chemical control of Botryosphaeria panicle and shoot blight of pistachio in California. California pistachio industry annual report 1994–1995, 73–76. Fresno: California Pistachio Commission.

Michailides, T. J., D. P. Morgan, and G. Weinberger. 1995. Chemical control of Botrytis shoot blight with benomyl by ground or air application. California pistachio industry annual report 1994–1995, 77. Fresno: California Pistachio Commission.

Michailides, T. J., R. E. Rice, and J. M. Ogawa. 1987. Succession and significance of several hemipterans attacking a pistachio orchard. Journal of Economic Entomology 80:398–406.

Michailides, T. J., B. L. Teviotdale, and G. Weinberger. 1999. Botryosphaeria blight of pistachio: Identification and control manual. Fresno: California Pistachio Commission.

Michailides, T. J., D. P. Morgan, Z. Ma, Y. Luo, D. Felts, M. A. Doster, and H. Reyes. 2005. Conventional and molecular assays aid diagnosis of crop diseases and fungicide resistance. California Agriculture 59(2): 115–123.

Mila, A. L., and T. J. Michailides. 2006. Use of Bayesian methods to improve prediction of panicle and shoot blight severity of pistachio in California. Phytopathology 96:1142–1147.

Mila, A., G. F. Driever, D. P. Morgan, and T. J. Michailides. 2005. Effects of latent infection, temperature, precipitation, and irrigation on panicle and shoot blight of pistachio in California. Phytopathology 95:926–932.

Morgan, D. P., G. F. Driever, D. Felts, W. H. Krueger, and T. J. Michailides. 2009. Evaluation of two disease warning systems for Botryosphaeria panicle and shoot blight of California pistachio and efficient control based on early-season sprays. Plant Disease 93:1175–1181.

Munkvold, G. P., and J. J. Marois. 1994. Eutypa dieback of sweet cherry and occurrence of *Eutypa lata* perithecia in the Central Valley of California. Plant Disease 78:200–207.

Ntahimpera, N., G. F. Driever, D. Felts, D. P. Morgan, and T. J. Michailides. 2002. Dynamics and pattern of latent infection caused by *Botryosphaeria dothidea* on pistachio buds. Plant Disease 86:282–287.

Pantidou, M. L. 1973. Fungus-host index for Greece. Athens: Benaki Phytopathological Institute.

Parmeter, J. R. Jr. 1970. Biology and pathology of *Rhizoctonia solani*. Berkeley: University of California Press.

Pearson, T. C., M. A. Doster, and T. J. Michailides. 2001. Automated detection of pistachio defects by machine vision. Applied Engineering in Agriculture 17:729–732.

Pryor, B. M., and T. J. Michailides. 2002. Morphological and molecular characterization of *Alternaria* isolates associated with Alternaria late blight of pistachio. Phytopathology 92:406–416.

Rice, R. E., J. K. Uyemoto, J. M. Ogawa, and W. M. Pemberton. 1985. New findings on pistachio problems. California Agriculture 39:15–18.

Rumbos, I. C. 1986. Isolation and identification of *Eutypa lat*a from *Pistacia vera* in Greece. Journal of Phytopathology 116:352–357.

Simmons, E. G. 1967. Typification of *Alternaria, Stemphylium,* and *Ulocladium*. Mycologia 59:67–92.

Sneh, B., L. Burpee, and A. Ogoshi. 1991. Identification of *Rhizoctonia* species. St. Paul, MN: APS Press.

Swart, W. J., and W. M. Bates. 1995. First report of stem canker caused by *Botryosphaeria obtusa* on pistachio. Plant Disease 79:1036–1038.

Swart, W. J., and J. T. Blodgett. 1998. First report of *Botryosphaeria dothidea* basal canker of pistachio trees in South Africa. Plant Disease 82:960.

Tzavella-Klonari, K. 1979. Studies on the biology, epidemiology and pathogenesis of the disease caused by the fungus *Septoria pistaciarum* Caracciolo on pistachio nut tree. Vol. KA. (Ph.D. thesis.) School of Agricultural Sciences, Aristotle University of Thessaloniki.

Wasfy, E. H., I. A. Ibrahim, and H. M. Elarosi. 1974. New Alternaria disease of pistachio in Egypt. Phytopathologique Mediterranea 13:110–111.

Young D. J., and T. J. Michailides. 1989. First report of Septoria leaf spot of pistachio in Arizona. Plant Disease 73:9.

Part 7 Vertebrate Pest Management

27 Vertebrate Pest Management

Roger A. Baldwin and Terrell P. Salmon

California ground squirrels, pocket gophers, meadow voles, rabbits, wild pigs, coyotes, and birds can be significant pests in pistachio orchards. They cause damage by feeding on nuts, tree roots, and bark and can stunt tree growth or even kill young trees. Burrows, mounds, and depressions created by ground squirrels, pocket gophers, and wild pigs interfere with orchard maintenance, irrigation, and harvesting operations. In addition, rodents, rabbits, and coyotes gnaw on drip irrigation lines.

The extent of vertebrate pest problems in a pistachio orchard is determined in large part by the orchard's location. Problems are often worse in orchards adjacent to rangeland, waterways, or other areas where pests are not controlled. Orchard management activities also have some effect; for example, flood irrigation and orchard floor cultivation may discourage some rodent populations.

The most successful approach to dealing with rodent and rabbit problems is to manage these populations so they do not reach levels where significant damage occurs. This preventive approach is of particular importance in organic orchards, where options for reducing destructive populations are limited to labor-intensive techniques such as shooting and trapping.

Management programs for vertebrate pests involve four basic steps:

- Correctly identify the species causing the problem. Signs such as tracks, feces, plant damage, and burrows may be used to identify which species is responsible for damage.
- Assess potential management options to determine practical control methods.
- Take early action and use control methods appropriate for the orchard and time of year, with due consideration for the environment and nontarget species.
- Establish a monitoring system to detect reinfestation to determine when additional corrective measures or controls are necessary.

Regular monitoring of vertebrate pests in and around orchards should form the basis of a management program. Historical records of pest population levels, control measures implemented, and results from these control measures can be used to help determine the best management approach.

Vertebrate control equipment and supplies (baits, fumigants, etc.) are available at local retail outlets such as farm supply and hardware stores. In addition, many agricultural commissioners' offices sell some rodent control materials. For further information on sources of control materials, consult your local UC Farm Advisor or county agricultural commissioner.

Control Actions

For most vertebrate pests, more than one control method is available to reduce damaging populations. Table 27.1 lists control options for important vertebrate pests in pistachio orchards. Details on how to use these controls are given in the sections that follow. It is a good idea to consult your county agricultural commissioner before you use any controls to find out which procedures work best in your location and what restrictions apply to their use. The timing of control actions is often critical and is determined in large part by the life cycle of the target pest. Become familiar with the biology of the vertebrate pests affecting your orchards and the available control options so you will be able to plan the most cost-effective management strategies.

Take steps when preparing and planting an orchard to reduce potential problems with

Table 27.1. Control methods for vertebrate pests of pistachios

Pest	Control method								
	Habitat modification	Trapping	Baiting	Fencing	Tree protectors	Repellents	Frightening	Shooting	Fumigating
ground squirrels	×	×	×					×	×
pocket gophers	×	×	×						×
voles	×		×		×				
rabbits	×	×*	×	×	×	×		×	
birds	×						×	×*	

Note: *Useful for some species.

vertebrates. Properly installed fencing protects young trees against rabbits and deer. Deep plowing and disking destroy or disperse resident vole populations and also destroy large parts of the burrow systems of pocket gophers, reducing the risk of reinvasion. A vegetation-free zone around an orchard reduces the risk of invasion by voles, although the zone size has not been researched. Take steps to eliminate pocket gopher and ground squirrel populations before planting a new orchard; it is much easier to control them prior to planting than afterward. Trapping, baiting, shooting, or burrow fumigation can be used, depending on the pest involved and the situation. Once trees are planted, tree guards protect against damage by rabbits and, to some degree, voles. Once the orchard is in place, develop and implement a management program to address any vertebrate pest problems that may arise.

Biological Control

Vertebrate populations are affected most by the availability of food and cover, not by diseases in the populations or the presence of predators. Although a number of predators such as hawks, owls, foxes, coyotes, and snakes feed on some vertebrate pest species, the role of predators in rodent control is limited by

- their tendency to modify their diet according to the relative abundance of prey species
- the high reproductive rate of small rodents, which allows rodent populations to compensate for loss to predation
- predator-avoidance strategies developed by prey species

Predators seldom keep vertebrate pests from reaching damaging levels and should not be relied on to prevent vertebrate damage to crops, though they might be considered a small component of an integrated management program. Growers in favorable settings can enhance the orchard habitat for predators, for example, by installing nest boxes or raptor perches; however, there is little or no data to suggest that the installation of nest boxes or raptor perches reduces vertebrate pest damage. One study showed that despite raptors' quick acceptance of perches and an increase in their numbers in study areas, rodent numbers did not change.

Legal Aspects

Under the California Fish and Game Code, if California ground squirrels, meadow voles, pocket gophers, or black-tailed jackrabbits are causing or about to cause a crop depredation, they may be lethally removed at any time by the owner or tenant of a property. Cottontail and brush rabbits may also be lethally removed when causing damage, but not before damage occurs. Trapping is often used to control vertebrate pests, but in California trapping most mammals, even for pest purposes, requires a trapping license issued by the California Department of Fish and Wildlife. A trapping license is not needed for rats, mice, moles, voles, and gophers. Live (cage) traps are sometimes used for managing vertebrate pests; be aware that under the California Fish and Game Code, it is illegal to trap and relocate any wildlife, including rodents. Live-trapped pest animals should be euthanized humanely. Methods considered humane by the American Veterinary Medical Association include gassing with carbon dioxide and shooting. Drowning is not an approved method of euthanasia.

Only pesticides that are registered with the Environmental Protection Agency's Department of Pesticide Regulation (DPR) can legally be used for vertebrate pest control. Registered materials are listed in DPR's database, available at the DPR website, cdpr.ca.gov/. You may also contact your agricultural commissioner for information on registered materials.

In some areas of California, pistachio orchards are located within the range of federally and state-protected endangered species. Species likely to be of concern include the San Joaquin kit fox, several species of rare kangaroo rats, and, where burrow fumigants are used, the blunt-nosed leopard lizard. See the composite map showing ranges of these species (fig. 27.1). Special guidelines apply to the use of toxic baits and fumigants for vertebrate pest control in these areas. Your county agricultural commissioner has the latest detailed maps that show the ranges of endangered species and the latest information on restrictions that apply to pest control activities in those areas. You can also get more information on endangered species regulations from the DPR website, cdpr.ca.gov/docs/endspec/.

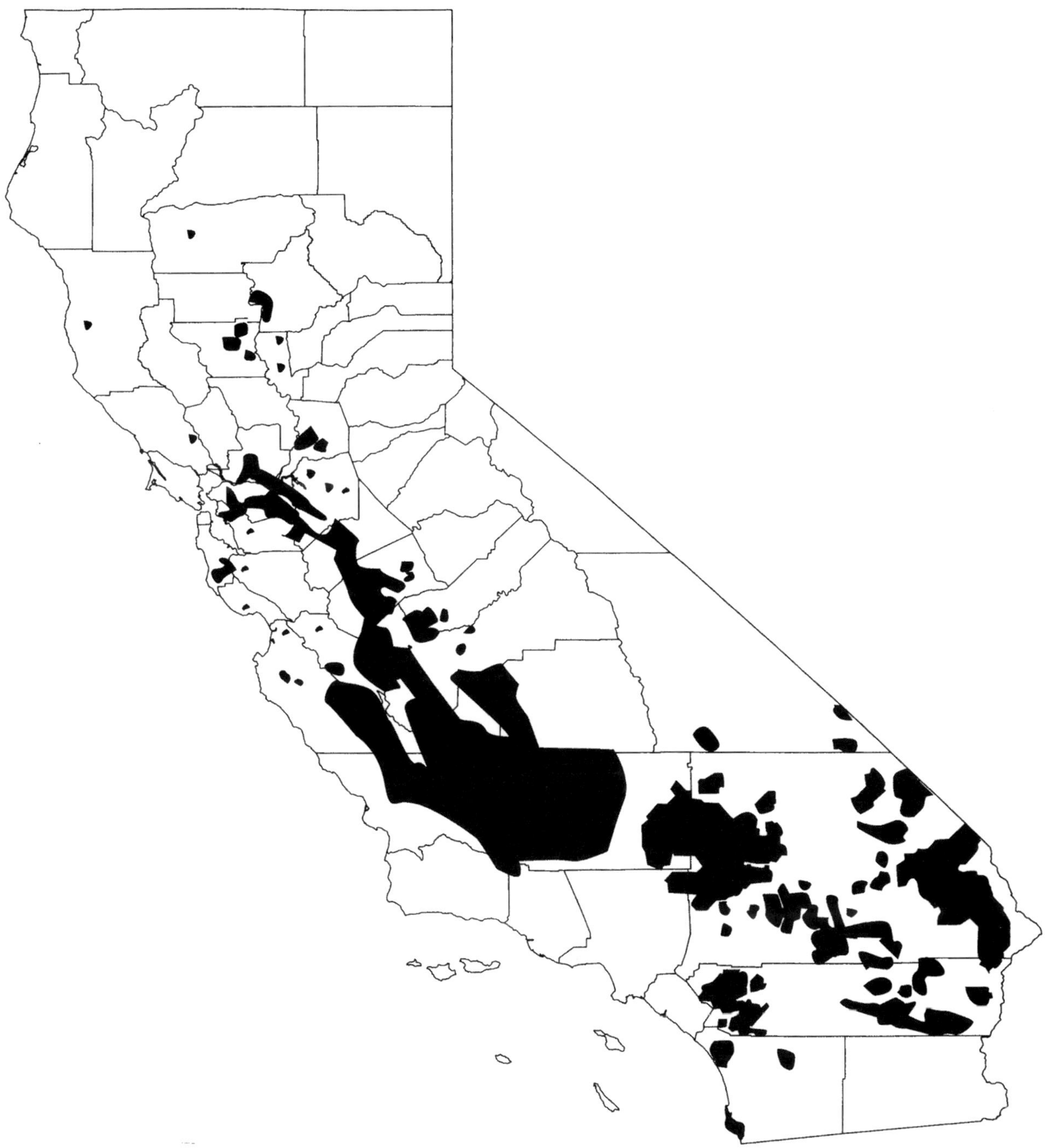

Figure 27.1. Distribution of endangered vertebrate species in California. This is a composite of current maps showing approximate locations of endangered species that may affect pest management options in or near pistachio orchards. Check with your county agricultural commissioner or the DPR website, cdpr.ca.gov/docs/endspec/, for the latest information regarding restrictions that may apply to a specific location.

Major Vertebrate Pests of Pistachios

California Ground Squirrel (*Otospermophilus beecheyi*)

California ground squirrels feed on nuts and damage trees by chewing bark off tree limbs or trunks. Burrow systems in the orchard may interfere with orchard management activities and divert irrigation water. Problems also occur when ground squirrels chew on and damage drip lines (fig. 27.2). In some areas, ground squirrels pose a health risk to humans through the spread of bubonic plague.

Figure 27.2. Damage to microsprinkler irrigation tubing caused by California ground squirrels. *Photo:* R. A. Baldwin.

Figure 27.3. The California ground squirrel has mottled brown or gray fur and a slightly bushy tail; it can be a significant pest of pistachios. *Photo:* J. K. Clark.

Figure 27.4. Ground squirrel burrows in orchard. *Photo:* R. A. Baldwin.

The adult ground squirrel has a head and body 9 to 11 inches long and a slightly bushy tail (fig. 27.3). The fur is mottled dark and light brown or gray. Ground squirrels dig burrows along ditches and fencerows, around buildings, and within and bordering many agricultural crops. They tend to avoid flood-irrigated orchards, thick chaparral, dense woods, very moist areas, and lands that are under complete and frequent cultivation. They will travel 100 yards or more to feed in adjacent crops. In orchards, they often dig entrances to their burrow systems near the base of trees (fig. 27.4). Ground squirrels live in colonies, which may grow very large if left uncontrolled. Each ground squirrel burrow system can have several openings, with scattered soil in front. Individual ground squirrel burrows are 5 to 30 feet long and 2.5 to 4 feet below the surface, and burrow openings are about 4 to 6 inches wide. Burrows provide a place to retreat, sleep, hibernate, rear young, and store food.

Ground squirrels are active during the day, with peak activity occurring in the morning and late afternoon. Ground squirrels are primarily herbivorous. In spring, ground squirrels consume a variety of green grasses and forbs. When these begin to dry and produce seed in early summer, the squirrels switch to seed, grains, and nuts. Ground squirrels hibernate during winter, emerging in late January or February. Some squirrels (usually the young of the year) are active year-round. Many adults go into a temporary dormant state, called estivation, during the hottest parts of the year. Females have one litter in the spring, averaging eight young. The young emerge from the burrow after about 6 weeks. Figure 27.5 illustrates periods of activity for the California ground squirrel.

Management guidelines

Ground squirrel activity should be monitored periodically (at least once a month) both inside and on the perimeter of an orchard in spring, summer, and autumn when squirrels are active. Midmorning is usually the best time of day to observe squirrel activity. Controls should be implemented as soon as activity is first observed to keep numbers from increasing. Ground squirrels are extremely adaptable, so habitat modification only has limited benefit in a management program. Burrow fumigants, poison baits, and traps currently are the most effective control methods. Where ground squirrels are a major problem, keep annual records of the dates that squirrels emerge from hibernation and when first young are seen aboveground; changes in the general number of squirrels; and control methods used,

Figure 27.5. Activity periods and preferred food sources for the California ground squirrel, which is critical information for choosing the most-effective control action for ground squirrels and the proper timing for control. Activity periods vary somewhat from one growing area to another depending on local climate.

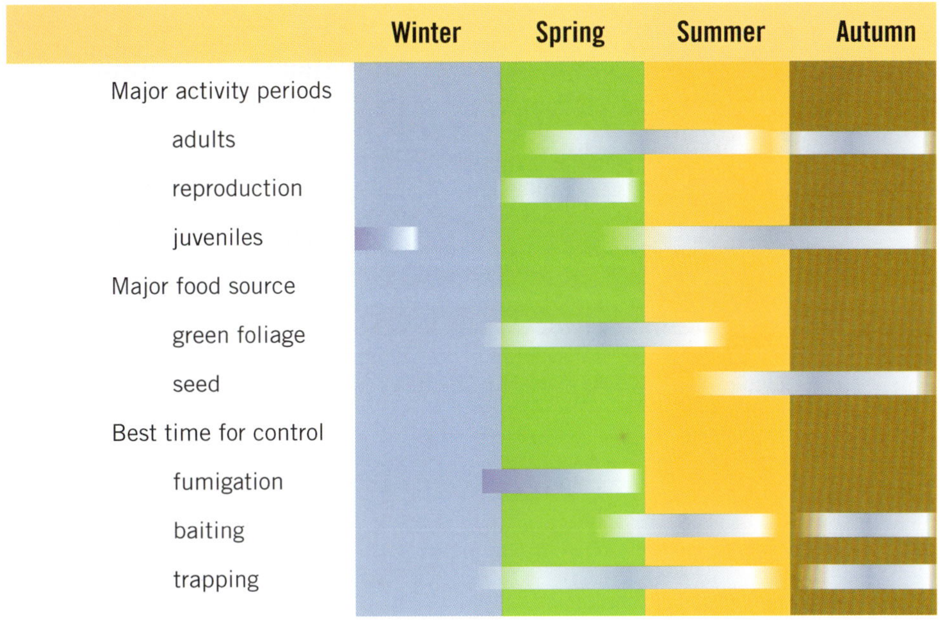

dates of use, and their effect. Use these records as the basis for future management decisions.

The type of control action needed for ground squirrels depends primarily on their activity patterns and feeding preferences during the time of year when control is to be undertaken (see fig. 27.5). The choice of control action is also influenced by the location of the infestation and the number of squirrels present. From a biological point of view, the most effective time of year to control ground squirrels is in early spring, when adults have emerged but before they have reproduced. For best control, use burrow fumigation about 3 weeks after the first squirrels emerge from hibernation. Because squirrels are feeding nearly exclusively on green vegetation early in the season, poisoned grain baits are generally not effective until late spring and early summer. In late spring or summer if squirrels are moving from adjacent lands into the orchard to feed, baiting or trapping along the perimeter offers the most effective control if access to the neighboring property is not possible. While baiting can be done in summer and autumn, squirrels are usually eating nuts at this time, leading to poor bait acceptance.

Habitat modification. Removing brush piles, stumps, and debris in and around an orchard may help limit buildup of squirrel populations and makes it easier to monitor squirrel activity. Squirrels may quickly reinvade abandoned burrow systems. Deep plowing (ripping) along orchard perimeters destroys burrow entrances and helps slow the rate of invasion.

Trapping. Trapping is an effective control method. However, it is quite labor intensive and thus is most practical for small infestations. Trapping can be implemented at any time of year when squirrels are active, but it is most effective in spring before reproduction increases the population. The most commonly used traps are kill traps such as Conibear and box traps.

Conibear traps are placed unbaited over the burrow entrance to capture squirrels as they leave the burrow (fig. 27.6). Within the range of the San Joaquin kit fox, the trap must be placed in a covered box with an entrance no larger than 3 inches wide to exclude foxes. Box traps are also effective for ground squirrels and are baited with walnuts, almonds, oats, barley, or melon rinds (fig. 27.7). The bait is placed inside the trap, either behind or tied to the trigger. Traps should be prebaited for several days before setting the trigger mechanism. Once ground squirrels have become accustomed to taking bait, the trigger mechanism can be set. As with all traps, steps should be taken to minimize the risk of capturing nontarget wildlife and pets. For example, traps can be placed inside boxes with 3-inch openings, large enough for ground squirrels to enter but not other nontarget species.

Fumigation. Fumigants can be very effective against ground squirrel populations and can be used prior to squirrels reproducing and during times

Figure 27.6 A Conibear trap is placed in the entrance to a ground squirrel burrow. Secure the trap to a stake. *Photo:* J. K. Clark.

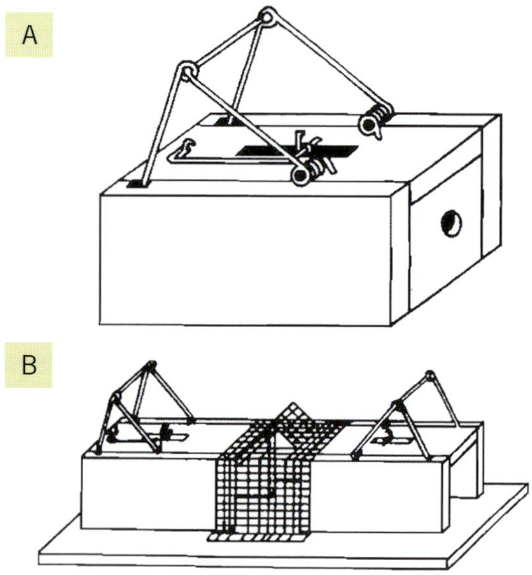

Figure 27.7 Single box traps (A) can be used for ground squirrels. To use a pair of traps, remove the backs, connect the two traps with wire mesh, and attach them to a board (B).

when poison baits are relatively ineffective. The best time to fumigate burrows is in late winter or early spring, after squirrels have emerged from hibernation but prior to breeding. Because ground squirrels seal themselves within nesting chambers, fumigation is not effective when squirrels are hibernating or estivating. For fumigation, it is important that the soil is moist so that gases are retained in high concentrations within the burrow; dry soil allows gases to escape. When using a fumigant, be sure to treat all active burrow systems in and around the orchard. Recheck all areas a few days after fumigation and retreat any burrows that have been reopened. Do not use fumigants in burrows that extend beneath occupied buildings, and carefully follow the label directions for fumigant use.

Gas cartridges provide an easy and relatively safe way to fumigate ground squirrel burrows. They are available commercially and from some agricultural commissioners' offices. Use one or two cartridges for each burrow that shows signs of activity. More than two may be needed for a large burrow system. Quickly push the ignited cartridge, fuse-end first, into the burrow using a shovel handle or stick, and seal the burrow entrance with soil. Watch nearby burrow entrances; treat and seal any that begin to leak smoke. The larger and more complex the burrow system, the more fumigant needed.

Aluminum phosphide is a very effective fumigant when used early in the spring when soil moisture is high. However, aluminum phosphide is a restricted-use material; a permit is required for purchase or use. Application personnel must be trained in the material's proper use and on its potential hazards. When aluminum phosphide pellets come into contact with moist soil in the burrow, they produce phosphine gas, which is highly toxic to any animal. When using aluminum phosphide, treat every active burrow, plug the entrance with a wad of newspaper, and cover with soil. Do not add water or moist paper to the burrow during the application process.

Poison baits. Poison baits are usually the most cost-effective way of controlling ground squirrels, especially large populations and over large areas. Bait consists of grain or pellets treated with a poison registered for ground squirrel control. To be effective, the bait must be used at a time of year when ground squirrels are active and feeding on seed (usually late spring to early summer and autumn; see fig. 27.5). Before using baits, place small amounts of untreated grain (breakfast oats work fine) near burrows in the morning and check in the afternoon to see if the squirrels eat it. If the grain is taken, proceed with baiting. Otherwise wait for several days to a week and try again. As nuts reach maturity, bait acceptance may decrease, given the squirrels' strong preference for nuts. Therefore, nuts may need to be harvested before reinstituting baiting programs. However, this may be too late for grower needs, which reinforces the point that ground squirrel control is best implemented before squirrel densities become too high.

All poison baits registered for ground squirrel control are restricted-use materials; they include the acute toxin zinc phosphide and anticoagulant baits (diphacinone and chlorophacinone). Zinc phosphide can be applied as spot treatments to control ground squirrels during the nonbearing season or outside of orchards. For spot treatments, 1 tablespoon of bait is scattered over an area of 2 to 3 square feet around each burrow opening. Alternatively, the bait may be broadcast over a larger area using a mechanical spreader at the rate of 6 to 10 pounds of bait per swath acre. Bait shyness can occur with zinc phosphide baits when squirrels ingest a sublethal dose, become sick and learn to avoid the bait during future applications. This can result in low effectiveness of zinc phosphide baiting programs. Prebaiting the area with untreated grain 2 to 3 days prior to the application of zinc phosphide may reduce the chances of bait shyness and improve the effectiveness of baiting programs. Control with zinc phosphide is usually achieved within 48 hours of bait application.

With anticoagulant rodenticides, ground squirrels must ingest several doses of bait over a period of several days. Control is slower than with zinc phosphide, but there is less chance of squirrels becoming bait-shy. Another advantage is the availability of an antidote (vitamin K_1) in the event of accidental poisoning of nontarget species. Anticoagulants can be applied in bait stations, as spot treatments near burrows, or broadcast over larger areas. Be sure to read the label directions carefully to determine what application method is appropriate.

Bait stations are commonly used in and around orchards to provide bait for squirrels. Various bait station designs can be used; they all let squirrels in while excluding larger animals. One design uses openings that are 3 or 4 inches in diameter and incorporates baffles to keep the bait inside the station (fig. 27.8). Within the range of the San Joaquin kit fox, entrances to bait stations must not be greater than 3 inches wide. Special types of bait stations must be used within the ranges of endangered kangaroo rat species to ensure that these species are excluded. Contact your local agricultural commissioner or see the DPR website, cdpr.ca.gov/docs/endspec/, for the latest recommendations on use of poison baits and bait stations in areas that are within the range of endangered species.

Bait stations should be placed near runways or burrows and should be secured so that they are not easily tipped over. If squirrels are moving into an orchard from adjacent areas, place bait stations along the perimeter of the orchard where squirrels are invading, one station every 100 feet. Use more stations when the number of squirrels is high. Check bait stations daily at first, then as often as needed to keep the bait replenished. If bait feeding is interrupted, the bait's effectiveness is greatly reduced. Be sure to pick up any bait that is spilled and replace bait that is wet or moldy. Successful baiting usually takes 2 to 4 weeks. Continue to supply bait until feeding ceases and no more squirrels are observed; then properly dispose of unused bait.

Spot treatments and broadcast applications are permissible in orchards during the dormant season (read the product label for further specifications) and can be used at any time in noncrop areas. Spot treatments are most economical and effective for small populations. For spot treatments, scatter bait evenly over 40 to 50 square feet near active burrows (the amount will vary depending on the product used). Reapply in 4 or 5 days (the label will specify the proper application interval for each specific product) for a total of two applications to ensure squirrels are exposed to a continual supply of bait. Scattering takes advantage of the ground squirrels' natural foraging behavior and minimizes the risk to nontarget species. Do not pile bait or spread it on bare ground; this increases the hazard to livestock and nontarget wildlife. Bait may also be applied using a mechanical broadcaster such as a seed spreader mounted on the back of an ATV (all-terrain vehicle). Usually, squirrels retreat to their burrows when sick, and will die there. Dispose of any visible carcasses to prevent poisoning predators or scavengers. Burying is a good method as long as carcasses are buried deep enough to discourage scavengers.

Pocket Gophers (*Thomomys* spp.)

Pocket gophers are potentially serious orchard pests, especially in young orchards. They feed on the bark of tree crowns and roots, girdling and killing young trees and reducing the vigor of older trees (fig. 27.9). Gophers may also damage drip irrigation lines. Burrow systems may divert irrigation water, causing stress to young trees, and possibly increasing soil erosion. Mounds of soil associated with burrows can interfere with harvesting, mowing, and other orchard management activities.

Figure 27.8. Four-inch plastic pipe can be used to construct a bait station for ground squirrels. The pipe must have its entrance restricted to 3 inches when used within the range of the San Joaquin kit fox. Place baffles inside the pipe to keep bait inside the station. *Photo:* J. K. Clark.

Figure 27.9. Some of the roots of this tree have been eaten by gophers, greatly reducing the upper portion of the root system. Note the underground tunnel next to the trunk. *Photo:* J. K. Clark.

Figure 27.10. Adult pocket gopher emerging from a burrow. Pocket gophers are seldom seen aboveground. *Photo:* J. K. Clark.

Figure 27.11. Typical fan-shaped pocket gopher mound. The burrow opening is plugged. *Photo:* R. A. Baldwin.

An adult pocket gopher is 6 to 8 inches long and has a stout yellowish- or grayish-brown body, a short tail, and small ears and eyes (fig. 27.10). External fur-lined cheek pouches open outside the lips on each side of the mouth and are used for carrying food. Pocket gophers are rarely seen aboveground, spending most of their time in a system of tunnels they construct 6 to 18 inches below the surface. Gophers are solitary except during breeding and when young are being raised. A single burrow system can cover hundreds of square feet and consists of main tunnels with lateral branches used for feeding or to push excavated soil to the surface. The fan-shaped mounds that are formed over the openings of lateral tunnels are the most obvious signs of gopher infestation (fig. 27.11). Pocket gophers plug burrow openings with soil so that the tunnel system is completely closed and the temperature and humidity in the burrow remain fairly constant. Gophers feed primarily on the roots of herbaceous plants, but they may also dig lateral tunnels to feed on aboveground vegetation within a few inches of the tunnel opening. Once they have finished feeding, they plug these holes with soil.

In orchards and other irrigated lands, gophers may breed throughout the year, with a breeding peak in late winter or early spring. Females usually produce one or two litters per year, with litters averaging about five young. When the young are mature, they leave the burrow to find a favorable location to establish their own burrow. Some may take over burrow systems that recently have been vacated. Buildup of gopher populations in an orchard is favored by the presence of cover crops, especially those that contain clovers, which are a favored food source. When cover crops dry up or are removed, gophers may feed on the bark of tree roots and crowns (see fig. 27.9). Cover crops also make detection of pocket gophers difficult.

Management guidelines

Pocket gophers are active throughout the year, and if left uncontrolled, a population can increase rapidly. Damage to orchard trees is almost always belowground and usually is not evident until trees show signs of stress such as yellow or brown leaves. It is therefore important to control gophers when activity is first detected. The typical fan-shaped mounds are a good indicator of gopher presence. These mounds are produced in greatest numbers in the spring and autumn when soil is moist. Monitor for activity monthly in the spring and pay close attention to orchard perimeters where gophers may move in from adjacent infested areas. Gophers patrol and expand their burrows on a regular basis, so their activity can be determined by first flattening the mound and then looking for new activity in 1 or 2 days. Mounds of moist soil are also indicative of recent gopher activity. Special attention must be given to orchards with abundant ground cover, given the increased probability of gophers in these areas and the greater difficulty in detecting gopher activity in dense vegetation. With persistent effort, pocket gophers can be controlled and even eliminated from an orchard. The most effective control methods are baiting, trapping, and fumigation.

Habitat modification. Pocket gophers should be controlled prior to planting an orchard. The land can be deep-ripped so that most burrow systems are destroyed and many gophers are killed. Poison baits or traps can then be used to eliminate any remaining gophers. In established orchards, clean cultivation of the orchard floor reduces the food supply and may destroy some burrows, making the orchard less favorable for gophers. Clean cultivation also makes it easier to monitor gopher activity. Flood-irrigated orchards may have fewer problems than orchards that are drip or sprinkler irrigated. Gophers should be controlled in adjacent areas to prevent orchards from being invaded.

Trapping. Traps are highly effective in controlling small infestations of pocket gophers but can be expensive and labor intensive over large acreage. Either pincher traps (e.g., Macabee and Gophinator) or box traps may be used (figs. 27.12 and 27.13). To place traps, use a probe near a fresh mound to locate the main tunnel, which is usually on the lower side of the mound and 8 to 12 inches below the surface. You will feel reduced resistance when the probe enters the burrow, and the probe will drop quickly about 2 inches when you find it. The number of traps required will depend on the number of directions the tunnel goes, but usually a tunnel requires only one or two traps. Wire traps to a stake. You may cover the hole to keep light out, but recent research shows that this is not necessary. If the trap is not visited within 24 hours, move it to a new location.

Fumigation. Gas cartridges are not effective against pocket gophers. In extensive burrow systems, it is difficult to maintain a lethal concentration of the fumigant in the system, and pocket gophers quickly seal off their tunnels when they detect poison gas. However, aluminum phosphide (a restricted-use material; a permit is required for purchase or use) can be highly effective. It is most effective in late winter or early spring when soil moisture is high (high soil moisture prevents toxic gases from escaping through the soil) and prior to the gopher's breeding period. Aluminum phosphide can also be effective later in the year when treatments follow irrigation. Follow label instructions and all safety precautions given.

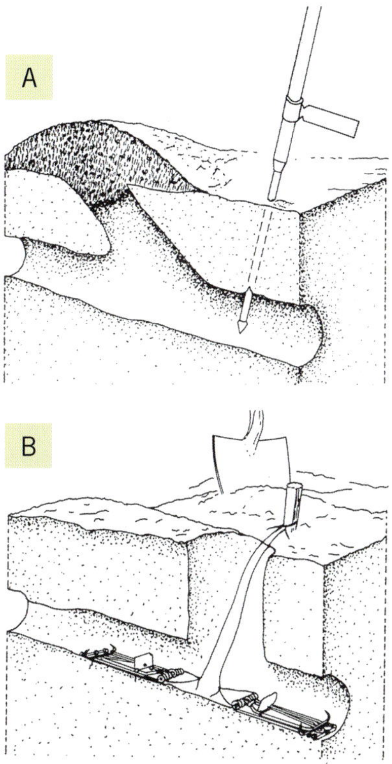

Figure 27.12. To place a Macabee- or Gophinator-type trap, use a probe to find the gopher tunnel (A). Begin probing 4 to 8 inches from the plug side of a mound. The probe will suddenly drop a few inches when you hit the main tunnel. Use a shovel to expose the main tunnel and place two traps in the tunnel, one in each direction (B). Tie the traps to a stake that is tall enough to be seen easily. Push each trap well back into the tunnel; the hole may be covered or left uncovered.

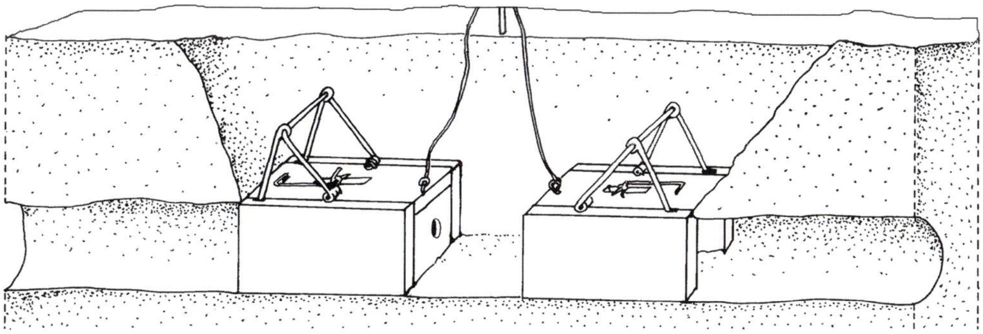

Figure 27.13. Place as many box traps as there are tunnels, and be sure to stake traps down. The opening may be covered or left uncovered depending on trapper preference.

To use aluminum phosphide, probe to find the main tunnel, widen the probe opening, drop two to four tablets (see label for specific number) into the burrow, and seal off the hole to prevent light from entering the tunnel. This process should be repeated for each burrow system, resulting in approximately two treatments per burrow system. As with any control method, keep monitoring for signs of renewed gopher activity and re-treat areas if you find new mounds.

Poison baits. Baits may be applied by hand or with a mechanical applicator. Strychnine, zinc phosphide, and anticoagulant baits are registered for control of gophers. Bait must be applied belowground. Follow bait label directions carefully for application methods and amounts.

With hand-baiting, locate the main tunnel of the burrow system by probing near fresh activity. Then enlarge the probe opening and place a small amount of bait in the burrow. This should be repeated at least two times for each burrow system. Hand-operated mechanical bait dispensers have a bait reservoir and bait-release mechanism and permit probing and bait dispensing in one operation. These devices are substantially faster than hand-baiting.

Mechanical burrow builders are tractor drawn and are an efficient way of controlling gophers over large areas (fig. 27.14). The device constructs an artificial burrow beneath the soil and deposits bait within it at preset intervals and quantities. The artificial burrows are constructed between tree rows and should be made to intercept some of the pocket gophers' natural burrows so that gophers will explore them and consume the bait. When using a mechanical burrow builder, periodically check to see that the artificial burrows are formed properly. Soil moisture must be right to produce a well-formed, smooth, artificial burrow. If possible, wait at least 10 days after baiting before running other equipment over treated areas. Mechanical burrow builders may not work well on rocky or sandy soils. In some mature orchards that have not been deep-cultivated, surface roots may be damaged by the burrow builder. After you have used a mechanical burrow builder, follow up with a program of trapping or hand-baiting to control remaining gophers. Burrow builders should not be used in areas with a small or localized gopher population. Creating new burrows throughout the orchard creates a larger tunnel system for easier reinvasion.

Figure 27.14. A mechanical burrow builder is pulled behind a tractor to dispense bait in artificial burrows. This device can be effective when large numbers of pocket gophers are present in an orchard. *Photo:* J. K. Clark.

Meadow Voles (*Microtus californicus*)

Meadow voles can cause serious damage to trees by feeding on bark at the base of the tree. Small trees are most susceptible to being completely girdled or killed by voles, but even large trees can suffer significant damage. Voles can also chew holes in irrigation lines. Voles are most likely to cause problems in orchards with year-round cover crops or where vegetation is allowed to build up around the base of trees. Vegetation management and, when necessary, the proper use of trunk guards on young trees, usually keep damage to a minimum. Baiting is used to control voles when populations reach damaging levels.

Meadow voles are small, stocky rodents with small ears and eyes, short legs, and short tails (fig. 27.15). Their coarse fur is usually dark brown or grayish brown. Adults are larger than a house mouse but smaller than a rat. They are active year-round and during the day and night. Females produce several litters per year, with peaks in reproduction in spring and autumn. Because voles mature rapidly and can produce many litters annually, their populations can increase rapidly. Vole populations tend to cycle, peaking every 4 to 7 years followed by a rapid decline in population size. Voles can be detected through the presence of narrow runways in grass or other ground cover and numerous shallow burrows with openings about 1.5 inches in diameter. Voles seldom travel far from their burrows and runways.

Management guidelines

In midwinter, begin to monitor monthly for active runways in cover crops or weedy areas in and around orchards. Look for fresh vole droppings and short pieces of clipped vegetation in runways. Look for burrow openings around the base of trees. Where burrow openings are present, check for tree damage. Characteristic damage is complete or partial girdling of tree trunks from just below the surface to as high as voles can reach on the trunk (usually not more than 5 in; fig. 27.16). In some situations, voles climb higher on young trees. Be sure to check carefully; once a decrease in tree vigor is detected, it may be too late to prevent significant injury to the tree. Carefully monitor fencerows, ditch banks, and other areas near the orchard where permanent vegetation is favorable for the buildup of vole populations.

Habitat modification. Controlling vegetative cover reduces or eliminates food and cover necessary for vole survival. Because voles do not travel more than a few feet from their burrows to obtain food, any significant destruction of food or cover results in burrow abandonment or mortality. Eliminating vegetative cover in adjacent areas reduces the potential for voles to invade the orchard. Clean cultivation or weed control within 3 feet of the trunks of trees makes the habitat unsuitable for voles. If you maintain ground cover in row middles, keep it mowed short.

Trunk guards. Cylindrical wire or plastic trunk guards can protect young trees from voles. An effective guard can be made of a cylinder of ¼- or ½-inch hardware cloth that is 24 inches wide and of sufficient diameter to allow for several years' growth of the tree trunk. Trunk guards must extend at least 6 inches below the soil surface to discourage voles from burrowing underneath. Plastic and other fiber materials also make effective trunk guards and are often less expensive, provide sunburn protection, and are more convenient to use. However, they may not provide as much protection against voles. If you use any of these other materials, be sure to check underneath the guards periodically to make sure voles are not getting to the trees. Good herbaceous control around the base of trees increases the effectiveness of tree guards.

Poison baits. Poison grain baits can be very effective in reducing meadow vole populations within orchards and can be used to control vole populations in adjacent areas before they invade orchards. Single-dose baits (e.g., zinc phosphide) and multiple-dose ones (e.g., anticoagulants) are available, but there may be baiting restrictions in some areas to protect endangered species. It is extremely important to understand and follow the label directions for use.

For small infestations, scatter bait around infested areas, focusing on runways and around burrows. When allowed by the bait's label, broadcast baiting can be used to treat larger areas. For noncropland, apply bait in autumn or spring before the reproduction peak. However, application within an orchard is restricted to the nonbearing season (i.e., postharvest to fruit set). Bait acceptance will depend on the time of year and availability of alternative foods.

Figure 27.15. Voles are gray or brown, with a blunt nose, inconspicuous ears, and a short, slightly hairy tail. *Photo:* J. K. Clark.

Figure 27.16. Numerous vole burrow openings around the base of a tree with substantial bark damage. Removing vegetation from around the base of trees reduces the risk of girdling from voles. *Photo:* N. O'Connell.

Black-Tailed Jackrabbit (*Lepus californicus*) and Cottontail and Brush Rabbits (*Sylvilagus* spp.)

Jackrabbits, cottontails, and brush rabbits may damage nut orchards (figs. 27.17 and 27.18), although damage is almost always limited to orchards less than 4 years old. Rabbits may remove bark from trees and clip off branches within their reach to eat buds and young foliage. Trunk girdling is usually higher on the trunk than damage caused by meadow voles. The damage appears as vertical lines or grooves in the bark, often at 45-degree angles. Rabbits may also gnaw on drip irrigation lines.

Although jackrabbits are the most common of rabbit-type pests, they are technically classified as hares. Jackrabbits are about the size of a large house cat, have very long ears, short front legs, and long hind legs. They live in the more open areas of the Central Valley, coastal valleys, and foothills. They make depressions underneath bushes or other vegetation, where they remain secluded during the day. Young are born fully haired, with open eyes, and become active within a few days. Cottontails and brush rabbits are smaller and have shorter ears. Young are born blind and hairless, and they stay in the nest for several weeks.

Rabbits usually breed, bear young, and live outside orchards. They are active all year but damage trees primarily in winter and early spring, when other sources of food are limited. Common control methods for rabbits include fencing, tree guards, repellents, baiting, trapping, and shooting, depending on the species.

Figure 27.17. Jackrabbits are gray brown, with very long ears and large hind legs. *Photo:* J. K. Clark.

Figure 27.18. Cottontail and brush rabbits are smaller than jackrabbits and have shorter ears. *Photo:* J. K. Clark.

Management guidelines

Inspect young trees periodically for damage. If you find damage, look for droppings and tracks that indicate rabbits are the cause. Monitor the orchard perimeter in the early morning, or late evening or at night using a spotlight, to identify the rabbit species responsible for the damage, the approximate number of depredating individuals, and the location where rabbits are entering the orchard. The method of rabbit control should depend on the urgency of the problem and the situation; depredating populations are best managed before severe damage occurs.

Habitat modification. Rabbits usually invade orchards from adjacent fields, but unless the land is under the grower's direct management, modification of the outlying habitat is usually impractical. The removal of rabbit foods such as orchard cover crops and weeds may reduce the number of rabbits that visit the orchard and make them easier to detect; however, because removal of vegetative cover may temporarily increase tree damage (as trees would be the only food source left for rabbits), habitat modification to reduce damage within the orchard is rarely practical.

Exclusion. Where rabbits are a constant and continuing threat to young trees, fencing the entire orchard may be an effective management approach. Make the fence of woven wire or poultry netting at least 4 feet high and with a mesh diameter of 1 inch or less. Bury the mesh 6 inches deep with a 6-inch lip turned out at a 90-degree angle at the bottom to prevent rabbits from burrowing underneath. However, the cost of a rabbit fence may be prohibitive for a large orchard, given the relatively short-term need for rabbit fencing.

Individual tree guards may provide a cheaper alternative and offer a practical way to prevent damage to replants in an established orchard. Guards may be solid or made of net or mesh materials such as metal, hardware cloth, plastic, or sturdy fibrous sheets. Make the cylinders at least 2.5 feet high to prevent jackrabbits from reaching foliage and limbs by standing on their back legs (fig. 27.19).

Poison baits. Poison baits offer a practical and economical way to control large areas with heavy rabbit infestations. Multiple-dose anticoagulant baits for rabbit control are available from many county agricultural commissioners' offices. Follow label directions carefully. Place baits (1 to 5 lb) in bait stations specifically designed for rabbits. Locate them near trails and secure them so they cannot easily be tipped over. Use as many stations as is necessary to ensure that all rabbits have easy access to bait, spacing them 50 to 200 feet apart along the perimeter where rabbits are entering the orchard. Inspect bait stations every morning for the first several days to keep bait supplies replenished; it may take this long before rabbits become accustomed to feeding at the stations. Increase either the amount of bait in the stations or the number of stations if all the bait is consumed in a single night. Replace bait that becomes wet or moldy. It usually takes 2 to 4 weeks or longer before results are seen with anticoagulant bait. Continue baiting until feeding ceases and you no longer observe rabbits.

Search for carcasses daily in early morning and remove and bury or otherwise dispose of them to reduce secondary poisoning hazards to scavengers. Do not place bait where humans and livestock can access it. Protect diurnal seed-eating birds by covering or removing the bait during daylight hours. As always, follow baiting restrictions when endangered species are potentially present within the baiting area.

Repellents. Chemical repellents may provide temporary relief from rabbits. Spray or brush the repellent on tree trunks during the dormant season or on foliage or trunks during the growing season; proper application times are listed on the label. Additional applications may be necessary to renew repellency lost through rain or sprinkler irrigation or to protect new growth.

Trapping. Box traps can provide effective control of cottontails and brush rabbits when populations are small. Traps are ineffective for controlling jackrabbits, as they typically do not enter traps.

Shooting. When low numbers of rabbits are present and causing damage, shooting can be an effective control for all three types of rabbits. For best results, patrol systematically in the early morning or late evening.

Wild Pigs *(Sus scrofa)*

Wild pigs include both domestic hogs that have escaped (also known as feral hogs) and wild boar (fig. 27.20). Wild pigs primarily cause damage by rooting up plants and insects. This rooting, combined with wallowing, creates shallow depressions in the soil, which interfere with the application of irrigation water and increase soil erosion. Removal of cover vegetation can reduce the potential for wild pig damage by eliminating their primary food source. In some cases, heavy exclusionary fences may keep pigs out, though fences may not exclude very determined pigs. Shooting and trapping can be effective for controlling pigs. Consult your local agricultural commissioner for details and regulations on controlling wild pigs in your area.

Figure 27.19. Wire mesh cylinders can be used to prevent rabbit damage to trees. They prevent vole damage as well if the mesh is small enough and buried several inches underground. *Photo:* J. K. Clark.

Figure 27.20. Wild pigs often travel in groups and are commonly found in brushy areas in the foothills. Water is an important resource for wild pigs. *Photo:* Courtesy Sedgwick Reserve, University of California Santa Barbara Natural Reserve System.

Figure 27.21. Coyotes are medium-size members of the dog family. Damage to pistachio orchards is primarily from destruction of microsprinkler systems. *Photo:* Courtesy of the Regents of the University of California.

Coyote *(Canis latrans)*

Coyotes occasionally cause damage in pistachio orchards by chewing on irrigation lines (fig. 27.21). Coyotes are classified as nongame mammals and can be taken whenever they are causing a problem. Contact your agricultural commissioner for details about regulations and control options.

Birds

Most significant avian pests of pistachios belong to the family Corvidae and include ravens *(Corvus corax)*, crows (*Corvus brachyrhynchos*; fig. 27.22), scrub jays (*Aphelocoma coerulescens*; fig. 27.23), and yellow-billed magpies (*Pica nuttalli*; fig. 27.24). Other depredating birds you might find in orchards include European starlings *(Sturnus vulgaris)*, blackbirds (*Agelaius* spp., *Euphagus* spp.), and house finches *(Carpodacus mexicanus)*; however, they typically only cause minimal damage in pistachio orchards.

Corvids can subsist on many kinds of food, including carrion, although their typical diet includes insects, acorns, wild berries, and fruit. Cultivated fruit and nuts are often easy to obtain, and birds take full advantage of these food sources. Scrub jays instinctively store food in the autumn and carry away more fruit and nuts than they can eat. Thus, a single bird can account for considerable crop loss. Jays usually are solitary birds and occur singly or in pairs; rarely do they flock, although they do occasionally when food is abundant. In contrast, crows and ravens join into flocks for most of the year and establish a central roost, which can consist of thousands of birds. Each day, crows fly from the roost to feeding areas. Magpies generally occur in smaller groups of a few to several dozen birds. Corvids cause the greatest amount of damage during the period from shell split to harvest. After harvest, these birds can remove mummy nuts left in the orchard; however, allowing undisturbed access after harvest may contribute to increased bird damage the following year by improving survival and encouraging nearby nesting in spring.

Regular monitoring helps determine when damage starts so you can initiate control actions early. Birds are much more difficult to control once they have become used to feeding in an orchard. Become familiar with damaging species so you can distinguish them from the many nonpest birds that may frequent orchards. To monitor the occurrence of bird damage, it is easier to watch for movement of birds into or within the orchard than to see the damage itself. Observing movements into the orchard also allows you to determine the source of bird invasions and then to scare off invaders before their feeding patterns become established. With crows, for example, scout birds often fly out in search of food sources during early morning and return to the flock to guide the other birds to the food sources. If scout birds can be deterred from an orchard, the flock will choose another food source.

Figure 27.22. Crows are large black birds. They roost in large numbers and can be serious pests in pistachio orchards. *Photo:* J. K. Clark.

Figure 27.23. The head, wings, and tail of the scrub jay are blue; the breast and back are gray. A single scrub jay can cause considerable crop loss. *Photo:* W. P. Gorenzel.

Bird counts can help you decide when to take control actions and whether the controls you have used are having an effect. Once nuts begin to develop, plan to drive through orchards twice a week during the time of day you expect most bird activity and count the birds you see. Where birds are moving into an orchard from adjacent habitat, they can be counted by a stationary observer. Handheld counters are useful for making these counts. Bird counts will tell you how severe the problem is and if it is increasing. Keep records of bird species, counts, and locations for each season; they will help you plan control actions. This way you can have control devices on hand and in working order when they are needed.

Management guidelines

Remove brush piles in or near orchards; they offer refuge and loafing sites for birds and may exacerbate problems. Removal of brushy vegetation adjacent to orchards may also reduce bird problems.

Frightening. Noisemakers and visual repellents are the primary means of controlling bird damage in pistachio orchards. To be effective, several different methods should be used in rotation and started as soon as birds appear. Once birds have become used to feeding in an orchard, they are much harder to frighten away. If your orchard is near an urban area, check to see if there are any local ordinances that restrict the use of noisemaking devices. Monitor bird activity while using frightening devices and switch to a different type when birds appear to be getting used to the technique you are using.

Roving patrols that fire shell crackers, bird bombs, or whistler bombs are among the most effective ways to frighten birds from orchards. Stationary noisemakers such as gas cannons (propane exploders) and electronic noisemakers are most effective if you use at least one device for every 5 acres and elevate them above the level of the tree canopy. Move the devices to new locations every 3 to 5 days so birds take longer to get used to them. The effectiveness of most frightening devices is enhanced with occasional shooting, which reinforces the frightening stimulus.

The most commonly used visual repellents are reflective streamers and large scare-eye balloons. Attach balloons to poles so they are above the tree canopy and hang reflective streamers in trees. Use visual repellents in combination with noisemakers to increase their effectiveness. Use one type of visual repellent with each type of noisemaker and switch to a different visual repellent when you switch noisemakers.

The most effective bird frightening program makes use of as many different kinds of noisemakers and visual repellents as practical. Shooting is usually used to reinforce the various frightening methods. The following is an example of a diverse 4-week frightening program:

- First week: Patrol orchard firing cracker shells.
- Second week: Use stationary propane exploders.
- Third week: Use stationary propane exploders and also patrol orchard firing cracker shells.
- Fourth week: Install electronic noisemakers with scare-eye balloons or reflective streamers near trees where damage is most severe.

Shooting. Birds such as scrub jays and magpies that usually invade orchards in small numbers can often be controlled by shooting. A depredation permit is required if you want to shoot scrub jays or ravens. No permit is required to shoot crows or magpies that are causing damage. However, it is a good idea to check with authorities as regulations may change. Where permissible, shooting occasionally at a few birds increases the effectiveness of other noisemaking techniques because the birds begin associating loud noises with the real hazard of firearms.

Figure 27.24. Yellow-billed magpies can be serious pests for pistachio growers. They have distinct black and white markings on the body, a very long tail, and a bright yellow beak. *Photo:* J. K. Clark.

Part 8 Physiological Disorders

28 Alternate Bearing, Nut Blanking, and Shell Splitting

Louise Ferguson and Vito S. Polito

Female pistachio trees display three physiological characteristics that directly affect marketable fruit yield, though they occur at different stages of fruit development. The first is the alternate-bearing habit, marked by an annual fluctuation of large crops with small crops. The second is the production of blanks (i.e., fully expanded shells containing either an unfertilized embryo or a fertilized embryo that aborted soon after fertilization or failed to fully develop to maturity). And the third is the production of nonsplit nuts, which contain a partially or fully developed kernel but fail to split along the lateral shell suture. All three phenomena appear to be correlated with crop load and are therefore probably related to carbohydrate limitations (Sajjadinia et al. 2010; Spann et al. 2008; Vemmos et al. 2012) or to embryo formation and development (Shuraki and Sedgley 1996). Thus far, little is known about the specific physiological mechanism of each phenomenon, nor have control methods been developed.

Alternate Bearing

Pistachios bear their nuts laterally on 1-year-old shoots. Once female pistachios are mature, virtually all the axillary lateral buds on the current year's shoot growth are inflorescence buds. However, when budded on rootstocks with the more vigorous *Pistacia integerrima* parentage, the second and third flush may produce vegetative buds on the more distal shoot growth. The single apical vegetative bud that produces the current season's vegetative shoot growth bears the preformed buds for the following season's crop and the potential for neoformed buds—that is, buds formed the same season (Spann et al. 2008). Therefore, the individual bearing shoots within a tree are producing the current year's crop as well as generating the growth and buds for the subsequent year's crop. This bearing habit exacerbates the alternate-bearing pattern of increasingly larger and smaller crops that is common to all pistachio cultivars. The simultaneous demand of maturing the current crop on 1-year-old wood and producing the shoot growth bearing inflorescence buds for the subsequent crop exerts a high carbohydrate demand on the tree's available resources. The visible manifestation of this alternate-bearing pattern is easily observed during the growing season. When multiple clusters of filled split nuts are produced on the 1-year-old wood, the inflorescence buds on the current year's shoot growth will abscise by mid-August. The observable sequence of alternate bearing and possible mechanisms are discussed below.

Within the inflorescences on 1-year–old wood, the individual pistachio nuts begin growth by expanding only the nutshells in mid-April, reaching full size by mid- to late May. The nutshells then begin to harden by thickening and changes in composition (Allan et al. 2015, unpublished data). When the nutshell is about 50 percent hardened, the kernel within the nutshell begins enlarging, generally in late June. As the kernel enlarges, the basal inflorescence buds on the growth beyond the nut clusters die and drop, or abscise, sequentially from the shoot's base to the tip. Only the inflorescence buds abscise. The axillary and apical vegetative buds do not abscise. The loss of inflorescence buds is greater during heavy crop years and on shoots with a smaller leaf surface area to crop load ratio. Therefore, following a heavy crop year, for any individual shoot the new growth generally bears little or no fruit, as the buds abscised the previous season. As trees, orchards, and even industries mature, the synchrony within each appears to increase. In California, the industry-wide synchrony from 1979 through 2015 was remarkably regular (Anon. 2015), though slightly mitigated at times by new acreage entering bearing. This synchrony suggests that climatic patterns play a major role in initiating alternate bearing. Thus far, however, the role of weather in alternate bearing (other than major winter or spring freezes) has not been demonstrated.

Recent research has demonstrated a transient carbohydrate decrease in branches with heavy crop loads. It has also shown that new extension shoot growth precedes the abscission of the developing inflorescence buds on the current year's shoot growth (Spann et al. 2008). This finding, combined with multiple earlier studies demonstrating changes in shoot and leaf carbohydrate and plant hormone levels before and during inflorescence bud abscission, suggests, but does not prove, that a temporary competition for the stored carbohydrates in fruit-bearing shoots initiates a hormonal signal that causes the bud abscission on current year's shoot growth that results in alternate bearing (Caruso et al. 1992; Crane et al. 1971, 1972, 1976, 1977; Marra et al. 2009; Nzima et al. 1997; Takeda and Crane 1980; Takeda et al. 1980; Vemmos et al. 2012; Wolpert and Ferguson 1990). Alternate bearing is not unique to pistachio trees. Several types of fruit trees alternate bear. However, thus far only pistachios and another apically dominant alternate-bearing species, *Olea europaea,* appear to alternate bear through the mechanism of premature inflorescence bud abscission during the previous year (Chao 2014).

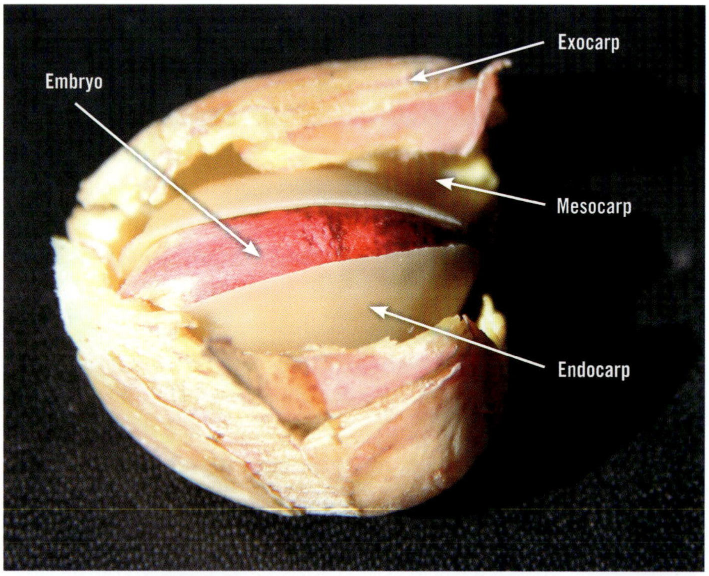

Figure 28.1. A fully mature and naturally ripened pistachio nut (embryo or seed) with the shell (endocarp) naturally split, and the hull (internal mesocarp and external exocarp) puffed and turning red. *Photo:* L. Ferguson.

Alternate bearing has not been demonstrated to be harmful to the tree, but it is disruptive to commercial farming operations and marketing. Attempts to alleviate alternate bearing by foliar nutritional and growth regulator sprays have not been successful. However, heavy mechanical pruning, particularly topping, after a heavy crop can successfully decrease alternate bearing for as long as 7 years (Ferguson et al. 1995). This would be considered rejuvenation pruning. The topping and hedging does not directly affect the mechanism of alternate bearing. It alters the ratio of fruit bearing to nonbearing vegetative shoots by precipitating strong vegetative shoot growth. This decreases yield the first 2 years after mechanical pruning and decreases the amplitude of alternate bearing in the subsequent years. If more moderate mechanical topping and hedging is started in young orchards (those that are 6 to 10 years old), before alternate bearing is severe, it prevents severe alternate bearing from developing (Dennis Elam, personal communication, 2004). Currently, pruning is the only method available to mitigate or prevent alternate bearing.

Blank Nuts

A pistachio fruit consists of a nutmeat (embryo, seed, or kernel) enclosed in a thin, hard shell (endocarp) surrounded by a fleshy hull (mesocarp and exocarp, or collectively the pericarp) (fig. 28.1). The fruit grows from the pistil of the female flower. The pistil has a single ovary at its base. The ovary forms the fruit wall, including the shell and the hull. Within the ovary is a single ovule. The ovule, which contains the female germ cell (egg), develops into the edible kernel. Extending from the ovary is a three-part style. Each of the three parts of the style terminates in a stigma. As the flower opens, the stigmatic surfaces become receptive to pollen.

Fruit set typically follows successful pollination. Pollen is released from staminate flowers on male trees and is carried by the wind to the stigmas of pistillate flowers on female trees. When a pollen grain lands on a receptive stigma surface, it germinates to form a pollen tube. The pollen tube is an elongated cell that grows through stigma and style tissue to the basal ovary and into the ovule. As it does so, it carries male germ cells to the egg cell, where fertilization is completed. Multiple pollen tubes germinate and grow through the style, but only one successfully reaches the ovule. That pollen tube enters the ovule and releases its contents to complete fertilization. Fertilization is the fusion of a male germ cell with the egg cell. Thus, the reproductive process leading to fruit set in pistachio has three parts: pollination, involving the transfer of pollen to the stigma; pollen tube growth, the transfer of male germ cells through the stigma and style to the ovule; and fertilization, the fusion of the male and female germ cells in the ovule.

The fusion of male and female germ cells produces a single-celled zygote that eventually grows to form an embryonic plant. This embryonic plant comprises the kernel. The process begins slowly, however. The first division of the zygote does not occur for several weeks after flowering, during which time the ovary, or shell, grows to its final size. After ovary growth is complete, the kernel grows to fill it. This is an unusual pattern of growth, and the differential timing of pistachio ovary and kernel development has implications for alternate bearing, blank nut production, and shell splitting.

Pistachio pollination is unique in that the pollen tube penetrates a tissue surrounding the egg called the chalaza. This structure is located on the opposite side of the ovule from the micropyle, and it delays the union of the ovule and pollen (Grundwag 1975). During the delay, the pollen travels from the chalaza to the micropyle, and the stimulation of that movement stimulates the growth of the pericarp (Grundwag 1975; Martinez-Palle and Herrero1995). Polito (1998) also demonstrated that pollen stimulation could cause parthenocarpy, or fruit set, without successful pollen tube development or subsequent successful fertilization of the ovule or both. Blank nuts result when there is fruit set and ovary growth but the embryo fails to grow, leaving the nutshell empty, or blank. Blanking can occur during two different phases of pistachio nut development, nut setting and nut filling. Shuraki and Sedgley (1996) described two mechanisms of blank nut production, embryo abortion and kernel development. They described five major causes of the incomplete embryo development. The most frequent causes were the flower ovary lacking an embryo sac, spontaneous degeneration of the embryo sac early in the pollination, and later degeneration in the nucellar or cotyledeon structures. Another more common pollination problem in many fruit is the failure of the pollen tube to penetrate the ovule; and, finally, a less common occurrence is the endosperm cells failing to multiply. Without proper ovule structure or successful fertilization, the resulting fruitlet would be a blank. The degeneration of the fertilized embryo, nucellus, or cotyledeon would result in a partially filled, or defect, fruit. Examples of blank nuts produced during different stages of development can be seen in figure 28.2.

Blank Production During Nut Set

The first empty shells (blanks) are produced as a result of events that occur at fruit set. Pollination occurs but fertilization fails either because the pollen tube does not complete its growth to the ovule or the ovule is not viable when the pollen tube arrives. The stimulus of pollination and pollen tube growth is sufficient to induce fruit set and shell formation, but failure of fertilization means no embryo forms, so there is no kernel to fill the shell. This phenomenon is called parthenocarpy and is the basis for the production of several types of seedless fruit, including some seedless citrus varieties. Parthenocarpy is fairly common; however, it is normally found in fruit that have many seed rather than in single-seeded fruit such as pistachio.

Experimental evidence suggests that the pollination-induced parthenocarpy discussed above is one cause of blank pistachio nuts. In one study, flowers were pollinated with pollen that had been exposed to a high dose of gamma radiation. The radiation treatment was at a threshold level that permitted pollen germination but inhibited full pollen tube growth. These experiments resulted in a high percentage of blank nuts. One explanation is that the pollination stimulus, which was not eliminated in the irradiated pollen, is sufficient to set fruit, most likely by triggering a hormonally mediated signal that leads to fruit set.

Figure 28.2. Various stages of blank nuts. (A) Partially grown embryo. (B) Embryo that has minimal nut growth. (C) Note the red color development, puffy and tattered shell, and yellow color at the abscission zone of the nut, as opposed to the adjacent nut (D) with little hull color development or puff. The former contains a fully matured and split nut. The latter contains only a desiccated embryo. *Photo:* L. Ferguson.

One of the causes of blanking may occur during nut set. Some evidence suggests that it may be associated with poor boron nutrition. Boron is involved in several important aspects of plant reproductive biology, including pollen germination on the pistil, pollen tube growth, and ovule longevity, all of which may have a role in pollination-induced parthenocarpy. Research has demonstrated boron leaf levels below 120 ppm dry weight (August leaf sample) are associated with an increased percentage of blank nuts at harvest (Brown et al. 1993).

Blanking During Nut Fill

Blanks also develop in July during kernel enlargement, when a percentage of fertilized embryos fail to enlarge to fill the shells. It is not known what determines that percentage, but it is suspected that the tree's stored carbohydrate capacity initially determines the percentage of filled nuts. It has been demonstrated that thinning a cluster prior to nut growth results in a higher percentage of filled nuts on the thinned cluster. However, the thinned and unthinned clusters have virtually the same absolute number of filled nuts, though the thinned clusters may have slightly larger nuts. This response to thinning also has been demonstrated on a whole-tree scale in pruning experiments (Ferguson et al. 1995).

Effect of Alternate Bearing on Blanking

Production of blank nuts is strongly affected by alternate bearing. The percentage of blank nuts appears to be consistently higher in the low crop year, the off-year (table 28.1). The greater percentage of blank nuts in the off-year further corroborates, though does not prove, the theory that the carbohydrate status of a tree the previous year determines the crop load it can set and carry through maturation to shell split. Markedly different tree crop loads within a crop year (see table 28.1) had virtually the same percentage of blank nuts. For example, in 1989 the control trees had 0.2 pound of crop per tree and 19.1 percent blanks, while the hedged and topped trees had 11.4 pounds of crop and 16.0 percent blanks. There was no statistically significant difference in these percentages despite the over 50-fold difference in crop loads. It appears that, other than maintaining boron leaf levels above 120 parts per million, there is little that can be done to affect the percentage of blank nuts.

Effect of Temperature on Nut Growth

As with all growth processes in plants, growth of a pistachio fruit is a function of heat. Recent research has demonstrated that the three major pistachio cultivars grown in California (Kerman, Golden Hills, and Lost Hills) have slightly different heat unit requirements, calculated as growing degree-days (GDDs) above 50°F, for shell growth, shell hardening, and kernel growth, commonly referred to as Stages I, II, and III (Allan 2015). Kerman and Lost Hills require approximately 756 and 751 GDDs, respectively, to complete Stage I, full shell volume, while Golden Hills requires about 6 percent less, approximately 705 GDDs. However, for Stage II shell hardening, Kerman requires 2,583 GDDs, as opposed to Golden Hills, which requires 2,830 GDDs, and Lost Hills, which requires 3,157 GDD. This may explain why the Lost Hills cultivar has weak shell-hinge strength, which allows kernels to drop out of intact shells during processing. The shell may not be fully hardened at harvest, and as a result, the hinge widens and drops out the kernel. Kernel growth, Stage III, does not begin measurable expansion growth until 931 GDDs for Golden Hills, 982 GDDs for Lost Hills, and 1,000 GDDs for Kerman. Kernel growth is completed earliest for Golden Hills at 1,904 GDDs, 2,021 GDDs for Lost Hills, and 2,111 GDDs for Kerman. By calculation, Golden Hills requires 207 GDDs, almost 10 percent less than Kerman, and Lost Hills 90 GDDs fewer. Thus, both can be harvested earlier than Kerman, even if the shells are not fully hardened. Little is known about the effects of GDDs on pollination, fertilization, early fruit growth, and final nut size. Therefore, little is known about the effects of temperature on blank nut production.

Nonsplit Nuts

The edible pistachio, unlike the species used for rootstocks, is characterized by splitting of the nutshell at maturity (see fig. 28.1). A nonsplit nut is not fully mature, as the final dehiscence, or longitudinal splitting, of the shell has not occurred (as shown in figure 28.3). Splitting begins about the end of July, at least 1 month before fruit maturity, and continues through mid-September, progressing simultaneously with nutmeat maturation. Final nutmeat maturity is indicated by "puffing," or separation of the hull from the shell. This is accompanied by a breakdown of chlorophyll pigments in the hull, allowing the red pigments to become visible. Figure 28.1 shows the hull of a fully mature nut with a split shell. Mature nuts detach from the fruit pedicel with a light shaking of the rachis, and the nut hull slips off easily with a light pressure exerted at the stem end of the nut.

Table 28.1. Effect of one-time mechanical hedging and topping* on yield and percentage of blank, split, and nonsplit nuts

Year and treatment	Tree yield (lb/tree)	Blank nuts (%)	Split nuts (%)	Nonsplit nuts (%)
1985 (off-year)				
control	6.4 a†	8.1 a	86.4 a	3.2 a
hedge	4.6 b	7.2 b	86.1 a	3.0 a
top	3.7 b	8.4 a	84.9 a	2.9 a
hedge and top	1.8 c	8.1 a	85.8 a	2.7 a
average		8.0	85.8	3.0
1986 (on-year)				
control	49.2 a	2.1 a	63.4 a	30.4 a
hedge	44.6 a	2.5 a	65.5 a	28.1 a
top	39.9 b	2.9 a	64.2 a	30.1 a
hedge and top	28.1 c	2.6 a	62.5 a	29.4 a
average		2.5	63.9	29.5
1987 (off-year)				
control	3.5 c	10.1 ab	83.3 ab	3.7 a
hedge	7.6 b	8.5 b	83.8 a	3.1 a
top	11.7 a	11.5 a	80.3 b	3.7 a
hedge and top	14.2 a	10.2 ab	83.1 ab	3.6 a
average		10.1	82.6	3.5
1988 (on-year)				
control	34.1 a	2.7 a	72.9 a	20.3 a
hedge	26.3 b	2.8 a	72.6 a	21.3 a
top	24.5 b	2.2 a	72.3 a	22.4 a
hedge and top	26.2 b	2.2 a	72.6 a	22.7 a
average		2.5	72.6	21.6
1989 (off-year)				
control	0.2 c	19.1 a	47.6 c	19.1 a
hedge	0.6 c	17.7 b	61.2 b	13.5 c
top	6.0 b	14.1 c	63.6 a	15.1 b
hedge and top	11.4 a	16.0 a	64.4 a	12.8 c
average		16.7	59.2	15.1
1990 (on-year)				
control	37.1 a	11.2 a	67.5 a	21.1 a
hedge	35.3 a	9.3 a	68.5 a	22.1 a
top	30.4 b	11.7 a	63.5 a	22.8 a
hedge and top	27.2 b	13.8 a	67.4 a	18.7 a
average		11.5	66.7	21.2
1991 (off-year)				
control	3.0 c	19.2 a	59.0 a	21.1 b
hedge	4.2 c	16.0 a	57.1 ab	26.0 b
top	17.2 b	15.1 a	50.1 bc	34.0 a
hedge and top	25.8 a	17.0 a	47.1 c	35.1 a
average		16.8	53.3	29.1

Notes: *Control trees received regular annual hand-pruning.

†Values within a crop year followed by the same letter are statistically equal.

Figure 28.3. Filled nuts with unsplit shells. These nuts can be artificially split and sold in-shell or removed and sold as shelling stock. *Photo:* L. Ferguson.

Pistachio nuts may split along the longitudinal ridges of the shell and at the tip of the shell. Splitting can occur in any combination of one or both of the longitudinal ridges, with or without the tip, or at the tip alone. Investigation of the anatomical structure of the longitudinal- and tip-split regions indicates that these parts of the shell differ from one another structurally, suggesting that different mechanisms may be involved in shell separation at each site (Polito and Pinney 1999).

Shell split is dependent on nutmeat growth and development within the shell. Figure 28.4 shows the relationship between shell split and kernel growth. Note that kernel growth begins after the shell has reached its full size. The first split shells are seen about the time the kernel has grown to fill the shell and exerts physical pressure on it. At this time, the shell is fully lignified (hardened), and the cells that form the regions where longitudinal splitting occurs are dead. This fact seems to rule out the possibility that biochemical factors are involved as a controlling mechanism in shell splitting, making it unlikely that a chemical agent to enhance split percentages will be discovered.

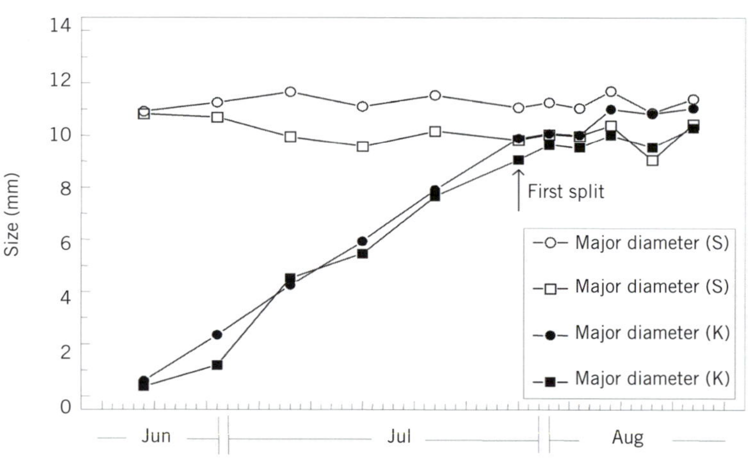

Figure 28.4. Kernel (K) and shell (S) growth are indicated by the major (in the plane of the longitudinal split lines) and minor (perpendicular to the plane of the longitudinal split lines) diameters. Data points represent the means of 20 samples. *Source:* Polito and Pinney 1999.

There is a correlation between kernel size to shell size ratios and longitudinal splitting (Polito and Pinney 1999). Table 28.2 shows the relationship of kernel diameters to inner shell diameters for six samples of fresh (not dried to 5 to 7% moisture), filled nuts. In each case, the ratio of kernel to shell size is greater for fully split nuts than it is for tip-split or nonsplit nuts. Interestingly, there is no statistical difference between tip-split and nonsplit nuts, a finding that is consistent with anatomical indications that tip splitting and longitudinal splitting involve different mechanisms. Furthermore, for fully split nuts, the ratios of kernel to shell minor diameters (i.e., perpendicular to the plane of the longitudinal ridges) are consistently greater than one. This means that for fully split nuts, kernel size is greater than shell size in the direction where force against the inside of the shell would tend to drive the shell halves apart.

These correlations implicate mechanical force generated by the growing kernel against the shell as the mechanism for shell splitting. Deficit irrigation experiments supported this concept. When irrigation stress was applied during Stage I, the period of rapid endocarp (nutshell) expansion, the result was smaller nutshells and a higher percentage of split nuts (Goldhamer et al. 1985; Goldhamer and Kanber 2012). Commercially, this method of obtaining a higher percentage of split nuts is not feasible, as a smaller crop of larger split or shelled nuts is currently more profitable than a higher percentage of smaller split nuts. If it is the growth of the kernel that drives shell splitting, then factors that enhance kernel size relative to shell size should lead to more split nuts. Crop load is one such factor; irrigation management is another. Both can affect the split nut percentages.

There is an inverse relationship between tree crop load, the percentage of nuts with split shells, and the percentage of blank nuts (table 28.1). As crop load increases, the percentage of nuts with split shells decreases, and the percentage of blank nuts decreases. Thus, in on-years the marketable crop is decreased by nonsplits, and in off-years it is decreased by blanks. Further, it is interesting to note that the percentage of nonsplit nuts and blanks is much more strongly correlated with on- and off-years than with individual tree crop load. A review of the 7 years of crop production in trees with markedly

Table 28.2. Ratio of kernel to shell (K:S) diameters, at their widest point, for undried, filled nuts

	Major diameter (K:S ratio)			Minor diameter (K:S ratio)		
Sample*	Full split	Tip split	Nonsplit	Full split	Tip split	Nonsplit
1	0.97a†	0.90b	0.89b	1.06a	0.90b	0.96b
2	0.98a	0.85b	0.85b	1.07a	0.90b	0.90b
3	0.95a	0.88b	0.83b	1.06a	0.94b	0.92b
4	1.02a	—	0.90b	1.10a	—	0.97b
5	0.94a	—	0.83b	1.02a	—	0.88b
6	0.97a	—	0.81b	1.02a	—	0.90b

Notes: *For each sample, n = 20; samples 4 through 6 had fewer than 20 tip-split nuts.
†For each diameter ratio for each sample, values with different letters differ significantly ($p < 0.05$).

different crop loads in table 28.1 shows that only when the crop load is as little as 0.2 pound per tree versus 11.4 pounds per tree (1989), or 3.0 versus 25.8 pounds per tree (1991), are significant differences produced in the percentage of split shells. This suggests that the ability of the tree to support and mature a crop is a stronger factor in determining the percentage of split nuts than the actual pounds of crop on the tree in a season. Or, it suggests that the carbohydrate status of a tree produced the previous season and entering the current season plays a large role in its ability to successfully mature a crop.

Effects of Female Cultivar, Pollen Source, and Rootstock on Shell Split

Recent experimental results have demonstrated that both the female cultivar and the male pollen source can affect nutshell split. First, it is well established that variability in shell split percentages and widths exists among edible pistachio cultivars. This was one of the primary selection criteria for new cultivars Golden Hills and Lost Hills as well as for Kerman. Second, research has demonstrated that the Peters and Ask male pollen produces a higher percentage of split nutshells than pollen of Atlantica males. There is no evidence of rootstock producing differences in shell split percentages. In rootstock trials conducted in California, no significant differences in shell split percentages have been detected among Kerman scions on *P. atlantica, P. integerrima,* and the hybrids of these two rootstocks (Ferguson et al. 2002).

Effects of Preharvest Production Practices on Shell Split

Some production practices affect shell split. In decreasing order of impact, these factors are harvest time, irrigation management, boron nutrition, and dormant pruning.

Research results demonstrated that the highest percentages of shell split are achieved when harvest is delayed until the maximum number of nuts display hull dehiscence, or separation, from the nutshell. Dehiscence is signaled primarily by hull color change to red on the surface exposed to the sun, followed by the hull puffing, or appearing loose, followed by tattering of the exocarp (see figure 28.2). In practice, the progress of hull dehiscence is evaluated by observing the early hull color change, then randomly sampling trees for split nuts. A 100-nut sample is collected from around the tree, with the knowledge that nuts in the upper southwest quadrant will mature first, and the percentage of nuts with easily removed hulls and the split shells is determined. This is done daily until the increase in the percentage of split nuts levels off.

Maximizing shell split must be balanced against the threat of *Amyelois transitella,* navel orangeworm (NOW), infestation. NOW infestation levels dictate insecticide application timing. These insecticides have preharvest intervals. Other factors influencing the shell split percentage are the availability of a harvesting contractor and the potential for shell staining from hull degradation. Delaying harvest until maximum nutshell split percentages are achieved may result in NOW infestation and shell staining.

Research has demonstrated that irrigation stress during Stage I of nut growth from mid-May through the end of June increases the percentage of shell split. Irrigation stress during Stage III, from mid-August through early September, when shells are actively splitting, significantly decreases the percentage of split nuts (Goldhamer et al. 1985; Goldhamer and Kanber 2012).

An early study demonstrated that if boron is deficient, a late dormant spray of 2 to 5 pounds of Solubor (20% boron active ingredient) per acre, applied at inflorescence bud swell, significantly increased the percentage of split nuts (Brown et al. 1993). However, if boron levels are above 120 parts per million, split percentages are not increased with additional boron applications.

Dormant pruning has a negligible effect on split percentages. Dormant pruning produced the alterations in yield presented in table 28.1. These data indicate that only when dormant pruning produced differences that varied significantly in crop load per tree were there significant differences in the percentage of splits, as in 1989 and 1991. Further, pruning treatments done 4 and 7 years earlier produced these differences. Later research demonstrated approximately half a pistachio tree's inflorescence buds can be removed and the tree will compensate by setting more nuts per cluster with the same percentage of split nuts as the unpruned controls. Thus, dormant pruning has a limited effect on shell split percentages; unless the pruning is quite severe, it does not impact tree crop load and therefore will not impact shell split percentages.

It is logical to expect that severe salinity, particularly high sodium and chloride levels, could decrease shell split, primarily through osmotic effects producing the same effects as deficit irrigation during Stage III. However, a long-term rootstock salinity tolerance trial did not demonstrate a significant effect on shell split percentages with soil water extract electroconductivities (EC_e) of up to 12 dS/m (Sanden et al. 2004).

Effects of Postharvest Factors on Shell Splitting

No postharvest practice has yet been demonstrated to significantly impact the split percentage. However, pistachio shells have a high moisture content. When subjected to heat during postharvest transport, preprocessing delays, and drying, the shell moisture content decreases, shrinking shells about the nut and increasing split width. Thus, pistachios can leave the field with a lower percentage of wide splits than when they arrive at the processor. During processing, an increase in split width can occur very early in drying and increase as dryer temperatures rise from 125° to 190°F. This increase can result in nut kernels dropping out of the shell. However, postharvest heat has not been demonstrated to increase the split percentages.

In conclusion, controllable factors that affect shell split include harvest timing, irrigation, and boron nutrition. To obtain the best split shell percentages with the Kerman cultivar, harvest timing should be based on the appearance of hull color and a sampling for the split percentage. Trees should not be water stressed from mid-August through September, and boron levels should be maintained above 120 parts per million by dry weight of a July leaf sample.

References

Allan, C. J., C. H. Crisosto, and L. Ferguson. 2015. Determining thermal unit requirements of three developmental stages of *Pistacia vera* spp. fruit growth. Master's thesis, Department of Plant Sciences, University of California Davis.

Anonymous. 2015. Pistachio bearing acreage, production and yield per acre. Administrative Committee for Pistachios website, http://www.acpistachios.org/statistics.shtml.

Brown, P., L. Ferguson, and G. Picchioni. 1993. Boron nutrition of pistachio: Final report. In California pistachio industry annual report 1993, 57–59. Fresno: California Pistachio Commission.

Caruso, T., L. Di Marco, and A. Raimondo. 1992. Effects of debudding and defruiting on alternate bearing in pistachio (*Pistacia vera* L.). Fruit Varieties Journal 46:170–174.

Chao, Yi-yun. 2014. Alternate bearing in olive (*Olea europaea* L.) Master's thesis, Department of Plant Biology, University of California, Riverside.

Crane, J. C., and I. Al-Shalan. 1977. Carbohydrate and nitrogen levels in pistachio branches as related to shoot extension and yield. Journal of the American Society for Horticultural Science 102:326–329.

Crane, J. C., and M. M. Nelson. 1971. The unusual mechanism of alternate bearing in the pistachio. American Society for Horticultural Science 6:484–490.

———. 1972. Effects of crop load, girdling and auxin application on alternate bearing of the pistachio. American Society for Horticultural Science 97:337–339.

Crane, J. C., P. B. Catlin, and I. Al-Shalan. 1976. Carbohydrate levels in pistachio as related to alternate bearing. Journal of the American Society for Horticultural Science 101:371–374.

Ferguson, L., J. Maranto, and R. H. Beede. 1995. Mechanical topping mitigates alternate bearing of 'Kerman' pistachios (*Pistacia vera* L.). HortScience 30(7): 1369–1372.

Ferguson, L., R. H. Beede, H. Reyes, and P. Metheney. 2002. California pistachio rootstock evaluations. Proceedings of the 3rd International Symposium on Pistachios and Almonds. Acta Horticulturae 591:63–66.

Goldhamer, D. A., and R. Kanber. 2012. Pistachio. In Crop yield response to water. Rome: United Nations FAO Irrigation and Drainage Paper No. 66.

Goldhamer, D. A., R. K. Kjelgren, L. Williams, and R. Beede. 1985. Water use requirements of pistachio trees and response to water stress. Proceedings of the National Conference on Advances in Evapotranspiration. American Society of Agricultural Engineers: 216–223.

Grundwag, M. 1975. Seed set in some *Pistacia* L. (Anacardiaceae) species after inter- and intra-specific pollination. Israel Journal of Botany 24:205–211.

Marra, F. P., E. Barone, L. M. Mantia, and T. Caruso. 2009. Toward the definition of a carbon budget model: Seasonal variation and temperature effect on respiration rate of vegetative and reproductive organs of pistachio trees *(Pistacia vera)*. Tree Physiology 29:1095–1103.

Martinez-Palle, E., and M. Herrero. 1995. The ponticulus: A structure bridging pollen tube access to the ovule in *Pistacia vera*. Sexual Plant Reproduction 8:217–222.

Nzima, M. D., G. C. Martin, and C. Nishijima. 1997. Seasonal changes in total nonstructural carbohydrates within branches and roots of naturally "off" and "on" 'Kerman' pistachio trees. Journal of the American Society for Horticultural Science 122(6): 856–862.

Polito, V. S. 1998. Seedlessness and parthenocarpy in *Pistacia vera* L. (Anacardiaceae): Temporal changes in patterns of vascular transport to ovules. Annals of Botany 83:363–368.

Polito, V. S., and K. Pinney. 1999. Endocarp dehiscence in pistachio (*Pistacia vera* L.). International Journal of Plant Sciences 160:827–835.

Sajjadinia, A., A. Ershadi, H. Hokmabadi, M. Khayat, and M. Gholami. 2010. Gas exchange activities and relative water content at different fruit growth and developmental stages of ON and OFF cultivated pistachio trees. Australian Journal of Agricultural Engineering 1:1–6.

Sanden, B. L., L. Ferguson, H. C. Reyes, and S. C. Grattan. 2004. Effect of salinity on evapotranspiration and yield of San Joaquin Valley pistachios. Proceedings of the 4th International Symposium on Irrigation of Horticultural Crops. Acta Horticulturae 664:583–589.

Shuraki, Y. D., and M. Sedgley. 1996. Fruit development of *Pistacia vera* (Anacardiaceae) in relation to embryo abortion and abnormalities at maturity. Australian Journal of Botany 44:35–45.

Spann, T. M., R. H. Beede, and T. M. DeJong. 2008. Seasonal carbohydrate storage and mobilization in bearing and non-bearing pistachio *(Pistacia vera)* trees. Tree Physiology 28:207–213.

Takeda, F., and J. C. Crane. 1980. Abscissic acid in pistachio as related to inflorescence bud abscission. Journal of the American Society for Horticultural Science 105:573–576.

Takeda, F., K. Ryugo, and J. C. Crane. 1980. Translocation and distribution of 14C-photosynthates in bearing and non-bearing pistachio branches. Journal of the American Society for Horticultural Science 105:642–644.

Vemmos, S. N., A. Papagiannopoulou, and S. Coward. 2012. Effect of shoot girdling on photosynthetic capacity, leaf carbohydrate, and bud abscission in pistachio (*Pistacia vera* L.). Photosynthetica 50(1): 35–48.

Wolpert, J. A., and L. Ferguson. 1990. Inflorescence bud retention in 'Kerman' pistachio: Effect of defruiting date and branch size. HortScience 25(8): 919–921.

Measurement Conversion Table

U.S. customary	Conversion factor for U.S. customary to metric	Conversion factor for metric to U.S. customary	Metric
Length			
inch (in)	2.54	0.394	centimeter (cm)
foot (ft)	0.3048	3.28	meter (m)
yard (yd)	0.914	1.09	meter (m)
mile (mi)	1.61	0.62	kilometer (km)
Area			
acre (ac)	0.4047	2.47	hectare (ha)
square foot (ft^2)	0.0929	10.764	square meter (^{2})
square mile (mi^2)	2.59	0.386	square kilometer (km^2)
Volume			
fluid ounce (fl oz)	29.57	0.034	milliliter (ml)
gallon (gal)	3.785	0.26	liter (l)
inch (in)	305	0.00328	millimeter
acre-inch (ac-in)	102.8	0.0097	cubic meter (m^3)
acre-foot (ac-ft)	1,233	0.000811	cubic meter (m^3)
cubic foot (ft^3)	28.317	0.353	liter (l)
cubic yard (yd^3)	0.765	1.307	cubic meter (m^3)
gallon per acre (gal/ac)	9.36	0.106	liter per hectare (l/ha)
gallon per ton (gal/T)	4.173	0.239	liter per metric ton (l/t)
Mass			
ounce (oz)	28.35	0.035	gram (g)
pound (lb)	0.454	2.205	kilogram (kg)
ton (T)	0.907	1.1	metric ton (t)
pound per acre (lb/ac)	1.12	0.89	kilogram per hectare (kg/ha)
ton per acre (T/ac)	2.24	0.446	metric ton per hectare (t/ha)
Pressure			
pound per square inch (psi)	6.89	0.145	kilopascal (kPa)
pound per square inch (psi)	0.0689	14.4	bar
Temperature			
Fahrenheit (°F)	°C = (°F - 32) ÷ 1.8	°F = (°C × 1.8) + 32	Celsius (°C)

Index

An *italic* "f" after a page number, such as 272*f*, refers to a photo or an illustration; an *italic* "t," as in 11*t*, refers to a table.

C

M

N

O

P

Q

R

T